AF673377

TERI ENERGY & ENVIRONMENT DATA DIARY AND YEARBOOK 2015/16

TERI's Governing Council

TERI ENERGY & ENVIRONMENT DATA DIARY AND YEARBOOK 2015/16

The Energy and Resources Institute

ISBN 978-81-7993-583-5

Suggested citation
TERI (The Energy and Resources Institute). 2016. *TERI Energy & Environment Data Diary and Yearbook 2015/16*. New Delhi: TERI

Published by
The Energy and Resources Institute (TERI)
TERI Press
Darbari Seth Block
IHC Complex, Lodhi Road
New Delhi – 110 003
India

Tel. 2468 2100 or 4150 4900
Fax 2468 2144 or 2468 2145
India +91 • Delhi (0) 11
Email teripress@teri.res.in
Website www.teriin.org

Printed in India

NCL...TOWARDS ENVIRONMENTAL EXCELLENCE

Northern Coalfields Limited, popularly known as NCL, is situated in the Central east part of India with 10 large opencast mines located in Singrauli district of Madhya Pradesh and Sonebhadra district of Uttar Pradesh. NCL is a coal-producing Miniratna company, which supplies coal to the major thermal power plants of the region and is a major source of energy to the nation. Approximately, 12 percent of the country's coal production is contributed by NCL and the company operates in areas that have abundant resource of power grade coal. The company believes in producing coal with utmost care to the environment with a vision of sustainable development.

NCL is committed to protect the environment through prevention, mitigation of pollution, proper recycling and disposal of wastes, conservation of bio-diversity and bringing awareness amongst the stakeholders for continual improvement in environmental performances following the best practices, including social and lifestyle upgradation. This is the vision of the corporate environment policy of our holding company, Coal India Limited and NCL, as well.

The Company has planned for integrated reclamation by internal dumping & backfilling, which reclaims the land to near original profile and the landscape, in future, shall merge with the existing terrain. It therefore gives a lot of importance to aesthetics, ecological restoration, topsoil utilization, water management and overall improvement of the ecology.

The flora and fauna of the mining areas is planned and restored effectively and at times the vegetation and land use has improved as compared to the original land use by bio-reclamation of mined out land, done through proper selection of useful and native species. The planned reclamation has improved the bio-diversity of many areas.

The Company has accorded high priority to Environmental Management, Ecology Development and Pollution Control. It is one of the best environmentally managed Coal Producing Companies in India. It is also an Integrated Management System certified company complying with ISO 14001, ISO 9001 and OSHAS 18001.

There are 10 fully operational Effluent Treatment Plants with sufficient capacity of treatment, which is 50% more than present requirement, for pollution control of water and its reuse. The colony water is treated in Sewage Treatment Plants in most of the projects with the latest state of the art system for management of domestic waste.

NCL has developed useful water storage ponds and water bodies for rainwater harvesting which are being used by the local people for different purposes. Catch drains and garland drains are provided to deal with surface runoff.

The reclamation work, both physical and biological, in NCL is one of the best in Coal India with bio matting used in Jayant, Dudichua and Nigahi. The number of trees planted by NCL is more than 2.26 crores, which is the highest amongst all subsidiary companies of Coal India Limited, with more than 1,374 per employee. In recent years more stress was given to bio-reclamation and eco-restoration. The reclamation work done in the mines like Jayant and Nigahi is one of the best mining reclamation models in the country.

The ambient air quality is monitored for all the mines on fortnightly basis and 98% of the results indicate that the Ambient Air Quality is within the norms. In recent years,

there is a reduction of 3% in the SPM values on an average basis in NCL; this is because of large-scale tree plantation and improved pollution control by increasing the frequency of water sprinkling and improved roads of the company.

NCL has also started a demonstration scheme on Ecological Restoration of mined out areas in Krishnashila and Nigahi projects on 10 hectares with the help of the Forest Research Institute, Dehradun. The new environmental initiatives include mist spray system, continuous fixed air quality monitoring systems, solar power plants and solar panel heaters, mobile road sweepers and a model Eco Park. Study of carbon footprint and environmental economics for the company is also in the offing. The endeavor to green NCL shall continue in future and the company shall strive to be one the best Green Coal companies of the nation for fulfilling the dream of a better life and sustainable environment

Century by CREST, Chandigarh

CREST scored a Century by installing 100 Rooftop Solar Power Plants with overall installed capacity of 5.3MWp

These 100 Solar Power Plants include followings:

- Installed & Commissioned 1480 kWp Rooftop Solar Power Plants in 35 Govt. Schools out of 125 Govt. Schools and will install on all schools by March, 2017.
- Installed & Commissioned 2.58MWp Rooftop Solar Power Plants on 10 Govt. Colleges.
- Largest Rooftop Solar PV Plant of 1 MW capacity Installed & Commissioned at Punjab Engineering College (PEC), Chandigarh.

CREST have received following prizes from the MNRE, on the First Foundation Day of Association of Renewable Energy Agencies of State (AREAS) on 27th August 2015 at Bengaluru, Karnataka for its excellent performance in the field of renewable energy.

1. Prize for second highest grid connected solar rooftop power capacity addition in the country.
2. Prize for second highest cumulative grid connected renewable power capacity amongst the UTs in the country.
3. Prize for second for installation of renewable power applications in its office building at Chandigarh.

Chandigarh Renewal Energy, Science & Technology Promotion Society (CREST)
(Under the Aegis of Department of Science & Technology, Chandigarh Administration),
Ist Floor, Paryavaran Bhawan, Sector-19-B, Chandigarh. Ph. 0172- 2771919.
www.crestchd.org.in, email: crestchandigarh@gmail.com

Contents

Preface xiii
About TEDDY 2015/16 xv
Contributors xvii
Acknowledgements xix
List of tables xxi
List of figures xxix
List of maps xxxiii

1 Energy and environment: an overview 1
2 Commercial energy balance tables and conversion factors 13

Energy supply

3 Coal and lignite 25
4 Petroleum and natural gas 73
5 Power 99
6 Renewable energy sources and technologies 163

Energy demand

7 Agriculture 197
8 Industry 211
9 Transport 229
10 Household energy 271

Local and global environment

11 Environment 297
12 Climate change 343

Energy and environment goals

13 Sustainable development goals and implications for India 369
14 Air pollution and health 375

Index 393

Preface

India with its current and projected future economic growth has established an identity of a stable emerging economy, but this growth is accompanied by the increase in energy demand and related environmental degradation. Although energy access has improved in India, there still exists a substantial population, particularly in rural India, that depends on non-commercial fuels as a source of cooking, which have adverse effects on air quality and human health. The year 2015 marked an important milestone for the future of the world with the launch of new sustainable development goals (SDGs) and adoption of an ambitious Paris Agreement on climate change. The SDGs have specific goals related to energy. Goal 7 of the SDGs focuses on ensuring access to affordable, reliable, sustainable, and modern energy for all. Besides, energy is also implicitly addressed in other SDGs.

A certain level of climate change is inevitable as reiterated by the Fifth Assessment Report (AR5) of the Intergovernmental Panel on Climate Change (IPCC) and it would pose risks for a range of ecosystems in India and the livelihoods of people, particularly the poor. The twenty-first session of the Conference of the Parties held in Paris produced an ambitious agreement which recognized the need to hold the global temperature rise to "well below 2°C". The Paris Agreement rests on nationally determined contributions at its core. India's intended nationally determined contributions (INDCs) have been put forward with the intent of reducing its emission intensity by 33–35 per cent by 2030 (over the 2005 level). These INDCs are built on India's goal of installing 175 GW of renewable power capacity by 2022. Reliable data would be required to ensure the optimization of the pathway for the development, transition, and monitoring of progress towards global goals and commitments. It is, therefore, imperative to build capacities for national data collection, disaggregation, and analysis.

The Energy and Resources Institute (TERI) takes pride in presenting the thirtieth edition of its flagship publication *TEDDY (TERI Energy & Environment Data Diary and Yearbook)*, which is the only comprehensive energy and environment yearbook in India. *TEDDY* serves as a ready reckoner for energy and environment related information and statistics sourced from government agencies, policy documents, and other secondary sources.

In this publication, energy supply is discussed in chapters on coal and lignite, petroleum and natural gas, power, and renewable energy sources and technologies. The section on energy demand covers extensively the major energy-consuming sectors of agriculture, industry, transport, and households. The section on local and global environment examines local environmental issues, such as the comprehensive state of environment assessment, forestry, biodiversity, and climate change, and their respective impacts on society. The publication provides a review of government policies that have implications for these energy and environment sectors. It also contains detailed data tables that provide enhanced information on the sectors covered in the publication. An updated commercial energy flow in India with explanations prepared by TERI researchers is an important aspect of this publication, as it predicts the flow of net available energy and its consumption by demand sectors in the future. The Green Focus at the end of each chapter provides a description of sustainable development practices or initiatives related to each discussed sector. *TEDDY* has been envisaged to be at the forefront of global and local policy discourses. From this perspective, it also features a section on sustainable development goals as well as air pollution and health.

TERI Energy & Environment Data Diary and Yearbook has been a much sought after reference document across the world and it has been cited in government policy documents, scholarly articles, journals, and other peer-reviewed books in India and abroad.

At TERI, we attempt to bring constant refinement to *TEDDY* and are hopeful that as all the previous issues, *TEDDY 2015/16* would also be of immense value to policy makers, business organizations, academic institutions, and research scholars. Comments and suggestions from readers shall be appreciated as these would help us in improving subsequent issues of this publication.

Advisory and Editorial Team
TERI Energy & Environment Data Diary and
Yearbook 2015/16

About TEDDY 2015/16

TERI Energy & Environment Data Diary and Yearbook (TEDDY) is an annual publication brought out by The Energy and Resources Institute (TERI) since 1986. The publication provides updated information on the energy supply sectors (coal and lignite, petroleum and natural gas, power, and renewable energy sources), energy-consuming sectors (agriculture, industry, transport, residential, and commercial sectors), and environment (local and global). The entire publication uses interactive graphs, figures, maps, and tables to explain facts, which makes the book an interesting read. The publication also provides a review of the government policies that have implications for energy and environment.

The data in the yearbook provides the latest available information at the time of compilation of the chapters. The analytical narrative supporting the data has been well researched by sector experts at TERI. Care has been taken to ensure that continuity of information is maintained so that the readers can understand and analyse the trends and patterns of change in the energy and environment sector over a period of time.

The launch of *TEDDY 2014/15* witnessed the presence of eminent people, including Mr Prabir Sengupta (Distinguished Fellow, TERI), Mr Chandrashekhar Dasgupta (Distinguished Fellow Emeritus, TERI), Dr Prodipto Ghosh (Distinguished Fellow, TERI), Mr Anil Kumar Jain (Advisor-Energy, National Institution for Transforming India), Dr Leena Srivastava (Vice Chancellor, TERI University), and Dr P C Maithani (Director, Ministry of New and Renewable Energy).

The overall structure of this year's *TEDDY* is the same as the previous editions, with some improvisations. For the reference of our readers, the following table lists the chapters of the publication.

Chapter 1	:	Energy and environment: an overview
Chapter 2	:	Commercial energy balance tables and conversion factors
Chapter 3	:	Coal and lignite
Chapter 4	:	Petroleum and natural gas
Chapter 5	:	Power
Chapter 6	:	Renewable energy sources and technologies
Chapter 7	:	Agriculture
Chapter 8	:	Industry
Chapter 9	:	Transport
Chapter 10	:	Household energy
Chapter 11	:	Environment
Chapter 12	:	Climate change
Chapter 13	:	Sustainable development goals and implications for India
Chapter 14	:	Air pollution and health

The thirtieth edition of the publication, *TEDDY 2015/16*, comes with several interesting features. The Green Focus at the end of each chapter highlights sustainable initiatives and successful practices, which are of current interest in the sectors discussed under the sections on energy supply, energy demand, and local and global environment. The publication also features a section with a discussion on sustainable development goals (SDGs) and air pollution and health.

We are hopeful that like all the previous editions of the publication, *TEDDY 2015/16* would be of immense value to our readers including policy makers, business organizations, academic institutions, research scholars, and development practitioners. Suggestions and comments from the readers shall be highly appreciated as these would help us in improving subsequent issues of this publication.

Contributors

Overall guidance

R K Pachauri
Prabir Sengupta

Editors

Ashutosh Senger
Shailly Kedia
Rinki Jain

Chapter Authors and Reviewers

Energy and environment: an overview
Authors : Ashutosh Senger and Rinki Jain
Reviewers : Shailly Kedia and Suneel Pandey

Commercial energy balance tables and conversion factors
Author : Nishant Jain
Reviewer : Rinki Jain

Coal and lignite
Authors : Swati Mitchelle DSouza and Jonathan Donald Syiemlieh
Reviewer : Nitya Nanda

Petroleum and natural gas
Authors : Siddharth Singh and Madhura Joshi
Reviewer : Pravin Kumar Agarwal

Power
Authors : S Narayan Kumar and Ashutosh Senger
Reviewer : K Ramanathan

Renewable energy sources and technologies
Authors : Shreya Agarwal, Nagendra Kumar, Abhishek Agrawal, Sarvesh Devraj, Anshu Yadav, and Ankit Narula
Reviewers : Shirish Garud and Ujjwal Bhattacharjee

Agriculture
Authors : Manish Anand, Shilpanjali Deshpande Sarma, and Prakashkiran S Pawar
Reviewer : Nutan Kaushik

Industry
Author : Sachin Kumar
Reviewer : Prosanto Pal

Transport
Author : Sangeetha Ann Wilson
Reviewer : Shri Prakash

Household energy
Authors : Martand Shardul and Bigsna Gill
Reviewer : S Arun

Environment
Authors : Jai Malik and Ashutosh Senger
Reviewers : Suresh Chauhan, Siddharth Edake, and R Suresh

Climate change
Authors : Swati Agarwal, Shivani Wadehra, and Saurabh Bhardwaj
Reviewer : Manish Kumar Shrivastava

Sustainable development goals and implications for India
Authors : Shailly Kedia and Ipsita Satpathy
Reviewer : Chandrashekhar Dasgupta

Air pollution and health
Authors : Rinki Jain and Karnika Palwa
Reviewers : Meena Sehgal, Sumit Sharma, and Shailly Kedia

Maps

R K Joshi

Data compilation

M Jayaraman

Executive support

P D Tiwari

Documentation

Library and Information Centre, TERI

Publishing: copy edit, graphics, and production

Aman Sachdeva
Anupama Jauhry
Sachin Bhardwaj
Santosh Gautam
Shashi Bhushan
Shilpa Mohan
Sushmita Ghosh

Information dissemination

Gitesh Sinha
Kakali Ghosh
Sangeeta Gupta
Sanjeev Sharma

CD publishing

Rajiv Sharma

Acknowledgements

The thirtieth edition of *TERI Energy & Environment Data Diary and Yearbook (TEDDY)* is a product of collective effort by the research and publishing team at The Energy and Resources Institute (TERI). The publication provides the latest updated information and statistics about the energy and the environment in India.

Throughout the preparation of the publication, the team gained guidance and support from Dr R K Pachauri, Executive Vice Chairman, TERI; without his leadership this publication would not have been possible. We especially thank Mr Prabir Sengupta, Distinguished Fellow, TERI for his constant encouragement, support, and guidance for this publication.

We extend our gratitude to all the authors and reviewers of the publication who have done due fact checking and put together research narratives for the chapters. We greatly acknowledge the colleagues at TERI Press for their constant support in bringing out the publication. Finally, we thank Mr P D Tiwari for extending all possible executive support.

We are confident that this publication meets the highest standards of scholarship and would be immensely useful to our readers including policy makers, business organizations, academic institutions, and research scholars working in areas of energy and environment.

List of tables

Energy and environment: an overview

1 Final commercial energy consumption (in MTOE) in India by sector 2

Commercial energy balance tables and conversion factors

1 Main conversions used in the petroleum industry 13
2 Heat units and others expressed 14
3 Categorization of coking coal grades 14
4 Categorization of non-coking coal grades 14
5 Standard useful heat values 14
6 Calorific values of different petroleum products 15
7 Conversion factors of petroleum products 15
2.1 Commercial energy balance (million tonnes of oil equivalent): 2012/13 18
2.2 Commercial energy balance (million tonnes of oil equivalent): 2011/12 19
2.3 Commercial energy balance (million tonnes of oil equivalent): 2010/11 20
2.4 Commercial energy balance (million tonnes of oil equivalent): 2009/10 21
2.5 Commercial energy balance (million tonnes of oil equivalent): 2008/09 22

Coal and lignite

1 Coal production in 2011/12, 2012/13, 2013/14, and 2014/15 by company 28
2 Coal consumption pattern (in MT) 30
3 Coal handling by ports (MT) 35
4 Coal supply and demand (in MT) 36
5 Companies participating in the initial rationalization of linkages 37
3.1 Coal resources of states (million tonnes) as on 1 April 2015 44
3.2 Coal resources (million tonnes) by category as on 1 April 2015 44
3.3 Lignite resources of states and union territories (million tonnes) as on 1 April 2015 . 45
3.4 Lignite mined out (in million tonnes) till 31 March 2015 45
3.5 Coal production (million tonnes) in various states: 2000/01–2014/15 46
3.6 All-India production (million tonnes) of coal, lignite, and other derivatives: 2000/01–2014/15 47
3.7 Coal production (million tonnes) of various companies: 2000/01–2014/15 48
3.8 Coal dispatches (million tonnes) in various states: 2000/01–2014/15 50
3.9 Coal dispatches (million tonnes) of various companies: 2000/01–2014/15 51
3.10 Supply of indigenous coal ('000 tonnes) to steel plants: 2008/09–2013/14 53
3.11 Availability and consumption (million tonnes) of raw coal by sector: 2000/01–2014/15 54
3.12 Indigenous coal supply plan (BE) and achievement 2014/15 55
3.13 Output per man-shift (tonnes) in opencast and underground mining: 2000/01–2014/15 56
3.14 Inter-modal split of coal movement (million tonnes): 2000/01–2014/15 56
3.15 Traffic handled (million tonnes) at major ports: 2007/08–2014/15 57

3.16 Customs duty structure on imported coal: 2010/11 58
3.17 Import (million tonnes) of coking and non-coking coal: 1991/92–2014/15 58
3.18 Comparison of gross calorific value and useful heat value for non-coking coal (with 6% moisture content) 59
3.19a Categorization of coal and coke 60
3.19b Grades of coal 61
3.20 Pricing of non-coking coal based on gross calorific value 62
3.21 Revised rates of royalty (₹/tonne) 64
3.22 Royalty paid (₹ million) on coal to the states: 1991/92–2012/13 65
3.23 Average pithead unit value (₹/tonne) of coal: 1974/75–2008/09 66
3.24a Allocation of coal and lignite blocks during 2014/15 67
3.24b Allotment of captive coal blocks as on 31 March 2014 by year and sector 68
3.25 Annual plan and actual expenditure (₹ million): 1993/94–2015/16 69

Additional tables in the CD

3.26a Raw coal feed and clean coal production (million tonnes) of washeries: 1995/96–2006/07
3.26b Coking coal washery performance ('000 tonnes): 2007/08–2013/14
3.26c Performance of non-coking coal washeries owned by collieries: 2007/08–2013/14
3.27 Production (million tonnes) from coal washeries: 1990/91–2006/07
3.28 Coal pithead stock (million tonnes) of various companies: 1995/96–2014/15
3.29 Coal production (million tonnes) in various companies: 1993/94–2014/15
3.30 Supply of indigenous coal ('000 tonnes) to steel plants: 2000/01–2007/08

Petroleum and natural gas

1 Trend in onshore-offshore crude oil production (in MT) 74
2 Production of crude oil by region (in MT) 75
3 Top five countries from which India imported crude oil in 2014/15 (in MT) 75
4 Capacity, length, and operators of major crude pipelines 77
5 Installed capacity and crude throughput 79
6 Gross refining margins (in $/bbl) 80
7 Production of petroleum products (in '000 tonnes) 80
8 Consumption of petroleum products (in '000 tonnes) 81
9 Imports of petroleum products (in '000 tonnes) 81
10 Exports of petroleum products (in '000 tonnes) 82
11 Length and capacity of major petroleum product pipelines 83
12 Under-recoveries to oil companies (in ₹ crore) by product 84
13 Fiscal subsidy on PDS kerosene and domestic LPG (in ₹ crore) 84
14 Share of under-recoveries (in ₹ crore) 84
15 Production of natural gas (in BCM) 85
16 Production of onshore natural gas by region (in BCM) 85
17 Production of offshore natural gas by region (in BCM) 85
18 Existing and planned LNG terminals in India 87
19 Details of existing natural gas pipelines 90
20 Gas pipelines under construction 91
21 Offtake of natural gas by industry (in MCM) 91
22 Coalbed methane production in India (in mmscm) 93

Additional tables in the CD

4.1 Hydrocarbon discoveries by ONGC, OIL, PSC regime during 2013/14

4.2 Oil and natural gas reserves: 2005–2015

4.3 Production ('000 tonnes) of crude oil: 2000/01–2014/15

4.4 Production (million cubic metres) of natural gas: 2000/01–2014/15

4.5 Reserve replacement ratio and life of reserves for crude oil (million tonnes) and natural gas (billion cubic metres): 1995/96–2014/15

4.6 Installed refining capacity and refinery crude processing ('000 tonnes): 2007/08–2014/15

4.7 Production ('000 tonnes) of petroleum products: 2008/09–2014/15

4.8 Imports ('000 tonnes) and value (₹ million) of crude oil and petroleum products: 2007/08–2014/15

4.9 Exports ('000 tonnes) and value (₹ million) of crude oil and petroleum products: 2007/08–2014/15

4.10 Sales/consumption of petroleum products ('000 tonnes): 2007/08–2014/15

4.11 Sector-wise consumption of natural gas (2008/09–2014/15) (in MMSCM)

4.12 Subsidies on major petroleum products (₹ million): 1995/96–2008/09

4.13 Compressed natural gas activities in India as on 31 March 2015

4.14 Exploratory and developmental drilling by the Oil and Natural Gas Corporation Ltd and Oil India Ltd: 2005/06–2014/15

4.15 Refinery crude throughput and production ('000 tonnes) of petroleum products: 2005/06–2014/15

4.16 Average annual compound growth rates of consumption of petroleum products (%)

4.17 Consumption ('000 tonnes) of major petroleum products, by state (public sector undertakings sales only): 2014/15

4.18 Development of spot prices of key Organization of the Petroleum Exporting Countries, crudes ($ per barrel): 2006–14 (September)

4.19 Realization of excise and customs duties from crude oil and petroleum products (₹ million): 2005/06–2014/15

4.20 Consumer prices of natural gas (₹/1000 cubic metres) (excluding royalty, taxes, and all statutory levies)

4.21a Tenth Five-year Plan outlay of public sector undertakings under the purview of the Ministry of Petroleum and Natural Gas (₹ crore)

4.21b Twelfth Five-year Plan outlay and actual expenditure for petroleum and natural gas sectors (₹ crore)

4.21c Twelfth Five-year Plan outlay and actual expenditure for petroleum and natural gas sectors (₹ crore)

Power

1 Transmission and distribution network (in ckt km) 102

2 Capacity addition target and achievement for the Twelfth Five-year Plan (MW) by sector and source 104

3 Current status of smart grid pilot projects in India 107

5.1a Installed capacity (MW) as on 31 March 2015 113

5.1b Growth of installed electrical energy generation capacity (MW), utilities by mode .. 119

5.2 Annual gross generation in utilities (GWh): 2010/11–2012/13 120

5.3 Annual plant load factor (%) of thermal/nuclear electricity generation plants, by state/union territory: 2012/13 121

5.4 Installed capacity (MW) and generation (GWh): 1990/91–2012/13 (with respect to non-utilities of installed capacity 1 MW and above) 123

5.5 Length of transmission and distribution lines (ckt km) by region and state/union territory as on 31 March 2013 125

5.6 Transmission and distribution losses as a percentage of availability in states/ union territories: 2001/02–2012/13 127

5.7a Aggregate technical and commercial losses (%) for SEBs and PDs selling directly to the consumers: 2011/12–2013/14 129
5.7b Aggregate technical and commercial losses (%) for all discoms of unbundled states: 2011/12–2013/14 130
5.7c Aggregate technical and commercial losses (%) for utilities selling directly to consumers: 2009/10–2011/12 134
5.8 Energy sales to ultimate consumer utilities (GWh): 2012/13 138
5.9 Number of consumers and connected load: 2000/01–2012/13 140
5.10 Per capita consumption (kWh) of electricity: 2000/01–2012/13 141
5.11 Power supply position: peak demand versus peak met (2014/15) 143
5.12 Power supply position: requirement versus availability (April 2014–March 2015) ... 145
5.13a Status of consumer metering: 2006/07 147
5.13b Number of consumers (in lakh): 2009/10–2013/14 149
5.14 List of existing inter-regional links 150
5.15 Details of nuclear power projects as on December 2015 152
5.16 Progress report of village electrification as on 31 October 2015 153
5.17 Electrification of rural households 155
5.18 Status of hydroelectric potential development, in terms of installed capacity of above 25 MW as on 31 October 2015 157
5.19 Profit and loss of power utilities (₹ million) by state: 2011/12–2013/14 159
5.20 Status of reforms and restructuring in power sector 161

Additional tables in the CD

5.21 Plan outlay and expenditure (₹ million) for the power sector
5.22 Electricity generation and consumption (GWh) in selected industries having captive power plants of 1 MW and above: 2012/13

Renewable energy sources and technologies

1 Status of bio-power (grid-connected/off-grid) and family–size biogas plants in India (till August 2015) 165
2 State-wise list of biomass gasifier plants installed in India with their capacity 166
3 List of solar photovoltaic systems with different capacity and financed by MNRE ... 167
4 Details of some successfully operated biogas-based power plants with performance 170
5 Biomass co-generation projects installed (MW capacity) in different states during 2014–15 170
6 JNNSM targets and achievements 173
6.1a Cumulative deployment of various renewable energy systems/devices in India as on 31 December 2014 183
6.1b State-wise renewable energy potential (MW) 184
6.2 Wind power cumulative installed capacity (MW) in India, by state: 2006–15 185
6.3 Wind power installed capacity (MW) in India, by state: 2002/03–2014/15 186
6.4 Wind power potential in India 187
6.5 Potential, installed, and under implementation projects in small hydropower sector (up to 25 MW), by state (as on 31 December 2014) 188
6.6 National Biogas and Manure Management Programme (plants in number) 189
6.7 State-wise cumulative commissioned biomass power and bagasse cogeneration projects 191
6.8 Tentative state-wise breakup of renewable power target to be achieved by 2022 192

Agriculture

1 Electricity consumption in the agriculture sector 200
2 Number of diesel and electric pumps used in India 202
3 Production, imports, and consumption of fertilizers (in thousand tonnes of nutrients) 202
4 Water requirement for irrigation in India (in BCM) 205
5 Efficiencies of various facilities for utilization of irrigation water 205
6 Comparison of water use in traditional and drip irrigation systems for different crops 206
7 State-wise potential and actual area under micro-irrigation in India 206

Industry

1 Annual growth rate of industrial production (%) 212
2 Production of aluminium by primary aluminium producers 213
3 Specific energy consumption for 2014 213
4 Crude steel capacity and production in India (in MTPA) 214
5 Percentage share of different process routes in crude steel production 214
6 Installed capacity of various fertilizers (in MTPA) 215
7 Installed capacity of fertilizer plants (urea) in India 216
8 Production of urea, diammonium phosphate, and complexes (in MT) 217
9 Production of spun yarn (in million kg) 218
10 Capacity share of major cement plants (2012/13) 219
11 General trend of SEC in Indian cement plants 219
12 Product-wise installed capacity and production of alkali chemicals 220
13 Average operating SEC of the chlor-alkali sector 220
14 Production of paper from different raw materials in 2010/11 (in MT) 221
15 Installed capacity and production of selected pulp and paper mills for 2010/11 222
16 Distribution of enterprises by main source of energy 222
8.1 Installed capacity ('000 tonnes), production ('000 tonnes), and capacity utilization (%) in the major steel industry: 2007/08–2014/15 226
8.2 Installed capacity (lakh tonnes), production (lakh tonnes), and capacity utilization (%) in the fertilizer industry: 2000/01–2014/15 227
8.3 Import (lakh tonnes) of fertilizers: 2000/01–2014/15 227
8.4 Production of aluminium (in tonnes): 2006/07–2014/15 227
8.5 Number of working, medium, small, and micro enterprises/industries: 1992/93–2013/14 228

Transport

1 Type and length of different road categories in India (as on 31 March 2013) 231
2 Status of NHDP phases 231
3 Total number of registered motor vehicles in India (2009–12) (in million) 232
4 Length of traction (in km) by gauge (as on 31 March 2014) 235
5 Details of National Waterways in India 238
9.1 Key data on transport: 2000/01–2013/14 244
9.2 Gross domestic product in the transport sector (million rupee): 1999/2000–2011/12 246
9.3 Consumption ('000 tonnes) of petroleum, oil, and lubricants in the transport sector: 2007/08–2014/15 247
9.4 Total length (kilometre) of national highways (as on 31 March): 2000–13 248

9.5 Registered motor vehicles (thousand): 1951–2013 ... 250
9.6 Registered motor vehicles (number) in India: 31 March 2013 ... 251
9.7 Indian shipping and gross registered tonnage (million): 2000–14 ... 253
9.8 Capacity available (million tonnes) at major ports, by commodity: 2000–14 ... 253
9.9 Cargo traffic handled (million tonnes) at major ports, by commodity: 2005/06–2014/15 ... 254
9.10 Performance indicators of selected ports: 2009/10–2013/14 ... 255
9.11 Performance of the railways: 2003/04–2013/14 ... 256
9.12 Railway traffic: 2000/01–2013/14 ... 257
9.13 Electrification of railway routes and tracks (kilometre): 2000/01–2013/14 ... 257
9.14 Energy intensity in railways (consumption per thousand gross tonne-km): 2010/11–2013/14 ... 258
9.15 Trends in relative rail and road traffic: 2000/01–2012/13 ... 259
9.16 Consumption of fuel/energy in Indian Railways: 2009/10–2013/14 ... 260
9.17 Traffic trends in terms of aircraft, passenger, and cargo movements: 2002/03–2014/15 ... 261
9.18 Growth of civil aviation 2000/01–2014/15 ... 262
9.19 Physical performance of Air India Limited: 2011/12–2014/15 ... 263
9.20 Share of different modes of transport in gross domestic product at basic and constant prices 2011/12–2013/14 ... 264
9.21 Production of motor vehicles (number) in India: 2003/04–2013/14 ... 265
9.22 Total and surfaced road length (kilometre) by categories in India as on 31 March 2013 ... 266
9.23 Total road length (kilometre) and percentage share of each category of road (1951–2013) ... 267
9.24 Length (kilometre) of national highways in India: 2005–13 (as on 31 March) ... 269

Additional tables in the CD

9.25 Registered motor vehicles (in number) in million plus cities as on 31 March 2013
9.26 Vessel traffic at major ports (in number), by category: 2000/01–2013/14
9.27 Cargo traffic handled ('000 tonnes) at minor ports, by principal commodity: 2003/04–2013/14

Household energy

1 Distribution of households on the basis of energy source for lighting ... 272
2 Source of energy for domestic cooking in India ... 273
3a Percentage distribution of households by primary energy source for cooking (2001/02–2011/12) ... 274
3b Percentage distribution of households by primary energy source for lighting (2001/02–2011/12) ... 275
4a Distribution of rural households (per 1000) of each household type by primary source of energy for cooking ... 277
4b Distribution of urban households (per 1000) of each household type by primary source of energy for cooking ... 277
5 Progress in village electrification in India ... 279
6 Electrical energy sales to ultimate consumers (utilities): 2012/13 ... 280
7 Consumption of LPG and SKO by the domestic sector ... 281
10.1a State-wise distribution of households based on source of energy for lighting ... 285
10.1b State-wise distribution of households by type of fuel used for cooking ... 287

10.2a Source of energy for domestic lighting in rural India 289
10.2b Source of energy for domestic lighting in urban India 291
10.3a Source of energy for lighting in slum households 293
10.3b Source of energy for cooking in slum households 293

Environment

1 Forest cover of India in 2013 305
2 Number of threatened species in India 307
3 Details of protected areas in India (as on 6 October 2015) 307
11.1a Ambient SO_2 levels in highly polluted cities ($\mu g/m^3$): 1995–2012 311
11.1b Ambient NOx levels in highly polluted cities ($\mu g/m^3$): 1995–2012 319
11.1c Ambient RSPM levels in highly polluted cities ($\mu g/m^3$): 1999–2012 327
11.2 Number of air monitoring stations in India violating the 24 hourly standards for a particular per cent of times in a year (2001–10) 335
11.3 Concentration of pollutants as a result of burning biofuels 337
11.4 New National Ambient Air Quality Standards (NAAQS) for India 338
11.5 Interventions adopted to control vehicular emissions in the period 1991–2010 339
11.6 Available water resources: summary for India and world 341
11.7 Municipal solid waste generation in India (2009–12) 342

Climate change

1 Probability statements regarding RCP scenarios 349
2 Level of CO_2 emissions and per capita CO_2 emissions in the world 350
3 Emission trend across four major CO_2-emitting countries 351
4 Emissions related to energy 351
5 Global emissions and emission gap under the implementation of INDCs 351
6 CO_2 emissions in India (in $MtCO_2$) 352
7 CO_2 emissions within energy subsectors 353
8 GHG emission trends in India by gases (million tonnes of CO_2 equivalent) 354
9 Total approved CDM projects in 2015 355
10 National Action Plan on Climate Change 357
11 Current status of state action plans on climate change 358
12.1 Carbon dioxide emissions across world economies 363

Sustainable development goals and implications for India

1 Sustainable Development Goals for Agenda 2030 371
2 Efforts on reporting energy and environment data/statistics 373

Air pollution and health

1 Percentage of total global deaths due to causes attributed to air pollution 379
2 Total number of YLL, YLD, and DALY (in million) for 2000 and 2012 for male and female of all age groups for the diseases attributable to air pollution 380
3 Indicative estimate of the economic costs of health impacts of ambient air pollution in OECD countries, India, and China in 2005 and 2010 384

List of figures

Energy and environment: an overview

1 Commercial energy flow in India for 2012/13 ... 5

Coal and lignite

1 All-India installed capacity (as on 31 March 2015) ... 26
2 Coal reserves in India (MT) (as on 1 April 2015) ... 26
3 Raw coal production over 10 years ... 28
4 Lignite production over 10 years ... 29
5 Production of coal from opencast and underground mining (in per cent) from 2005/06 to 2014/15 ... 32
6 Import of coking and non-coking coal (in MT) ... 35
7 Coal transportation by various modes for 2014/15 ... 36
8 Accident and death rates in coal mines in India ... 40

Petroleum and natural gas

1 Quantity and value of crude oil imports in India ... 76
2 Imports of crude oil and petroleum products *vis-a-vis* total imports ... 76
3 Natural gas production in India by sector (in BCM) ... 86
4 Natural gas imports (in MT) ... 86
5 Geology of natural gas resources ... 93

Power

1 Installed capacity by source (as on 31 March 2015) ... 100
2 Installed capacity by sector: 2013/14 versus 2014/15 ... 100
3 Trend in electricity generation by source (from 2008/09 to 2014/15) ... 100
4 Electricity generation by sector (as on 31 March 2015) ... 101
5 Installed capacity of captive generation by source: 2011/12 versus 2012/13 ... 101
6 Captive generation by source ... 101
7 Trend in annual plant load factor of central, state, and private thermal stations from 2007/08 to 2014/15 ... 102
8 System losses by region: 2011/12 versus 2012/13 ... 102
9 Trend in peak deficit from 2009/10 to 2014/15 ... 103
10 Trend in energy deficit: 2009/10 to 2014/15 ... 103
11 Consumer-connected load by category (as on 31 March 2013) ... 103
12 Share of different segments within the total short-term transaction for 2014/15 ... 106
13 Trends in the price of short-term transactions of electricity (2014/15) ... 106

Renewable energy sources and technologies

1 Grid-connected renewable energy installed capacity and power sector in India ... 164

Agriculture

1 Foodgrain production of India ... 198
2 HSD and LDO consumption in agriculture ... 199

3 Number of tractors sold (from 1985/86 to 2014/15) 201
4 Number of power tillers sold (from 1985/86 to 2014/15) 201
5 Production of urea, diammonium phosphate, and other complex fertilizers (in MT) 203
6 Ultimate irrigation potential in India 203

Industry
1 Energy consumption trends in urea plants 217

Transport
1 Consumption of high-speed diesel in different transport modes (in '000 tonnes) 230
2 Registered motor vehicles (in thousand) for 1951–2012 233
3 Share of consumption of different petroleum products in the road transport sector 233
4 Trend of originating passenger and passenger kilometre growth in railways (in million) 235
5 Trend of originating freight traffic and freight tonne kilometre growth in railways (in million) 235
6 Growth in the consumption of electricity and high-speed diesel by the railways for locomotive services 236
7 Consumption of petroleum products in shipping (tonnes, %) during 2011/12 238
8 Trend in the growth of domestic passengers and cargo in the aviation sector 238

Household energy
1 Population without access to electricity and reliance on traditional biomass 272
2a Distribution of urban households (per 1000) of each household type by the primary source of energy for lighting 276
2b Distribution of households (per 1000) of each rural household type by the primary source of energy for lighting 276
3 Energy balance on a net calorific value basis—residential consumption in India 278
4 Consumption of electricity in India by sector (2013/14) 278
5 Consumption of electricity by the domestic sector (in '000 GWh) 281

Environment
1 Urban population as a percentage of the total population of some selected countries 298
2 Percentage of locations in residential/industrial/commercial/rural areas exceeding NAAQS 298
3 Top 10 cities with highest NO_2 values (annual average) in 2012 299
4 Top 10 cities with the highest PM10 values (annual average) in 2012 299
5 Prevalence of active tuberculosis, ARI, and partial/complete blindness in India by type of cooking fuel used 300
6 BOD trend of waterbodies in India 301
7 Total coliform trend of waterbodies in India 301
8 Faecal coliform trend of waterbodies in India 302
9 Top 10 states generating the highest municipal solid waste per day (2009–12) 303

Climate change
1 All-India annual mean temperature anomalies for 1901–2013 (based on the 1961–90 average) 344
2 Spatial patterns of linear trends of (a) maximum and (b) minimum temperatures 345
3 Spatial patterns of trends (°C per 100 years) in mean annual temperature anomalies (1901–2014) 346

4 Sub-divisional trends of (a) seasonal and (b) monsoon rainfall for 1901–2003 348
5 Time series of active (upper panel) and break (lower panel) days during the monsoon season 349
6 Over 1000 scenarios from the IPCC *Fifth Assessment Report* 350
7 India's primary energy demand by fuel type and related CO_2 emissions in the INDC scenario 353

Sustainable development goals and implications for India

1 Key milestones leading to Agenda 2030 370

Air pollution and health

1 Annual mean concentration of PM_{10} ($\mu g/m^3$) in regions of the world........ 376
2 Annual mean concentration of $PM_{2.5}$ ($\mu g/m^3$) in regions of the world 377
3 Leading causes of global deaths and premature deaths........ 378
4 Total global deaths attributable to household and ambient air pollution in 2012 (by region) 379
5 The number of global deaths in children due to acute lower respiratory infections 380
6 Predicted average gain in life expectancy with a reduction in average annual levels of $PM_{2.5}$ 10 $\mu g/m^3$ in 25 European cities 381
7 Total deaths from ambient particulate matter and ozone pollution in India........ 382
8 Percentage distribution of deaths from ambient particulate matter pollution in India 382
9 Percentage by state of medically certified deaths due to diseases of the respiratory system in 2011........ 383
10 Total cases by state of ARI infection deaths due to diseases of the respiratory system in 2011........ 383

List of maps

Coal and lignite

1 Gross geological coal resources by state as on 1 April 2015 27

Petroleum and natural gas

1 Petroleum refineries and major crude oil pipelines in India 78
2 LNG terminals and natural gas pipelines in India 89

Power

1 Number of LEDs sold till 26 October 2015 109

Renewable energy sources and technologies

1 Installed capacity of grid-connected solar power plants 174
2 Geothermal map of India 176

Agriculture

1 State-wise irrigation potential in India ('000 hectares) 204

Environment

1 Forest cover of India 306
2 United Nations' list of protected areas of India in 2014 308

ENERGY AND ENVIRONMENT: AN OVERVIEW 1

Introduction

India is currently the fourth largest consumer of primary energy in the world after China, United States of America, and Russian Federation (BP 2015). The total primary energy supply in the country for 2012/13 was around 320 million tonnes of oil equivalent (MTOE). Coal continues to be the dominant fuel, with a share of 57% in the total primary energy supply. India's human development index for 2014 was 0.609, which placed the country in the medium human development category and at 130th position among the 188 countries (UNDP 2015). Given the correlation between energy access and human development, access to energy is at the core of India's policy making.

The Environmental Performance Index (EPI)[1], developed for 178 countries, placed India at the 155th position in the world in terms of the highest average exposure to fine particulate matter ($PM_{2.5}$) (EPI 2014). Occupying 2.4% of the world's land area, India is home to one-sixth of the total plant species in the world (MoSPI 2013). Out of the 34 biodiversity hotspots in the world, four are present in India: the Himalayas, the Western Ghats, the North East, and the Nicobar Islands. India's environmental jurisprudence is evolving with the intent to strike a balance between economic growth and environment. With growing energy and resource demand and the growing emphasis on environmental consequences, there is a need for a coordinated and effective institutional framework for sustainable development in India.

The *Fifth Assessment Report* of the Intergovernmental Panel on Climate Change (IPCC) predicts water stress and food shortage in South Asia in the future (IPCC 2014). For countries such as India, sustainable development is the basis for assessing climate change responses. The Intended Nationally Determined Contribution (INDC) submitted by the Government of India includes adaptation, mitigation, requirement for finance, technology transfer, and capacity building. As per the submitted INDC, India aims to reduce the emission intensity per unit of its gross domestic product (GDP) by 33%–35% by 2030 from the 2005 level and achieve 40% of electric power installed capacity from non-fossil fuels by 2030 (GoI 2015). At the 21st Conference of Parties of the United Nations Framework Convention on Climate Change held in Paris, India emphasized on climate justice, carbon space for equitable growth and the principle of Common but Differentiated Responsibility.

The Government of India has revised the renewable energy capacity target to 175 000 MW till 2022, comprising 100 000 MW solar, 60 000 MW wind, 10 000 MW biomass, and 5000 MW small hydropower (SHP). The National Smart Grid Mission in the power sector has been established to plan and monitor the implementation of policies and programmes related to smart grid activities in India (MoP 2015). Clean energy cess has been increased from ₹100 to ₹200 per tonne of coal to finance clean environment initiatives. The Government of India has established the National Adaptation Fund on Climate Change (NAFCC) with the objective to assist state and union territories that are particularly vulnerable to the adverse effects of climate change in meeting the cost of adaptation (PIB 2015a).

[1] A metric that ranks countries on the basis of their performance on high-priority environmental issues in two broad policy areas: protection of human health against environmental harm and protection of ecosystems.

Table 1 Final commercial energy consumption (in MTOE) in India by sector

Sector	1980/81	1985/86	1990/91	1995/96	2000/01	2005/06	2009/10	2010/11	2011/12	2012/13
Agriculture	1.6 (2.3%)	2.4 (2.6%)	4.9 (3.9%)	8.4 (5.3%)	15.2 (7.9%)	15.1 (6.9%)	23.14 (7.32%)	18.70 (5.9%)	21.79 (6.17)	22.53
Industry	36.9 (53.7%)	49.2 (53.0%)	62.9 (50.4%)	77.5 (48.6%)	77.4 (40.4%)	96.2 (44.4%)	137.98 (46.62%)	146.72 (46.7%)	160.09 (45.35)	171.21
Transport	17.4 (25.3%)	21.7 (23.4%)	28 (22.4%)	37.2 (23.4%)	33.5 (17.5%)	36.5 (16.8%)	55.34 (17.5%)	63.39 (20.2%)	76.46 (21.66)	80.30
Residential and commercial	5.6 (8.1%)	8.9 (9.6%)	12.6 (10.1%)	15.3 (9.6%)	24.1 (12.6%)	32.6 (15.1%)	43.43 (13.73%)	44.09 (14%)	48.7 (13.79)	48.94
Other energy uses*	1.9 (2.8%)	2.7 (2.9%)	3.9 (3.1%)	6.8 (4.3%)	13.4 (7.0%)	18.7 (8.6%)	30.25 (9.56%)	14.33 (4.6)	15.07 (4.27)	9.43
Non-energy uses**	5.3 (7.7%)	7.9 (8.5%)	12.6 (10.9%)	14.1 (8.8%)	28 (14.6%)	17.5 (8.1%)	26.15 (8.27%)	27.17 (8.6%)	30.9 (8.75)	25.04
Total	68.7 (100%)	92.8 (100%)	124.9 (100%)	159.3 (100%)	191.6 (100%)	216.6 (100%)	316.29 (100%)	314.4 (100%)	353.01 (100%)	357.45

*This comprises energy spent in other activities, which include miscellaneous uses and mining.
**Non-energy uses exist only for naphtha and natural gas sectors since both these fuels are consumed as feedstock in fertilizers and petrochemicals.
MTOE – million tonnes of oil equivalent
Note Figures in parentheses indicate the percentage share of each sector.
Sources TERI (various years); CEA (2013); MoPNG (2013); MoC (2014)

Energy supply and demand

The energy sector in India is closely linked to the country's economic activity. The past and current patterns of energy generation and consumption are bound to determine the economic future and well-being of the country. The final commercial energy consumption in India during the last three decades is given in Table 1.

Sources of energy supply in India

Coal and lignite

In terms of the total commercial energy supplied, as well as the installed power generation capacity, coal forms the largest source of energy in India with 60% of the installed capacity in the country. Coal production has witnessed an upward trend. In 2014/15, the production of raw coal was 612 MT at a year-on-year growth of 8.25%, which was higher than the production of the last 4 years combined. According to the estimates of the Coal Controller's Organisation, coal production has seen a considerable jump from 407 MT in 2005/06 to 612 MT in 2014/15. The two most prevalent methods of coal mining in India are opencast mining and underground mining. The share of opencast mining in total raw coal production grew to 92.09% in 2014/15; however, the share of underground mining fell from 14.98% in 2005/06 to 7.91% in 2014/15.

Coal import has risen in India, especially in non-major ports. The capacity of ports to handle coal traffic, better evacuation facilities in non-major ports, and proximity of thermal power plants to the ports are believed to be some of the factors for the rise of coal import at these ports. There have been important developments related to coal pricing. The Cabinet Committee of Economic Affairs approved the proposed pass-through mechanism in June 2013, allowing power producers to pass on higher cost of imported coal to consumers, contrary to the previous planned price-pooling mechanism. Other topics, including the Coal Mines (Special Provision) Act, 2015, policies and regulatory issues, coal transportation, and environmental and socio-economic challenges associated with coal mine exploration, are discussed in Chapter 3.

Petroleum and natural gas

Production of crude oil witnessed a decline of 0.53% from 37.7 MT in 2013/14 to 37.5 in 2014/15 (MoPNG 2015). Production in offshore fields saw

Commercial energy flows in India: Explanation for the Sankey diagram (Figure 1)

1. The net availability of natural gas refers to that availability of gas net of flaring and liquefied petroleum gas (LPG) extraction. The formula for estimating the net availability of natural gas in million tonnes of oil equivalent (MTOE) is explained in Table A.

Table A Formula for estimating net availability of natural gas

Production	36.61
+ Imports	16.20
– LPG extraction from natural gas	2.41
– Flaring of natural gas	0.81
– Petroleum refining	8.10
– Own use	3.29
Net availability of natural gas	38.20

Source MoPNG (2013)

2. The net availability of coal and lignite refers to the total availability of coal and lignite net of own use and washery rejects. The formula for the net availability of coal and lignite in MTOE is given in Table B.

Table B Formula for estimating net availability of coal and lignite

Production of coal and lignite	209.58
+ Imports of coal	94.74
– Exports of coal	1.47
+ Stock changes	3.88
– Own use*	0.16
– Coal washery rejects	3.98
Net availability of coal and lignite	302.59

*Own use refers to the amount of coal consumed in collieries for their own consumption.

Source MoC (2013)

3. The net availability of petroleum products refers to the crude oil converted to petroleum products and available for final consumption. The formula for the net availability of petroleum products in MTOE is given in Table C.

Table C Formula for estimating net availability of petroleum products

Crude throughput	219.21
– Refinery boiler fuel	18.35
+ Imports of petroleum products	16.48
– Exports of petroleum products	65.91
+ Stock changes	-25.42
+ LPG extracted from natural gas	2.41
Net availability of petroleum products	128.4

Source MoPNG (2013)

4. Energy loss from generation is the sum of conversion losses in power generation (this takes into account the thermal efficiency of different power plants), auxiliary consumption in power stations, and transmission and distribution losses (Table D).

Table D Calculation of energy loss from generation

Conversion losses in power generation	132.50
+ Auxiliary consumption in power stations	5.85
+ Transmission and distribution losses	18.25
Total energy loss from power generation	156.60

Note Final consumption of electricity across commercial, residential, agricultural, industrial, transport, and other energy and non-energy sectors is 61.58 MTOE (CEA 2014).

5. The net availability of nuclear, renewable, and hydropower sources is the total electricity generated from these sources. Nuclear, renewable, and hydropower sources generated 2.83, 5.09, and 9.79 MTOE of energy, respectively (CEA 2014).
6. Final consumption refers to the energy available for the final demand sectors: residential, commercial, agricultural, industrial, transport, and other energy and non-energy sectors. Table E gives the calculation for estimating the final energy consumption for each of these six sectors. Final energy consumption is the sum of consumption of coal and lignite, natural gas, petroleum products, and electricity. The data have been sourced from CEA (2014); MoC (2013); and MoPNG (2013). The final energy consumption across various sectors in India for 2012/13 is given in Table E.

Table E Final energy consumption across various sectors in India

Sector	Coal and lignite	Natural gas	Petroleum products	Electricity/power	Final consumption
Agricultural	–	0.16	9.69	12.68	22.53
Industrial	118.68	0.97	29.39	22.17	171.21
Transport	–	0.33	78.75	1.21	80.30
Residential	–	2.49	23.01	15.80	41.30
Commercial	–	0.02	1.36	6.26	7.64
Other energy uses	6.73	0.61	5.35	3.46	16.16
Non-energy uses	–	12.49	12.56	–	25.04

a major decline. Oil and Natural Gas Corporation's plan to increase production using improved oil recovery and enhanced oil recovery techniques seems to be the reason behind the increased offshore production of 0.7 MT from its fields. India imported 189 MT of crude oil in 2014/15 (MoPNG 2015). The demand for petroleum products is increasing in India. Considering the increase in consumption, the exports of petroleum products fell by 4 MT in 2014/15. While India is a net importer of crude oil, it is also one of the largest exporters of petroleum products in the world.

The production of petroleum products marginally increased from 220.19 MT in 2013/14 to 221 MT in 2014/15. Total consumption of petroleum products rose by 6.4 MT to 182.7 MT in 2014/15 over the 2013/14 levels. Natural gas production in India declined for the fifth consecutive year, with a production of 33.65 billion cubic metres (BCM) in 2014/15 as against 35.4 BCM in 2013/14.

Chapter 4 discusses pricing of petroleum products, natural gas imports, natural gas pricing, liquefied natural gas terminals, and other topics. The chapter also discusses unconventional fossil fuels and shale gas.

Power

As on 31 March 2015, the total installed capacity of the power sector in India was 271 GW, representing an increase of about 11% during 2014/15. With thermal power having the largest share of 84% in the total installed capacity (this excludes renewables), nearly 87% of the total thermal installed capacity in the country is accrued to coal. The installed capacity of captive generation has increased over the past few years. In the case of captive plants of 1 MW capacity, the installed capacity has risen from 39 375 MW in 2011/12 to 40 726 MW in 2012/13 (CEA 2014).

The electricity generated from captive generating units during 2012/13 was 144 billion units (BU) as against 134 BU in 2011/12. Here again, coal accounts

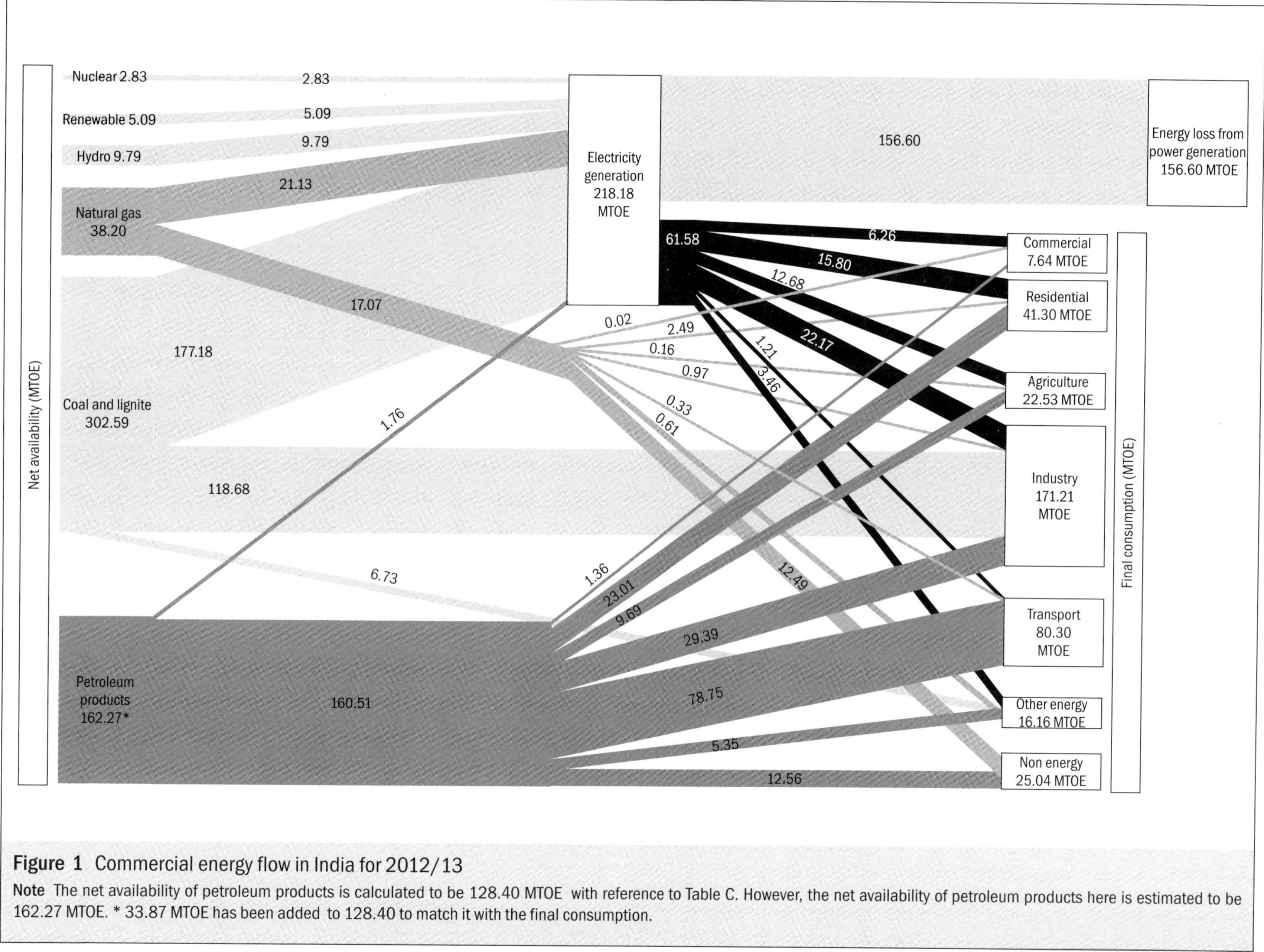

Figure 1 Commercial energy flow in India for 2012/13

Note The net availability of petroleum products is calculated to be 128.40 MTOE with reference to Table C. However, the net availability of petroleum products here is estimated to be 162.27 MTOE. * 33.87 MTOE has been added to 128.40 to match it with the final consumption.

for a major share of 84% in total electricity generation from captive plants, followed by gas, diesel, wind, and hydropower (CEA 2014). India has an extensive network of transmission and distribution of power. About 3 860 110 circuit km of lines operate at different voltages, and the length of distribution lines up to 500 V is 5 220 446 circuit km. The transmission and distribution lines grew by 4% during 2012/13 (CEA 2014). The electricity lost in transformation, transmission, and distribution system, including unaccounted electricity, was around 23% of the total available electricity in 2012/13 as against 23.65% of the total available electricity (CEA 2014). For 2012/13, the all-India energy sales were 708 BU. The industrial sector has been the highest consumer, accounting for 35% of the total electricity sold by utilities; this is followed by the domestic sector (26%) and agriculture (21%). Although there has been improvement in power supply position, the Indian power sector continues to face energy and peak shortages. Other topics, including policies, programmes, regulations, and power sector reforms, are discussed in Chapter 5.

Renewable energy sources

Of the total installed power generation capacity, the share of renewables is around 13.2%. The off-grid applications of renewable energy systems in India include biomass gasification, biogas technology, solar photovoltaic, solar cooking, solar industrial process heat and steam-generating systems, and solar air heating. The main renewable energy technologies that are grid connected include biomass gasification, bagasse-based biomass cogeneration, non-bagasse-based biomass cogeneration, onshore wind energy, grid-connected solar photovoltaic, solar thermal, and SHP plants.

So far the largest renewable energy system in India is that of wind power, reaching 24 088.36 MW of the installed capacity. Also, India is ranked fifth in the world for its wind installations (REN21 2015). The Government of India has announced plans for attracting investments worth $100 billion in the renewable energy sector over the next few years, along with the proposed revision of the National Solar Mission targets to 100 GW, as compared to the previous target of 22 GW by 2022.[2] The National Wind Energy Mission is expected to be launched soon, and the government has announced the target of doubling the installed wind power capacity to 40 GW by 2019.

India has witnessed a fairly good growth in renewable energy owing to supportive government policies, increasing costs of conventional energy, reductions in the cost of renewable energy technology, and economies of scale in manufacturing. The status and potential of several renewable energy technologies in the context of India are discussed in Chapter 6.

Nuclear power

India has a booming and largely indigenous nuclear power programme, and the country expects to have 14 600 MWe of nuclear capacity on line by 2024. The country aims to supply 25% of electricity from nuclear power by 2050.[3] India's fuel scenario marked by the shortage of fossil fuels is driving the nuclear investment for electricity. India's nuclear power strategy depends on a three-stage development programme involving conventional nuclear reactors, fast breeder reactors, and thorium-based reactors. Successful transition to thorium-based reactors will enable us to explore India's vast thorium resources, thus becoming much more energy independent beyond 2050. In the past, the expansion of India's atomic energy capacity was limited due to the lack of domestic uranium and the difficulty in obtaining the international supply of uranium fuel because of restrictions imposed by the Nuclear Suppliers Group. However, with these restrictions being lifted now, faster expansion of nuclear generation capacity is expected.[4]

The current installed nuclear power capacity in India is 5780 MW, and it is expected to increase to 10 080 MW by 2019. India is ranked 12th in terms of power generation from nuclear source as per data published in May 2015 by the Power Reactor Information System (PRIS) of the International Atomic Energy Agency (IAEA). India has signed nuclear agreements with the USA, France, Russia, Namibia, Mongolia, Republic of Korea, Argentine Republic, United Kingdom, Republic of Kazakhstan, Canada, Sri Lanka, and Australia (PIB 2015b).

[2] Details available at <www.pv-tech.org/news/india_to_boost_national_target_to_100gw_by_2022>, last accessed on 7 January 2016

[3] Details available at <www.world-nuclear.org/info/Country-Profiles/Countries-G-N/India/>, last accessed on 7 January 2016

[4] Details available at <http://planningcommission.nic.in/plans/mta/11th_mta/chapterwise/chap15_energy.pdf, last accessed on 7 January 2016

Hydropower

In India, hydropower projects are usually categorized into two segments: small and large hydro. Hydro projects up to 25 MW capacities have been categorized as SHP projects. The Ministry of Power is responsible for large hydro projects (above the station capacities of 25 MW), while the Ministry of New and Renewable Energy is responsible for SHP (up to 25 MW) projects. Power generated from LHP is expected to play an important role in supplementing the conventional power generation and meeting energy needs, especially for regions with good hydro potential. India's installed capacity of hydro is 42 283.42 MW.[5] As against the all-India target of 10 897 MW under the Twelfth Five-year Plan, 3311.02 MW of hydro capacity has been achieved (as on August 2015).

Sources of energy demand in India

Agriculture

Agriculture occupies a significant position in India's society and economy as it provides livelihood and employment to more than 54% of the population (Census of India 2011) and contributes around 13.9% to the GDP (MoA 2015). The agriculture sector has major implications for the overall demand and supply of energy in India. The share of total electricity consumption by the agriculture sector increased from 81 673 GWh in 2001/02 to 159 144 GWh in 2013/14 (CEA 2015). Energy use in agriculture at the farm level can be categorized as either direct or indirect. Direct energy use in agriculture is in the form of diesel and electricity to operate mobile and stationary equipment/machinery to prepare fields, plant and harvest crops, and transport inputs and outputs to and from markets. Indirect energy is consumed off the farm for the manufacture of inputs such as fertilizers and pesticides.

In the agriculture sector of India, high-speed diesel (HSD) is steadily replacing light diesel oil (LDO), which is used in stationary applications. Diesel is used in tractors, power tillers, and pump sets. Agriculture sector accounts for 0.63% of total HSD consumption in India. The energy scenario of India's agriculture sector, water efficiency in agriculture, agriculture planning, and other topics are discussed in Chapter 7.

Industry

The index of industrial production has shown a fluctuating trend over the last few years. Industrial growth peaked at 15.5% in 2007/08 (manufacturing growth rate was 18.4%) and then declined to 2.5% in 2008/09. It, however, recovered and further improved to 8.2% in 2010/11, followed by deceleration to 2.9% in 2011/12 and 1.1% in 2012/13. Industrial production again moderated, recording a negative growth of (−) 0.1% in 2013/14. The specific energy consumption (SEC) in 2014 for primary aluminium production for Asia, excluding China, was 14 714 kWh alternating kWhAC/tonne, as compared to the world average of 14289 kWhAC/tonne. There is a large scope for improving energy efficiency in aluminium production.

The steel sector contributes nearly 2% to India's GDP. India is the fourth largest producer of crude steel and the largest producer of direct reduced iron or sponge iron in the world. The domestic demand for total finished steel grew at the rate of 8% annually between 1991/92 and 2010/11. The average SEC of Indian steelmaking units is 7.18 Gcal/tcs (30 GJ/tcs), which is almost twice the world's leading plants.

Fertilizer production consumes approximately 1.2% of the world's total energy on an annual basis. The Indian fertilizer industry is comparable to the best in the world in terms of efficiency, capacity utilization, energy consumption, and utilities such as power and water. The new fertilizer plants are comparable to the best in the world and have been able to bring down the energy consumption in urea production from about 9% in 1987/88 to 6.04% in 2013/14. The textile industry is energy intensive with the wet processing or dyeing operation consuming almost 50% of the energy in a composite mill. Boilers and spinning plants consume most of the energy in the textile industry. With effective energy conservation measures, there is a large scope for saving energy in textile mills.

Energy cost accounts for 35%–45% of the production cost in cement plants; this makes cement industry energy intensive. The average electricity consumption is about 80 kWh per tonne of cement. The best thermal and electrical energy consumption currently achieved by the Indian cement industry is about 667 kcal/kg clinker and 67 kWh/tonne cement,

[5] Details available at <www.cea.nic.in/reports/monthly/executivesummary/2015/exe_summary-09.pdf>, last accessed on 30 September 2016

which are comparable to the best reported figures of 660 kcal/kg clinker and 65 kWh/tonne cement in developed countries such as Japan. In the chlor-alkali industry, the average operating SEC declined from 3351 kWh/tonne in 1990/91 to 2350 kWh/tonne in 2009/10 (SSEF 2013).

The pulp and paper industry is also highly energy intensive, and usually energy cost accounts for about 16%–25% of the total cost of production of paper. The SEC of integrated wood-based plants is 23–37 GJ/tonne compared to 18–22 GJ/tonne of paper internationally. With the adoption of energy conservation measures such as the delignification process in pulping and the use of pressure screens instead of centrifugal screens, the Indian pulp and paper industry can achieve better energy efficiency. Energy conservation measures for large industries, including aluminium, iron and steel, fertilizer, textile, cement, chlor-alkali, paper and pulp, micro, small, and medium enterprises, and various policy measures, are discussed in Chapter 8.

Transport

India's transport system is one of the world's largest transport systems, covering a land area of 3.3 million km^2 and serving a population of more than 1.21 billion (Census of India 2011). The transport sector contributed 5.7% to the country's GDP in 1999/2000, and the contribution further increased to 6.6% in 2010/11.[6] The sector is also a key energy-consuming sector; it consumed 34.24% of the total volume of petroleum products consumed in India in 2009/10 (MoPNG 2012). Of all the products consumed, HSD consumption was the maximum. The transport sector consumed 59.1% (33.74 MT) of the total HSD consumption (56.32 MT) in 2009/10 (MoPNG 2012).

India has the second largest road network (3.33 million km) in the world. Road transport contributed 4.8% to India's GDP in 2011/12. Within the transport sector, the road sector is the largest consumer of petroleum products. The sector now accounts for the largest share (72%) of HSD consumption in India (MoPNG 2012). Indian Railways is one of the largest rail networks in the world, with a network of 65 436 route km and a running track length of 89 236 km (MoR 2014). Rail transport is six times more energy efficient and four times more economical than road transport. The consumption of HSD in locomotive services increased from 2699.6 million litres in 2011/12 to 2780.9 million litres in 2013/14 (MoR 2015).

Direct coal consumption by the railways is almost negligible because of the phasing out of steam locomotives. The consumption of coal for locomotive services was 0.001 MT in 2013/14 (MoR 2015). Indian Railways has been striving to bring environment-friendly initiatives and save fuel over the years. Maritime transport handles more than 90% by volume and 70% by value of India's trade. During 2010/11, the shipping sector consumed 0.56 MT of HSD, 0.004 MT of LDO, and 0.78 MT of furnace oil (MoPNG 2012). Other topics, including energy consumption trends in various transport sectors and key legislations and regulations in the sector, are covered in detail in Chapter 9.

Household energy

Energy access is fundamental to the human development of a nation. More than 1200 million people worldwide lack access to affordable basic energy services for effective lighting and clean cooking. An estimated 791 million people still lack access to modern cooking methods (SE4ALL 2015). As per the recently released energy statistics of the Government of India, the domestic sector accounted for 22% of the total electricity consumed during 2013/14. According to the 68th round of National Sample Survey, electricity constitutes 50% of the household fuel expenditure for urban households and 22% for rural households. Based on the balances report of International Energy Agency, India's total residential consumption of energy for 2012 was close to 8.77% of the world's total residential energy consumption.

In India, 67.3% of households use electricity for lighting and 28.6% depend on LPG as fuel for cooking. Households in the rural areas of underdeveloped states are less reliant on electricity for lighting and are relatively more dependent on traditional biomass for cooking compared to the urban households of these states. However, according to a recent UN report, 50 million people in India have gained access to electricity. The consumption of superior kerosene oil is declining and the consumption of LPG for domestic use is increasing.

[6] Details available at < http://indiabudget.nic.in/es2011-12/echap-10.pdf>, last accessed on 7 January 2016

Other topics, including village electrification, sector-wise electricity consumption, policies and regulations for modern energy access in households, are discussed in Chapter 10.

Environment

As India progresses, the need to find solutions to local and global environmental problems has become a priority. In India, ambient air pollution has been identified as the fifth largest cause of mortality (Atkinson, Cohen, Mehta, *et al.* 2011). Respirable suspended particulate matter is of prime concern as it exceeds the ambient air quality standards at 83% of locations (CPCB 2014).

Water is essential for survival, and groundwater pollution is becoming a crucial issue. Groundwater meets the water requirement of 55% of irrigational, 85% of rural, and 50% of urban and industrial activities in the country. In India, 70% of surface water resources are contaminated by biological, toxic, organic, and inorganic pollutants (MoEF 2009). Water contamination weakens natural ecosystems that support human health, food production, and biodiversity (CAG 2012); therefore, water management has emerged as a key area of concern for the government.

As the urban population increased from 27.81% in 2001 to 31.16% in 2011 (Census of India 2001, 2011), the amount of solid waste generated has also increased. Moreover, till February 2009, there were about 36 165 hazardous waste generating industries, discharging about 6.2 MT of hazardous wastes every year. India is among the few countries that have developed a biogeographic classification for conservation planning and have mapped biodiversity-rich areas. As per the International Union for Conservation of Nature, India has 1039 threatened species. Forest cover in India has increased by 3775 km^2 since 2013 (FSI 2015). For further details regarding the state of environment of India and the various interventions adopted in this regard by the government, refer to Chapter 11.

Climate change has been accepted as a reality and is believed to be unequivocally and essentially driven by anthropogenic factors, as indicated in the *Fifth Assessment Report* of IPCC. The all-India annual mean temperatures have shown a significant warming trend of 0.51°C per 100 years (Kothawale, Munot, and Krishna Kumar 2010). The regional climate models for India, barring some parts in southern peninsular India, show a rise of 9%–16% in the future monsoon rainfall towards the end of the 21st century (MoEFCC 2012). The models project that even though the frequency of cyclones might decrease, the intensity of the cyclones occurring in the future would be higher than the present level. Absence of significant changes in the tracks of cyclones has been reported for the future climate simulations (MoEF 2012).

Carbon dioxide (CO_2) contributes more than 90% to the total greenhouse gas (GHG) emissions in India owing primarily to fuel combustion. Along with the rise in absolute CO_2 emissions, the world average of per capita emissions has risen from 4.88 tonnes per capita in 2012 to 5.10 tonnes per capita in 2013. India's per capita CO_2 emission is well below the global average at 1.6 tonnes CO_2 per capita.

It is estimated that with 6.4% GDP (real) growth per year, GHG emissions (excluding land use, land-use change, and forestry) are projected to reach 3.5 $GtCO_2e$ in 2020 and 5.0–5.1 $GtCO_2e$ in 2030. An inference can be drawn that a 40% increase in emissions from the 2010 levels by 2020 and a doubling of the 2010 levels by 2030 are expected in India. This growth in emissions is in sync with India's 2020 intensity pledge and INDC.

India has been actively participating in the multilateral negotiations under the United Nations Framework Convention on Climate Change (UNFCCC). India's INDC submitted to the UNFCCC in response to the Conference of Parties decisions 1/CP.19 and 1/CP.20 for 2021–30 communicates that India aims to reduce the overall emission intensity and improve energy efficiency of its economy over time and at the same time protect the vulnerable sectors of the economy and segments of the society. Observed climate trends, GHG emission trends, government responses to address climate change, and other topics are discussed in Chapter 12.

Energy and environment in context of Sustainable Development Goals

At the United Nations Sustainable Development Summit 2015, heads of state as well as government and high-ranking representatives adopted the new global Sustainable Development Goals (SDGs). The sustainable development agenda with 17 goals and 169 targets came into effect on 1 January 2016. The SDGs will guide the global and national development policy agendas over the next 15 years. A quick comparison between the Millennium Development Goals (MDGs) and SDGs shows that SDGs are universal in nature and have specific goals related to energy, which were missing in the MDGs. The SDGs include the topics that were missing in the MDGs, such as arid and semi-arid, oceanic, mountain, grassland, and arctic ecosystems and desertification. The reality of climate change also gets reflected within the SDGs.

In achieving the MDG targets, India has made significant progress in meeting goals such as those related to forestry and safe drinking water; however, it has not made sufficient progress on many environmental issues such as biodiversity loss and degradation of ecosystems. Therefore, India needs to develop appropriate and specific national indicators for the effective realization of SDG targets, which are measurable and accepted in entire India. Chapter 13 discusses India's energy and environment related goals, India's MDG experience, and the development of SDGs.

The build-up of primary and secondary pollutants has made ambient air pollution a prominent global threat to human health. According to the *Fifth Assessment Report* of the IPCC, "nearly all the non-CO_2 climate-altering pollutants are health damaging, either directly or by contributing to secondary pollutants in the atmosphere". The third goal of the 17 goals submitted to the UN General Assembly by the Open Working Group on Sustainable Development Goals aims at ensuring "healthy lives" and promoting "well-being for all at all ages". It has nine sub-goals, and the sub-goal 3.9 targets a substantial reduction in the number of deaths and illnesses from hazardous chemicals, as well as air, water, and soil pollution and contamination, by 2030. There is a need for reporting and monitoring the impacts of environmental pollution. Indoor air pollution has emerged as a serious concern in rural India because of the dependence on traditional biomass for cooking and space heating and on kerosene or other liquid fuels for lighting. This concern can be warranted by the fact that in rural India only 11.4% of households use LPG for cooking (Census of India 2011). Chapter 14 discusses the impact of ambient and household air pollution on health at the global and national levels along with recommendations on policy measures to reduce the health impact of air pollution in the national context.

References

Atkinson, R., A. Cohen, S. Mehta, and H. Anderson. 2011. Systematic review and meta-analysis of epidemiological time-series. ***Air Quality, Atmosphere and Health*** 383–391

BP (British Petroleum). 2015. BP Statistical Review of World Energy. Details available at http://www.bp.com/content/dam/bp/pdf/energy-economics/statistical-review-2015/bp-statistical-review-of-world-energy-2015-full-report.pdf

CAG (Comptroller and Auditor General). 2012. ***Performance Audit of Water Pollution in India, Report No. 21 of 2011/12***. New Delhi: CAG, Ministry of Environment, Forests, and Climate Change

CEA (Central Electricity Authority). 2013. ***All-India Electricity Statistics: General Review 2013***. New Delhi: CEA, Government of India

CEA (Central Electricity Authority). 2014. ***All-India Electricity Statistics: General Review 2014***. New Delhi: CEA, Government of India

CEA (Central Electricity Authority). 2015. ***All-India Electricity Statistics: General Review 2015***. New Delhi: CEA, Government of India

Census of India. 2001. Office of the Registrar General and Census Commissioner, India. New Delhi: Government of India

Census of India. 2011. Office of the Registrar General and Census Commissioner, India. New Delhi: Government of India

CPCB (Central Pollution Control Board). 2014. ***National Ambient Air Quality Status and Trends-2012***. New Delhi: CPCB

EPI (Environmental Performance Index). 2014. 2014 EPI results at-a-glance. Details available at http://epi.yale.edu/epi , last accessed on 17 November 2015

FSI (Forest Survey of India). 2015. ***India State of Forest Report 2015***. Dehradun: FSI

GoI (Government of India). 2015. India's Intended Nationally Determined Contribution. Details available at http://www4.unfccc.int/submissions/INDC/Published%20Documents/India/1/INDIA%20INDC%20TO%20UNFCCC.pdf, last accessed on 17 November 2015

IPCC (Intergovernmental Panel on Climate Change). 2014. ***Climate Change 2014: Impacts, Adaptation, and Vulnerability—Summary for Policymakers***. Geneva: IPCC

Kothawale, D. R., A. A. Munot, and K. Krishna Kumar. 2010. Surface air temperature variability over India during 1901–2007, and its association with ENSO. ***Climate Research*** 42: 89–104

MoA (Ministry of Agriculture). 2015. ***Annual Report 2014/2015***. New Delhi: MoA, Government of India

MoC (Ministry of Coal). 2013. ***Coal Directory of India 2012/13***. Kolkata: Coal Controller's Organisation

MoC (Ministry of Coal). 2014. ***Coal Directory of India 2012/13***. Kolkata: Coal Controller's Organisation

MoEF (Ministry of Environment and Forests). 2009. ***State of Environment Report***. New Delhi: MoEF, Government of India

MoEFCC (Ministry of Environment, Forests, and Climate Change). 2012. ***India's Second National Communication to the United Nations Framework Convention on Climate Change***. New Delhi: MoEFCC, Government of India. Details available at http://unfccc.int/resource/docs/natc/indnc2.pdf

MoP (Ministry of Power). 2015. National Smart Grid Mission. Details available at http://powermin.nic.in/upload/pdf/National_Smart_Grid_Mission_OM.pdf, last accessed on 17 November 2015

MoPNG (Ministry of Petroleum and Natural Gas). 2012. ***Basic Statistics on Indian Petroleum and Natural Gas***. New Delhi: MoPNG, Government of India

MoPNG (Ministry of Petroleum and Natural Gas). 2013. ***Indian Petroleum and Natural Gas Statistics 2012–13***. New Delhi: MoPNG, Government of India

MoPNG (Ministry of Petroleum and Natural Gas). 2015. ***Petroleum and Natural Gas Statistics 2014–15***. New Delhi: MoPNG, Government of India

MoR (Ministry of Railways). 2014. ***Indian Railways Annual Statistical Statements 2012–13***. New Delhi: MoR, Government of India

MoR (Ministry of Railways). 2015. ***Indian Railways Annual Statistical Statements 2013–14***. New Delhi: MoR, Government of India

MoSPI (Ministry of Statistics and Programme Implementation). 2013. ***Compendium of Environmental Statistics in India 2013***. New Dellhi: MoSPI, Government of India

PIB (Press Information Bureau). 2015a. Government has established National Adaptation Fund on Climate Change: Javadekar. Details available at http://pib.nic.in/newsite/PrintRelease.aspx?relid=124326, last accessed on 17 November 2015

PIB (Press Information Bureau). 2015b. India ranks at 12th position in terms of power generation from nuclear source as per data published in May 2015 by PRIS of IAEA. Details available at http://pib.nic.in/newsite/PrintRelease.aspx?relid=123553, last accessed on 24 November 2015

REN21 (Renewable Energy Policy Network for the 21st Century). 2015. ***Renewables 2015 Global Status Report: REN21***. Paris: REN 21

SE4ALL (Sustainable Energy for All). 2015. ***Progress towards Sustainable Energy: Global Tracking Framework 2015 Key Findings***. Vienna: SE4ALL

SSEF (Shakti Sustainable Energy Foundation). 2013. ***Technology Compendium on Energy Saving Opportunities Chlor-Alkali Sector***. New Delhi: SSEF

UNDP (United Nations Development Programme). 2015. ***Human Development Report 2015***. New York: UNDP

COMMERCIAL ENERGY BALANCE TABLES AND CONVERSION FACTORS

2

The energy balance tables prepared by TERI provide an account of different kinds of commercial energy sources. Columns represent various energy sources and rows represent the origin and utilization of these sources. All figures are represented by the common unit million tonnes of oil equivalent (MTOE). Original units have been converted into MTOE units based on the calorific values of the respective energy sources. The main conversion factors are given in Tables 1 and 2.

Columns

Columns 2 to 7: Sources of primary energy

Sources of primary energy can be referred to as (commercially traded) resources that occur "as is" in nature and are used to produce energy from coal, water in motion (hydroelectricity), uranium (nuclear energy), wind, biomass, natural gas, crude oil, solar radiation, geothermal resources, and so on. Coal and lignite, hydropower, nuclear resources, renewable energy sources, natural gas, and crude oil are the main sources of primary energy.

Column 2: Coal and lignite

The values in Column 2 pertain to coking coal, non-coking coal, and lignite. In India, coking coal is categorized based on ash content. Non-coking coal, which used to be categorized based on the useful heat value (UHV), is now graded based on the gross calorific value (GCV). The categorization of different grades of coking and non-coking coal is given in Tables 3 and 4.

Table 1 Main conversions used in the petroleum industry

Item	Conversion factor	
Crude oil	1 tonne	= 7.33 barrels
		= 1.165 m^3 (kilolitres)
	1 barrel	= 0.136 tonne
		= 0.159 m^3 (kilolitre)
	1 m^3	= 0.858 tonnes
		= 6.289 barrels
	1 MT	= 1.111 BCM natural gas
		= 39.2 BCF natural gas
		= 0.805 MT (LNG)
		= 40.4 trillion BTU
Natural gas	1 BCM	= 35.3 BCF natural gas
		= 0.90 MT crude oil
		= 0.73 MT LNG
		= 36 trillion BTU
		= 6.29 million barrels of oil equivalent
LNG	1 MT	= 1.38 BCM natural gas
		= 48.7 BCF natural gas
		= 1.23 MT crude oil
		= 52 trillion BTU
		= 8.68 million barrels of oil equivalent
CNG	1 kg	= 1.244 standard cubic metres of natural gas
		= 1.391 litres of petrol
		= 1.399 litres of HSD oil

BTU – British thermal units; BCF – billion cubic feet; BCM – billion cubic metre; CNG – compressed natural gas; HSD – high-speed diesel; LNG – liquefied natural gas; MT – million tonnes

Source MoPNG (2009)

Table 2 Heat units and others expressed

Items	Conversion (MTOE)
Heat units	
10 million BTU	0.25
100 million therms	0.25
10 000 teracalories	1.00
Solid fuels	
1 MT of coal	0.67 (international)[#]
	0.34 (domestic weighted average)*
1 MT of lignite	0.33
Natural gas	
1 BCM	0.90
Town gas	
1 BCM	0.42
Electricity	
12 000 million kWh	0.86

BCM – billion cubic metre; BTU – British thermal units; kWh – kilowatt-hour; MT – million tonnes; MTOE – million tonnes of oil equivalent

[#]*BP Statistical Review of World Energy*

*Calculations used for this energy balance

Note According to the *BP Statistical Review of World Energy*, 1 MT of oil produces 4400 million units (kWh) of electricity in a modern power station.

Source MoPNG (2009)

Table 3 Categorization of coking coal grades

Coking coal grade	Ash content (%)
Steel Grade I	Less than 15
Steel Grade II	Between 15 and 18
Washery Grade I	Between 18 and 21
Washery Grade II	Between 21 and 24
Washery Grade III	Between 24 and 28
Washery Grade IV	Between 28 and 35
	Ash + moisture content (%)
Semi-coking Grade I	Less than 19
Semi-coking Grade II	Between 19 and 24

Source CCO (2008)

Calorific values for domestic coking coal and non-coking coal have been worked out based on the weighted average of UHVs, weights being the production figures. The UHVs considered for lignite, imported coking coal, imported non-coking coal, and coal for export are provided in Table 5.

Table 4 Categorization of non-coking coal grades

Old grading based on useful calorific value (kcal/kg)	New grading based on gross calorific value (kcal/kg)
A (6 200 and above)	G1 (above 7 000)
	G2 (between 6 701 and 7 000)
	G3 (between 6 401 and 6 700)
B (5 600 to 6 200)	G4 (between 6 101 and 6 400)
	G5 (between 5 801 and 6 100)
C (4 940 to 5 600)	G6 (between 5 501 and 5 800)
D (4 200 to 4 940)	G7 (between 5 201 and 5 500)
	G8 (between 4 901 and 5 200)
E (3 360 to 4 200)	G9 (between 4 601 and 4 900)
	G10 (between 4 301 and 4 600)
F (2 400 to 3 360)	G11 (between 4 001 and 4 300)
	G12 (between 3 700 and 4 000)
G (1 300 to 2 400)	G13 (between 3 400 and 3 700)
	G14 (between 3 101 and 3 400)
Non-coking coal ungraded	G15 (between 2 801 and 3 100)
	G16 (between 2 501 and 2 800)
	G17 (between 2 201 and 2 500)

Source CCO (2015)

Table 5 Standard useful heat values

Item	Useful heat value (kcal/kg)
Lignite	3 000
Coking coal (imported)	6 500
Non-coking coal (imported)	6 500
Coal (exported)	6 000

Columns 3 to 5: Hydropower, nuclear power, and renewable energy sources

Data for hydropower and energy from renewable sources pertain to both utilities and captive power plants. Currently, nuclear power is generated only in utilities.

Some of the standard conversion factors used for transforming energy units into a common unit are as follows:

1 gigawatt-hour (GWh) = 0.0036 petajoules (PJ)

1 MTOE = 41.868 PJ

1 kWh electricity = 860 kcal

Column 6: Natural gas

Although the calorific value of natural gas in India ranges from 8000 kcal/m^3 to 9480 kcal/m^3, we have assumed a calorific value of 9000 kcal/m^3 for the study related to energy balances.

Column 7: Crude oil

1 MT of crude oil = 1 MTOE

Columns 8 to 16: Sources of secondary energy

Secondary energy sources include the various products obtained from the refining of crude oil, such as liquefied petroleum gas (LPG), naphtha, motor gasoline, aviation turbine fuel (ATF), kerosene, high-speed diesel (HSD), light diesel oil, fuel oil,[1] and other petroleum products. The calorific values of different petroleum products are given in Table 6.

Column 17: Total petroleum products

The figures for total petroleum products are the sum totals of the figures for the various energy sources produced as a result of refining crude oil. The conversion factors for the main petroleum products are given in Table 7. Since these conversion factors are approximations and the actual factors depend on the densities of various products, the conversion factor for "other petroleum products" is adjusted to equate the crude throughput (less refinery fuel) with the total production of petroleum products.

Column 18: Thermal power

Thermal power includes power generated from (1) coal and lignite, (2) natural gas, and (3) petroleum products.

Column 19: Total power

Total power includes (1) thermal power, (2) hydropower, (3) nuclear power, and (4) power from renewable energy sources.

Column 20: Total energy

Column 20 is the summation of all forms of energy outlined earlier.

Table 6 Calorific values of different petroleum products

Petroleum product	Calorific value (kcal/kg)
LPG	11 300
Naphtha	10 750
Kerosene	10 450
ATF	10 650
Fuel oil	9 850
Motor gasoline	10 700
HSD	10 350

ATF – aviation turbine fuel; HSD – high-speed diesel; LPG – liquefied petroleum gas
Source MoPNG (2009)

Table 7 Conversion factors of petroleum products

Product	TOE/tonne	Barrel/tonne
Refinery gas	1.15	8.00
Ethane	1.13	16.85
LPG	1.13	11.60
Aviation gasoline	1.07	8.90
Motor gasoline	1.07	8.53
Jet gasoline	1.07	7.93
Jet kerosene	1.07	7.93
Other kerosene	1.05	7.74
Naphtha	1.08	8.50
Gas/diesel oil	1.04	7.46
Heavy fuel oil	0.96	6.66
Petroleum coke	0.74	5.50
White spirit	0.96	7.00
Lubricants	0.96	7.09
Bitumen	0.96	6.08
Paraffin waxes	0.96	7.00
Non-specified products	0.96	7.00

LPG – liquefied petroleum gas; TOE – tonnes of oil equivalent
Source MoPNG (2009)

Rows

Rows 3 to 5: Production, import, and export

Rows 3–5 refer to the indigenous production, import, and export of a particular source of energy.

[1] Fuel oil includes furnace oil and low sulphur heavy stock (LSHS)/ hot heavy stock (HHS).

Row 6: Stock changes

A stock build is shown as a negative number and a stock draw as a positive number.

Row 7: Availability

The availability of primary sources of energy has been worked out by adding imports and stock changes to production and subtracting exports from it.

Row 8: Petroleum refining

Row 8 accounts for the generation of secondary fuels as a result of crude oil refining.

Row 9: Own use

Any fuel used for its own production/extraction is reflected in Row 9.

Row 10: LPG extraction

Row 10 refers to the extraction of LPG from natural gas. The same value occurs in both the columns—natural gas and LPG.

Row 11: Power generation

Row 11 refers to the amount of power generated from each fuel.

Row 12: Conversion loss in power generation

Conversion loss in power generation is calculated based on the efficiency of inputs (that is, electricity generated per unit of input) in different types of power stations.

Row 13: Auxiliary consumption in power stations

The percentage of auxiliary consumption in different types of power stations has been applied to power generation less conversion loss in power stations in each case.

Row 14: Transmission and distribution losses

Row 14 refers to the transmission and distribution (T&D) losses in power stations (utilities only).

Row 15: Flaring of natural gas

Row 15 accounts for the amount of natural gas that escapes and flares up during the process of extraction.

Row 16: Coal washery rejects

Raw coal feed to washeries is converted into washed coal and middlings, which are sent to power plants or industries. However, a certain amount of coal comes out as rejects, which are not usable.

Row 17: Conversion

Conversion is defined differently for different fuels. In the case of coal, conversion implies the amount of coal used for power generation. Conversion of natural gas includes the utilization for its own use, for LPG extraction, and for power generation as well as its flaring. Conversion in the case of crude oil is the amount of crude oil sent for petroleum refining (which is the same as its availability). The conversion value for sources of secondary energy equals their production (from crude oil or natural gas) minus their use in power generation. In the case of thermal power and total power, conversion refers to losses in power generation, that is, conversion loss, auxiliary consumption, and T&D losses. Conversion of total energy also includes flaring of natural gas and coal washery rejects.

Row 18: Net availability

Row 18 refers to the amount of each energy source available (after consumption in power plants, own use, losses, and so on) for consumption across various economic activities.

Rows 19 to 26: Consumption

Rows 19–26 display the total consumption of each energy source and its break-up by sector (that is, agriculture, industry, transport, residential, commercial, other energy uses, and non-energy uses).[2]

[2] In the absence of disaggregated sectoral data, sales of naphtha by private parties have been distributed between power generation and petrochemicals as per the weighted average of direct sales of naphtha to power plants and petrochemicals sector in respective years.

References

CCO (Coal Controller's Organisation). 2008. ***Coal Directory of India 2006/07***. Kolkata: CCO

CCO (Coal Controller's Organisation). 2010. ***Coal Directory of India 2008/09 (Part-I: Coal Statistics)***. Kolkata: CCO

CEA (Central Electricity Authority). 2008. ***All India Electricity Statistics: general review 2008***. New Delhi: CEA

CEA (Central Electricity Authority). 2009. ***All India Electricity Statistics: general review 2009***. New Delhi: CEA

MoPNG (Ministry of Petroleum and Natural Gas). 2009. ***Indian Petroleum and Natural Gas Statistics 2007/08***. New Delhi: MoPNG, Government of India

MoPNG (Ministry of Petroleum and Natural Gas). 2010. ***Indian Petroleum and Natural Gas Statistics 2008/09***. New Delhi: MoPNG, Government of India

Bibliography

CCO (Coal Controller's Organisation). 2009. ***Coal Directory of India 2007/08***. Kolkata: CCO

CCO (Coal Controller's Organisation). 2011. ***Coal Directory of India 2009/10 (Part-I: Coal Statistics)***. Kolkata: CCO

CCO (Coal Controller's Organisation). 2012. ***Coal Directory of India 2010/11 (Part-I: Coal Statistics)***. Kolkata: CCO

CCO (Coal Controller's Organisation). 2013. ***Coal Directory of India 2011/12***. Kolkata: CCO

CCO (Coal Controller's Organisation). 2014. ***Coal Directory of India 2012/13***. Kolkata: CCO

CCO (Coal Controller's Organisation). 2015. ***Coal Directory of India 2013/14***. Kolkata: CCO

CEA (Central Electricity Authority). 2010. ***All India Electricity Statistics: general review 2010***. New Delhi: CEA

CEA (Central Electricity Authority). 2011. ***All India Electricity Statistics: general review 2011***. New Delhi: CEA

CEA (Central Electricity Authority). 2012. ***All India Electricity Statistics: general review 2012***. New Delhi: CEA

CEA (Central Electricity Authority). 2013. ***All India Electricity Statistics: general review 2013***. New Delhi: CEA

CEA (Central Electricity Authority). 2014. ***All India Electricity Statistics: general review 2014***. New Delhi: CEA

MoPNG (Ministry of Petroleum and Natural Gas). 2008. ***Indian Petroleum and Natural Gas Statistics 2006/07***. New Delhi: MoPNG, Government of India

MoPNG (Ministry of Petroleum and Natural Gas). 2011. ***Indian Petroleum and Natural Gas Statistics 2009/10***. New Delhi: MoPNG, Government of India

MoPNG (Ministry of Petroleum and Natural Gas). 2012. ***Indian Petroleum and Natural Gas Statistics 2010/11***. New Delhi: MoPNG, Government of India

MoPNG (Ministry of Petroleum and Natural Gas). 2013. ***Indian Petroleum and Natural Gas Statistics 2012/13***. New Delhi: MoPNG, Government of India

MoPNG (Ministry of Petroleum and Natural Gas). 2014. ***Indian Petroleum and Natural Gas Statistics 2013/14***. New Delhi: MoPNG, Government of India

TABLE 2.1 Commercial energy balance (million tonnes of oil equivalent): 2012/13

	Primary energy									Secondary energy												
	Coal and lignite				Hydro-power	Nuclear power	Renewable energy sources	Natural gas	Crude oil	LPG	Naphtha	Motor gasoline	Aviation turbine fuel	Kerosene	High-speed diesel	Light diesel oil	Fuel oil	Other petroleum products	Total petroleum products	Thermal power	Total power	Total energy
	Coking coal	Non-coking coal	Lignite	Total																		
Supply																						
Production	21.46	174.19	13.93	209.58	9.79	2.83	5.09	36.61	37.86												17.70	301.75
Imports	23.11	71.64		94.74				16.20	184.80	7.11	1.87	0.16	0.00	0.00	0.65	0.00	1.03	5.67	16.48		0.41	312.21
Exports	0.03	1.43		1.47						0.23	9.30	17.82	4.97	0.02	23.25	0.01	5.69	4.63	65.91		0.01	67.37
Stock changes	1.29	2.72	-0.13	3.88					-3.44	-0.43	1.97	2.29	-0.15	-0.36	-0.82	0.01	-2.78	-25.14	-25.42			-24.99
Availability	45.82	247.12	13.80	306.73	9.79	2.83	5.09	52.81	219.21													521.60
Petroleum refining								8.10	219.21	8.69	18.66	32.23	10.73	8.22	94.27	0.38	14.45	46.67	234.31			234.31
Own use	0.02	0.13		0.16				3.29	18.35													
LPG extraction								2.41		2.41									2.41			
Power generation		167.03	10.16	177.18	9.79	2.83	5.09	21.13			0.42				0.22	0.14	0.98		1.76	200.07	217.78	217.78
Conversion loss in power generation		111.91	6.81	118.71				13.27							0.08	0.05	0.37		0.51	132.50	132.50	132.50
Auxiliary consumption in power stations		4.61	0.28	4.89	0.06	0.28	0.34	0.22							0.00	0.00	0.05		0.06	5.17	5.85	5.85
Transmission and distribution losses																			0.00		18.25	18.25
Flaring of natural gas								0.81											0.00			0.81
Coal washery rejects	1.10	2.88		3.98																		3.98
Conversion				177.18				35.73	219.21	11.10	18.24	32.23	10.73	8.22	94.05	0.25	13.47	46.67	234.96	137.66	156.59	161.39
Net availability				125.41	9.73	2.54	4.75	17.07	200.86	17.56	12.77	16.85	5.61	7.84	70.63	0.25	6.03	22.57	160.10	62.41	61.58	364.17
Consumption				125.41				17.07		17.56	12.77	16.85	5.61	7.84	70.63	0.25	6.03	22.57	160.10		61.58	364.17
Agriculture								0.16		0.00					9.61	0.00	0.08		9.69		12.68	22.53
Industry	33.52	82.45	2.72	118.68				0.97		0.61	0.00				2.84	0.07	3.28	22.57	29.39		22.17	171.21
Transport								0.33		0.24		16.85	5.61		55.78	0.00	0.27		78.75		1.21	80.30
Residential								2.49		15.33				7.68					23.01		15.80	41.30
Commercial								0.02		1.32				0.04					1.36		6.26	7.64
Other energy uses	2.27	1.32	3.14	6.73				0.61		0.05	0.22			0.12	2.39	0.17	2.41		5.35		3.46	16.16
Non-energy uses								12.49		0.00	12.56								12.56			25.04

LPG – liquefied petroleum gas
Note Conversion factor for coal used for power generation adjusted.

TABLE 2.2 Commercial energy balance (million tonnes of oil equivalent): 2011/12

	Primary energy									Secondary energy												
	Coal and lignite				Hydro-power	Nuclear power	Renewable energy sources	Natural gas	Crude oil	LPG	Naphtha	Motor gasoline	Aviation turbine fuel	Kerosene	High-speed diesel	Light diesel oil	Fuel oil	Other petroleum products	Total petroleum products	Thermal power	Total power	Total energy
	Coking coal	Non-coking coal	Lignite	Total																		
Supply																						
Production	21.47	182.84	12.70	217.01	11.23	2.78	4.51	42.80	38.09												18.52	316.42
Imports	20.67	46.18		66.84				11.96	171.73	5.74	2.12	0.70	0.00	0.59	1.09	0.00	1.08	4.36	15.68		0.45	266.22
Exports	0.06	1.15		1.21						0.20	10.90	15.54	4.86	0.04	21.21	0.09	7.58	2.90	63.31		0.01	64.52
Stock changes	0.67	- 1.30	-0.13	-0.76					1.60	1.02	10.39	2.11	0.04	0.23	1.64	0.01	-3.32	-5.64	6.49			7.34
Availability	42.75	226.57	12.57	281.88	11.23	2.78	4.51	54.77	211.42													525.45
Petroleum refining								5.08	211.42	8.29	18.46	28.77	10.71	7.81	84.78	0.48	18.68	21.83	199.81			199.81
Own use	0.02	0.14		0.16				3.26	15.42													
LPG extraction								2.50		2.50									2.50			
Power generation		154.35	8.75	163.10	11.23	2.78	4.51	23.90			0.24				0.17	0.12	1.00		1.54	188.55	207.06	207.06
Conversion loss in power generation		107.02	6.07	113.09				12.44							0.10	0.07	0.57		0.74	126.27	126.27	126.27
Auxiliary consumption in power stations		3.92	0.22	4.15	0.08	0.28	0.28	0.29							0.00	0.00	0.04		0.04	4.48	5.12	5.12
Transmission and distribution losses																			0.00		17.92	17.92
Flaring of natural gas								0.97											0.00			0.97
Coal washery rejects	1.54	0.95		2.49																		2.49
Conversion				163.10				35.71	211.42	10.79	18.22	28.77	10.71	7.81	84.60	0.36	17.67	21.83	183.10	130.75	149.31	152.77
Net availability				116.13	11.15	2.50	4.23	19.05	196.00	17.35	19.84	16.04	5.90	8.60	66.13	0.28	7.86	17.65	159.63	57.80	58.19	353.01
Consumption				116.13				19.05		17.35	19.84	16.04	5.90	8.60	66.13	0.28	7.86	17.65	159.63		58.19	353.00
Agriculture								0.16		0.00					9.44	0.00	0.07		9.51		12.12	21.79
Industry	39.66	69.20	2.94	111.80				1.90		0.69	0.00				2.99	0.10	3.99	17.65	25.42		20.97	160.09
Transport								1.25		0.25		16.04	5.90		51.44	0.00	0.36		73.99		1.22	76.46
Residential								2.61		15.05				8.41					23.46		14.71	40.78
Commercial								1.05		1.18				0.06					1.24		5.62	7.92
Other energy uses	1.53			4.33				0.84		0.17	0.18			0.13	2.25	0.17	3.45		6.35		3.55	15.07
Non-energy uses								11.24		0.00	19.66								19.66			30.90

LPG – liquefied petroleum gas
Note Conversion factor for coal used for power generation adjusted.

TABLE 2.3 Commercial energy balance (million tonnes of oil equivalent): 2010/11

	Primary energy									Secondary energy												
	Coal and lignite				Hydro-power	Nuclear power	Renewable energy sources	Natural gas	Crude oil	LPG	Naphtha	Motor gasoline	Aviation turbine fuel	Kerosene	High-speed diesel	Light diesel oil	Fuel oil	Other petroleum products	Total petroleum products	Thermal power	Total power	Total energy
	Coking coal	Non-coking coal	Lignite	Total																		
Supply																						
Production	20.59	180.92	11.32	212.82	9.85	2.26	3.45	46.17	37.71	–	–	–	–	–	–	–	–	–	–	–	15.56	312.27
Imports	12.66	32.13	–	44.79	–	–	–	8.05	163.59	5.07	2.22	2.26	0.00	1.43	2.07	0.00	0.71	4.39	18.14	–	0.48	234.58
Exports	0.07	2.58	–	2.64	–	–	–	–		0.18	11.47	14.53	4.77	0.04	21.07	0.09	6.47	9.47	68.07	–	0.01	70.72
Stock changes	-0.62	-2.19	-0.01	-2.82	–	–	–	–	4.85	0.33	1.90	-0.51	-0.02	-0.18	0.31	-0.02	-3.50	-0.20	-1.90	–	–	0.13
Availability	32.57	208.28	11.30	252.15	9.85	2.26	3.45	54.22	206.15	–	–	–	–	–	–	–	–	–	–	–	–	476.26
Petroleum refining	–	–	–	–	–	–	–	2.90	206.15	8.52	18.85	27.96	10.19	8.05	80.78	0.55	19.70	21.83	196.44	–	–	196.44
Own use	0.03	0.14	–	0.16	–	–	–	1.63	15.87	–	–	–	–	–	–	–	–	–	–	–	–	–
LPG extraction	–	–	–	–	–	–	–	2.45	–	2.45	–	–	–	–	–	–	–	–	2.45	–	–	–
Power generation	–	141.91	8.16	150.07	9.85	2.26	3.45	25.56	–	–	0.55	–	–	–	5.06	0.13	1.24	–	6.98	182.61	198.18	198.18
Conversion loss in power generation	–	–	–	104.50	–	–	–	14.99	–	–	–	–	–	–	2.91	0.08	0.71	–	3.70	123.19	123.19	123.19
Auxiliary consumption in power stations	–	–	–	3.78	0.07	0.00	0.22	0.27	–	–	–	–	–	–	0.09	0.00	0.04	–	0.13	4.18	4.47	4.47
Transmission and distribution losses	–	–	–	–	–	–	–	–	–	–	–	–	–	–	–	–	–	–	0.00	–	16.73	16.73
Flaring of natural gas	–	–	–	–	–	–	–	0.87	–	–	–	–	–	–	–	–	–	–	0.00	–	–	0.87
Coal washery rejects	1.44	1.20	–	2.64	–	–	–	–	–	–	–	–	–	–	–	–	–	–	–	–	–	2.64
Conversion	–	–	–	150.07	–	–	–	33.42	206.15	10.97	18.30	27.96	10.19	8.05	75.72	0.42	18.46	21.83	191.91	127.36	144.38	147.89
Net availability	–	–	–	99.28	9.78	2.26	3.24	20.80	190.28	16.19	10.94	15.19	5.41	9.26	57.03	0.31	9.20	16.55	140.08	55.25	54.27	314.42
Consumption	–	–	–	99.28	–	–	–	20.80	–	16.19	10.94	15.19	5.41	9.26	57.03	0.31	9.20	16.55	140.08	–	54.27	314.42
Agriculture	–	–	–	–	–	–	–	0.17	–	0.00	–	–	–	–	7.59	0.00	0.07	–	7.66	–	10.87	18.70
Industry	32.73	61.68	2.34	96.74	–	–	–	1.31	–	0.67	0.00	–	–	–	6.65	0.12	4.10	16.55	28.10	–	20.57	146.72
Transport	–	–	–	–	–	–	–	0.36	–	0.00	–	15.19	5.41	–	40.54	0.00	0.75	–	61.89	–	1.15	63.39
Residential	–	–	–	–	–	–	–	0.33	–	15.45	–	–	–	9.11	–	–	–	–	24.56	–	13.35	38.24
Commercial	–	–	–	–	–	–	–	0.36	–	0.00	–	–	–	0.00	–	–	–	–	0.00	–	5.50	5.85
Other energy uses	–	–	–	2.53	–	–	–	1.87	–	0.07	0.17	–	–	0.15	2.25	0.17	4.28	–	7.09	–	2.84	14.33
Non-energy uses	–	–	–	–	–	–	–	16.39	–	0.00	10.78	–	–	–	–	–	–	–	10.78	–	–	27.17

LPG – liquefied petroleum gas
Note Conversion factor for coal used for power generation adjusted.

TABLE 2.4 Commercial energy balance (million tonnes of oil equivalent): 2009/10

	Primary energy						Secondary energy												
	Coal and lignite	Hydropower	Nuclear power	Renewable energy sources	Natural gas	Crude oil	LPG	Naphtha	Motor gasoline	Aviation turbine fuel	Kerosene	High-speed diesel	Light diesel oil	Fuel oil	Other petroleum products	Total petroleum products	Thermal power	Total power	Total energy
Supply																			
Production	212.24	8.96	1.60	3.23	42.75	33.69	–	–	–	–	–	–	–	–	–	–	–	13.79	302.47
Imports	47.61	–	–	–	8.86	159.26	3.09	1.75	0.41	0.00	1.03	2.64	0.00	0.77	4.98	14.68	–	0.46	230.40
Exports	1.47	–	–	–	–	–	0.15	10.65	10.45	4.89	0.05	19.11	0.00	4.97	2.78	53.04	–	0.01	54.51
Stock changes	-3.86	–	–	–	–	-0.18	0.21	1.52	-0.37	-0.09	-0.19	-1.09	0.00	-2.30	-4.41	-6.71	–	–	-10.75
Availability	254.52	8.96	1.60	3.23	51.60	192.77	–	–	–	–	–	–	–	–	–	–	–	–	467.60
Petroleum refining	–	–	–	–	–	192.77	9.14	18.39	24.11	9.90	8.93	75.85	0.45	17.61	21.15	185.53	–	–	185.53
Own use	0.26	–	–	–	–	11.42	–	–	–	–	–	–	–	–	–	–	–	–	–
LPG extraction	–	–	–	–	2.53	–	2.53	–	–	–	–	–	–	–	–	2.53	–	–	–
Power generation	156.50	8.96	1.60	3.23	19.23	–	–	1.37	–	–	–	3.96	0.16	0.71	–	6.20	181.93	195.72	195.72
Conversion loss in power generation	108.53	–	–	–	11.15	–	–	–	–	–	–	2.06	0.08	0.37	–	2.51	122.20	122.20	122.20
Auxiliary consumption in power stations	3.96	0.06	0.00	0.20	0.20	–	–	–	–	–	–	0.08	0.00	0.03	–	0.11	4.27	4.53	4.53
Transmission and distribution Losses	–	–	–	–		–	–	–	–	–	–	–	–	–	–	0.00	–	16.66	16.66
Flaring of natural gas	–	–	–	–	0.89	–	–	–	–	–	–	–	–	–	–	0.00	–	–	0.89
Coal washery rejects	2.06	–	–	–		–	–	–	–	–	–	–	–	–	–	–	–	–	2.06
Conversion	156.50	–	–	–	22.65	192.77	11.68	17.02	24.11	9.90	8.93	71.88	0.29	16.90	21.15	181.87	126.46	143.39	146.34
Net availability	95.69	8.90	1.60	3.02	28.95	181.35	17.36	9.64	13.72	4.93	9.72	54.33	0.29	10.41	18.93	139.32	55.47	52.78	316.75
Consumption	95.69	–	–	–	28.95	–	17.36	9.64	13.72	4.93	9.72	54.33	0.29	10.41	18.93	139.32	–	52.33	316.29
Agriculture	–	–	–	–	0.15	–	0.00	–	–	–	–	11.60	0.01	1.11	–	12.72	–	10.27	23.14
Industry	87.43	–	–	–	0.42	–	0.59	0.00	–	–	–	4.43	0.14	4.70	18.93	28.80	–	21.34	137.98
Transport	–	–	–	–	0.00	–	0.25	–	13.72	4.93	–	34.45	0.02	0.91	–	54.28	–	1.07	55.34
Residential	–	–	–	–	0.22	–	15.38	–	–	–	9.51	–	–	–	–	24.89	–	12.20	37.30
Commercial	–	–	–	–	0.00	–	0.99	–	–	–	0.07	–	–	–	–	1.06	–	5.07	6.13
Other energy uses	8.26	–	–	–	4.19	–	0.15	7.46	–	–	0.14	3.85	0.13	3.69	–	15.41	–	2.38	30.25
Non-energy uses	–	–	–	–	23.97	–	0.00	2.18	–	–	–	–	–	–	–	2.18	–	–	26.15

LPG – liquefied petroleum gas

Note Conversion factor for coal used for power generation adjusted.

TABLE 2.5 Commercial energy balance (million tonnes of oil equivalent): 2008/09

	Primary energy						Secondary energy												
	Coal and lignite	Hydropower	Nuclear power	Renewable energy sources	Natural gas	Crude oil	LPG	Naphtha	Motor gasoline	Aviation turbine fuel	Kerosene	High-speed diesel	Light diesel oil	Fuel oil	Other petroleum products	Total petroleum products	Thermal power	Total power	Total energy
Supply																			
Production	192.60	9.48	1.28	2.44	29.56	33.51	–	–	–	–	–	–	–	–	–	–	–	13.21	268.87
Imports	31.13	–	–	–	–	132.78	2.67	5.40	0.42	0.00	1.49	2.89	0.00	1.57	4.51	18.95	–	0.51	190.34
Exports	0.97	–	–	–	–	0.00	0.12	8.17	5.81	3.94	0.08	15.21	0.00	5.87	0.27	39.47	–	0.01	40.44
Stock changes	0.01	–	–	–	–	-5.51	-0.05	1.79	0.29	0.05	-0.28	-2.27	0.00	-1.95	-2.69	-5.11	–	–	-10.60
Availability	222.76	9.48	1.28	2.44	37.80	160.77	–	–	–	–	–	–	–	–	–	–	–	–	408.16
Petroleum refining	–	–	–	–	–	160.77	7.91	15.94	17.14	8.60	8.59	65.09	0.60	16.98	10.25	151.08	–	–	151.08
Own use	0.31	–	–	–	–	11.91	–	–	–	–	–	–	–	–	–	–	–	–	–
LPG extraction	–	–	–	–	2.44	–	2.44	–	–	–	–	–	–	–	–	2.44	–	–	–
Power generation	135.54	9.48	1.28	2.44	11.34	–	–	3.38	–	–	–	3.66	0.18	2.11	–	9.33	156.21	169.42	169.42
Conversion loss in power generation	92.59	–	–	–	6.00	–	–	–	–	–	–	1.51	0.08	0.87	–	2.45	101.04	101.04	101.04
Auxiliary consumption in power stations	3.55	0.08	0.18	0.00	0.15	–	–	–	–	–	–	0.07	0.00	0.04	–	0.11	3.81	4.07	4.07
Transmission and distribution losses	–	–	–	–	–	–	–	–	–	–	–	–	–	–	–	0.00	–	15.50	15.50
Flaring of natural gas	–	–	–	–	0.97	–	–	–	–	–	–	–	–	–	–	0.00	–	–	0.97
Coal washery rejects	1.72	–	–	–	–	–	–	–	–	–	–	–	–	–	–	–	–	–	1.72
Conversion	135.54	–	–	–	14.76	160.77	10.35	12.56	17.14	8.60	8.59	61.43	0.41	14.87	10.25	144.20	104.85	120.61	123.30
Net availability	85.19	9.40	1.10	2.44	23.04	148.86	12.84	11.58	12.05	4.71	9.72	46.84	0.37	8.72	11.80	118.63	51.36	49.31	276.17
Consumption	85.19	–	–	–	23.04	–	12.84	11.58	12.05	4.71	9.72	46.84	0.37	8.72	11.80	118.63	–	45.36	272.23
Agriculture	–	–	–	–	0.14	–	0.00	–	–	–	–	10.49	0.01	0.17	–	10.66	–	9.27	20.07
Industry	85.19	–	–	–	5.32	–	0.57	0.08	–	–	–	3.69	0.17	3.65	11.80	19.96	–	16.85	127.32
Transport	–	–	–	–	0.00	–	0.00	–	12.05	4.71	–	28.71	0.03	0.75	–	46.24	–	1.02	47.26
Residential	–	–	–	–	0.09	–	12.02	–	–	–	9.54	–	–	–	–	21.56	–	11.18	32.84
Commercial	–	–	–	–	0.00	–	0.00	–	–	–	0.04	–	–	–	–	0.04	–	4.60	4.65
Other energy uses	–	–	–	–	1.38	–	0.25	5.65	–	–	0.13	3.96	0.16	4.16	–	14.31	–	2.45	18.13
Non-energy uses	–	–	–	–	16.11	–	0.00	5.85	–	–	–	–	–	–	–	5.85	–	–	21.96

LPG – liquefied petroleum gas

Note Conversion factor for coal used for power generation adjusted.

ENERGY SUPPLY

- COAL AND LIGNITE
- PETROLEUM AND NATURAL GAS
- POWER
- RENEWABLE ENERGY SOURCES AND TECHNOLOGIES

COAL AND LIGNITE 3

AT A GLANCE

- Coal remained the dominant fuel in India in 2015, comprising 56.48% of the total consumption of fuel.
- The Coal Mines (Special Provision) Act, 2015 laid guidelines for the competitive bidding of captive coal blocks and also laid provisions for the commercialization of coal mining.
- In 2013/14, the production of coal from captive coal blocks was more than the target for the actual production.
- Coal India Ltd constituted 80.7% of the total coal production, followed by Singareni Collieries Company Ltd (8.58%) and the private sector (7.41%) in 2013/14.
- Coal imports have grown at a compound annual growth rate of 25% over the past 5 years, from 68.91 million tonnes (MT) in 2010/11 to 212.103 MT in 2014/15.

Introduction

Coal remains the mainstay of India's energy mix with more than 60% of the installed capacity of the country (Figure 1). This trend is expected to continue as the share of coal in the primary commercial energy supply mix is projected to remain more than 50% even by 2030 (TERI 2015).

Some major changes in the coal sector were witnessed last year with new policies being implemented or considered. Chief among those changes was the Coal Mines (Special Provisions) Act, 2015, which laid guidelines for the provisions competitive bidding of captive coal blocks and provisions for commercialization of coal mining. Although the provisions for commercialization have not been implemented in practice, they have paved the way for opening the sector after more than 40 years. Another major change was increase in supply for matching demand. Coal demand (coking and non-coking) in India grew by 30% from FY10 to FY15, while coal production increased by half of the demand in the same period, leading to a rise in imports. As such, the Ministry of Coal (MoC) has set a target of increasing coal production by 1.5 billion tonnes (BT) till 2020 and is looking for ways to achieve this target. Such an ambitious target would mean a renewed focus on underground mining and also increased production from captive blocks. The MoC has also initiated a plan to rationalize coal linkages to deal with inadequate or delayed supply and optimize transportation costs. This chapter discusses each of these developments and highlights the challenges involved in achieving them.

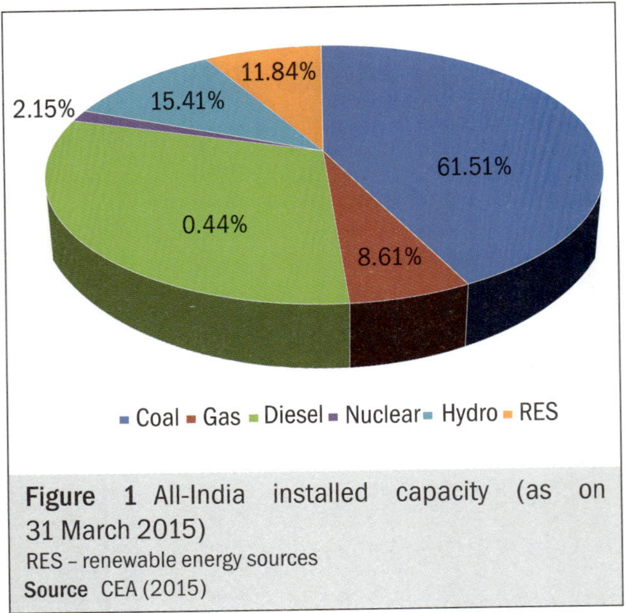

Figure 1 All-India installed capacity (as on 31 March 2015)
RES – renewable energy sources
Source CEA (2015)

Coal reserves

Coal resources in India stood at 306 596 million tonnes (MT) in April 2015 (CCO 2015), measured up to a maximum depth of 1200 m. Of this, 43% constitute proved geological reserves, while indicated geological reserves have risen to 47% and inferred geological resources constitute the remaining 10% (Figure 2). The increase in proved reserves on a year-on-year basis is mainly due to the discovery of coal resources in Ib River (2059.16 MT), Mand Raigarh (1690.36 MT), Talcher (896.35 MT), Hasdeo-Arand (432.56 MT), Wardha Valley (151.93 MT), Nand-Bander (133.98 MT), Barjora (86.52 MT), Godavari Valley (77.27 MT), Bishrampur (62.07 MT), Ramgarh (45.52 MT), West Bokaro (40.36 MT), and Raniganj (29.05 MT) coalfields (CMPDI 2015). But these coal resources, measured as per the Indian Standard Procedure, include the coal that has been mined out and also the coal blocks that are located in protected areas or beneath towns and cities or waterbodies and, therefore, cannot be mined. As per the United Nations Framework Classification, India had 19.8 BT of extractable coal reserves till 2013 (these are only for the coal blocks under Coal India Ltd (CIL) and exclude 119 assigned blocks) (CMPDI 2014).

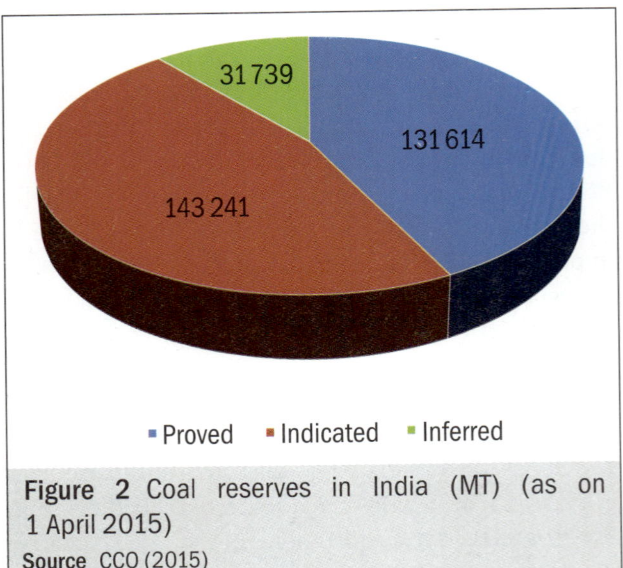

Figure 2 Coal reserves in India (MT) (as on 1 April 2015)
Source CCO (2015)

Map 1 shows the estimated gross geological coal resources by state as on 1 April 2015. Of the 306 BT of the total coal resources, about 131 BT have been extracted till date, according to an MoC statement (CCO 2015).

Coal mining

Coal

In 2014/15, India produced 612 MT of raw coal at the year-on-year growth rate of 8.25%, which was higher than the production of the last 4 years combined. While CIL and Singareni Collieries Company Ltd (SCCL) accounted for 80.7% and 8.58% of the raw coal production, respectively, the private sector mined 7.41% of coal (Table 1).

Trends in raw coal production during the past 10 years are shown in Figure 3. Coal production rose from 407 MT in 2005/06 to 612 MT in 2014/15, with the jump from 2013/14 to 2014/15 being the highest (8%). Coal reserves are concentrated mainly in Jharkhand (26%), Odisha (25%), Chhattisgarh (18%), West Bengal (10%), Madhya Pradesh (9%), and Maharashtra (4%) (CCO 2015).

With coal demand expected to reach 1200 MT by 2019/20, CIL is expected to chip in at least 1000 MT, of which 908 MT is the expected contribution from unidentified projects (PIB 2015a). CIL envisages taking up 129 new coal mining projects, of which 21 have been approved and 20 given nod in principle. A contribution of 87 MT is envisaged from new projects in 2016/17 and 182 MT in 2019/20 (PIB 2015b). Moreover, if CIL and other coal producers maintain an annual growth rate of 8.25%, they will be able to meet the target of 910.28 MT by 2020.

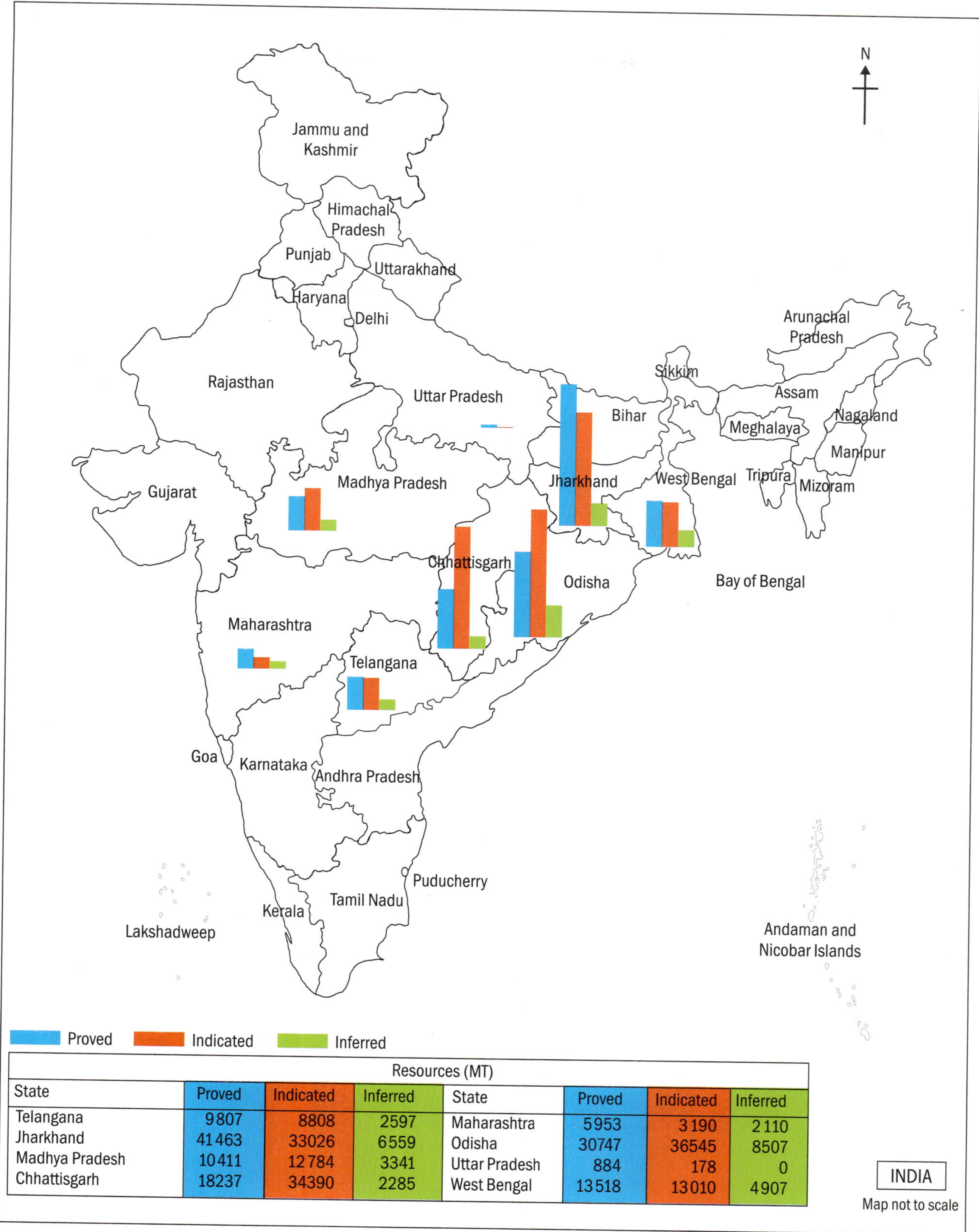

Resources (MT)							
State	Proved	Indicated	Inferred	State	Proved	Indicated	Inferred
Telangana	9807	8808	2597	Maharashtra	5953	3190	2110
Jharkhand	41463	33026	6559	Odisha	30747	36545	8507
Madhya Pradesh	10411	12784	3341	Uttar Pradesh	884	178	0
Chhattisgarh	18237	34390	2285	West Bengal	13518	13010	4907

Map 1 Gross geological coal resources by state as on 1 April 2015

Table 1 Coal production in 2011/12, 2012/13, 2013/14, and 2014/15 by company

Company	Production in 2011/12 (in MT)	Production in 2012/13 (in MT)	Production in 2013/14 (in MT)	Production in 2014/15 (in MT)
CIL	435.838	452.200	462.413	494.234
SCCL	52.211	53.190	50.469	52.536
Total public (including CIL, SCCL, and other PSUs)	503.843	521.677	528.081	567.033
Private	36.107	34.725	37.685	45.402
Total	539.950	556.402	565.766	612.435

CIL – Coal India Ltd; PSU – public sector undertaking; SCCL – Singareni Collieries Company Ltd
Source CCO (2015)

Lignite production

Lignite production increased from 30.2 MT in 2005/06 to 48.2 MT in 2014/15 (Figure 4). As on 1 April 2015, the estimated lignite reserve in the country is around 44 114 MT. The largest deposits of lignite are located in Tamil Nadu (80%). Other states where lignite deposits are located include Rajasthan, Gujarat, Jammu and Kashmir, and Puducherry.

Captive coal blocks

The Coal Mines (Special Provision) Act, 2015 laid provisions for the auction and commercialization of coal mining. Under the provisions of this act, the companies should meet the following criteria to be eligible for participating in the bidding process:

- A government company or a joint venture (JV) formed by such a company or between a state government or the central government
- A company or a JV formed by two or more companies having common specified end use
- A company engaged in a specified end use having a coal linkage or one that has made an application for a coal linkage
- A government company or a JV formed by such a company with any other company with a common specified end use

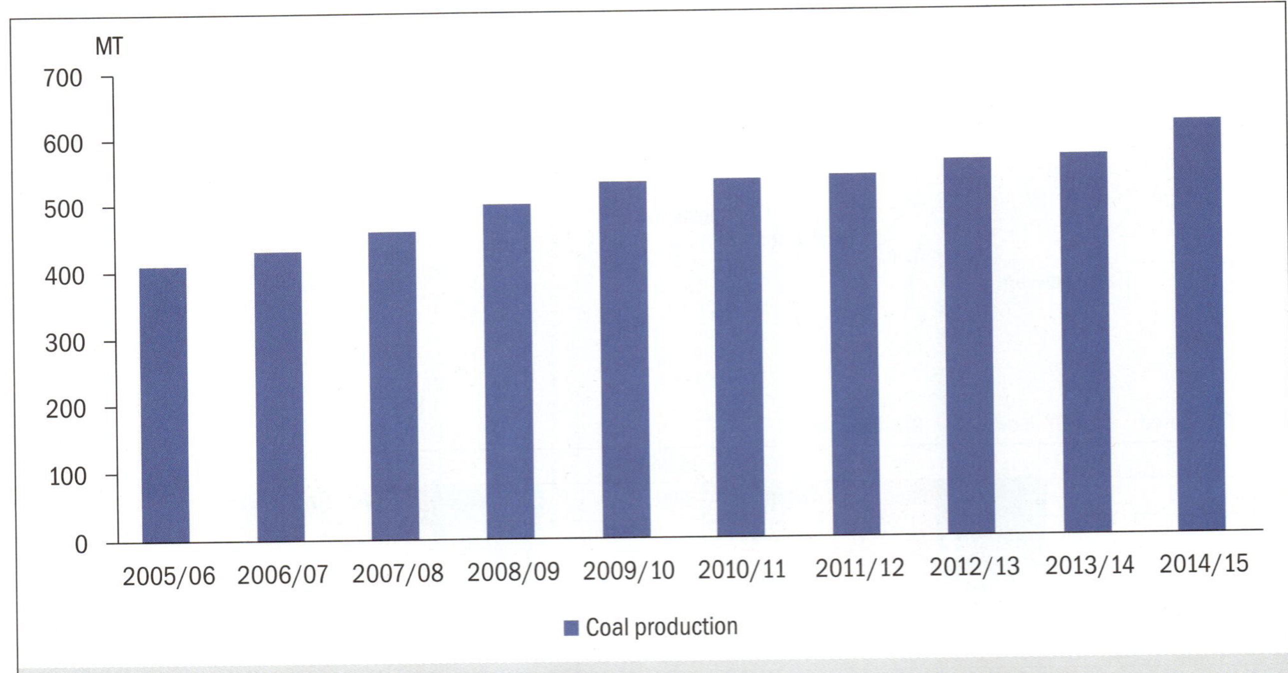

Figure 3 Raw coal production over 10 years
Source CCO (2015)

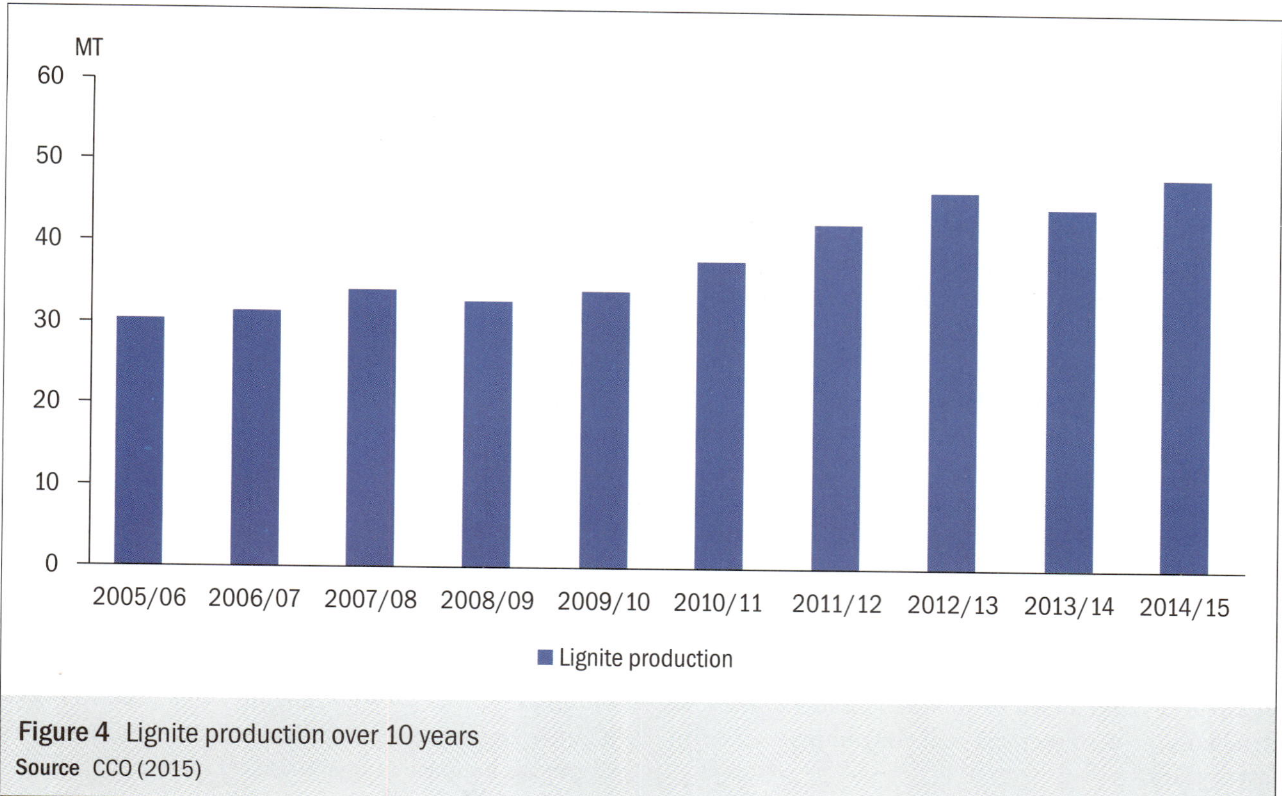

Figure 4 Lignite production over 10 years
Source CCO (2015)

The genesis of this act lies in the Supreme Court judgement that in 2014 upheld the views of the Comptroller and Auditor General (CAG) and questioned the legality of the captive coal block allocation process. The court declared the allocation of 204 coal blocks (of the 218 blocks allocated since 1993) illegal and arbitrary (Manohar Lal Sharma vs The Principal Secretary & Ors 2014). It cancelled the allocation and ordered the government to re-auction the blocks using a transparent auction-based mechanism. Accordingly, the coal blocks for the regulated sector (electricity) were auctioned under the reverse bidding process, while those for the non-regulated sector (steel, cement, fertilizers) were auctioned under the forward bidding process. The regulated sector would follow the process of reverse auction whereby the company that wins the block is the one that has offered the lowest bid to the CIL ceiling price for a similar grade. The non-regulated sectors would follow a forward auction process whereby the company that offers the most to the notified price for a similar coal grade would win the block. Such a company would have to pay an upfront payment of 10% of the intrinsic value of the mine or the value computed by compounding the annuitized reserve price over the mine's life, whichever is higher. Moreover, for blocks reserved for the power sector, a fixed reserve price of ₹100/tonne must be paid to the state government as per the actual production over and above the royalty.

Apart from questioning the allocation process, which led to an estimated loss of ₹185 000 crore to the exchequer, the CAG in its report also questioned the lack of production from the allocated blocks (CAG 2012). This is because production from captive blocks did not match the expected targets. For example, the Eleventh Five-year Plan estimated a production target of 104 MT from 93 blocks in 2011/12. However, only 29 of these blocks produced about 36 MT by the end of the year. Moreover, the trend of production from these blocks has always been well below the expected targets. Till March 2015, 56 blocks were auctioned or allocated, and production rose from 39 MT in 2013/14 to 52.77 MT in 2014/15.

Coal consumption

Globally, coal consumption grew by 0.4% in 2014, which is well below the annual growth of 2.9%, driven by the falling consumption in China, Ukraine, and the

United Kingdom. In comparison, India experienced the largest volumetric increase ever and in the world at 11.1% (BP 2015). Coal consumption in India has been increasing steadily. In the past 6 years, demand increased from 606 MT in 2009/10 to 787 MT in 2014/15. These levels are expected to touch 980 MT by 2016/17 (Planning Commission 2012). Much of this increase is expected to come from the power sector. For example, in the last 9 years, the power sector (utilities and captive) has seen a compound annual growth rate (CAGR) of 7.5%, with demand rising from 336 MT in 2006/07 to 601 MT in 2014/15 (Table 2). In the same period, consumption of coking coal grew at a CAGR of 5.86% from 35.17 MT to 55.46 MT. The non-coking coal segment, with a 93% share in the total consumption, increased from 428.7 MT in 2006/07 to 731 MT in 2014/15. The other segments, which include fertilizers, LTC/soft coke, and colliery consumption, saw a rise in consumption from 72.91 MT in 2006/07 to 103.85 in 2014/15, with a CAGR of 4.52%. Table 2 gives the detailed year-wise sectoral coal consumption over the last 9 years.

Coal pricing

The gross calorific value (GCV) pricing system is an international pricing standard that came into effect on 1 January 2012 in the Indian coal industry. The GCV pricing system classifies coal into 17 grades at an interval of 300 kcal/kg. This system facilitates a narrower bandwidth for coal pricing and, therefore, closely reflecting the quality and price of coal (see Annexure 1). This would enable pricing of coal produced at different mines under a standard formula based on the total heat value of coal, with an exception of coal produced at the Rajmahal mines of Eastern Coalfields Ltd, where an additional amount of ₹390 per tonne is charged over and above the prescribed notified rate.

Coal India Ltd revised and rationalized the basic notified prices of all grades of non-coking coal,[1] except G1, G2, and G5 (see Annexure 1), which came into effect on 28 May 2013 and was expected to bring additional revenue of ₹2500 crore in a full year and ₹2100 crore for the remaining months of the fiscal year 2013/14. Higher grades of coal, G3 and G4, with a GCV ranging between 6000 and 6300 will see a 12% reduction in prices. To neutralize the loss of revenue on this account, CIL has approved a 10% increase in the prices of a large segment of coal from G6 to G17, with a GCV ranging from 2200 to 6000. The two moves taken together translate into a 4.75% aggregate increase in coal prices (Dutta 2013).

The Cabinet Committee of Economic Affairs approved the proposed pass-through mechanism on 21 June 2013, which allows power producers to pass on the higher cost of imported coal to consumers as against the averaging of prices of imported and domestic coal under the price-pooling mechanism planned earlier. The quantum of coal to be imported has not yet been worked out, but as per MoC

Table 2 Coal consumption pattern (in MT)

Sector	2006/07	2007/08	2008/09	2009/10	2010/11	2011/12	2012/13	2013/14	2014/15	CAGR (%)
Steel and coke	35.17	39.02	37.66	39.39	40.00	46.67	52.30	53.98	55.46	5.86
Power (utilities)	307.92	332.40	362.08	380.13	405.00	460.00	512.00	535.00	551.60	7.55 (including captive power)
Power (captive)	28.13	29.31	32.94	38.47	40.00	40.00	43.00	46.14	50.00	7.55 (power utilities)
Cement	19.74	21.27	20.09	20.80	25.98	28.89	30.24	30.00	26.12	3.56
Others	72.91	82.29	96.26	103.46	113.80	120.47	135.30	104.57	103.85	4.52
Total	463.87	504.29	549.03	582.25	624.78	696.03	772.84	769.69	787.03	

Source CCO (2015); MoC (2011)

[1] Details available at <http://pib.nic.in/newsite/PrintRelease.aspx?relid=101114>, last accessed on 30 January 2014.

Annexure 1 Gradation of non-coking coal with revised prices

Pithead run of mine price for non-coking coal with effect from 28 May 2013			
Grades	kcal/kg	Power utilities (including IPP), fertilizers, and defence	Sectors other than power utilities (including IPP), fertilizers, and defence
		₹/tonne	₹/tonne
G1	Exceeding 7000	*	*
G2	Exceeding 6700 but not exceeding 7000	4870	4870
G3	Exceeding 6400 but not exceeding 6700	3890	3890
G4	Exceeding 6100 but not exceeding 6400	3490	3490
G5	Exceeding 5800 but not exceeding 6100	2800	2800
G6	Exceeding 5500 but not exceeding 5800	1600	2150
G7	Exceeding 5200 but not exceeding 5500	1400	1890
G8	Exceeding 4900 but not exceeding 5200	1250	1690
G9	Exceeding 4600 but not exceeding 4900	970	1310
G10	Exceeding 4300 but not exceeding 4600	860	1160
G11	Exceeding 4000 but not exceeding 4300	700	950
G12	Exceeding 3700 but not exceeding 4000	660	890
G13	Exceeding 3400 but not exceeding 3700	610	820
G14	Exceeding 3100 but not exceeding 3400	550	740
G15	Exceeding 2800 but not exceeding 3100	510	680
G16	Exceeding 2500 but not exceeding 2800	450	610
G17	Exceeding 2200 but not exceeding 2500	400	540

GCV – gross calorific value; IPP – independent power producer

*For GCV exceeding 7000 kcal/kg, the price shall be increased by ₹150/tonne, over and above the price applicable for GCV band exceeding 6700 but less than 7000 kcal/kg, for increase in GCV for every 100 kcal/kg or part thereof. For Western Coalfields Ltd, there will be a 10% add on over and above the price mentioned above for GCV bands not exceeding 5800 kcal/kg and below. An additional amount of ₹390/tonne to be charged over and above the notified price in respect of the coal produced from Rajmahal mine of Eastern Coalfields Ltd.

estimates, if CIL imports 15% of coal, this would result in an increase in electricity tariff across the country by 15–17 paisa per unit.

Imported coal price

In 2015, steam coal prices decreased in South Africa and Australia, while prices were volatile in Indonesia with a general oversupply of coal in the market. Prices of South African coal (6000 kcal/kg NAR[2]) decreased to $50.70 per tonne free on board (FOB) on 22 September from $52.60 per tonne FOB[3] on 31 August. Prices of Australian coal (6300 kcal/kg gross as received [GAR][4]) witnessed a drop to $55 per tonne FOB on 22 September from the 31 August rate of $58.45 per tonne FOB. Prices of Indonesian coal (5900 kcal/kg GAR) marginally decreased from $51.75 on 31 August to $50.35 per tonne FOB on 22 September. A similar trend was observed for the 5000 kcal/kg GAR category Indonesian coal, where the price fell from $39 per tonne on 31 August to $38.15 per tonne FOB.

Coal prices across Europe decreased in the first two weeks of July 2015 due to receding demands and reduced imports from major Asian buyers. The world's biggest coal importer, China, is witnessing a reduced demand in coal imports with rising

[2] This is the basis that determines the amount of usable energy coal contains and is the difference between the as-received energy and the amount of energy to vaporize the moisture in the coal formed by combustion of hydrogen.

[3] It is the rate at which coal is sold and it excludes ocean freight and insurance but includes loading costs.

[4] It is defined as the measure of heat value, which takes into account the presence of moisture in coal.

environmental costs and reducing purchasing power. Coal imports in China during January–August 2015 witnessed a 31.3% fall over the same period last year. India and China are focusing on hiking their domestic production, which has affected global coal imports. In the short term, the global coal market is expected to witness continued price pressures and may see improvement by the end of 2015 with potential price support as producers adjust their supply rates (Mjunction 2015).

Mining techniques

Two methods of coal mining are prevalent in India, namely, opencast (OC) mining and underground (UG) mining. The share of OC mining in the total raw coal production grew from 85.02% in 2005/06 to 92.09% in 2014/15. The share of UG mining, however, fell from 14.98% in 2005/06 to 7.91% in 2014/15 (Figure 5). Moreover, the output per manshift for UG mines was 0.78 MT in 2014/15 compared to 13.06 MT for OC mines. This trend is worrying given the rising demand for coal and the inability of coal mining via OC mines to fulfil this demand. Production from UG mines has been stagnating over the years due to overemphasis on OC mining. This has inversely led to the use of technology that has not been updated, leading to inefficient and labour-intensive mining.

Coal beneficiation

Coal beneficiation refers to activities that improve the quality of coal. Coal washing is one of the most prevalent techniques adopted under beneficiation. The washing process consists of removing impurities (mainly ash, sulphur, and rock) from raw coal. The extent to which coal is cleaned before use depends on the purpose for which it is mined and on the specifications it must meet for its use. The Ministry of Environment, Forests, and Climate Change (MoEFCC) restricts the use of coal containing more than 34% ash content in power stations located 500 km away from pitheads. Moreover, given the estimated increase in coal production, it has become imperative to increase the number of coal washeries.

The installed capacity of washeries for thermal coal was about 131.24 MT per year as on 31 March 2014, with a total of 52 PSUs and private coal washeries. Coking coal and non-coking coal amounting 26.69 MT per year and 101.55 MT per year, respectively, were registered as washed coal as on 31 March 2014 (Annexure 2) (MoSPI 2015).

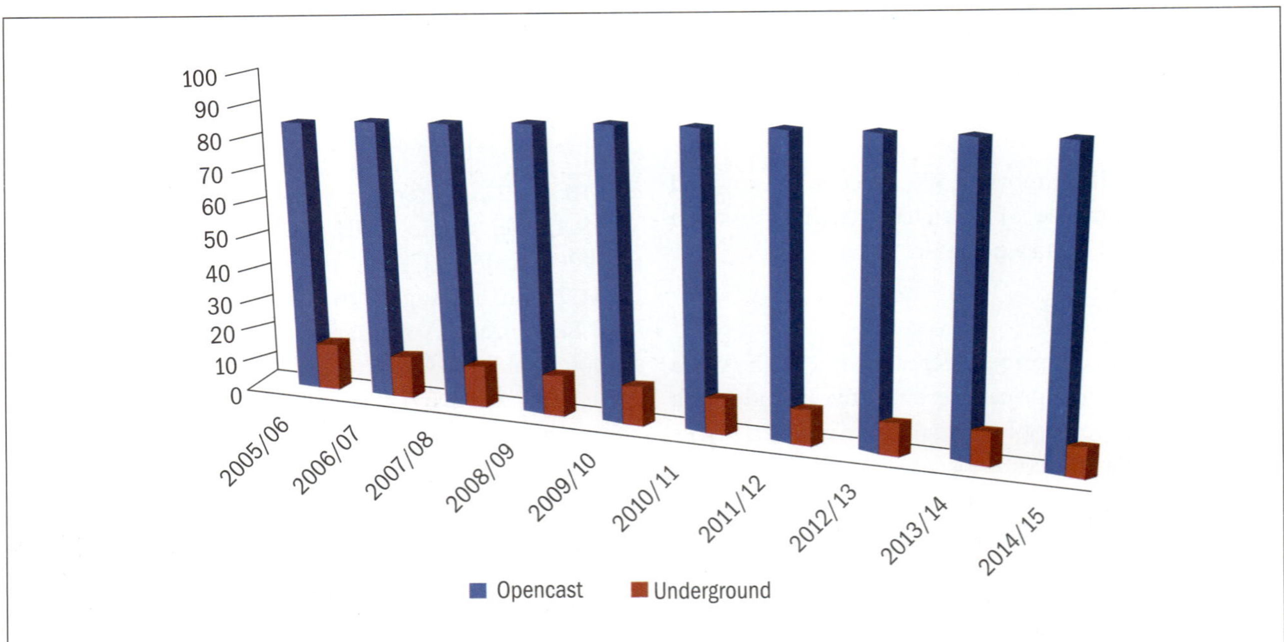

Figure 5 Production of coal from opencast and underground mining (in per cent) from 2005/06 to 2014/15
Source CCO (2015)

Annexure 2 Installed capacity of coal washeries in India as on 31 March 2014

Sl No.	Washery and operator	State of location	Capacity (MTY)
			31 March 2014*
	Coking coal		
1	Dudga-II, CIL	Jharkhand	2.00
2	Bhojudih, CIL	West Bengal	1.70
3	Patherdih, CIL	Jharkhand	1.60
4	Moonidih, CIL	Jharkhand	1.60
5	Sudamdih, CIL	Jharkhand	1.60
6	Mahuda, CIL	Jharkhand	0.63
7	Kathara, CIL	Jharkhand	3.00
8	Swang, CIL	Jharkhand	0.75
9	Rajrappa, CIL	Jharkhand	3.00
10	Kedla, CIL	Jharkhand	2.60
11	Nandan, CIL	Madhya Pradesh	1.20
	(A) CIL		19.68
12	Durgapur, SAIL	West Bengal	1.50
13	DCOP, DPL	West Bengal	1.35
14	Chasnala, IISCO	Jharkhand	1.50
15	Jamadoba, TISCO	Jharkhand	0.90
16	West Bokaro-II, TISCO	Jharkhand	1.80
17	West Bokaro-III, TISCO	Jharkhand	2.10
18	Bhelatand	Jharkhand	0.86
	(B) PSU and private		10.01
	Total (A + B)		29.69
	Non-coking coal		
1	Dugda-I, CIL	Jharkhand	2.50
2	Madhuban, CIL	Jharkhand	2.50
3	Gidi, CIL	Jharkhand	2.50
4	Piparwar, CIL	Jharkhand	6.50
5	Kargali, CIL	Jharkhand	2.72
6	Bina, CIL	Uttar Pradesh	4.50
	(A) CIL		21.22
7	Dipka, Aryan Coal Beneficiation Pvt. Ltd	Chhattisgarh	12.00
8	Gevra, Aryan Coal Beneficiation Pvt. Ltd	Chhattisgarh	5.00
9	Panderpauni, Aryan Coal Beneficiation Pvt. Ltd	Maharashtra	3.00
10	Chakabuwa, Aryan Energy Pvt. Ltd	Chhattisgarh	4.00
11	Indaram, Aryan Coal Beneficiation Pvt. Ltd	Andhra Pradesh	–
12	Talcher, Aryan Energy Pvt. Ltd	Odisha	2.00

Contd...

Annexure 2 *Contd...*			
Sl No.	Washery and operator	State of location	Capacity (MTY)
			31 March 2014*
13	Wani, Kartikay Coal Washeries Pvt. Ltd (Aryan)	Maharashtra	2.50
14	Korba, ST-CLI Coal Washeries Ltd	Chhattisgarh	5.20
15	Ramagundam, Gupta Coalfields and Washeries Ltd	Andhra Pradesh	2.40
16	Sasti, Gupta Coalfields and Washeries Ltd	Maharashtra	2.40
17	Wani, Gupta Coalfields and Washeries Ltd	Maharashtra	1.92
18	Umrer, Gupta Coalfields and Washeries Ltd	Maharashtra	0.75
19	Bhandara, Gupta Coalfields and Washeries Ltd	Maharashtra	0.75
20	Gondegaon, Gupta Coalfields and Washeries Ltd	Maharashtra	2.40
21	Majri, Gupta Coalfields and Washeries Ltd	Maharashtra	2.40
22	Bilaspur, Gupta Coalfields and Washeries Ltd	Chhattisgarh	3.50
23	Ghugus, Gupta Coalfields and Washeries Ltd	Maharashtra	2.40
24	Talcher, Global Coal Mining (Pvt.) Ltd	Odisha	2.50
25	Ib Valley, Global Coal Mining (Pvt.) Ltd	Odisha	3.25
26	Ramagundam, Global Coal Mining (Pvt.) Ltd	Andhra Pradesh	1.00
27	Wani, Bhatia International Ltd	Maharashtra	3.73
28	Ghugus, Bhatia International Ltd	Maharashtra	4.00
29	Jharsuguda, Bhatia International Ltd	Odisha	1.50
30	Tamnar, Jindal Steel and Power Ltd	Chhattisgarh	6.00
31	Wani, Indo Unique Flame Ltd	Maharashtra	2.40
32	Nagpur, Indo Unique Flame Ltd	Maharashtra	0.60
33	Punwat, Indo Unique Flame Ltd	Maharashtra	2.40
34	Dharamsthal, BLA Industries	Madhya Pradesh	0.33
	(B) Private		80.33
	Total (A + B)		101.55
	Gross total (coking + non-coking)		131.24

CIL - Coal India Limited; IISCO - India Iron & Steel Company; MTY - metric tonnes per year; PSU - public sector undertakings; SAIL - Steel Authority of India Ltd; TISCO - Tata Iron and Steel Company

*Provisional

Source MoSPI (2015)

Coal imports, port capacity, and infrastructure

The demand for coal during 2014/15 was 787 MT, whereas the indigenous availability was 612.43 MT. This gap was filled with imports, which rose from 166.85 MT in 2013/14 to 212.10 in 2014/15 (Figure 6). Of this, non-coking coal constituted almost 80% of the imports due to increasing requirements from the power sector. Coal imports grew at a CAGR of 25% during the last 5 years from 68.91 MT in 2010/11 to 212.103 MT in 2014/15.

Moreover, coal imports at non-major ports have been steadily increasing as compared to imports at major ports. For example, in 2014/15 alone, combined imports from Mundra, Dahej, Dhamra, Gangavaram, Krishnapatnam, and SEZ Mundra (92.04 MT) exceeded imports from major ports of Chennai Sea, Cochin Sea, Ennore Sea, Kandla Sea, Kolkata

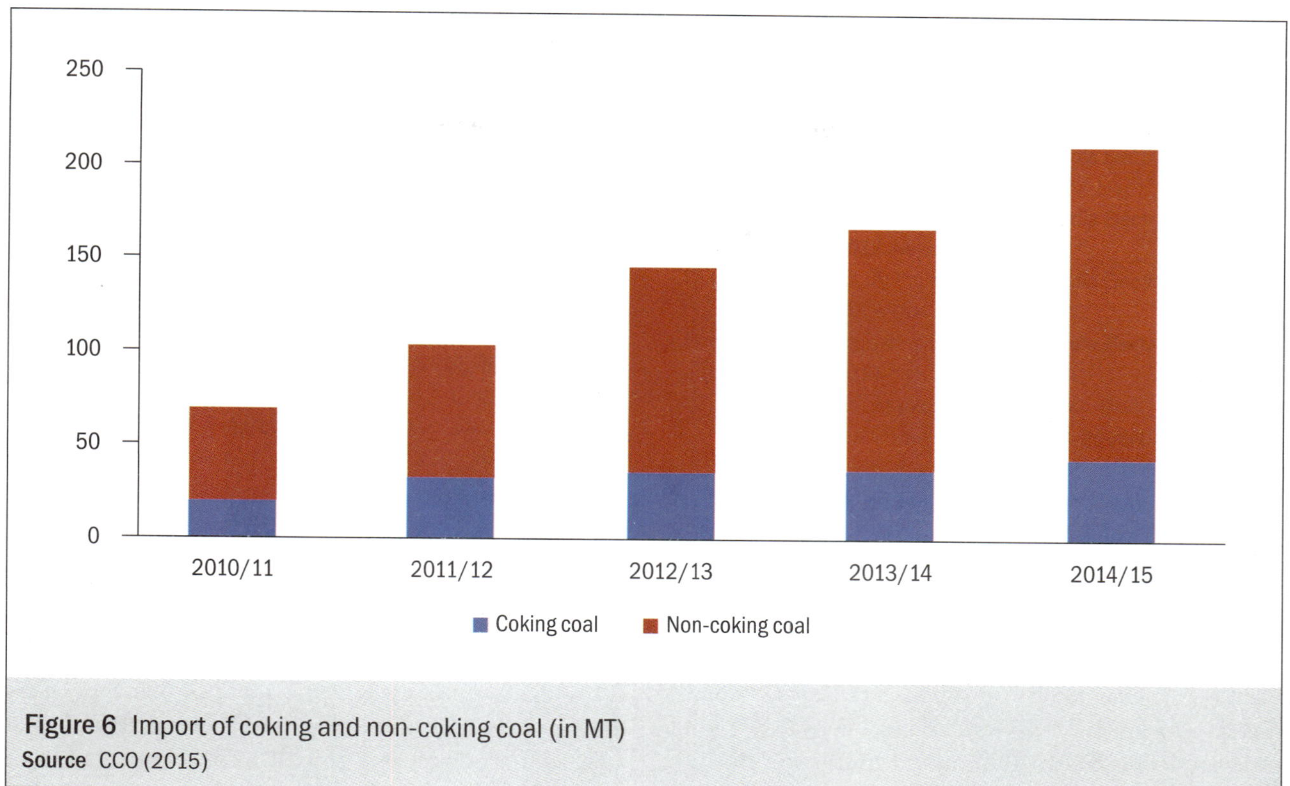

Figure 6 Import of coking and non-coking coal (in MT)
Source CCO (2015)

Sea, Marmagao Sea, Mumbai Sea, Mangalore Sea, Paradip Sea, Tuticorin Sea, and Visakhapatnam Sea (89.35 MT). Table 3 gives a detailed break-up of the amount of coal handled at major and non-major ports in India from 2011/12 to 2014/15.

Some of the reasons for this rise have been the capacity of ports to handle coal traffic, better evacuation facilities from the non-major ports, and proximity of thermal power plants (TPPs) to the ports. In 2012/13, the capacity utilization of thermal coal at the major ports handling imported coal remained high, with Paradip recording a utilization of 130.53%, closely followed by Haldia at 92% and Ennore at 71%, well above the internationally recommended norm of 70% and below (MoS 2013).

Table 3 Coal handling by ports (MT)

Year	Major ports	Non-major ports	Total coal handled
2011/12*	58.44	44.66	102.85
2012/13*	67.27	78.51	145.78
2013/14**	71.91	94.94	166.85
2014/15#	89.35	122.74	212.10

Source *CCO (2013), **CCO (2014), #CCO (2015)

In the same year, a commodity-wise analysis showed that the weighted average pre-berthing detention (the time taken by a ship from its arrival till it starts moving for a berth) came to 2.2 days for coking coal and 0.91 days for thermal coal. The average turnaround time (total time spent by a ship since its entry till its departure) at the major ports was 4.29 days, while a commodity-wise analysis showed that the weighted average turnaround time for bulk carriers carrying coking coal was 6.4 days and that for thermal coal was 3.3 days (MoS 2013). Another factor that has enabled growth in coal imports at non-major ports is the proximity of these ports to TPPs. The majority of the coal imported at Mundra and Dahej feeds the TPPs located in these areas. These ports also have better infrastructural facilities for cargo handling systems, draft, and IT, as well as better investment in warehousing facilities. Coal import figures have already exceeded the projections made in the Twelfth Five-year Plan. As per the Twelfth Five-year Plan, imports are expected to rise to 185.50 MT in 2016/17, with the assumption of the total projected coal demand of 980.50 MT and the projected domestic supply of 795 MT (Table 4).

Table 4 Coal supply and demand (in MT)

Source	2012/13 (budgeted estimate)	2012/13 (actual)	2013/14 (budgeted estimate)	2013/14 (actual)	2014/15 (budgeted estimate)	2015/16** (actual)	Twelfth Plan (projection for 2016/17)
Coal India Ltd	470	464.95	492.00	471.09	520.00	488.92	615.00
Singareni Collieries Company Ltd	53.1	52.08	56.00	47.89	55.50	52.66	57.00
Others	57.2	50.58	66.55	52.03	68.25	66.05	123.00
Total indigenous supply	580.3	567.60	614.55	571.00	643.75	607.63	795.00
Actual import		145.79		168.50*		212.103	
Demand projected	772.84		769.69		787.03		980.50
Gap projected to be met by import	192.54		155.14		143.28		185.50

Note Demand used to be assessed by the erstwhile Planning Commission at the beginning of each annual plan.
*Provisional figures up to January 2014
**CCO (2015)
Source MoC (2014)

Transportation

Coal is evacuated through railways, roads, merry-go-rounds (MGRs), belts, conveyors, and ropeways. The Indian Railways accounts for close to half (49%) of the total share of raw coal dispatched from the pitheads to the centres of consumption. Figure 7 shows the trend in the mode-wise dispatch of coal during 2014/15 across CIL, SCCL, and public and private entities. The figure shows that roads account for 28%, followed by MGR at 15% in 2014/15. The envisaged coal movement matrix in the terminal year of the Twelfth Five-year Plan (2016/17) constitutes 58% share of rail, 25% share of road, 11% share of MGR, and 6% share of belt/rope. This includes the movement of 800 MT of indigenous coal and coal products and 265.50 MT of imported coal. The plan for increased movement of imported coal is the reason for the increased share of railways in the Twelfth Five-year Plan (FICCI 2013).

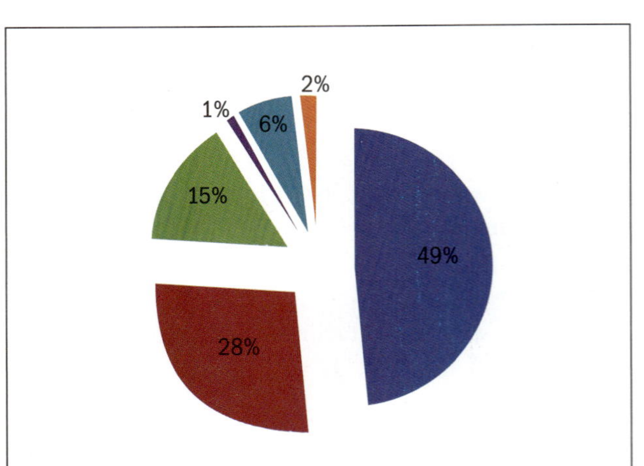

Figure 7 Coal transportation by various modes for 2014/15
MGR – merry-go-rounds
Source CCO (2015)

In India, coal is mainly found in the east and central-east, while the consumers are spread across the country. The major coal consumption points are located in north India, the western states of Gujarat and Maharashtra, and the southern region. With the upcoming centres of production for CIL coalfields, three rail lines have been identified in three potential coalfields to evacuate coal from pitheads: Jharsuguda–Barpali–Saradegna railway line (53 km) in Ib Valley, Odisha; Bhupdevpur–Korichapar–Dharamjaigarh railway line (180 km) in Mand Raigarh Coalfield, Chhattisgarh; and Tori–Shivpur–Kathautia railway line (90.7 km) in North Karanpura, Jharkhand. The Ministry of Railways envisages commissioning these three railway lines by June 2016, September 2016, and December 2016, respectively. These railway lines would help in increasing production to the extent of 200 MT from these coalfields (MoC 2013).

Although these railway lines will help in transporting coal, the problem of delay or lack of transportation infrastructure, which leads to inadequate supply to TPPs, may not be solved. The MoC plans to solve this problem in an ad hoc manner through the latest rationalization plan. A sub-committee constituted in 2014 looked into the rationalization of sources for the power sector and submitted a proposal for the rationalization of 21 power plants (MoC 2014c). As per this proposal, powerhouses are divided into four categories: pithead plant, hinterland plant, coastal plant—if a coastal plant is near the pithead it is classified as a pithead plant—and intermodal plant. These powerhouses would look at bilateral or multilateral swaps to minimize the landed cost of coal and optimize supply. The proposal projected that such a step would save around ₹4500–6000 crore annually. In the first stage, the rationalization process aims to swap between coastal and hinterland plants such that the plants closer to the coast would use higher quality imported coal and swap their domestic linkages with hinterland plants. Table 5 lists the companies participating in the rationalization process.

In the second stage, a multilateral approach would be followed whereby a final allocation matrix based on an optimization model would be achieved via the identification of clusters, resulting in a net reduction in the residual amount on a cumulative basis. Such a plan, while preparing for the existing linkages on the supply side, has not taken into account the nature of physical linkages, for example, adequate railway routes. Freight transportation has to compete with passenger trains on the main trunk routes. Some of the main trunk routes are running at a capacity utilization factor of more than 100%. On the flip side, considering the production cycle in a colliery, the present capacity at the siding routes often remains underutilized for almost seven months a year (summer and monsoon). This eventually creates a suboptimal situation wherein the railway authorities find it difficult to adjust train schedules and arrange additional wagons during the five months of peak transportation. Such a situation needs to be avoided by finding a mechanism to create and finance additional infrastructure.

Policies and regulations

The MoC governs the coal sector in India. CIL, a public sector entity, is the world's largest coal producer and produces about 81% of India's coal. In general, the central and state governments jointly manage resources in India. Coal mines were nationalized in the early 1970s. By an amendment to the Coal Mines (Nationalization) Act, 1973, two exceptions were introduced in 1976: (1) captive mining by private companies engaged in the production of iron and steel and (2) sub-lease of coal mining to private parties in isolated small pockets not amenable to economic development and not requiring rail transport.

Private coal mines in India were nationalized in two phases during 1971–73. The Coal Mines Nationalization Act of 1973 was amended in June 1993 to allow coal mining for captive consumption for generating power, washing of coal obtained from a mine, and other end uses to be notified by the government from time to time.

Important acts for coal sector

- The Mines and Minerals (Development and Regulation) Amendment Act, 2015
- The Coal Bearing Areas (Acquisition and Development) Act (CBAADA), 1957

Table 5 Companies participating in the initial rationalization of linkages

State	Participating companies	Logistics savings (crore/year)
Tamil Nadu	TANGEDCO and NTPC	370
Gujarat	GSECL and NTPC	720
Maharashtra	MAHAGENCO, Indiabulls Power, Adani Power, NTPC, NSPCL	303
Haryana	HPGCL, APCPL, DVC, Lanco, Anpara	281
Rajasthan	RRVUNL and NTPC	423
Punjab	PSPCL, DVC, UPRVUNL	175
Total		2272

APCPL – Aravali Power Company Private Ltd; DVC – Damodar Valley Corporation; GSECL – Gujarat State Electricity Corporation Ltd; HPGCL – Haryana Power Generation Corporation Ltd; MAHAGENCO – Maharashtra State Power Generation Company Ltd; NSPCL – NTPC-SAIL Power Company Ltd; NTPC – National Thermal Power Corporation; PSPCL – Punjab State Power Corporation Ltd; RRVUNL – Rajasthan Rajya Vidyut Utpadan Nigam Ltd; TANGEDCO – Tamil Nadu Generation and Distribution Corporation; UPRVUNL – Uttar Pradesh Rajya Vidyut Utpadan Nigam Ltd

Source MoC (2015b)

- The Mineral Concession Rules (MCR), 1960 (amended in 2000)
- The Coal Mines (Nationalization) Act, 1973 (amended in 1976 and 1993)
- The Coal Mines (Conservation and Development) Act, 1974 (amended in 1986)
- The Colliery Control Order, 1945 (amended in 2000)
- The Colliery Control Rules, 2004
- Auction by Competitive Bidding of Coal Mines (Amendment) Rules, 2012
- The New Coal Distribution Policy, 2007 with Presidential directives in 2012 and second directive in 2013
- Draft Coal Regulatory Authority Bill, 2013
- The Coal Mines (Special Provisions) Act, 2015

Policy and regulatory issues

Although the legislative framework governing the coal sector in India is well established, poor or no implementation of the existing laws have been observed. For instance, environmental clearances based on environmental impact assessment under the Environmental Protection Agency is subject to various flaws in implementation. Also, the system of mandatory compensatory afforestation as introduced under the Forest Conservation Act has not been fully achieved in practice.

Delay in clearances is also a major hurdle in coal mining. The MoC requires OC mines to commence production within 36–48 months and in the case of UG mines within 48–54 months. This timeline is rarely met since overlapping jurisprudence between state and central agencies delays clearances, often leading to delays of 10 years in some cases. Currently, various agencies at the central, state, and local levels are responsible for the management of the sector. However, roles and responsibilities of these jurisdictions in some cases are not clear. For instance, the MoC plays a conflicting role as both policy maker and administrator. This role is further highlighted considering the latest developments in the sector.

As mentioned earlier, the Coal Mines (Special Provisions) Act, 2015 that laid guidelines for the competitive bidding of coal block auctions has raised a few queries. The primary query is the viability of the pricing strategy as a result of the reverse bidding process. All the coal blocks auctioned for the regulated sector went from reverse bidding to forward bidding as per the notification of the nominated authority stating that bidding should be based on forward auctions if the bid price for a block is zero. In a forward auction, companies that win the bids cite 100% discount to the ceiling price and are willing to pay the government over and above the price at which they won the bid. With power tariffs being the sum of variable and fixed costs, fuel cost (which is part of the variable cost) is zero for these companies and they have to depend on the fixed cost to absorb the impact. On the contrary, discoms, especially those that have not signed long-term power purchase agreements (PPAs), may pay higher, based on higher fixed costs. For those that have signed long-term PPAs, there could be prolonged negotiations with gencos that attempt to revise fixed costs, thus leading to uncertainty. A few other issues that have cropped up related to the auction process include the need for dual bids and the legitimacy of allowing subsidiaries to participate in the same bid.

The 2015 act also laid provisions for the commercialization of coal mining. Providing guidelines for commercialization without putting in place a pricing policy, having an independent regulator, or resolving the existing supply–demand issues (problems with flexible spending accounts, coal mafia in transportation, and black marketing) would derail the actual process of transformation in the sector.

Another problem that needs to be addressed is the regulatory structure for coal pricing. At present, CIL (having a monopoly in the market) sets the price for each grade of coal. This structure will not be suitable when private players start coal mining. Therefore, the MoC needs to initiate a process of changing the pricing structure either by establishing a coal regulator that will decide pricing or by linking pricing to international markets. The former option has been under debate for a numbers of years, with coal producers unwilling to allow an external regulatory authority to set prices. The latter option, however, may not be viable considering that market prices could lead to higher price volatility. This would be unviable for the power sector, which is the primary consumer of coal and is a regulated market.

Environmental challenges

The consumption of coal has an adverse impact on environment and ecology. The fuel chain of coal begins with the exploration and development of mines, coal mining, beneficiation, transportation, combustion, and finally mine closure. Every stage of this fuel chain has a number of risks in terms of social displacement, rehabilitation and resettlement, livelihood impacts, land degradation, water pollution, air pollution (emission of SO_x, NO_x, CO, CO_2, SPM, PM_{10}, $PM_{2.5}$, and other harmful gases), destruction of forests and biodiversity, and health and safety issues.

Opencast mining has a sustained impact on the affected area. It destroys aquifer layers, changes the local lithology, creates voids, generates prominent forms of anthropogenic relief, and pollutes environmental elements, culminating into land degradation (Singh, Singh, and Singh 2007; Tudor and Catalin 2013). Moreover, unscientific techniques used in illegal mining have led to subsidence (localized or over an extended area) in large areas of Raniganj and Jharia. In Raniganj, mining activities have degraded a sizeable area of land and endangered about 42 localities. In fact, as per the MoC, the subsided area in the Jharia coalfield covers about 25.69 km^2, while the area affected in the Raniganj coalfield is about 8.62 km^2 (MoC 2014a). Moreover, since coal reserves are primarily located in forest areas, mining automatically leads to destruction of forests, which has a resounding impact on biodiversity. For instance, between 1973 and 2007, the forest cover in the Angul–Talcher region in Odisha reduced by 11% because of coal mining (Singh, Singh, and Singh 2010). The forest cover in the Damodar Valley Coalfield, which was once 65%, today stands at only 0.05% (Priyadarshi 2012).

Mining also has an impact on the water quality of the region around coalfields. For instance, the Damodar River has become one of the most polluted rivers in India on account of intensive mining in the coal mining belt (Dhar and Tiwary 1994). A study showed that a single coal washery discharged about 45 tonnes of fine coal in the river every day and as many as 11 coal washeries were located in the region with an installed capacity of 20.52 MT annually (Priyadarshi 2012). Coal transportation via roadways or railways also has environmental consequences. Moreover, since coal in India has high ash content, disposal of fly ash puts stress on land and water resources, with power plants using more than 1000 acres of land for storing ash. Fly ash production has witnessed a steady increase from 1 MT in 1947 to about 40 MT in 1994, about 112 MT in 2005 (Kumar, Mathur, and Sinha 2005), and 163.56 MT in 2012/13 (CEA 2014). Combustion of coal also releases a significant amount of greenhouse gases, such as carbon dioxide, other local pollutants, such as sulphur dioxide (SO_2), nitrogen oxide (NO_x), particulate matter (PM), and various trace metals, such as mercury, which get dispersed over large areas through stacks. Guidelines have so far existed for PM; however, a draft notification policy by the MoEFCC is looking to establish emission standards for the rest.

The problem is that there is no one policy that seeks to mitigate the environmental challenges emanating from coal. Each of the issues highlighted earlier has separate acts in place, which are under the jurisdiction of different ministries. While a single policy may be too radical to contemplate, more coordination between the different ministries and also better implementation of the acts could lead to a more efficient management of the system and thus mitigate the harshest consequences of coal mining.

Socio-economic challenges

Displacement and loss of livelihood

By conservative estimates, between 1950 and 1990, the coal mining sector is reported to have displaced around 2.0–2.5 million people in India (Terminski 2012). This problem is compounded by the rampant use of OC technology, which uses more land compared to UG mines but provides less employment and increases mining-induced displacement (Singh 2007). For example, an OC mine, which produces 40 MT of coal during its lifetime (about 15 years), will scar an area of about 25 km^2 (Herbert and Lahiri-Dutt 2004). Add to this is the problem of impact on livelihood. A livelihood study (Mishra 2009) conducted in the Ib Valley Coalfield in Odisha assessed the impact of different forms of capital on livelihood. The study found that while financial capital in terms of mean per capita income rose, physical capital in terms of basic infrastructure remained below par. Moreover, the health facilities developed by companies in the region were not available for the larger community.

Health and safety issues

Coal mining is a highly regulated industry due to the presence of many inherent operational and occupational hazards. Mine workers are exposed to a host of respiratory and cardiopulmonary ailments, thus increasing the risk of premature death at various stages of mining. The Coal Mine Safety Legislation in India is one of the most comprehensive and pervasive statutory frameworks for ensuring occupational health and safety. Compliance of these safety statutes is mandatory.

According to a 2010 CAG study on the corporate social responsibility activities of CIL, less than 8% of the contract workers are subjected to regular health check-ups though all are expected to be monitored regularly. This highlights the concern for inadequate attention to labour safety. Similarly, the safety of mine workers is also a prime concern since the number of fatalities is unacceptably high. Figure 8 shows the 10-year average rate of accidents and deaths per 1000 persons employed in coal mines in India since 1951. It can be seen that the rate of accidents has fallen from 0.61 during 1951–60 to 0.22 during 2011–14. The death rate also witnessed a fall from 0.82 during 1951–60 to 0.23 during 2011–14.

Illegal mining

After the nationalization of India's coal industry, only the central and state governments are authorized to mine coal; however, illegal coal mining and diversion are serious issues in India and have been highlighted by the Parliamentary Standing Committee on Coal and Steel (Standing Committee on Coal and Steel, Fifteenth Lok Sabha, 2012). As per a knowledge paper by FICCI, nearly one-thirds of mined coal is lost to illegal hoarding and selling (FICCI 2013). XLRI Jamshedpur and Indian School of Mines Dhanbad estimated a loss of ₹34 crore as royalty to the Jharkhand Government in 2006 due to illegal mining (MoC 2015a). The illegal production and marketing of coal have resulted in the destruction of environmental commons, which has also caused various social and environmental externalities.

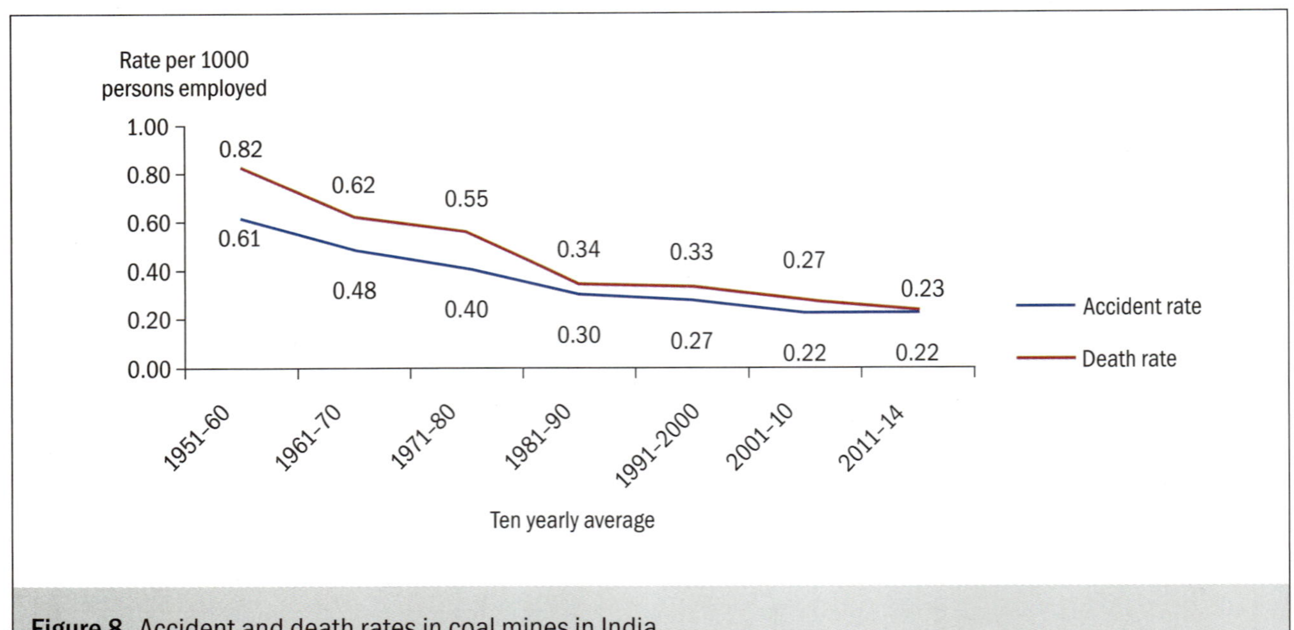

Figure 8 Accident and death rates in coal mines in India
Source MoLE (2015)

Increase in clean energy cess on coal

The Union Budget 2015/16 has proposed increasing the clean energy cess to ₹200 per tonne from ₹100 per tonne of coal to finance clean environment initiatives, thus incentivizing the clean energy sector. This cess will provide funds for projects based on renewable energy and green energy initiatives. The collected cess will contribute ₹12 000 crore annually to the National Clean Energy Fund (NCEF) (MoF 2015a). The NCEF will help in financing projects under the National Solar Mission and provide subsidy for off-grid solar applications. Soft loans are also provided by the Indian Renewable Energy Development Agency to clean energy projects. The Economic Survey 2014/15 released by the Ministry of Finance has suggested a fivefold hike in coal cess to around ₹500 per tonne from the existing level to help curb the annual carbon emissions of 214 million tonnes, which is 11% of India's annual emissions (MoF 2015b). The fivefold increase in cess will also help to bring domestic and international coal prices at par.

References

BP. 2015. ***BP Statistical Review of World Energy***. Details available at https://www.bp.com/content/dam/bp/pdf/energy-economics/statistical-review-2015/bp-statistical-review-of-world-energy-2015-full-report.pdf

CAG (Comptroller and Auditor General). 2012. ***Performance Audit of Allocation of Coal Blocks and Augmentation of Coal Production.*** New Delhi: CAG, Government of India

CCO (Coal Controller's Organisation). 2013. ***Provisional Coal Statistics 2012–13.*** Kolkata: CCO

CCO (Coal Controller's Organisation). 2014. ***Provisional Coal Statistics 2013–14.*** Kolkata: CCO

CCO (Coal Controller's Organisation). 2015. ***Provisional Coal Statistics 2014–15.*** Kolkata: CCO

CEA (Central Electricity Authority). 2014. ***Report on Fly Ash Generation at Coal/Lignite Based Thermal Power Stations and its Utilization in the Country for the year 2011–12 and 2012–13.*** New Delhi: CEA

CEA (Central Electricity Authority). 2015. ***Executive Summary: Power Sector.*** New Delhi: CEA, Government of India

CMPDI (Central Mine Planning and Design Institute). 2014. ***UNFC Classification of Coal Resources of CIL Blocks as on 1.04.2013.*** Ranchi: CMPDI, Government of India. Details available at www.cmpdi.co.in/unfc_code.php, last accessed on 19 November 2015

CMPDI (Central Mine Planning and Design Institute). 2015. ***Inventory of Coal Resources in India.*** Ranchi: CMPDI, Government of India. Details available at www.cmpdi.co.in/coalinventory.php#, last accessed on 10 November 2015

Dhar, B. and R. Tiwary. 1994. Environmental pollution from coal mining activities in Damodar River Basin, India. ***Mine Water and the Environment*** 13: 1–10

Dutta, I. 2013. Coal India revises prices. ***The Hindu,*** 28 May

FICCI (Federation of Indian Chambers of Commerce and Industry). 2013. ***Knowledge Paper for National Conference on Re-energising Indian Coal Sector: Interventions through Policy, Competition and Technology.*** New Delhi: FICCI

Herbert, T. and K. Lahiri-Dutt. 2004. Coal sector loans and displacement of indigenous populations: Lessons from Jharkhand. ***Economic and Political Weekly*** XXXIX(23): 2403–2409

Hsu, A., J. Emerson, M. Levy, A. de Sherbinin, L. Johnson, O. Malik, J. Schwartz, and M. Jaiteh. 2014. ***The 2014 Environment Performance Index.*** New Haven, CT: Yale Center for Environment law and Policy

Kumar, V., M. Mathur, and S. Sinha. 2005. A case study: manifold increase in fly ash utilization in India. Details available at http://c-farm.org/001_A_CASE_STUDY.pdf

Manohar Lal Sharma vs The Principal Secretary & Ors, 120, 463, 515, 283 of 2012 (Supreme Court September 2014)

Mishra, P. P. 2009. Coal mining and rural livelihoods: Case of the Ib Valley Coalfield, Orissa. ***Economic and Political Weekly,*** XLIV(44): 117–123

Mjunction. 2015. ***Coal Insights,*** 8(11): 32

MoC (Ministry of Coal). 2011. ***Report of the Working Group on Coal and Lignite for Formulation of Twelfth Five Year Plan (2012–2017).*** New Delhi: MoC, Government of India

MoC (Ministry of Coal). 2013. New coal distribution policy (F.No. 23011/90/2013-CPD). Government of India. Details available at http://powermin.nic.in/upload/pdf/Impact_on_tariff_in_the_conclude_PPAs.pdf, last accessed on 5 January 2015

MoC (Ministry of Coal). 2014a. ***Annual Report, 2013–14.*** New Delhi: MoC, Government of India

MoC (Ministry of Coal). 2014b. ***Approach Paper for Auctioning of Coal Mines.*** New Delhi: MoC, Government of India

MoC (Ministry of Coal). 2014c. Minutes of the workshop on rationalization of coal linkages. Details available at www.coal.nic.in/sites/upload_files/coal/files/webform/notices/261214c.pdf, last accessed on 28 January 2015

MoC (Ministry of Coal). 2015a. ***Committee on Public Undertakings, 2014–2015, Fourth Report, Sixteenth Lok Sabha.*** New Delhi: MoC, Government of India

MoC (Ministry of Coal). 2015b. ***Report of the New Inter-Ministerial Task Force on Rationalization of Sources of Coal to Power Plants.*** New Delhi: MoC, Government of India

MoF (Ministry of Finance). 2015a. ***Budget 2015–2016.*** New Delhi: MoF, Government of India

MoF (Ministry of Finance). 2015b. ***Economic Survey, 2014–15.*** New Delhi: MoF, Government of India

MoLE (Ministry of Labour and Employment). 2015. ***Standard Note, 01.01.2015.*** New Delhi: Directorate General of Mines Safety, MoLE, Government of India

MoS (Ministry of Shipping). 2013. Basic port statistics of India 2012–13. Details available at Ministry of Shipping: http://shipping.nic.in/showfile.php?lid=1676, last accessed on 29 January 2015

MoSPI (Ministry of Statistics and Programme Implementation). 2015. ***Energy Statistics 2015.*** New Delhi: Central Statistics Office, MoSPI

PIB (Press Information Bureau). 2015a. CIL gearing up for one billion tonne coal production mark. Details available at http://pib.nic.in/newsite/pmreleases.aspx?mincode=42, last accessed on 21 October 2015

PIB (Press Information Bureau). 2015b. Proposals to open new coal mines. Details available at http://pib.nic.in/newsite/pmreleases.aspx?mincode=42, last accessed on 21 October 2015

PIB (Press Information Bureau). 2015c. Year end review: major FDI policy changes. Details available at http://pib.nic.in/newsite/PrintRelease.aspx?relid=114271 GoI.

Planning Commission. 2012. ***Twelfth Five-year Plan (2012–2017).*** New Delhi: Planning Commission, Government of India

Priyadarshi, N. 2012. Effects of mining on environment in the state of Jharkhand, India. Details available at http://nitishpriyadarshi.blogspot.in/2012/05/effects-of-mining-on-environment-in.html

Singh, G. 2007. Mitigating environmental and social impacts of mining in India. ***Mining Engineers' Journal*** pp. 8–24

Singh, P., R. Singh, and G. Singh. 2007. Evaluation of land degradation due to coal mining: a vibrant issue. Paper presented at the ***First International Conference on MSECCMI,*** New Delhi, India

Singh, P. K., R. Singh, and G. Singh. 2010. Impact of coal mining and industrial activities on land use pattern in Angul-Talcher region of Orissa, India. ***International Journal of Engineering Science and Technology*** 2(12): 7771–7784

TERI (The Energy and Resources Institute). 2015. ***Energy Security Outlook: Defining a Secure and Sustainable Energy Future for India.*** New Delhi: TERI

Terminski, B. 2012. Mining-induced displacement and resettlement: social problem and human rights issue (a global perspective). Details available at http://indr.org/wp-content/uploads/2013/04/B.-Terminski-Mining-Induced-Displacement-and-Resettlement.pdf

Tudor, G. and N. M. Catalin. 2013. Considerations regarding the risk of land degradation induced by mining operations in coal open pits. In ***13th SGEM GeoConference on Science and Technologies in Geology, Exploration and Mining, SGEM2013 Conference Proceedings,*** 16–22 June, Vol 1, pp. 357–362

Bibliography

Bose, P. R. 2013. Thermal coal imports rise 28% on strong demand. ***The Hindu Business Line***, 27 November

BP. 2014. ***BP Statistical Review 2014: India in 2013***. Details available at www.bp.com/content/dam/bp/pdf/Energy-economics/statistical-review-2014/BP-Statistical-Review-of-World-Energy-2014-India-insights.pdf, last accessed on 25 November 2014

Mjunction. 2013. Minor ports take major steps. ***Coal Insights***, 8(3)

TABLE 3.1 Coal resources of states (million tonnes) as on 1 April 2015

State	Proved	Indicated	Inferred	Total
Andhra Pradesh	0	1 149	432	1 581
Arunachal Pradesh	31	40	19	90
Assam	465	47	3	515
Bihar	0	0	160	160
Chhattisgarh	18 237	34 390	2 285	54 912
Jharkhand	41 463	33 026	6 559	81 048
Madhya Pradesh	10 411	12 784	3 341	26 536
Maharashtra	5 953	3 190	2 110	11 253
Meghalaya	89	17	471	576
Nagaland	9	0	307	315
Odisha	30 747	36 545	8 507	75 799
Sikkim	0	58	43	101
Telangana	9 807	8 808	2 597	21 212
Uttar Pradesh	884	178	0	1 062
West Bengal	13 518	13 010	4 907	31 435
Total	131 614	143 242	31 740	306 596

Note Singrimari Coalfield of Assam (non-coking) is included in Gondwana Coalfield, not considered in tertiary coalfields. Inferred includes inferred mapping.
Source CCO (Coal Controller's Organisation). 2015. ***Provisional Coal Statistics 2014/15***. Kolkata: CCO, Ministry of Coal, Government of India

TABLE 3.2 Coal resources (million tonnes) by category as on 1 April 2015

Type of coal	Proved	Indicated	Inferred	Total
Coking				
Prime coking	4 614.35	698.71	0	5 313.06
Medium coking	13 388.90	12 113.56	1 879.47	27 381.93
Blendable/semi-coking	482.16	1 003.68	221.68	1 707.52
Subtotal	18 485.41	13 815.95	2 101.15	34 402.51
Non-coking	112 534.85	129 325.81	28 839.75	270 700.41
Tertiary coal	593.81	99.34	799.49	1 492.64
Total	131 614.07	143 241.10	31 740.39	306 595.56
Percentage share	42.93	46.72	10.35	100.00

Source Details available at <http://www.cmpdi.co.in/coalinventory.php>, last accessed on 12 November 2015

TABLE 3.3 Lignite resources of states and union territories (million tonnes) as on 1 April 2015

State/union territory	Proved	Indicated	Inferred	Total
Puducherry	0	405.61	11.00	416.61
Tamil Nadu	3 735.23	22 900.05	8 573.62	35 208.90
Rajasthan	1 168.53	2 670.84	1 887.34	5 726.71
Gujarat	1 278.65	283.70	1 159.70	2 722.05
Jammu and Kashmir	0	20.25	7.30	27.55
Kerala	0	0	9.65	9.65
West Bengal	0	1.13	1.64	2.77
Total	6 182.41	26 281.58	11 650.25	44 114.24

Source CCO (Coal Controller's Organisation). 2015. ***Provisional Coal Statistics 2014/15***. Kolkata: CCO, Ministry of Coal, Government of India

TABLE 3.4 Lignite mined out (in million tonnes) till 31 March 2015

State	District	Mine	Mined out 2014/15	As on 1 April 2015
Tamil Nadu	Cuddalore	Neyveli – Mine I and Expansion	9.050	324.94
		Neyveli – Mine IA	2.910	35.8
		Neyveli – Mine II and Expansion	13.220	252.73
		Total	25.180	613.470
Gujarat	Kutch	Panandhro	2.790	109.92
		Matanomadh	3.210	23.62
		Umarsar	0.010	0.01
	Bharuch	Rajpardi	0.610	15.25
	Bhavnagar	Surkha North	1.200	8.19
		Alapar (Khadsaliya)	0.190	2.49
	Surat	Tadakeshwar	0.880	11.06
		Vatsan	0.640	23.24
		Mangrol	0.360	6.35
		Valia	2.410	3.01
		Total	12.300	203.140
Rajasthan	Bikaner	Barsingsar	1.350	5.59
		Gurha East	1.000	4.21
	Barmer	Giral	0.430	10.08
		Sonari	0.670	1.56
		Kapurdhi	7.000	16.22
	Nagaur	Kasnau/Matasukh	0.300	1.55
		Total	10.750	39.210
Grand total			48.230	855.820

Source Details available at <http://www.portal.gsi.gov.in/gsiDoc/pub/lignite%20resource_2015.pdf>, last accessed on 13 November 2015

TABLE 3.5 Coal production (million tonnes) in various states: 2000/01–2014/15

State	2000/01	2001/02	2002/03	2003/04	2004/05	2005/06	2006/07	2007/08	2008/09	2009/10	2010/11	2011/12	2012/13	2013/14	2014/15
Telangana	30.274	30.811	33.236	33.854	35.303	36.138	37.707	40.604	44.542	50.429	51.333	52.211	53.190	50.469	52.536
Arunachal Pradesh	–	–	–	–	–	–	–	0.079	0.142	0.251	0.299	0.222	0.073	0	0
Assam	0.660	0.640	0.633	0.733	0.628	1.101	1.050	1.101	1.009	1.113	1.101	0.602	0.605	0.664	0.779
Chhattisgarh	64.699	53.621	56.758	61.505	69.253	76.358	83.241	90.172	101.922	109.953	113.825	113.958	117.830	127.095	134.764
Jammu and Kashmir	0.033	0.035	0.025	0.019	0.023	0.019	0.016	0.017	0.011	0.023	0.023	0.020	0.019	0.019	0.013
Jharkhand	33.520	76.813	78.628	79.526	78.038	85.423	88.764	90.895	96.272	105.917	108.949	109.566	111.274	113.091	124.147
Madhya Pradesh	69.927	44.156	45.736	49.826	52.511	55.579	59.726	67.841	71.325	74.074	71.104	71.123	75.948	75.590	87.609
Maharashtra	28.754	30.830	31.359	32.912	34.529	36.119	36.215	36.403	38.705	41.005	39.336	39.159	39.134	37.223	38.257
Meghalaya	4.065	5.149	4.406	5.439	5.345	5.566	5.787	6.541	5.489	5.767	6.974	7.206	5.640	5.732	5.732
Odisha	44.803	47.805	52.229	60.147	66.604	70.540	81.160	89.482	98.402	106.409	102.565	105.476	110.132	112.917	123.627
Uttar Pradesh	16.863	16.533	17.783	15.791	16.804	15.721	12.228	11.426	12.029	13.968	15.526	16.178	16.090	14.721	14.957
West Bengal	20.098	21.394	20.479	21.494	23.577	24.475	24.938	22.521	22.905	23.133	21.659	24.230	26.467	28.244	30.014
All-India	313.696	327.787	341.272	361.246	382.615	407.039	430.832	457.082	492.753	532.042	532.694	539.951	556.402	565.765	612.435

Source CCO (Coal Controller's Organisation). 2015. ***Provisional Coal Statistics 2014/15***. Kolkata: CCO, Ministry of Coal, Government of India

TABLE 3.6 All-India production (million tonnes) of coal, lignite, and other derivatives: 2000/01–2014/15

Item	2000/01	2001/02	2002/03	2003/04	2004/05	2005/06	2006/07	2007/08	2008/09	2009/10	2010/11	2011/12	2012/13	2013/14	2014/15 (P)
Coking coal	30.900	28.688	30.195	29.401	30.224	31.511	32.097	34.455	34.809	44.413	49.547	51.660	51.582	56.818	57.451
Non-coking coal	282.796	299.119	311.077	331.845	352.391	375.528	398.735	422.627	457.948	487.629	483.147	488.290	504.820	508.947	554.984
Total coal	313.696	327.807	341.280	361.246	382.610	407.039	430.840	457.090	492.757	532.042	532.694	539.950	556.402	565.765	612.435
Lignite	24.247	24.800	26.018	27.958	30.411	30.228	31.285	33.980	32.421	34.071	37.733	42.332	46.453	44.271	48.257
Total coal and lignite	337.950	352.607	367.298	389.210	413.021	437.267	462.125	491.070	525.178	566.113	570.427	582.282	602.855	610.036	660.692
Soft coke	–	–	–	–	–	–	–	–	–	–	–	–	–	–	–
Hard coke	13.381	13.558	13.917	13.121	12.673	13.347	12.566	12.542	12.619	12.663	12.880	14.330	11.694	12.606	14.355
Washed coking coal	8.642	8.012	8.204	8.201	8.790	8.376	7.025	7.171	7.181	6.547	6.955	6.496	6.550	6.614	6.011
Washed non-coking coal	–	–	–	–	10.556	12.555	12.688	12.686	13.550	13.963	14.531	15.437	14.190	15.699	17.294
Middlings (coking)	6.279	4.908	4.961	4.849	5.168	5.582	5.876	6.150	5.294	4.642	4.643	3.674	5.464	4.913	4.721
Middlings (non-coking)	–	–	–	–	1.605	2.793	2.858	3.276	3.264	3.264	3.589	3.669	3.825	3.926	3.742
Total exports (excluding coke)	1.292	1.903	1.517	1.627	1.374	1.989	1.554	1.627	1.655	2.450	1.875	2.014	2.443	2.188	1.238
Total imports (excluding coke)	20.930	20.548	23.260	21.683	28.950	38.586	43.081	49.794	59.003	73.255	68.918	102.853	145.785	166.857	212.103

P– provisional

Source CCO (Coal Controller's Organisation). 2015. ***Provisional Coal Statistics 2014/15***. Kolkata: CCO, Ministry of Coal, Government of India

TABLE 3.7 Coal production (million tonnes) of various companies: 2000/01–2014/15

Company	2000/01	2001/02	2002/03	2003/04	2004/05	2005/06	2006/07	2007/08	2008/09	2009/10	2010/11	2011/12	2012/13	2013/14	2014/15
Coal India Ltd															
Eastern Coalfields Ltd	28.030	28.552	27.178	28.003	27.253	31.114	30.466	24.059	28.135	30.058	30.803	30.558	33.901	36.047	40.006
Bharat Coking Coal Ltd	25.969	25.252	24.153	22.682	22.314	23.311	24.205	25.215	25.514	27.512	29.004	30.207	31.211	32.612	34.512
Central Coalfields Ltd	31.746	33.806	36.975	37.338	37.390	40.513	41.319	44.146	43.236	47.083	47.521	48.004	48.061	50.022	55.652
South Eastern Coalfields Ltd	60.331	64.120	66.604	71.009	78.550	83.024	88.502	93.791	101.150	108.009	112.705	113.837	118.219	124.261	128.275
Mahanadi Coalfields Ltd	44.803	47.805	52.229	60.049	66.080	69.604	80.001	88.012	96.336	104.079	100.280	103.119	107.895	110.439	121.379
Western Coalfields Ltd	35.200	37.009	37.815	39.526	41.410	43.204	43.212	43.512	44.700	45.735	43.654	43.110	42.287	39.729	41.147
Northern Coalfields Ltd	41.400	42.462	45.101	47.033	49.950	51.518	52.158	59.623	63.650	67.670	66.253	66.401	70.021	68.639	72.484
North Eastern Coalfields	0.660	0.640	0.633	0.733	0.628	1.101	1.050	1.101	1.009	1.113	1.101	0.602	0.605	0.664	0.779
Total	268.139	279.646	290.688	306.373	323.575	343.389	360.913	379.459	403.730	431.259	431.321	435.838	452.200	462.413	494.234
Singareni Collieries Company Ltd	30.274	30.811	33.236	33.854	35.303	36.138	37.707	40.604	44.546	50.429	51.333	52.211	53.190	50.469	52.536
Jammu and Kashmir Minerals Ltd	0.033	0.035	0.025	0.019	0.023	0.019	0.016	0.017	0.011	0.023	0.024	0.020	0.019	0.019	0.013
Damodar Valley Corporation	0.374	0.335	0.171	0.381	0.401	0.368	0.385	0.385	0.267	0.141	0.311	0.328	0.203	0.054	0.066
Bengal Emta Coal Mines Ltd	2.410	2.910	3.090	3.770	3.994	3.992	4.765	4.229	4.139	3.303	2.876	2.598	3.005	0	0
Jharkhand State Mineral Development Corporation Ltd	0.400	0.558	0.493	0.444	0.589	0.762	0.677	0.601	0.401	0.461	0.399	0.118	0	0	0
Indian Iron and Steel Company Ltd	1.242	1.140	0.820	0.770	0.892	0.658	0.695	1.021	1.018	1.298	1.082	0.598	0.715	0.622	0.626
Tata Steel Ltd	5.340	5.650	5.920	6.150	6.374	6.521	7.041	7.209	7.282	7.210	7.026	7.461	7.295	6.972	6.044
SAIL	–	–	–	–	–	–	–	–	–	0.063	0.014	0.040	0.102	0.069	0.025
WBPDCL	–	–	–	–	–	–	–	–	–	0.251	0.257	2.814	3.266	2.606	6.221
APMDTCL	–	–	–	–	–	–	–	0.079	0.142	0.063	0.299	0.221	0.073	0	0
ICML	–	–	–	–	2.401	2.622	2.546	2.754	2.984	3.213	2.929	3.745	3.129	2.708	3.492
JSPL	–	–	–	–	2.965	5.310	5.968	5.994	5.998	5.999	5.999	5.998	5.999	5.999	5.989
HIL	–	–	–	–	0.524	0.936	1.159	1.470	2.066	2.330	2.285	2.357	2.237	2.478	2.248
Meghalaya Coal	–	–	–	–	5.345	5.566	5.787	6.541	5.489	5.767	6.974	7.206	5.640	5.732	5.732
MIL	–	–	–	–	0.126	0.439	0.668	0.835	0.989	1.000	0.952	0.851	0.795	0.919	1.000
BLA Industries Pvt. Ltd	–	–	–	–	0.096	0.282	0.218	0.329	0.236	0.299	0.297	0.299	0.300	0.300	0.300
CML	–	–	–	–	0.008	0.012	0	0	0	0	0	0	0	0	0
PANEM Coal Mines Ltd	–	–	–	–	–	0.025	1.603	3.797	6.175	8.476	8.031	8.301	6.926	5.879	3.433
PIL	–	–	–	–	–	–	0.625	0.900	0.919	1.000	1.000	1.000	1.000	1.000	1.000

Contd...

TABLE 3.7 *Contd...*

Company	2000/01	2001/02	2002/03	2003/04	2004/05	2005/06	2006/07	2007/08	2008/09	2009/10	2010/11	2011/12	2012/13	2013/14	2014/15
JNL	–	–	–	–	–	–	0.059	0.279	0.396	0.560	0.406	0.480	0.480	0.446	0.703
JPL	–	–	–	–	–	–	–	0.578	4.893	6.045	5.688	5.250	5.250	6.226	6.248
SISCL	–	–	–	–	–	–	–	0.001	0.051	0.140	0.114	0.160	0.248	0.148	0.196
ESCL	–	–	–	–	–	–	–	–	0.013	0.062	0.034	0.106	0.099	0.461	0.436
KEMTA	–	–	–	–	–	–	–	–	0.991	2.252	2.275	2.189	2.506	2.502	2.478
UML	–	–	–	–	–	–	–	–	0.013	0.062	0.300	0.351	0.560	0.762	0.790
SEML	–	–	–	–	–	–	–	–	0.008	0.291	0.433	0.774	0.976	1.165	1.189
B S Ispat Ltd	–	–	–	–	–	–	–	–	–	–	0.015	0.003	0.062	0.081	0.031
DVC Emta Coal Mines Ltd	–	–	–	–	–	–	–	–	–	–	0.021	1.165	1.836	1.519	1.001
TUML	–	–	–	–	–	–	–	–	–	–	–	0.066	0.341	0.317	0.198
RRUVNL	–	–	–	–	–	–	–	–	–	–	–	–	0.293	1.197	3.443
WBMDTCL	–	–	–	–	–	–	–	–	–	–	–	–	0.348	0.726	1.042
SPL	–	–	–	–	–	–	–	–	–	–	–	–	0.225	1.695	9.406
Sova Ispat Alloys Ltd	–	–	–	–	–	–	–	–	–	–	–	–	0.089	0.276	0.400
MPSMCL	–	–	–	–	–	–	–	–	–	–	–	–	–	0.005	1.500
Total	308.212	321.085	334.443	351.761	382.616	407.039	430.832	457.082	492.757	531.997	532.695	542.548	559.407	565.766	

APMDTCL – Arunachal Pradesh Mineral Development and Trading Corporation Ltd; CML – Castron Mining Ltd; ESCL – Electro Steel Castings Ltd; HIL – Hindalco Industries Ltd; ICML – Integrated Coal Mining Ltd; JNL – Jayaswal Neco Ltd; JPL – Jindal Power Ltd; JSPL – Jindal Steel and Power Ltd; KEMTA – Karnataka Emta; MIL – Monnet Ispat Ltd; MPSMCL – Madhya Pradesh State Mining Corporation Ltd; PIL – Prakash Industries Ltd; RRUVNL – Rajasthan Rajya Vidyut Utpadan Nigam Ltd; SAIL – Steel Authority of India Ltd; SEML – Sarda Energy and Minerals Ltd; SISCL – Sunflag Iron and Steel Company Ltd; SPL – Sandhya Projects Ltd; UML – Usha Martin Ltd; TUML – Topworth Urja and Metals Ltd; WBPDCL – West Bengal Power Development Corporation Ltd; WBMDTCL – West Bengal Mineral Development and Trading Corporation Ltd

Source CCO (Coal Controller's Organisation). 2015. ***Provisional Coal Statistics 2014/15***. Kolkata: CCO, Ministry of Coal, Government of India

TABLE 3.8 Coal dispatches (million tonnes) in various states: 2000/01–2014/15

State	2000/01	2001/02	2002/03	2003/04	2004/05	2005/06	2006/07	2007/08	2008/09	2009/10	2010/11	2011/12	2012/13	2013/14	2014/15
Telangana	30.30	31.00	33.40	33.83	34.71	35.32	37.49	41.79	44.41	49.27	50.05	51.39	52.03	47.89	52.66
Arunachal Pradesh	–	–	–	–	–	–	–	0.08	0.13	0.23	0.25	0.322	0.055	0	0
Assam	0.80	0.60	0.60	0.87	0.57	1.17	1.18	1.20	0.84	1.07	1.10	0.80	0.62	0.58	0.73
Bihar	46.30	–	–	–	–	–	–	–	–	–	–	–	–	–	–
Chhattisgarh	22.60	54.80	58.80	61.92	70.15	75.00	80.53	90.79	103.02	106.92	109.56	114.61	121.06	124.67	129.39
Jammu and Kashmir	–	–	–	–	0.03	0.02	0.01	0.02	0.01	0.02	0.03	0.02	0.01	0.01	0.01
Jharkhand	30.40	75.40	75.80	78.88	76.61	79.67	84.29	88.90	95.41	99.86	106.64	109.79	119.28	116.80	122.69
Madhya Pradesh	70.40	41.70	45.00	48.90	51.69	54.95	60.00	68.34	72.04	73.48	69.44	69.56	60.41	63.10	74.24
Maharashtra	28.80	31.80	31.80	32.58	33.52	34.79	35.51	37.39	39.24	40.74	38.24	38.11	38.32	37.21	38.55
Meghalaya	4.10	5.20	4.40	5.44	5.35	5.57	5.79	6.54	5.49	5.77	6.97	7.21	5.64	5.73	5.73
Odisha	47.30	49.00	51.40	59.35	66.78	69.14	77.59	85.15	93.32	100.59	104.36	104.82	114.21	116.80	125.38
Uttar Pradesh	17.30	18.80	17.30	15.53	17.02	15.85	12.39	11.22	12.45	13.59	15.39	15.47	28.82	30.81	29.02
West Bengal	19.20	20.80	20.10	20.56	22.24	24.11	25.03	22.16	22.82	22.26	21.44	23.20	26.69	28.47	29.21
All-India	317.50	329.10	338.60	357.86	378.66	395.59	419.80	453.56	489.17	513.79	523.47	535.30	567.14	572.06	607.63

Source CCO (Coal Controller's Organisation). 2015. ***Provisional Coal Statistics 2014/15***. Kolkata: CCO, Ministry of Coal, Government of India

TABLE 3.9 Coal dispatches (million tonnes) of various companies: 2000/01–2014/15

Company	2000/01	2001/02	2002/03	2003/04	2004/05	2005/06	2006/07	2007/08	2008/09	2009/10	2010/11	2011/12	2012/13	2013/14	2014/15 (P)
Coal India Ltd															
Eastern Coalfields Ltd	27.774	27.804	26.754	26.910	26.671	28.203	29.342	25.020	27.849	28.827	29.364	30.491	35.544	35.974	38.220
Bharat Coking Coal Ltd	25.683	24.696	21.776	23.391	22.076	22.012	23.798	23.811	24.353	24.863	29.251	30.070	32.997	34.058	33.666
Central Coalfields Ltd	32.631	33.064	36.537	36.197	35.744	38.564	38.640	42.024	43.829	44.274	46.218	48.033	52.886	52.121	55.337
South Eastern Coalfields Ltd	60.443	64.860	68.267	70.872	78.760	81.161	86.149	94.976	102.994	105.870	109.000	115.139	121.973	122.013	123.209
Mahanadi Coalfields Ltd	47.302	49.030	51.360	59.352	66.290	68.214	76.421	83.634	91.284	98.147	102.087	102.521	111.959	114.342	122.996
Western Coalfields Ltd	35.213	38.027	38.111	39.159	40.280	41.718	42.146	44.885	45.349	45.499	42.551	41.959	41.539	39.939	41.240
Northern Coalfields Ltd	42.058	42.684	44.428	46.484	50.117	51.679	52.620	59.016	64.231	66.669	64.208	63.605	67.021	71.892	73.518
North Eastern Coalfields	0.760	0.620	0.640	0.870	0.568	1.170	1.182	1.200	0.835	1.071	1.102	0.800	0.618	0.577	0.733
Total	271.864	280.785	287.873	303.235	320.506	332.721	350.298	374.566	400.724	415.220	423.781	432.618	464.537	470.916	488.919
Singareni Collieries Company Ltd	30.314	31.043	33.367	33.829	34.707	35.321	37.487	41.793	44.410	49.266	50.046	51.389	52.025	47.892	52.662
Jammu and Kashmir Minerals Ltd	0.034	0.026	0.024	0.031	0.027	0.020	0.014	0.016	0.012	0.017	0.025	0.023	0.014	0.013	0.013
Damodar Valley Corporation	0.381	0.325	0.145	0.401	0.451	0.348	0.396	0.391	0.217	0.205	0.193	0.410	0.226	0.045	0.055
Bengal Emta Coal Mines Ltd	2.405	2.913	3.093	3.761	3.799	4.189	4.681	4.321	4.132	3.300	2.883	2.581	3.002	0	0
Jharkhand State Mineral Development Corporation Ltd	0.398	0.560	0.498	0.444	0.598	0.761	0.677	0.601	0.401	0.461	0.399	0.118	0	0	0.408
Indian Iron and Steel Company Ltd	1.258	1.133	0.864	0.765	0.886	0.655	0.689	1.015	1.015	1.291	1.089	0.598	0.716	0.622	0.619
WBPDCL	–	–	–	–	–	–	–	–	–	0.058	0.268	3.116	3.311	2.610	6.248
SAIL	–	–	–	–	–	–	–	–	–	0.226	0.014	0.040	0.097	0.075	0.025
APMDTCL	–	–	–	–	–	–	–	0.076	0.129	0.101	0.245	0.322	0.055	0	0
Tata Steel Ltd	5.338	5.658	5.918	6.140	6.377	6.506	7.041	7.215	7.282	7.221	7.026	7.438	7.314	6.969	6.061
Integrated Coal Mining Ltd	–	–	0.309	1.432	2.247	2.524	2.570	2.771	2.972	2.985	2.923	3.168	3.221	3.278	3.771
Jindal Steel and Power Ltd	1.416	1.549	2.111	2.424	3.012	5.310	5.968	5.994	5.997	5.999	5.995	5.993	5.999	5.999	5.989
Meghalaya Coal	4.065	5.149	4.406	5.439	5.345	5.566	5.787	6.541	5.489	5.767	6.974	7.206	5.640	5.732	5.732
Monnet Ispat Ltd	–	–	–	–	0.126	0.437	0.642	0.819	1.016	1.000	0.960	0.846	0.798	0.905	1.022
BLA Industries Pvt. Ltd	–	–	–	–	0.086	0.286	0.220	0.325	0.235	0.301	0.297	0.299	0.300	0.300	0.300
HIL	–	–	–	0.091	0.491	0.922	1.164	1.513	2.032	2.444	2.272	2.298	2.254	2.453	2.386
PSEB (Panem Coal Mines Ltd)	–	–	–	–	–	0.021	1.488	3.884	6.156	8.449	8.126	8.278	6.872	5.852	3.454

Contd...

TABLE 3.9 | *Contd...*

Company	2000/01	2001/02	2002/03	2003/04	2004/05	2005/06	2006/07	2007/08	2008/09	2009/10	2010/11	2011/12	2012/13	2013/14	2014/15 (P)
PIL	–	–	–	–	–	–	0.625	0.900	0.918	1.000	1.000	1.000	1.000	1.000	1.000
JNL	–	–	–	–	–	–	0.053	0.258	0.391	0.519	0.477	0.457	0.479	0.394	0.777
JPL	–	–	–	–	–	–	–	0.567	4.606	5.249	5.249	5.249	5.088	6.223	5.726
SIL	–	–	–	–	–	–	–	0.001	0.048	0.137	0.102	0.164	0.244	0.159	0.196
ESCL	–	–	–	–	–	–	–	–	0.012	0.029	0.022	0.037	0.074	0.370	0.717
UML	–	–	–	–	–	–	–	–	0.013	0.058	0.300	0.351	0.564	0.759	0.794
KPCL (KEMTA)	–	–	–	–	–	–	–	–	0.964	2.216	2.368	2.205	2.515	2.472	2.413
SEML	–	–	–	–	–	–	–	–	0.001	0.273	0.431	0.784	0.893	1.117	1.266
BSIL	–	–	–	–	–	–	–	–	–	–	–	0.006	0.019	0.027	0.136
TUML	–	–	–	–	–	–	–	–	–	–	–	0.039	0.367	0.300	0.208
DVC EMTA	–	–	–	–	–	–	–	–	–	–	–	1.169	1.844	1.523	1.006
SPL	–	–	–	–	–	–	–	–	–	–	–	–	0.081	1.845	9.261
SOVA Ispat Alloys Ltd	–	–	–	–	–	–	–	–	–	–	–	–	0.086	0.279	0.400
RRUVNL	–	–	–	–	–	–	–	–	–	–	–	–	0.293	1.197	3.443
WBMDTCL	–	–	–	–	–	–	–	–	–	–	–	–	0.265	0.734	1.119
MPSMCL											–	–	–	0	1.504
Total	317.473	329.141	338.608	357.992	378.658	395.587	419.800	453.567	489.172	513.792	523.465	538.202	570.193	572.060	607.630

APMDTCL – Arunachal Pradesh Mineral Development and Trading Corporation Ltd; CML – Castron Mining Ltd; ESCL – Electro Steel Castings Ltd; HIL – Hindalco Industries Ltd; ICML – Integrated Coal Mining Ltd; JNL – Jayaswal Neco Ltd; JPL – Jindal Power Ltd; JSPL – Jindal Steel and Power Ltd; KEMTA - Karnataka Emta; MIL – Monnet Ispat Ltd; MPSMCL- Madhya Pradesh State Mining Corporation Ltd; P – provisional; PIL – Prakash Industries Ltd; RRUVNL – Rajasthan Rajya Vidyut Utpadan Nigam Ltd; SAIL – Steel Authority of India Ltd; SEML – Sarda Energy and Minerals Ltd; SISCL – Sunflag Iron and Steel Company Ltd; SPL – Sandhya Projects Ltd; UML – Usha Martin Ltd; TUML – Topworth Urja and Metals Ltd; WBPDCL – West Bengal Power Development Corporation Ltd; WBMDTCL – West Bengal Mineral Development and Trading Corporation Ltd

Source CCO (Coal Controller's Organisation). 2015. ***Provisional Coal Statistics 2014/15***. Kolkata: CCO, Ministry of Coal, Government of India

TABLE 3.10 Supply of indigenous coal ('000 tonnes) to steel plants: 2008/09–2013/14

Steel plant	2008/09			2009/10			2010/11			2011/12			2012/13			2013/14		
	Prime coking coal	Medium coking coal	Total	Prime coking coal	Medium coking coal	Total	Prime coking coal	Medium coking coal	Total	Prime coking coal	Medium coking coal	Total	Prime coking coal	Medium coking coal	Total	Prime coking coal	Medium coking coal	Total
Steel Authority of India Ltd																		
Bhilai Steel Plant	462	688	1150	446	539	985	558	432	990	404	431	835	374	364	738	309	506	815
Bokaro Steel Plant	369	379	748	414	350	764	469	382	851	358	237	595	388	216	604	233	151	384
Durgapur Steel Plant	234	325	559	285	236	521	265	206	471	244	174	418	294	111	405	217	169	386
Indian Iron and Steel Company Ltd	605	37	642	554	49	603	524	25	549	348	81	429	298	16	314	322	0	322
Durgapur Projects Ltd	0	25	25	–	–	–	–	–	–	–	–	–	–	–	–	–	–	–
Rourkela Steel Plant	275	269	544	289	230	519	410	145	555	444	111	555	420	127	547	292	172	464
Total	1945	1723	3668	1988	1404	3392	2226	1190	3416	1798	1034	2832	1774	834	2608	1373	998	2371
Tata Steel Ltd	–	–	0	445	1679	2124	830	1954	2784	815	2038	2853	906	2348	3254	1027	2569	3596
Visakhapatnam Steel Plant	177	473	650	0	327	327	0	377	377	0	485	485	0	518	518	0	396	396
Grand total	2122	2196	4318	2433	3410	5843	3056	3521	6577	2613	3557	6170	2680	3700	6380	2400	3963	6363

Source CCO (Coal Controller's Organisation). 2015. ***Coal Directory of India 2013/14***. Kolkata: CCO, Ministry of Coal, Government of India

TABLE 3.11 Availability and consumption (million tonnes) of raw coal by sector: 2000/01–2014/15

Item	2000/01	2001/02	2002/03	2003/04	2004/05	2005/06	2006/07	2007/08	2008/09	2009/10	2010/11	2011/12	2012/13	2013/14	2014/15
Availability															
Production (raw coal)	313.70	327.79	341.27	361.25	382.62	407.04	430.83	457.08	492.76	532.04	532.69	539.95	556.40	565.77	612.44
Imports: coking coal	11.06	11.11	12.95	12.99	16.93	16.89	17.88	22.03	21.08	24.69	19.48	31.80	35.56	36.87	43.72
Imports: non-coking coal	9.87	9.44	10.31	8.69	12.03	21.70	25.20	27.77	37.92	48.57	49.43	71.05	110.23	129.99	168.39
Imports: coke	2.42	2.28	2.25	1.89	2.84	2.62	4.69	4.25	1.88	2.36	1.49	2.37	3.08	4.17	3.29
Exports: coking coal	0.62	0.88	0.16	0.16	0.24	0.05	0.11	0.04	0.11	0.27	0.11	0.10	0.06	0.01	0.04
Exports: non-coking coal	0.67	1.02	1.35	1.47	1.13	1.94	1.45	1.59	1.55	2.18	1.76	1.92	2.39	2.18	1.20
Exports: coke	0.01	0.01	0	0.20	0.16	0.16	0.08	0.10	1.34	0.13	0.73	0.61	1.20	0.15	0.10
Pithead stock (opening)	29.32	21.43	18.12	19.39	21.29	23.97	34.33	44.35	46.78	47.32	64.86	72.19	74.04	63.05	55.51
Pithead stock (closing)	21.43	18.12	19.39	21.29	23.97	34.33	44.35	46.78	47.32	64.86	72.19	74.04	63.05	55.51	59.45
Available for industrial consumption after adjustment of off-take	319.67	330.94	340.09	359.32	379.84	396.66	420.79	454.49	490.02	514.55	524.13	535.88	570.23	571.98	608.21
Total availability	343.64	352.02	363.99	381.10	410.20	435.73	466.96	506.97	550.11	587.53	593.17	640.69	712.62	741.99	822.56
Consumption/supply (raw and imported coal)															
Steel and washery	30.73	30.04	30.60	29.67	34.43	32.42	34.90	39.02	38.85	37.05	39.02	16.05	15.88	23.13	22.65
Railways	–	–	–	–	–	–	–	–	–	–	–	–	–	–	–
Power (utility and captive)	252.94	265.19	267.90	279.96	305.35	316.49	331.58	360.74	381.06	378.24	382.12	403.91	444.29	427.23	469.72
Cement	15.33	14.85	16.36	16.63	18.10	18.08	19.67	21.35	19.85	14.36	14.18	13.40	13.55	11.96	11.36
Fertilizers and chemicals	4.56	4.12	4.22	3.75	3.98	3.09	2.97	2.94	3.09	2.63	3.45	2.78	2.85	2.82	2.69
Sponge iron	3.71	4.40	6.17	7.59	10.99	14.70	17.47	20.92	13.94	17.26	17.02	21.28	20.83	15.12	14.68
Textiles	1.04	0.94	0.72	0.52	0.46	0.29	0.30	0.37	2.53	0.27	0.28	–	0.30	0.36	0.42
Paper	2.71	2.78	2.79	2.51	2.61	2.77	2.62	2.64	2.16	2.34	2.43	–	2.13	1.67	1.54
Brick	3.04	5.57	2.99	5.56	1.96	1.16	0.58	0.00	0.18	0.49	0.27	–	1.98	5.99	0.11
Other industries	23.05	24.31	28.60	31.77	28.35	43.20	51.24	53.77	84.86	61.16	64.70	77.74	67.96	82.98	84.46
Colliery consumption	2.20	1.80	1.48	1.33	1.18	1.07	0.99	0.93	0.85	0.76	0.62	0.58	0.47	0.64	0.58
Total consumption	339.31	353.98	361.83	379.28	407.41	433.26	462.32	502.66	547.37	514.56	524.09	535.73	570.23	571.89	608.21

Source CCO (Coal Controller's Organisation). 2015. ***Provisional Coal Statistics 2014/15***. Kolkata: CCO, Ministry of Coal, Government of India

TABLE 3.12 Indigenous coal supply plan (BE) and achievement: 2014/15

Sector	CIL			SCCL			Others		
	BE	Actual	Achievement (%)	BE	Actual	Achievement (%)	BE	Actual	Achievement (%)
Steel (indigenous)	7.12	15.07	212	–	–	–	3.16	7.22	228
Power (utilities)	405.00	374.78	93	35.00	39.21	112	26.89	4.51	17
Power (captive)	36.20	13.99	39	4.10	2.66	65	0.50	36.88	7376
Sponge iron/CDI	9.29	4.11	44	0.80	0.35	43	15.07	10.22	68
Cement	7.20	6.10	85	7.20	4.99	69	0.95	0.27	29
Others	54.79	3.93	7	8.35	3.97	48	6.96	79.39	1141
Colliery consumption	0.42	0.51	122	0.06	0.06	105	–	0.00	–
Total	520.02	418.49	80	55.51	51.24	92	53.53	138.49	261

BE – budgeted estimates; CDI – coal dust injection: CIL – Coal India Ltd; SCCL – Singareni Collieries Company Ltd

Source CCO (Coal Controller's Organisation). 2015. ***Provisional Coal Statistics 2014/15***. Kolkata: CCO, Government of India

TABLE 3.13 Output per man-shift (tonnes) in opencast and underground mining: 2000/01–2014/15

Company	2000/01	2001/02	2002/03	2003/04	2004/05	2005/06	2006/07	2007/08	2008/09	2009/10	2010/11	2011/12	2012/13	2013/14	2014/15
Overall															
CIL	2.30	2.45	2.67	2.82	3.05	3.26	3.54	3.79	4.09	4.48	4.74	4.92	5.32	5.79	6.19
SCCL	1.50	1.67	1.89	1.81	1.62	1.74	1.91	2.10	3.01	3.36	3.59	3.94	3.14	3.86	4.20
Opencast															
CIL	5.92	6.08	6.30	6.67	7.18	7.51	8.00	8.60	8.95	9.48	10.06	10.40	11.68	13.16	13.06
SCCL	5.94	6.74	7.67	7.67	8.83	9.60	9.50	10.76	10.60	10.71	11.98	13.26	11.87	11.10	12.14
Underground															
CIL	0.63	0.64	0.69	0.68	0.69	0.71	0.71	0.73	0.76	0.78	0.77	0.75	0.77	0.76	0.78
SCCL	0.79	0.85	0.86	0.86	0.85	0.89	0.90	1.02	1.05	1.08	1.10	1.10	1.13	1.12	1.10

CIL – Coal India Ltd; SCCL – Singareni Collieries Company Ltd

Source CCO (Coal Controller's Organisation). 2015. ***Provisional Coal Statistics 2014/15***. Kolkata: CCO, Ministry of Coal, Government of India

TABLE 3.14 Inter-modal split of coal movement (million tonnes): 2000/01–2014/15

Mode	2000/01	2001/02	2002/03	2003/04	2004/05	2005/06	2006/07	2007/08	2008/09	2009/10	2010/11	2011/12	2012/13	2013/14	2014/15
Rail	167.66	187.62	190.04	181.34	184.12	188.68	204.51	214.91	218.72	226.06	230.68	250.80	277.16	274.27	297.48
Road	57.39	44.31	48.63	72.28	83.97	92.22	101.18	122.23	146.96	159.22	168.87	165.53	161.30	164.86	167.98
Merry-go-round	71.12	74.63	76.18	85.32	85.36	86.11	87.93	86.39	92.80	97.32	93.28	86.84	94.59	88.63	93.52
Belt	8.85	7.27	7.69	10.43	13.82	16.68	17.22	17.78	24.04	23.60	24.02	23.07	23.45	30.46	33.56
Ropeway	9.49	6.40	6.09	5.80	8.09	8.38	8.96	8.95	6.65	7.10	6.02	5.88	5.39	4.64	4.43
Other means	2.98	3.34	0	2.73	3.29	3.51	0	3.31	0.00	0.50	0.38	3.05	7.87	8.40	10.66
Total	317.47	323.57	328.62	357.90	378.66	395.59	419.80	453.57	489.17	513.79	523.25	535.15	569.77	571.25	607.63

Source CCO (Coal Controller's Organisation). 2015. ***Provisional Coal Statistics 2014/15***. Kolkata: CCO, Ministry of Coal, Government of India

TABLE 3.15 Traffic handled (million tonnes) at major ports: 2007/08–2014/15

Port	2007/08		2008/09		2009/10		2010/11		2011/12		2012/13		2013/14		2014/15 (P)	
	All commodities	Only coal	All commodities	Only coal	All commodities	Only coal	All commodities	Only coal	All commodities	Only coal	All commodities	Only coal	All commodities	Only coal	All commodities	Only coal
Chennai/Ennore	68.72	12.80	68.99	14.39	71.76	12.34	72.47	11.39	70.66	14.07	71.29	14.93	78.44	22.48	82.79	24.35
Haldia/Kolkata	57.33	7.27	54.05	7.84	46.42	7.56	47.54	8.23	43.25	7.29	39.93	6.488	41.385	7.186	46.292	7.306
Jawaharlal Nehru	55.84	–	57.28	–	60.76	–	64.31	–	65.73	–	64.49	–	62.35	–	63.80	–
Kandla	64.92	1.18	72.23	1.87	79.50	3.23	81.88	3.49	82.50	4.23	93.62	4.44	87.01	6.35	92.50	9.97
Kochi	15.81	0.25	15.23	0.26	17.43	0.15	17.87	0.04	20.09	0.03	19.85	0.03	20.89	–	21.60	0.10
Mormugao	35.13	4.24	41.68	4.56	48.85	4.74	50.02	6.57	39.00	6.83	17.69	7.37	11.74	7.52	14.71	8.57
Mumbai	57.04	2.95	51.88	3.27	54.54	3.75	54.59	3.87	56.19	4.32	58.04	4.02	59.19	4.46	61.66	4.78
New Mangalore	36.02	1.69	36.69	1.93	35.53	2.81	31.55	2.86	32.94	4.02	37.04	6.91	39.37	8.35	36.57	8.18
Paradip	42.44	18.08	46.41	20.16	57.01	19.82	56.03	19.48	54.25	21.88	56.55	26.02	68.00	32.07	71.01	38.01
Tuticorin (V O Chidambaranar)	21.48	6.11	22.01	5.88	23.79	5.60	25.73	5.35	28.11	6.05	28.26	6.66	28.64	6.64	32.41	8.61
Visakhapatnam	64.60	10.35	63.91	11.00	65.50	11.72	68.04	11.46	67.42	10.06	59.04	9.80	58.50	9.67	58.00	8.85
All ports	519.31	64.93	530.36	71.16	561.09	71.71	570.03	72.73	560.13	78.79	545.79	86.66	555.50	104.73	581.34	118.73

P - provisional

Source Details available at <http://ipa.nic.in/oper.htm>, last accessed on 19 November 2015

TABLE 3.16 Customs duty structure on imported coal: 2010/11

Import of coal	Current duty (%)
Coking coal	
Ash content less than 12%	0
Ash content greater than 12%	5
Non-coking coal	5
Coke	5

Note With effect from 1 July 2010, it has been decided by the Government of India to bring the levy of Clean Energy Cess of ₹50 per tonne.
Source MoC (Ministry of Coal). 2013. ***Coal Directory of India 2011/12***. New Delhi: MoC, Government of India

TABLE 3.17 Import (million tonnes) of coking and non-coking coal: 1991/92–2014/15

Year	Coking coal	Non-coking coal	Total
1991/92	5.270	0.660	5.930
1992/93	6.320	0.420	6.740
1993/94	6.820	0.570	7.390
1994/95	10.150	1.240	11.390
1995/96	9.370	3.140	12.510
1996/97	10.620	2.560	13.180
1997/98	11.740	4.700	16.440
1998/99	10.020	6.510	16.530
1999/2000	10.992	8.708	19.700
2000/01	11.063	9.867	20.930
2001/02	11.107	9.441	20.548
2002/03	12.947	10.313	23.260
2003/04	12.992	8.691	21.683
2004/05	16.925	12.025	28.950
2005/06	16.891	21.695	38.586
2006/07	17.877	25.204	43.081
2007/08	22.029	27.765	49.794
2008/09	21.080	37.923	59.003
2009/10	24.690	48.565	73.255
2010/11	19.484	49.434	68.918
2011/12	31.801	71.052	102.853
2012/13	35.557	110.228	145.785
2013/14	36.872	129.985	166.857
2014/15	43.715	168.388	212.103

Source CCO (Coal Controller's Organisation). 2015. ***Provisional Coal Statistics 2014/15***. Kolkata: CCO, Ministry of Coal, Government of India

TABLE 3.18 Comparison of gross calorific value and useful heat value for non-coking coal (with 6% moisture content)

Grade	UHV (kcal/kg)		Ash (%)		GCV (kcal/kg)	
	From	To	From	To	From	To
A	–	>6200	–	<13.56	–	>6406
B	5600	6200	13.56	17.91	5997	6406
C	4940	5600	17.91	22.69	5447	5997
D	4200	4940	22.69	28.06	5042	5447
E	3360	4200	28.06	34.14	4469	5042
F	2400	3360	34.14	41.10	3814	4469
G	1300	2400	41.10	49.07	3064	3814

GCV – gross calorific value; UHV – useful heat value

Source TERI (Tata Energy Research Institute). 1997. ***Relative Economics of Fuels in Power Generation***. New Delhi: TERI

TABLE 3.19a Categorization of coal and coke

Class	Grade	Properties
Non-coking coal produced in all states other than Assam, Arunachal Pradesh, Meghalaya, and Nagaland	A	UHV > 6200 kcal/kg
	B	UHV 5600–6200 kcal/kg
	C	UHV 4940–5600 kcal/kg
	D	UHV 4200–4940 kcal/kg
	E	UHV 3360–4200 kcal/kg
	F	UHV 2400–3360 kcal/kg
	G	UHV 1300–2400 kcal/kg
Non-coking coal produced in the states of Arunachal Pradesh, Assam, Meghalaya, and Nagaland	Not graded	
Coking coal		
Steel	I	Ash content <15%
Steel	II	Ash content 15%–18%
Washery	I	Ash content 18%–21%
Washery	II	Ash content 21%–24%
Washery	III	Ash content 24%–28%
Washery	IV	Ash content 28%–35%
Semi-coking and weakly coking coals		
Semi-coking	I	Ash + moisture content<19%
Semi-coking	II	Ash + moisture content 19%–24%
Long-flame coal produced in collieries in all the states except Arunachal Pradesh, Assam, Meghalaya, Nagaland, and the Singareni Collieries in Andhra Pradesh	A	UHV > 6200 kcal/kg
	B	UHV 5600–6200 kcal/kg
	C	UHV 4940–5600 kcal/kg
	D	UHV 4200–4940 kcal/kg
Hard coke		
By-product	Premium	Ash content < 25%
By-product	Ordinary	Ash content 25%–30%
Beehive	Premium	Ash content < 27%
Beehive	Superior	Ash content 27%–31%
Beehive	Ordinary	Ash content 31%–36%

UHV – useful heat value

Notes

1. Coking coals have been classified thus by the erstwhile Coal Board under the Coal Mines (Conservation, Safety, and Development) Act, 1952 (12 of 1952) (since repealed) or such coals as have been declared or may be declared as coking coal by the Central Government under the Colliery Control Order, 1945, or the Coal Mines (Conservation and Development) Act, 1974 (28 of 1974), and the rules made under both the aforesaid acts.
2. Semi-coking coals and weakly coking coals are classified as blendable coals by the erstwhile Coal Board under the Coal Mines (Conservation, Safety, and Development) Act, 1952 (12 of 1952) (since repealed), or as may be declared as "semi-coking" or "weakly coking" coals by the Central Government under the Colliery Control Order, 1945 or the Coal Mines (Conservation and Development) Act, 1974 (28 of 1974) and the rules made under both the aforesaid acts.
3. Coals other than coking, semi-coking, and weakly coking coals are non-coking coals.
4. UHV is defined by totalling the formula: UHV in kcal/kg = [8900 – 138 × (percentage of ash content + percentage of moisture content)]. In case of coal with moisture less than 2% and volatile content less than 19%, the UHV shall be the value arrived as above, reduced by 150 kcal/kg for each 1% reduction in volatile content below 19% fraction pro-rata. Both moisture and ash shall be determined after equilibrating at 60% relative humidity and 40°C temperature as per relevant clauses of the Indian Standard Specification No. IS:1350–1959.
5. Ash percentage of coking coals and hard coke shall be determined after air-drying as per IS:1350-1959. If the moisture so determined is more than 2%, the determination shall be after equilibrating at 60% relative humidity at 40°C as per IS:1350-1959.
6. Long-flame coals shall be defined by the parameters as laid down in Table II of the Indian Standard Specification No. IS:770-1964, "General classification of coals (Revised)." The volatile matter (on unit coal basis) present in these coals exceeds 32%.
7. The above classification shall not apply to coals other than bituminous or sub-bituminous coals as specified under the Indian Standard Specification No. IS:77-1964.

Source MoC (Ministry of Coal). 1995. The Gazette of India. In ***Extraordinary***, Part II, Section 3, Sub-section (ii), No. 718. pp. 3-4. New Delhi: MoC, Government of India

TABLE 3.19b Grades of coal

Grades of coking coal			
Grade	Ash content		
Steel Grade - I	<15%		
Steel Grade - II	15%–18%		
Washery Grade - I	18%–21%		
Washery Grade - II	21%–24%		
Washery Grade - III	24%–28%		
Washery Grade - IV	28%–35%		
Grades of non-coking coal			
Grade	UHV (kcal/kg) = 8900 – 138 (A + M)	Corresponding ash% + moisture % at 60% RH and 40°C	Gross calorific value (kcal/kg) at 5% moisture level
A	>6200	<19.5	<6454
B	5600-6200	19.6-23.8	6049–6454
C	4940-5600	23.9-28.6	5597–6049
D	4200-4940	28.7-34.0	5089–5597
E	3360-4200	34.1-40.0	4324–5089
F	2400-3360	40.1-47.0	3865–4324
G	1300-2400	47.1-55.0	3113–3865
Grades of semi-coking and weakly coking coal			
Grade	Ash + Moisture content		
Semi-coking Grade I	<19%		
Semi-coking Grade II	19%–24%		
Grades of NEC coal			
Grade	UHV (kcal/kg)	Ash + Moisture content (%)	
A	6200-6299	18.85-19.57	
B	5600-6199	19.58-23.91	

UHV – useful heat value

Source Details available at <http://www.coal.nic.in>, last accessed on 1 November 2012

TABLE 3.20 Pricing of non-coking coal based on gross calorific value

Pithead run of mine price for non-coking coal applicable for all coal producing companies of CIL (Coal India Ltd) including NEC (North Eastern Coalfields)						
GCV bands (kcal/kg)	Pithead run of mine price for non-coking coal (₹/tonne)					
	Power utilities (including IPPs), fertilizer and defence sectors	Other sectors				
6700–7000	4870	4870				
6400–6700	4420	4420				
6100–6400	3970	3970				
5800–6100	2800	2800				
5500–5800	1450	1960				
5200–5500	1270	1720				
4900–5200	1140	1540				
4600–4900	880	1180				
4300–4600	780	1050				
4000–4300	640	870				
3700–4000	600	810				
3400–3700	550	740				
3100–3400	500	680				
2800–3100	460	620				
2500–2800	410	550				
2200–2500	360	490				
Coal prices of CIL for power utilities (including IPPs), fertilizer and defence sectors (₹/tonne)						
Coking coal (run of mine)						
Subsidiary	Steel Grade I	Steel Grade II	Washery Grade I	Washery Grade II	Washery Grade III	Washery Grade IV
Bharat Coking Coal Ltd (for 53 units)	3750	3140	2740	1980	1480	1370
Bharat Coking Coal Ltd (other than the 53 units)	–	–	2020	1680	1240	1150
Eastern Coalfields Ltd	–	–	2390	1990	1470	1370
Central Coalfields Ltd	–	–	1960	1620	1200	1120
Western Coalfields Ltd	–	–	1710	1410	1290	–
Semi-coking and weakly coking coal (run of mine)						
Subsidiary	Semi-coking Grade I			Semi-coking Grade II		
Eastern Coalfields Ltd (Raniganj)	2150			1790		
South Eastern Coalfields Ltd	1740			1450		
Direct feed coking coal (run of mine)	3720					
For direct feed cocking coal of collieries (ash exceeding 20% but not exceeding 21%)*						

Contd...

TABLE **3.20** | *Contd...*

Coal prices of CIL for consumers other than power utilities (including IPPs), and defence sectors (₹/tonne)						
Coking coal (run of mine)						
Subsidiary	Steel Grade I	Steel Grade II	Washery Grade I	Washery Grade II	Washery Grade III	Washery Grade IV
Bharat Coking Coal Ltd (for 53 units)	4880	4080	3560	2570	1920	1780
Bharat Coking Coal Ltd (other than the 53 units)	–	–	2630	2180	1610	1500
Eastern Coalfields Ltd	–	–	3110	2590	1910	1780
Central Coalfields Ltd	–	–	2550	2110	1560	1460
Western Coalfields Ltd	–	–	2220	1830	1680	–
Semi-coking and weakly coking coal (run of mine)						
Subsidiary	Semi-coking Grade I			Semi-coking Grade II		
Eastern Coalfields Ltd (Raniganj)	2800			2330		
South Eastern Coalfields Ltd	2260			1890		
Direct feed coking coal (run of mine)						
For direct feed coking coal of collieries (ash exceeding 20% but not exceeding 21%)[#]			4840			

*Bonus/penalty @ ₹130/tonne/per cent decrease/increase in ash
[#] Bonus/penalty @ ₹170/tonne/per cent decrease/increase in ash

Notes

1. For GCV exceeding 7000 kcal/kg, the price shall be increased by ₹150 per tonne over and above the price applicable for GCV band exceeding 6700 but not exceeding 7000 kcal/kg, for increase in GCV by every 100 kcal/kg or part thereof.
2. For Western Coalfields Ltd (WCL), there shall be a 10% add-on over and above the price mentioned above for GCV bands not exceeding 5800 kcal/kg and below.
3. All other elements of the ex-colliery delivered prices are presently applicable in terms of the last coal price notification circulated vide Ref. No. S&M: GM(F): Pricing: 1907 dated 26 February 2011 will continue to remain applicable.
4. For coal produced by the coal companies of CIL including NEC, other than non-coking coal, the prices as are presently applicable in terms of the coal price notification circulated vide Ref. No. CIL: S&M: GM(F): Pricing: 1907, dated 26 February 2011 will continue to remain applicable.

Source Details available at <http://www.coalindia.in/Documents/Revised_2nd_Ver_Coal_Price_for_uploading_310112.pdf>, last accessed on 1 November 2012

TABLE 3.21 Revised rates of royalty (₹/tonne)

Coal group	12 February 1981	1 August 1991		11 October 1994		2 September 1996		17 August 2002		1 August 2007		1 August 2012	
	All states	All states except Assam and West Bengal	Assam and West Bengal	All states except West Bengal	West Bengal	All states except Meghalaya and West Bengal	Meghalaya and West Bengal	All states except West Bengal	West Bengal	All states except West Bengal	West Bengal	All states except West Bengal	West Bengal
Group I coking coal SG I, SG II, WG I	7.00	150.00	7.00	195.00	7.00	195.00	7.00	250.00	7.00	180 + 0.05 P	7.00	180 + 0.05 P	7.00
Group I hand-picked coal (Assam, Arunachal Pradesh, Meghalaya and Nagaland)	7.00	150.00	7.00	150.00	6.50	150.00	150.00	250.00	7.00	180 + 0.05 P	7.00	180 + 0.05 P	7.00
Group II (coking coal WG II, WG III; SC I, SC II; non-coking A, B)	6.50	120.00	6.50	135.00	6.50	135.00	6.50	165.00	6.50	130 + 0.05 P	6.50	130 + 0.05 P	6.50
Group II ungraded run of mine coal (Assam, Arunachal Pradesh, Meghalaya, Nagaland)	6.50	120.00	6.50	120.00	–	120.00	–	–	–	–		–	
Group III (coking coal WG IV, non-coking coal Grade C)	5.50	75.00	5.50	95.00	5.50	95.00	5.50	115.00	5.50	90 + 0.05 P	5.50	90 + 0.05 P	5.50
Group IV (non-coking Grade D, Grade E)	4.30	45.00	4.30	70.00	4.30	70.00	4.30	85.00	4.30	70 + 0.05 P	4.30	70 + 0.05 P	4.30
Group V (non-coking Grade F, Grade G)	2.50	25.00	2.50	50.00	2.50	50.00	2.50	65.00	2.50	55 + 0.05 P	2.5	55 + 0.05 P	2.5
Group VI (coal produced in Andhra Pradesh)	5.00	70.00	–	75.00	–	75.00	–	90.00	–	As applicable for Grade II – Grade V			

P – price; SG – steel grade; WG – washery grade; SC – semi-coking

Notes

1. For the state of West Bengal, in addition to royalty, other charges are: rural employment (RE) cess 20%, Public Works Department (PWD) road cess ₹ 1, primary education (PE) cess 5%, Asansol Mines Board of Health cess ₹ 1, CST and SED are applicable as mentioned in notes 2 and 3.
2. For states except West Bengal, levy cess and other taxes specific to coal bearing lands, royalty allowed shall be adjusted for the local cesses or such taxes so as to limit the overall revenues to the formula-based yield.
3. The rates of royalty, which shall be a combination of specific and ad volerem rates of royalty shall be R = a + bP, where P (price) shall mean basic pithead price of run on mine coal and lignite as reflected in the invoice, excluding taxes, levies and other charges, and the value of a (fixed component) and b (variable or ad volerem component) would be as above.
4. Stowing excise duty applicable for all states.
5. Central sales tax @ 4% of (basic price + royalty + cess + surface transport charge from pit head to loading point).

Source CCO (Coal Controller's Organisation). 2013. ***Coal Directory of India 2011/12***. Kolkata: CCO, Ministry of Coal, Government of India

TABLE 3.22 Royalty paid (₹ million) on coal to the states: 1991/92–2012/13

Year	Andhra Pradesh	Assam	Jharkhand	Madhya Pradesh	Maharashtra	Odisha	Uttar Pradesh	West Bengal	Chhattisgarh	Total
1991/92	354	6	3 178	2 402	610	435	475	100	-	7 561
1992/93	1 000	5	5 043	3 845	1 089	631	668	87	-	12 369
1993/94	2 913	4	5 551	3 696	1 111	728	706	106	-	14 815
1994/95	1 445	4	6 131	4 389	1 597	1 041	874	96	-	15 577
1995/96	1 753	216	6 583	6 668	2 179	1 808	1 135	78	-	20 420
1996/97	1 971	99	6 828	6 841	2 003	1 854	1 214	109	-	20 919
1997/98	2 781	61	6 590	6 636	2 048	2 276	1 024	126	-	21 542
1998/99	2 002	52	6 079	6 751	1 846	2 289	868	98	-	19 984
2003/04	-	144	7 431	5 011	2 820	4 031	1 457	95	4 913	25 902
2004/05	-	127	7 359	5 296	2 930	4 604	1 599	93	5 335	27 343
2005/06[a]	-	150	7 814	5 645	3 074	4 551	1 738	105	5 489	28 566
2006/07[b]	-	124	3 461	3 774	2 008	3 400	415	80	3 844	17 106
2007/08	4 790	-	9 097	8 318	4 268	6 454	1 044	90	7 605	41 665
2008/09	5 607	-	10 672	9 502	5 018	7 731	1 149	94	8 942	48 715
2009/10	6 371	309	11 524	10 228	5 127	8 811	1 491	92	9 396	53 348
2010/11	7 800	283	12 736	8 637	4 998	9 324	1 677	97	10 243	55 795
2011/12	7 690	255	14 305	10 128	5 263	10 278	1 819	95	11 008	60 841
2012/13	11 429	427	19 161	14 883	7 042	12 251	2 370	112	16 242	83 915

[a]provisional
[b]provisional and up to November 2007
Source MoC (Ministry of Coal). 2014. ***Annual Report 2013/14***. New Delhi: MoC, Government of India

TABLE 3.23 Average pithead unit value (₹/tonne) of coal: 1974/75–2008/09

Year	Pithead unit value
1974/75	70.42
1979/80	118.95
1984/85	206.88
1989/90	278.51
1990/91	277.89
1991/92	354.90
1992/93	355.41
1993/94	433.21
1994/95	442.00
1995/96	446.23
1996/97	531.26
1997/98	599.22
1998/99	635.74
1999/2000	655.64
2000/01	648.77
2001/02	660.40
2002/03	708.75
2003/04	712.00
2004/05	788.00
2005/06	818.00
2006/07	818.00
2007/08	850.00
2008/09	850.00

Source CCO (Coal Controller's Organisation). 2010. ***Coal Directory of India 2008/09***. Kolkata: CCO, Ministry of Coal, Government of India

TABLE 3.24a Allocation of coal and lignite blocks during 2014/15

Sector	End use	Mode of allotment				
Coal blocks			Till 24 September 2014		Till 31 March 2015	
			No. of blocks	Geological reserves (MT)	No. of blocks	Geological reserves (MT)
PSU	Power	Government dispensation	44	15 914.64	18	3 795.29
	Commercial mining	Government dispensation	32	7 090.05	4	246.06
	Iron and steel	Government dispensation	2	360.23	1	251.88
	Total		78	23 364.92	23	4 293.23
Private companies	Power	Captive dispensation	16	2 576.87	5	466.13
	Power	Ultra mega power project	12	4 747.40	12	4 747.4
	Iron and steel	Captive dispensation	28	3 495.43	12	810.181
	Cement	Captive dispensation	2	225.39	2	225.39
	Small and isolated patch (commercial mining)	Captive dispensation	2	9.34	2	9.34
	Total		60	11 054.426	33	
All-India	Power		72	23 238.91	33	
	Iron and steel		30	3 855.66		
	Cement		2	225.39		
	Commercial mining		34	7 099.39		
	Grand total		138	34 419.346	56	10 551.671

Lignite blocks

Sector	End use	Mode of allotment	No. of blocks	Geological reserves (MT)
State PSU	Power	Government dispensation	10	1027.53
	Commercial	Government dispensation	9	462.40
	Subtotal		19	1489.93
Private	Power	Captive dispensation	5	99.50
	Commercial	Captive dispensation	1	7.80
	Subtotal		6	107.30
All-India	Power		15	1127.03
	Commercial		10	470.20
	Grand total		25	1597.23

PSU - Public Sector Undertakings

Source CCO (Coal Controller's Organisation). 2015. ***Provisional Coal Statistics 2015/15***. Kolkata: CCO, Ministry of Coal, Government of India

TABLE 3.24b Allotment of captive coal blocks as on 31 March 2014 by year and sector

Year of allotment	Power		Ultra mega power project		Iron and steel		Government commercial		Private commercial and cement		Coal to oil		Total	
	Coal blocks (numbers)	Geological reserves	Coal blocks (numbers)	Geological reserves	Coal blocks (numbers)	Geological reserves	Coal blocks (numbers)	Geological reserves	Coal blocks (numbers)	Geological reserves	Coal blocks (numbers)	Geological reserves	Coal blocks (numbers)	Geological reserves
1993	1	140.5											1	140.5
1994	1	22.6											1	22.6
1995	1	84.5											1	84.5
1996	1	125.7			3	535.0			2	9.3			6	670.0
1997	0	0											0	0
1998	3	434.8											3	434.8
1999	0	0			2	231.0							2	231.0
2000	0	0			1	156.0							1	156.0
2001	1	562.0			1	34.3							2	596.3
2002	1	92.3											1	92.3
2003	10	319.2			4	372.6	2	286.5					16	978.3
2004	4	2 143.5											4	2 143.5
2005	5	1 541.5			11	839.0	1	103.2					17	2 483.6
2006	10	5 100.1	6	1 635.0	3	144.7	15	3 768.5					34	10 648.3
2007	11	3 764.8	1	972.0	3	253.3	12	1 146.0	2	225.3			29	6 361.5
2008	5	1 320.5	1	100.0	1	250.0	1	84.3					8	1 754.8
2009	1	137.9	3	1 339.0	1	700.0							5	2 176.9
2010	0		1	800.0									1	800.0
2011	0						1	56.8					1	56.8
2012	0												0	0
2013	5	2 256.2											5	2 256.2
2014	0												0	0
Total	60	18 046.1	12	4 846.0	30	3 515.9	32	5 445.3	4	234.6			138	32 087.9

Notes 1. Geological reserves (in million tonnes) as estimated during allocation.
2. Out of total allocated 218 coal blocks, 80 have been deallocated by the Ministry of Coal, as on 31 March 2014. So, net allocated coal blocks are 138, as on 31 March 2014.

Source CCO (Coal Controller's Organisation). 2014. ***Provisional Coal Statistics 2013/14***. Kolkata: CCO, Ministry of Coal, Government of India

TABLE 3.25 Annual plan and actual expenditure (₹ million): 1993/94–2015/16

Year	Coal India Ltd	Singareni Collieries Company Ltd	Science and technology, regional exploration and others
1993/94			
Budget estimate	19 010.0	6 422.0	170.0
Revised	17 200.0	5 600.0	170.0
Actual expenditure	16 879.2	5 449.8	159.1
1994/95			
Budget estimate	20 629.0	4 834.0	235.0
Revised	18 500.0	5 500.0	229.9
Actual expenditure	16 045.6	5 165.8	48.4
1995/96			
Budget estimate	22 600.0	5 160.0	438.2
Revised	12 037.8	3 357.5	251.0
Actual expenditure	15 036.3	2 593.5	147.9
1996/97			
Budget estimate	21 430.0	5 389.7	528.0
Revised	13 108.3	3 250.0	280.0
Actual expenditure	15 304.5	2 500.6	280.7
1997/98			
Actual expenditure	18 245.5	2 084.8	303.0
1998/99			
Budget estimate	25 172.0	3 315.7	921.3
Revised	17 000.0	2 800.0	571.0
Actual expenditure	18 319.7	2 060.9	448.0
1999/2000			
Budget estimate	25 560.0	2 271.9	825.4
Revised	26 761.9	2 172.0	631.4
Actual expenditure	27 696.6	1 452.0	560.0
2000/01			
Budget estimate	27 903.2	3 278.2	3 612.2
Revised	17 670.0	2 350.0	2 385.0
Actual expenditure	10 593.2	1 941.0	2 818.1
2001/02			
Budget estimate	23 909.0	3 550.0	3 234.4
Actual expenditure	11 466.9	1 819.2	4 012.8

Contd...

TABLE **3.25** *Contd...*

Year	Coal India Ltd	Singareni Collieries Company Ltd	Science and technology, regional exploration and others
2002/03			
Budget estimate	21 900.0	4 050.0	3 117.7
Revised	17 000.0	2 270.0	2 630.7
Actual expenditure	11 921.7	1 394.9	1 574.3
2003/04			
Budget estimate	22 400.0	3 400.0	2 859.0
Revised	18 460.0	2 050.0	2 388.7
Actual expenditure	11 568.2	1 634.2	1 820.5
2004/05			
Budget estimate	23 100.0	3 250.0	2 233.2
Revised	18 773.5	2 750.0	2 756.7
Actual expenditure	11 883.1	2 748.7	1 898.3
2005/06			
Budget estimate	28 143.5	3 950.0	1 520.5
Revised	22 240.0	3 950.0	2 550.0
Actual expenditure	16 112.3	4 241.8	2 282.5
2006/07			
Budget estimate	30 637.0	5 770.9	1 687.1
Revised	24 200.0	5 770.9	3 000.0
Actual expenditure	20 596.9	4 486.3	2 564.0
2007/08			
Budget estimate	24 721.4	5 705.8	2 500.0
Revised	20 669.7	5 200.0	3 770.0
Actual expenditure	20 335.1	5 739.7	2 798.0
2008/09			
Budget estimate	32 147.0	6 653.0	3 000.0
Revised	27 550.0	6 653.0	2 100.0
Actual expenditure	25 071.7	6 504.4	1 979.1
2009/10			
Budget estimate	29 000.0	5 805.7	3 100.5
Revised	31 000.0	6 339.4	2 600.0
Actual expenditure	28 099.9	8 886.7	2 372.9
2010/11			
Budget estimate	38 000.0	13 349.3	4 000.0
Revised	36 150.0	11 245.7	4 705.2
Actual expenditure	25 397.2	6 438.1	3 749.6

Contd...

TABLE **3.25** *Contd...*

Year	Coal India Ltd	Singareni Collieries Company Ltd	Science and technology, regional exploration and others
2011/12			
Budget estimate	42 200.0	28 043.0	2 225.9
Revised	41 950.0	13 896.1	2 405.2
Actual expenditure	33 702.1	10 705.6	3 275.7
2012/13			
Budget estimate	42 750.0	32 203.3	5 440.0
Revised	39 000.0	32 203.3	4 160.0
Actual expenditure	29 152.3	20 472.6	2 126.9
2013/14			
Budget estimate	50 000.0	40 000.0	2 390.0
Revised	50 000.0	29 000.0	2 884.5
Actual expenditure	40 296.0	24 554.5	2 141.5
2014/15			
Budget estimate	52 250.0	38 500.0	2 884.5
Revised	52 250.0	37 600.0	2 384.5
2015/16			
Budget estimate	59 905.0	23 900.0	-

Source MoC (Ministry of Coal). 2015. ***Annual Report 2014/15***. New Delhi: MoC, Government of India

PETROLEUM AND NATURAL GAS

AT A GLANCE

- The production of crude oil declined marginally for the second year in a row, from 37.8 million tonnes (MT) in 2013/14 to 37.5 MT in 2014/15. However, crude oil imports did not increase significantly, although the value of imports fell significantly by 20.5%. The import dependency of crude oil now stands at 83.5%.
- India's refining capacity remained constant at 2015 MT per annum, while crude throughput increased marginally from 222.5 MT in 2013/14 to 223.2 in 2014/15.
- As diesel prices were fully decontrolled in October 2014, under-recoveries on the fuel were brought down to ₹10 935 crore in 2014/15, from ₹62 837 crore in the previous fiscal year.
- The production of natural gas declined in 2014/15 by about 5%, from 35.4 billion cubic metres (BCM) in 2013/14 to 33.65 BCM in 2014/15.
- Liquefied natural gas (LNG) imports grew from 13.03 MT in 2013/14 to 15.47 MT in 2014/15.
- The total natural gas supply (including LNG) in India in 2014/15 was around 53.98 BCM and imports formed around 37.7%.
- The government announced pooling of prices for the fertilizer sector, which came into effect on 1 July 2015.

Introduction

The domestic production of both oil and natural gas continued their declining trend in 2014/15. Crude oil production has been declining for 3 years, while natural gas production has been declining for the past 4 years. In particular, crude oil production from onshore blocks has declined, while that from offshore blocks has increased. In spite of the overall decline in the domestic production, imports of crude oil has remained constant. However, owing to the drop in international crude oil prices, the value of the imported crude oil fell by 20.5%. Further, the refining capacity remained constant in 2013/14, while refined crude throughput increased only marginally. Finally, as diesel under-recoveries were eliminated starting October 2014 after a series of price increases, the under-recoveries for the fuel in the financial year 2014/15 stood at ₹10 935 crore, down from ₹62 837 crore in 2013/14. Under-recoveries on domestic liquefied petroleum gas (LPG) and public distribution scheme (PDS) kerosene remain high, although the government hopes to cut leakages by the Direct Benefit Transfer of LPG (DBTL) subsidies and the "Give it up" campaign, urging wealthy citizens to forgo the subsidy.

The imports of liquefied natural gas (LNG) increased by 18.7% from 13.03 million tonnes (MT) in 2013/14 to 15.47 MT in 2014/15 and now constitute around 46% of the total consumption of natural gas in the country. The fall in the spot LNG prices in the Asian market has now made spot LNG more affordable. Owing to the terms agreed upon in the contract between Qatar and Petronet LNG (the largest supplier of long-term LNG for India), the cost of LNG from here has now become more than the spot LNG prices, forcing a reduction of 10% in its offtake (maximum allowed in the take or pay contract). LNG was delivered at a low, fixed cost between 2004 and 2008, after which it was linked to a five-year Japan Custom-cleared Crude (JCC) oil average. Barring Dahej, the other three LNG terminals operated at low capacities, either due to delay in the construction of supporting infrastructure (pipelines, breakwater facilities) or due to pricing issues. However, as the demand for natural gas and India's imports are projected to increase in the future, several LNG terminals are scheduled to come up between 2016 and 2020, which are currently at various stages of construction. It is essential that the supporting infrastructure is developed simultaneously to avoid underutilization of capacity. The government had also announced pooled pricing for the supply of natural gas to the fertilizer (urea) sector of India, and it came into force on 1 July 2015.

Oil

Crude oil

Domestic production

Oil production in India marginally fell to 37.5 MT in 2014/15 (Tables 1 and 2). Specifically, the drop in production was from onshore blocks. Offshore production, on the contrary, marginally improved from 2013/14. In particular, Oil and Natural Gas Corporation (ONGC) increased offshore production by 0.7 MT from its fields, including Mumbai High and Heera fields. This follows the company's plan to increase production using improved oil recovery and enhanced oil recovery techniques (The Hindu 2014). However, offshore production by ONGC is still well below its 2007/08 levels of 18 MT. Further, private onshore production fell for the first time since 2009/10, reflecting the marginal decline from Cairn India's Rajasthan fields.

Table 1 Trend in onshore-offshore crude oil production (in MT)

Year	Onshore				Offshore			Grand total
	ONGC	OIL	Private/JV	Total	ONGC	Private/JV	Total	
2014/15 (P)	6.1	3.4	9.1	18.5	16.2	2.7	18.9	37.5
2013/14	6.7	3.5	9.4	19.6	15.5	2.7	18.2	37.8
2012/13	6.9	3.7	8.8	19.4	15.6	2.8	18.4	37.9
2011/12	7.4	3.9	6.8	18.0	16.3	3.7	20.1	38.1
2010/11	7.5	3.6	5.4	16.4	17.0	4.3	21.3	37.7
2009/10	7.5	3.6	0.7	11.8	17.3	4.5	21.9	33.7
2008/09	7.6	3.5	0.2	11.3	17.8	4.4	22.2	33.5
2007/08	7.9	3.1	0.2	11.2	18.0	4.9	22.9	34.1
2006/07	8.1	3.1	0.2	11.3	18.0	4.7	22.7	34.0
2005/06	8.1	3.2	0.1	11.4	16.3	4.5	20.8	32.2
2004/05	8.3	3.2	0.1	11.6	18.2	4.2	22.4	34.0
2003/04	8.4	3.0	0.1	11.5	17.7	4.2	21.9	33.4
2002/03	8.5	3.0	0.1	11.5	17.6	4.0	21.6	33.0
2001/02	8.6	3.2	0.1	11.9	16.1	4.1	20.1	32.0
2000/01	8.4	3.3	0.1	12.0	16.6	3.8	20.4	32.4

JV – joint venture; MT – million tonnes; OIL – Oil India Ltd; ONGC – Oil and Natural Gas Corporation; P – provisional
Note The production figures have been rounded off to two decimal places.
Source MoPNG (2015b)

Table 2 Production of crude oil by region (in MT)

Year	Onshore					Offshore			Grand total
	Arunachal Pradesh	Andhra Pradesh/ Tamil Nadu	Assam/ Nagaland	Gujarat/ Rajasthan	Total	Private/JV	Mumbai High	Total	
2014/15 (P)	0.1	0.5	4.5	13.5	18.5	2.7	16.2	18.9	37.5
2013/14	0.1	0.5	4.7	14.2	19.6	2.7	15.5	18.2	37.8
2012/13	0.1	0.5	4.9	13.9	19.4	2.8	15.6	18.4	37.9
2011/12	0.1	0.6	5.0	12.3	18.0	3.7	16.3	20.1	38.1
2010/11	0.1	0.5	4.7	11.1	16.4	4.3	17.0	21.3	37.7
2009/10	0.1	0.5	4.7	6.4	11.8	4.5	17.3	21.9	33.7
2008/09	0.1	0.6	4.7	5.9	11.3	4.4	17.8	22.2	33.5
2007/08	0.1	0.6	4.4	6.2	11.2	4.9	18.0	22.9	34.1
2006/07	0.1	0.6	4.4	6.2	11.3	4.7	18.0	22.7	34.0
2005/06	0.1	0.6	4.5	6.3	11.4	4.5	16.3	20.8	32.2
2004/05	0.1	0.6	4.7	6.2	11.6	4.2	18.2	22.4	34.0
2003/04	0.1	0.7	4.6	6.1	11.5	4.2	17.7	21.9	33.4
2002/03	0.1	0.7	4.7	6.0	11.5	4.0	17.6	21.6	33.0
2001/02	0.1	0.7	5.1	6.0	11.9	4.1	16.1	20.1	32.0
2000/01	0.1	0.7	5.2	5.8	11.8	4.0	16.6	20.4	32.4

JV – joint venture; MT – million tonnes; P – provisional
Source MoPNG (2015b)

Import of crude oil

India's imports of crude oil remained at 189 MT in 2014/15 for the second year in a row. However, owing to low international crude oil prices, the value of these imports fell from ₹864 875 crore in 2013/14 to ₹687 350 crore in 2014/15 (Figures 1 and 2). The share of imported crude oil increased to 83.5% in 2014/15 from 74% in 2004/05. Further, the share of petroleum products in the overall imports in value terms was 23.6% in 2014/15, down from a high of 26.9% in 2013/14, when the value of the rupee fell to a low of ₹68 per dollar.

India continues to depend on the Middle East for its oil supplies, with Saudi Arabia being the largest supplier (Rajya Sabha 2015d). The top five countries that export crude oil to India also include Venezuela and Nigeria (Table 3).

The Government of India also announced a target of cutting import dependence on oil by 10% till 2022. To achieve this goal, the government has announced the following measures:

Table 3 Top five countries from which India imported crude oil in 2014/15 (in MT)

Rank	Country	2014/15 (April–December) (provisional)
1	Saudi Arabia	25.87
2	Venezuela	17.58
3	Iraq	17.52
4	Nigeria	14.43
5	Kuwait	12.96
Total		88.37

Note Data of imports from Reliance Industries Ltd are unavailable.
Source Rajya Sabha (2015f)

- Reassess hydrocarbon resources in all sedimentary basins in India
- Appraise unappraised areas within basins
- Draft a policy for the assessment of unexplored sedimentary basins

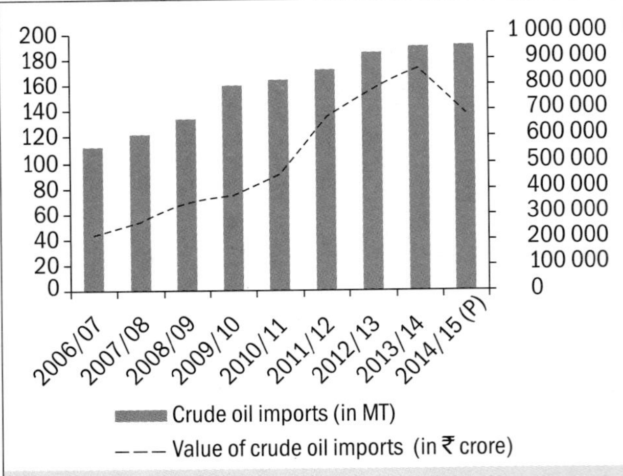

Figure 1 Quantity and value of crude oil imports in India
P - provisional
Source MoPNG (2015b)

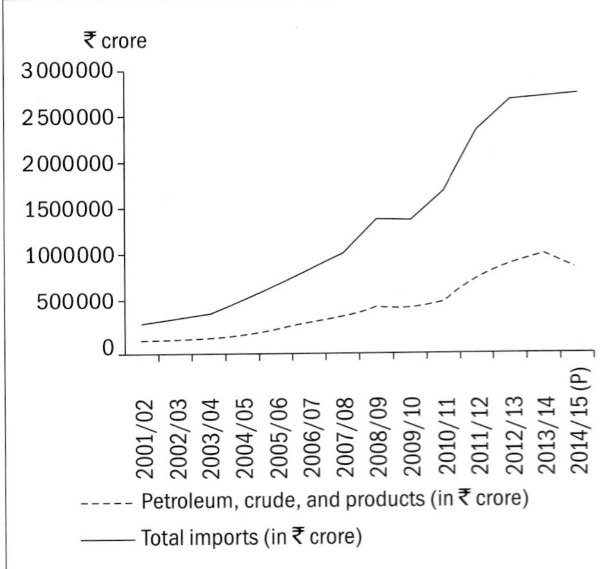

Figure 2 Imports of crude oil and petroleum products *vis-a-vis* total imports
P - provisional
Source RBI (2015)

- Promote the use of improved oil recovery and enhanced oil recovery techniques by national oil companies (NOCs)
- Create a policy framework to promote the early monetization of hydrocarbon discoveries
- Draft a policy for the exploration in the mining lease (ML) areas after the expiry of the exploration period
- Encourage the production of alternative energy sources such as coalbed methane (CBM), shale oil and gas, and gas hydrates (Rajya Sabha 2015e)
- Relax the production sharing contract (PSC) regime
- Announce a price premium for deep water, ultra-deep water, and high pressure, high temperature areas
- Provide a 40% subsidy in the North-East region to private companies (Rajya Sabha 2015g)

Crude oil pipelines

India has more than 9500 km of crude oil pipelines. Table 4 lists the operators, capacities, and lengths of major pipelines. See Map 1.

Strategic petroleum reserves

To enhance India's oil security, the government constituted the Indian Strategic Petroleum Reserves Ltd (ISPRL) for constructing strategic petroleum reserves (SPRs). The planned capacity of SPRs is 5.33 MT, and the SPRs are located in Visakhapatnam (1.33 MT), Mangalore (1.5 MT), and Padur (2.5 MT) (PIB 2015c). In the second phase of SPR construction, storage would be built in Chandikhol, Bikaner, and Rajkot. To fill the existing Visakhapatnam SPR with crude oil, the government has already allocated ₹4948 crore. Any remaining funds will be used for the remaining two SPRs. However, for the other two SPRs, the government will continue to explore alternative funding mechanisms. The annual operation and maintenance costs are estimated to be ₹47 crore for Visakhapatnam SPR and ₹179 crore for all of them put together. This includes the cost of their physical security (PIB 2015a).

Refining

Present capacity and throughput of existing refineries

Even as India is heavily dependent on crude oil imports, it has been a net exporter of petroleum products since 2001 (EIA 2014b). India's refining capacity in April 2015 stood at 215.07 MT per annum (MTPA), unchanged from the previous year (Table 5). India's refining capacity has come a long way since 2006/07, when it was 148.97 MTPA and it was only 62 MTPA in 1997/98.

Table 4 Capacity, length, and operators of major crude pipelines

Operator	Pipeline	Capacity (MTPA)	Length (km)
OIL	Duliajan-Digboi-Bongaigaon-Barauni	8.40	1193
IOCL	Salaya-Mathura-Panipat (including loop lines)	21.00	1870
IOCL	Paradip-Haldia-Barauni	11.00	1384
IOCL	Mundra-Panipat	8.4	1194
HMEL (JV with HPCL)	Mundra-Bathinda	9.00	1017
BORL (JV with BPCL)	Vadinar-Bina	6.00	937
Cairn	Barmer-Salaya	6.25	591
ONGC	Mumbai High-Uran trunk (offshore)	15.6	204
ONGC	Heera-Uran trunk (offshore)	11.5	81
ONGC	30″ BUT (OIL) (offshore)	18	203
Total	All pipelines, including those not listed above	146.6	9572.2

BORL - Bharat Oman Refineries Ltd; BPCL - Bharat Petroleum Corporation Ltd; BUT - Bombay High to Uran Trunk; HMEL - Hindustan Mittal Energy Ltd; HPCL - Hindustan Petroleum Corporation Ltd; IOCL - Indian Oil Corporation Ltd; JV - joint venture; MTPA - million tonnes per annum; OIL - Oil India Ltd; ONGC - Oil and Natural Gas Corporation

Note Smaller pipelines operated by ONGC have not been included in this table.

Source MoPNG (2015b)

No new refinery has been commissioned since 2012, when Hindustan Mittal Energy Ltd (Bathinda, Punjab) and Bharat Oman Refineries Ltd (Bina, Madhya Pradesh) started the refining activity. However, the capacity expansion has taken place in existing refineries. At the moment, new refineries are being constructed in Paradip, Odisha (by Indian Oil Corporation Ltd [IOCL] with a capacity of 15 MTPA); Cuddalore, Tamil Nadu (by Nagarjuna Oil Corporation Ltd with a capacity of 6 MTPA); Ratnagiri, Maharashtra (by Hindustan Petroleum Corporation Ltd [HPCL] with a capacity of 9 MTPA); and Barmer, Rajasthan (by HPCL with a capacity of 9 MTPA).

Gross refining margin

Gross refining margin (GRM) measures the profitability of refineries. GRM is the difference between the cost of crude oil and the average price realized on the refined product. GRMs in India fell across most refineries, other than Reliance Industries Ltd (RIL) Jamnagar, IOCL Guwahati, and IOCL Koyali (Table 6). Overall, IOCL was hit hardest, with an average GRM of only $0.27 per barrel (bbl). GRMs declined primarily because of inventory losses triggered by a decline in international crude prices. The price of the Indian basket of crude fluctuated between $104.5/bbl and $52.6/bbl at the end of 2014/15. Further, there was also pressure from increasing refining capacity in West Asia, USA, and China.

Petroleum products

Production

Although India's refining capacity has not increased since 2013, the production of petroleum products marginally increased to 221 MT (Table 7). The production of motor spirit and high-speed diesel (HSD) increased by 2 MT and 0.7 MT, respectively. However, the production of LPG, naphtha, furnace oil, and fuel oil declined.

Consumption

The domestic consumption of petroleum products rose by 6.4 MT to 182.7 MT in 2014/15 over the 2013/14 levels (Table 8). In particular, the consumption of naphtha, kerosene, light diesel oil, and furnace oil fell. The consumption of motor spirit, in particular, rose by 2 MT during the previous financial year.

Imports and exports of petroleum products

In light of the constant production and increasing consumption, the exports of petroleum products fell in 2014/15 by 4 MT and imports fell by 3.7 MT over

Map 1 Petroleum refineries and major crude oil pipelines in India
Source MoPNG (2015b)

Table 5 Installed capacity and crude throughput

Refinery and location	Refining capacity as on 1 April (MTPA)			Refining crude throughput (MT)				
	2013	2014	2015	2010/11	2011/12	2012/13	2013/14	2014/15 (P)
IOCL Guwahati, Assam	1.00	1.00	1.00	1.12	1.06	0.96	1.02	1.01
IOCL Barauni, Bihar	6.00	6.00	6.00	6.21	5.73	6.34	6.48	5.94
IOCL Koyali, Gujarat	13.70	13.70	13.70	13.56	14.25	13.16	12.96	13.29
IOCL Haldia, West Bengal	7.50	7.50	7.50	6.88	8.07	7.49	7.95	7.65
IOCL Mathura, Uttar Pradesh	8.00	8.00	8.00	8.88	8.20	8.56	6.64	8.52
IOCL Digboi, Assam	0.65	0.65	0.65	0.65	0.62	0.66	0.65	0.59
IOCL Panipat, Haryana	15.00	15.00	15.00	13.66	15.50	15.13	15.10	14.19
IOCL Bongaigaon, Assam	2.35	2.35	2.35	2.01	2.19	2.36	2.33	2.4
BPCL Mumbai, Maharashtra	12.00	12.00	12.00	13.02	13.36	13.08	12.68	12.82
BPCL Kochi, Kerala	9.50	9.50	9.50	8.70	9.47	10.11	10.29	10.36
HPCL Mumbai, Maharashtra	6.50	6.50	6.50	6.64	7.51	7.75	7.79	7.41
HPCL Visakh, Andhra Pradesh	8.30	8.30	8.30	8.20	8.68	8.03	7.78	8.77
CPCL Manali, Tamil Nadu	10.50	10.50	10.50	10.10	9.95	9.11	10.07	10.25
CPCL Narimanam, Tamil Nadu	1.00	1.00	1.00	0.70	0.61	0.64	0.56	0.53
NRL Numaligarh, Assam	3.00	3.00	3.00	2.25	2.83	2.48	2.61	2.78
ONGC Tatipaka, Andhra Pradesh	0.07	0.07	0.07	0.07	0.07	0.06	0.07	0.05
MRPL Mangalore, Karnataka	15.00	15.00	15.00	12.66	12.80	14.42	14.59	14.63
RIL Jamnagar, Gujarat	33.00	33.00	33.00	31.20	32.50	32.61	30.31	30.87
RIL (SEZ) Jamnagar, Gujarat	27.00	27.00	27.00	35.61	35.19	35.89	37.72	37.17
ESSAR Vadinar, Gujarat	20.00	20.00	20.00	14.87	13.50	19.77	20.20	20.49
BORL Bina, Madhya Pradesh	6.00	6.00	6.00	-	2.05	5.73	5.45	6.21
HMEL Bathinda, Punjab	9.00	9.00	9.00	-	-	4.90	9.27	-
Total	215.07	215.07	215.07	196.99	204.12	219.21	222.50	223.24

BORL - Bharat Oman Refineries Ltd; BPCL - Bharat Petroleum Corporation Ltd; CPCL - Chennai Petroleum Corporation Ltd; GGSR - Guru Gobind Singh Refinery; HMEL - Hindustan Mittal Energy Ltd; HPCL - Hindustan Petroleum Corporation Ltd; IOCL - Indian Oil Corporation Ltd; MRPL - Mangalore Refinery and Petrochemicals Ltd; MT - million tonnes; MTPA - million tonnes per annum; NRL - Numaligarh Refinery Ltd; ONGC - Oil and Natural Gas Corporation; P - provisional; RIL - Reliance Industries Ltd; SEZ - special economic zone

Source MoPNG (2015b)

the previous financial year. In particular, the import of LPG increased by 1.3 MT and the export of diesel fell by 1.9 MT. Among the various petroleum products, only the exports of petrol increased over the previous year. Tables 9 and 10 provide the imports and exports, respectively, of petroleum products in 2014/15.

Petroleum product pipelines

India has more than 14 000 km of petroleum product pipelines across the country. The capacity and length of the petroleum pipelines in the country classified according to operators are given in Table 11.

Pricing of petroleum products

By decontrolling the prices of petrol and diesel, the Government of India brought down the under-recoveries on the sale of these fuels to zero, starting from 25 June 2010 and 18 October 2014, respectively. Since then, the government is paying no subsidies on these fuels. As there were under-recoveries on diesel for more than 6 months in 2014/15, the government's share of subsidies was ₹10 935 crore, down from ₹62 837 crore in the previous fiscal year. This number will fall to zero as long as there are no under-recoveries on the fuel in 2015/16.

Table 6 Gross refining margins (in $/bbl)

Year	2010/11	2011/12	2012/13	2013/14	2014/15
IOCL (average of eight refineries)	5.72	3.63	3.16	4.24	0.27
BPCL (Mumbai)	4.23	1.73	4.67	3.95	3.97
BPCL (Kochi)	4.83	3.09	5.36	4.8	3.17
HPCL (Mumbai)	4.65	2.82	2.08	5.38	4.88
HPCL (Visakhapatnam)	5.81	2.95	2.08	1.5	1.12
CPCL (Chennai)	5.38	4.16	0.99	4.06	1.97
RIL	8.4	8.6	9.20	8.10	8.60
EOL	4.53	4.45	7.96	7.98	8.37
NRL	15.39	11.97	10.52	12.09	16.67
MRPL	5.9	5.96	2.45	2.67	-0.64
BORL	–	–	7.00	7.70	6.10

BORL – Bharat Oman Refineries Ltd; BPCL – Bharat Petroleum Corporation Ltd; CPCL – Chennai Petroleum Corporation Ltd; EOL – ESSAR Oil Ltd; HPCL – Hindustan Petroleum Corporation Ltd; IOCL – Indian Oil Corporation Ltd; MRPL – Mangalore Refinery and Petrochemicals Ltd; NRL – Numaligarh Refinery Ltd; RIL – Reliance Industries Ltd

Source PPAC (2015b)

Table 7 Production of petroleum products (in '000 tonnes)

Year	2008/09	2009/10	2010/11	2011/12	2012/13	2013/14	2014/15 (P)
LPG	9 158	10 334	9 708	9 547	9 825	10 030	9 840
MOGAS	16 020	22 537	26 138	27 186	30 118	30 275	32 325
Naphtha	16 452	18 788	19 196	18 825	19 018	18 505	17 391
Kerosene	8 391	8 703	7 809	7 861	7 971	7 418	7 559
ATF	8 071	9 304	9 589	10 065	10 088	11 220	11 103
HSD	62 905	73 298	78 057	82 880	91 103	93 759	94 428
LDO	606	472	590	502	400	423	358
FO	14 749	15 828	18 659	16 732	13 690	12 920	11 248
LSHS/HHS/RFO	2 935	2 518	1 860	1 701	1 364	485	671
Fuel oil	17 684	18 346	20 519	18 433	15 054	13 405	11 919
Lube oils	874	950	884	1 028	896	941	946
Bitumen	4 713	4 889	4 478	4 610	4 670	4 785	4 632
Petroleum Coke	4 241	3 709	2 711	7 837	10 943	12 068	12 448
Paraffin wax	69	64	61	47	52	55	51
Total wax	74	67	67	54	57	56	53
Others	5 964	12 478	15 081	14 381	17 598	17 871	18 137
Total	155 148	184 610	194 821	203 202	217 736	220 756	221 136

ATF – aviation turbine fuel; FO – furnace oil; HHS – hot heavy stock; HSD – high-speed diesel; LDO – light diesel oil; LPG – liquefied petroleum gas; LSHS – low sulphur heavy stock; MOGAS – Motor gasoline; P – provisional; RFO – residual fuel oil

Source MoPNG (2015b)

Table 8 Consumption of petroleum products (in '000 tonnes)

Year	2008/09	2009/10	2010/11	2011/12	2012/13	2013/14	2014/15 (P)
LPG	12 191	13 135	14 331	15 350	15 601	16 294	18 019
MS	11 258	12 818	14 194	14 992	15 744	17 128	19 075
Naphtha + NGL	13 911	10 134	10 676	11 222	12 289	11 305	10 939
SKO	9 303	9 304	8 928	8 229	7 502	7 165	7 087
ATF	4 423	4 627	5 078	5 536	5 271	5 505	5 578
HSD	51 710	56 242	60 071	64 750	69 080	68 364	69 404
LDO	552	457	455	415	399	386	365
Furnace oil	9 419	9 145	8 807	7 547	6 290	5 787	5 640
LSHS	3 169	2 484	1 982	1 759	1 366	449	377
Fuel oil	12 588	11 629	10 789	9 307	7 656	6 236	6 017
Lubes/Greases	2 000	2 539	2 429	2 633	3 196	3 305	2 964
Bitumen	4 747	4 934	4 536	4 638	4 676	5 007	4 983
Petroleum coke	6 166	6 586	4 982	6 138	10 135	11 756	14 407
Waxes	281	289	254	269	194	177	361
Others	4 469	5 111	4 315	4 654	5 315	5 781	5 789
Refinery fuel and losses	12 359	15 116	16 381	17 295	18 347	17 868	17 669
Total	145 958	152 924	157 422	165 427	175 404	176 274	182 656

ATF – aviation turbine fuel; HSD – high-speed diesel; LDO – light-diesel oil; LPG – liquefied petroleum gas; LSHS – low sulphur heavy stock; MS – motor spirit; NGL – natural gas liquids; SKO – superior kerosene oil

Source MoPNG (2015b)

Table 9 Imports of petroleum products (in '000 tonnes)

Year	2008/09	2009/10	2010/11	2011/12	2012/13	2013/14	2014/15 (P)
LPG	2 423	2 718	4 484	5 790	6 293	6 607	8 318
Petrol	397	385	1 702	654	147	235	372
Naphtha	5 023	1 734	2 063	2 091	1 735	1 026	858
Kerosene	1 448	985	1 381	564	0	0	30
Diesel	2 742	2 531	1 996	1 059	626	84	119
Lubes	986	1 419	1 291	1 434	1 468	1 674	1 779
Fuel oil	2 760	896	1 013	1 203	1 068	1 283	945
Bitumen	105	69	98	78	85	237	458
Others	2 702	3 928	3 352	2 977	4 352	5 571	7 544
Total	18 586	14 665	17 379	15 849	15 774	16 718	20 423

LPG – liquefied petroleum gas; P – provisional

Source MoPNG (2015b)

Table 10 Exports of petroleum products (in '000 tonnes)

Year	2008/09	2009/10	2010/11	2011/12	2012/13	2013/14	2014/15 (P)
LPG	109	131	154	174	200	227	254
Petrol	5440	9771	13578	14524	16657	15247	16048
Naphtha	7601	10043	10655	10139	8647	8322	7008
ATF	3701	4588	4478	4561	4664	5745	5520
Kerosene	77	46	33	34	23	15	15
Diesel	14720	18451	20335	20407	22464	26469	25559
LDO	0	41	98	84	9	30	6
Lubes	139	28	29	27	59	20	7
Fuel oil	6207	5186	6734	7895	5922	6159	4762
Bitumen	45	31	56	5	87	95	94
Others	905	2839	2927	2988	4675	5535	4653
Total	38944	51155	59077	60837	63408	67864	63928

ATF – aviation turbine fuel; LDO – light diesel oil; LPG – liquefied petroleum gas; P – provisional

Source MoPNG (2015b)

The government has also marginally increased taxes on petrol and diesel and announced to end subsidies on these products and increase taxation in the form of "green tax". The government claims that these measures would lead to a net reduction of 11 MT of CO_2 emissions in less than a year. This would constitute 0.6% of India's annual emissions (PIB 2015d).

Kerosene and domestic LPG, on the other hand, continue to be subsidized. In order to administer the domestic LPG subsidy, the government granted the DBTL subsidy in 2013. This direct transfer of subsidies was linked to Aadhaar (unique identification number) for 20 districts initially and was scaled up to 291 districts. DBTL was temporarily suspended in March 2014 because of the low penetration of Aadhaar in some districts.[1] The Dhande Committee was then set up to review the challenges faced in the implementation of DBTL, after which the PAHAL (DBTL) scheme was launched in November 2014 in 54 districts of India. It was expanded to entire India on 1 January 2015. Under PAHAL, customers can link their LPG costumer IDs to either Aadhaar or their bank accounts.

Further, in March 2015, the Prime Minister launched the "Give it up" campaign and appealed to the well-off people to voluntarily give up their LPG subsidy. While the estimates vary, the government claims that more than 2.8 lakh people from across the country have already given up LPG subsidy (PIB 2015g).

On the other hand, the pilot project on the direct benefit transfer of kerosene was carried out on a small scale (TERI 2014b). The Deendayal Upadhyaya Gram Jyoti Yojana aims to achieve the goal of 100% rural electrification by the end of FY17. With this in place, the government may phase out kerosene subsidies (The Financial Express 2015). The under-recovery on kerosene stood at ₹24804 crore in 2014/15, down from ₹30574 crore in 2013/14.

Tables 12, 13, and 14 give the under-recoveries to oil companies by product, fiscal subsidy on PDS kerosene and LPG, and the burden sharing of under-recoveries, respectively.

Ethanol blended petrol and bio-diesel

The Ethanol Blended Petrol (EBP) Programme mandates 5% blending of ethanol with petrol. To achieve this goal, the government has permitted up to 10% share of ethanol as per the specifications of

[1] In November 2013, the Supreme Court of India, upholding its interim order in September 2013 (which was in response to a public interest litigation), ruled that Aadhaar could not be made mandatory for receiving subsidies (IISD 2014).

Table 11 Length and capacity of major petroleum product pipelines

Operator	Pipeline	Capacity (MTPA)	Length (km)
GAIL	Jamnagar-Loni LPG pipeline	2.5	1414
GAIL	Vizag-Secunderabad LPG pipeline	1.3	618
IOCL	Koyali-Sanganer petroleum product pipeline	4.6	1287
IOCL	Mathura-Delhi-Bijwasan-Panipat petroleum product pipeline	3.7	258
IOCL	Barauni-Kanpur petroleum product pipeline	3.5	745
IOCL	Panipat-Delhi petroleum product pipeline	3	189
BPCL	Mumbai-Manmad-Manglya-Piyala-Bijwasan	6	1389
BPCL	Bina-Kota	4.4	259
HPCL	Mumbai-Pune-Solapur pipeline	4.3	508
HPCL	Visakh-Vijayawada-Secunderabad pipeline	5.4	572
HPCL	Mundra-Delhi pipeline	5	1054
HPCL	Ramanmandi-Bahadurgarh pipeline	4.7	243
PCCK	Cochin-Coimbatore-Karur pipeline	3.3	293
Total	All pipelines, including those not listed above	92.1	14463

BPCL – Bharat Petroleum Corporation Ltd; GAIL – Gas Authority of India Ltd; HPCL – Hindustan Petroleum Corporation Ltd; IOCL – Indian Oil Corporation Ltd; MTPA – million tonnes per annum; PCCK – Petronet Cochin Coimbatore Karur
Source MoPNG (2015b)

the Bureau of Indian Standards (BIS). To foster the availability of ethanol, in December 2014, the government fixed the price of ethanol between ₹48.5 and ₹49.5 per litre, depending on the distance of distillery from the depot (PIB 2015b). Further, ethanol produced from non-food feedstocks besides molasses to be procured subject to meet the relevant BIS specifications. In 2014, the government also began providing soft loans of up to 40% of the project cost to promote ethanol production (PIB 2014a).

However, there have been bottlenecks in procuring ethanol during the past few years. First, transportation procedures adopted by states are currently time consuming and complicated. Second, as sugarcane production is cyclical in nature, the supply of ethanol varies through the year. The Ministry of Consumer Affairs, Food and Public Distribution has urged state governments to take steps to dismantle the barriers that have impeded the successful implementation of EBP (PIB 2015h). On its part, the ministry has provided soft loans of ₹6000 crore and other financial assistance to sugar farmers. As a result of the steps taken by the government, the overall supply of ethanol, which was about 32 crore litres per year, shot up to 83 crore litres in 2015 (PIB 2015e).

Apart from EBP, the Ministry of Petroleum and Natural Gas (MoPNG) implemented the Bio-diesel Purchase Policy in 2006. Under this policy, bio-diesel meeting prescribed BIS specifications would be purchased at a uniform price for blending with HSD up to 5% by volume at 20 purchase centres across the country (PIB 2015i). Until August 2015, oil marketing companies had launched bio-diesel supplies in some retail outlets in Delhi, Vijayawada, Haldia, and Vizag (PIB 2015f).

Natural gas

Domestic natural gas production

Natural gas production fell again in 2014/15 by almost 5% from 35.4 BCM in 2013/14 to 33.65 BCM in 2014/15 (Table 15). There was some increase in production from the onshore fields in Rajasthan, Tripura, and to a lesser extent in Assam (Table 16); the offshore production of both public and private companies declined (Table 17).The steep rise in natural gas production starting 2009/10 and the unexpected fall after 2010/11 were due to the production from RIL-BP-Niko's KG-DWN-98/3 (popularly known as the KG-D6) block. Figure 3 shows the share of private and public sectors in the total natural gas production since 1999/2000.

Table 12 Under-recoveries to oil companies (in ₹ crore) by product										
Sensitive petroleum products	2005/06	2006/07	2007/08	2008/09	2009/10	2010/11	2011/12	2012/13	2013/14	2014/15 (P)
Petrol	2723	2027	7332	5181	5151	2227	-	-	-	-
Diesel	12647	18776	35166	52286	9279	34706	81192	92061	62837	10935
Domestic LPG	10246	10701	15523	17600	14257	21772	299970	39558	46457	36580
PDS kerosene	14384	17883	19102	28225	17364	19484	273520	29410	30574	24804
Total	40000	49387	77123	103292	46051	78190	1385410	161029	139868	72314

LPG – liquefied petroleum gas; PDS – public distribution system; P – provisional

Source MoPNG (2015b)

Table 13 Fiscal subsidy on PDS kerosene and domestic LPG (in ₹ crore)										
Year	2005/06	2006/07	2007/08	2008/09	2009/10	2010/11	2011/12	2012/13	2013/14	2014/15 (P)
PDS kerosene	1057	970	978	974	956	931	863	741	676	5
Domestic LPG	1605	1554	1663	1714	1814	1974	2137	1989	1904	18
Total	2662	2524	2641	2688	2770	2904	3000	2730	2580	23

LPG – liquefied petroleum gas; PDS – public distribution system; P – provisional

Source MoPNG (2015b)

Table 14 Share of under-recoveries (in ₹ crore)								
	Under-recovery and burden sharing (in ₹ crore)							
Year	2007/08	2008/09	2009/10	2010/11	2011/12	2012/13	2013/14	2014/15 (P)
Total under-recovery	77123	103292	46051	78190	138541	161029	139869	72314
Burden sharing through oil bonds/cash subsidy*	35290	71292	26000	41000	83500	100000	70772	22085
Borne by OMCs	16125	0	5621	6893	41	1029	2076	7407
Upstream companies#	25708	32000	14430	30297	55000	60000	67021	42822

*Since 2009/10, cash subsidy was provided by the government in place of oil bonds.

#Oil India Ltd, Oil and Natural Gas Corporation, and Gas Authority of India Ltd

OMC – oil marketing company; P – provisional

Source MoPNG (2015b)

Natural gas imports

The total global LNG trade reached 241.1 MT (around 332.7 BCM) in 2014, up 4.3 MT from 2013 (IGU 2015). The number of LNG exporting countries, between 2012 and 2014, has also risen from 17 to 19, though the overall supply remained stalled at the 2011 levels (BG Group 2015). At the same time, the number of international importers has gone up to 29 countries. For the first time, the Asian LNG market was larger than the European market in 2014, despite the demand in the Asian market being weaker than expected. The imports from India and China alone constituted just over 14% of the global trade (IGU 2015).

India was the fourth largest LNG importer in 2014/15, accounting for around 6.5% of the global imports. LNG imports of India grew from 13.03 MT in 2013/14 to 15.47 MT in 2014/15, representing an increase of 18.71% (Figure 4). The total natural gas supply (including LNG) in India during 2014/15 was 53.98 BCM, and imports constituted around 37.7% of this quantity. Owing to a fall in the spot LNG prices, which saw a low of around \$6–7 per

Table 15 Production of natural gas (in BCM)

Year	ONGC	OIL	Private/JV	Total
2014/15	22.03	2.72	8.9	33.65
2013/14	23.27	2.62	9.48	35.4
2012/13	23.55	2.64	14.49	40.68
2011/12	23.31	2.63	21.61	47.55
2010/11	23.09	2.35	26.77	52.21
2009/10	23.17	2.41	21.98	47.57
2008/09	22.49	2.26	8.09	32.84
2007/08	22.21	2.34	7.78	32.27
2006/07	22.25	2.27	7.04	31.56
2005/06	22.57	2.27	7.36	32.2
2004/05	22.99	2.01	6.78	31.77
2003/04	23.58	1.88	6.49	31.96
2002/03	24.24	1.74	5.41	31.4
2001/02	24.04	1.62	4.05	29.71
2000/01	24.02	1.86	3.6	29.48
1999/2000	23.25	1.73	3.47	28.45

JV - joint venture; OIL - Oil India Ltd; ONGC - Oil and Natural Gas Corporation
Source MoPNG (2015b)

million British thermal units (MBTU), the price of India's long-term LNG supply from Qatar at $13/MBTU became higher than spot cargoes. For the first time in 10 years, Petronet LNG asked for a 10% cut in the volumes delivered from Qatar (maximum allowed contractually). As spot LNG prices in Asia are expected to remain low, India is seeking to renegotiate the delivered LNG price from Qatar.

Table 17 Production of offshore natural gas by region (in BCM)

Year	ONGC Mumbai High	Private/JV offshore	Total
2014/15 (P)	17.27	7.59	24.86
2013/14	17.97	8.42	26.40
2012/13	18.10	13.70	31.80
2011/12	17.56	20.91	38.47
2010/11	17.59	26.05	43.64
2009/10	17.46	21.35	38.81

JV - joint venture; ONGC - Oil and Natural Gas Corporation; P - provisional
Source MoPNG (2015b)

Some of the countries from which India sourced its LNG imports in 2013/14 were Qatar, Nigeria, Yemen, Egypt, Norway, Algeria, and Equatorial Guinea (Lok Sabha 2014).[2]

The long-term contracts signed by different companies in India are as follows:

(i) Sabine Pass Liquefaction LLC, USA—3.5 MTPA for 20 years; signed by GAIL
(ii) Dominion Resources, USA—2.30 MTPA for 20 years; signed by GAIL
(iii) Gazprom Marketing and Trading, Singapore—2.5 MTPA for 20 years; signed by GAIL
(iv) Gas Natural Fenosa, Spain—0.75 MTPA for 3 years; signed by GAIL
(v) Gas de France, France—0.36 MTPA for 2 years; signed by GAIL
(vi) BG Group, United Kingdom—to initially supply 1.25 MTPA from 2015 with the potential increase to 2.5 MTPA for 20 years; signed by Gujarat State Petroleum Corporation (GSPC) (BG Group 2013)

Table 16 Production of onshore natural gas by region (in BCM)

Year	Andhra Pradesh	Arunachal Pradesh	Assam	Gujarat	Rajasthan	Tamil Nadu	Tripura	Total
2014/15	0.54	0.03	2.96	1.52	1.18	1.19	1.14	8.79
2013/14	1.17	0.04	2.87	1.66	0.98	1.30	0.82	9.01
2012/13	1.25	0.04	2.91	2.03	0.68	1.21	0.65	8.77
2011/12	1.25	0.04	2.91	2.03	0.69	1.21	0.65	8.77
2010/11	1.38	0.04	2.68	2.26	0.043	1.12	0.61	8.53
2009/10	1.47	0.04	2.72	2.45	0.24	1.18	0.65	8.76

Source MoPNG (2015b)

[2] This does not include the LNG imported by the Gujarat State Petroleum Corporation and Hazira LNG Pvt. Ltd.

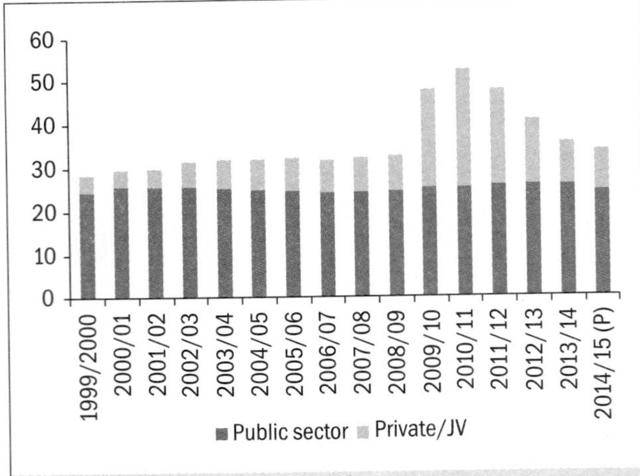

Figure 3 Natural gas production in India by sector (in BCM)
JV – joint venture; P – provisional
Source MoPNG (2015b)

(vii) Mobil Australia Resources Company (MARC), Australia's Gorgon project—1.44 MTPA for 20 years, delivering 1.44 MTPA for import of LNG. The supply is expected to start by 2016; signed by Petronet LNG (Petronet LNG 2015).

LNG terminals

Owing to declining domestic natural gas production, even if new discoveries are made and production increases, the gap between supply and potential demand will remain significant, and therefore larger quantities of natural gas will need to be imported. Thus, investment in the requisite infrastructure in the form of LNG terminals and pipeline networks is crucial. Currently, there are four commissioned terminals on the west coast of India: two in Gujarat, one in Maharashtra, and the recent one in Kochi (Table 18). To meet the rising demand, several more are under construction or in the planning stage (see Table 18).

Of the existing four terminals, Petronet LNG's terminal at Dahej is operated at more than 102.1% of its nameplate capacity (Petronet LNG 2015). The construction for its expansion to 15 MTPA is expected to be completed by 2016 and had achieved 56% progress till 31 March 2015. The company plans to expand this further to 17.50 MTPA. While the offtake of LNG has been lower due to the higher price of LNG from Qatar, the spot cargoes at Dahej have increased. The terminal at Kochi, commissioned in August 2013, is significantly underutilized (at 2.1% of its capacity in 2014/15) as Phase II of the pipeline network from the terminal is yet to be completed. Phase I has about 45 km of pipeline. From September 2014, Petronet LNG also started the supply of LNG by road tankers (small quantities) to industrial customers not connected by pipelines. So far is the supply only to a company in Trivandrum. The Hazira terminal operated at 60% of its capacity in 2014/15. Hazira LNG Pvt. Ltd, a joint venture (JV) between Royal Dutch Shell and Total SA,

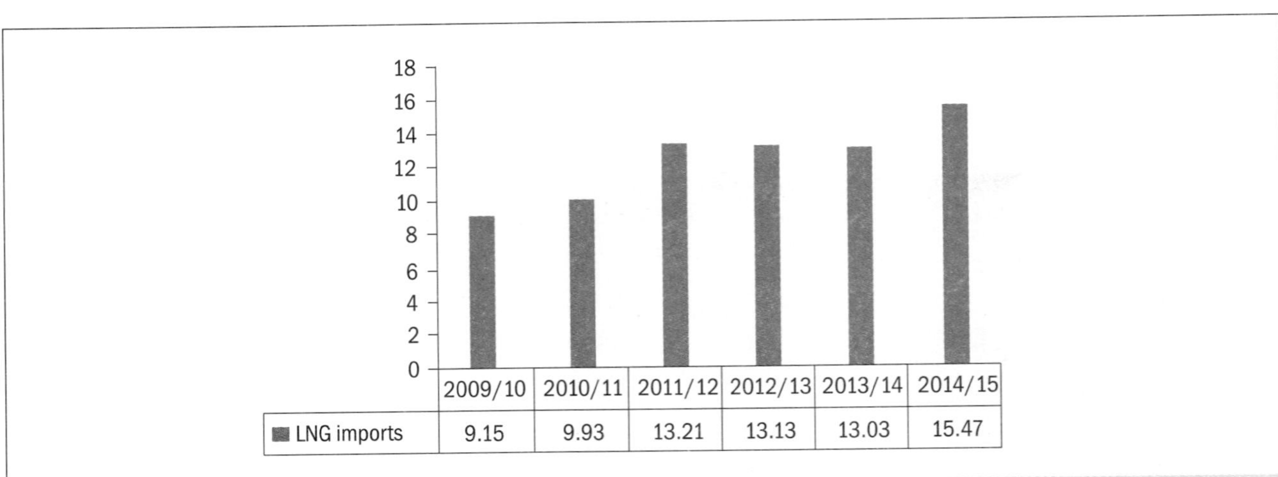

	2009/10	2010/11	2011/12	2012/13	2013/14	2014/15
LNG imports	9.15	9.93	13.21	13.13	13.03	15.47

Figure 4 Natural gas imports (in MT)
LNG – liquefied natural gas
Source MoPNG (2015b)

Table 18 Existing and planned LNG terminals in India

Project and developer	Location (state)	Capacity (MTPA)	Suppliers	Status
Dahej LNG terminal (Petronet LNG Ltd)	Dahej (Gujarat)	10 (operating at 11 MTPA; to be expanded to 15 MTPA by 2016; plans for further expansion to 17.50 MTPA)	RasGas (Qatar-based LNG supply company) and spot cargoes	Commissioned in February 2004 and commercial sales began in April 2004
Hazira LNG (Shell and Total)	Hazira (Gujarat)	5 (to be expanded to 7.5 MTPA by 2016/17)	Spot cargoes	Commissioned in April 2005
Dabhol terminal (Ratnagiri Gas and Power Pvt. Ltd)	Dabhol (Maharashtra)	5 (limited usage till breakwater facilities added by 2016/17)	Spot cargoes (long-term agreement for supply of LNG is yet to be finalized)	Commissioned in 2012
Kochi LNG (Petronet LNG Ltd)	Kochi (Kerala)	5	LNG sale and purchase agreement with Exxon Mobil Corporation for supply of approximately 1.5 MTPA of LNG from Gorgon LNG project, Australia	Commissioned in 2013
Ennore LNG (IOCL and CPCL)	Ennore (Tamil Nadu)	5	Yet to be finalized	Expected to be commissioned by 2016/17
Mundra LNG terminal (GSPC and Adani)	Mundra (Gujarat)	5	Yet to be finalized	Expected to be commissioned by 2016/17
Mangalore LNG (ONGC, BPCL, Mitsui)	Mangalore (Karnataka)	2-3	Yet to be finalized	Expected to be completed by 2018/19
APGDC-Kakinada FSRU terminal (APGIC, Shell, GDF Suez, and GAIL)	Kakinada (Andhra Pradesh)	3.5-5.0 (expandable to 10 MT)	Yet to be finalized	Expected to be completed by 2017/18
Gangavaram (Petronet LNG Ltd)	Gangavaram (Andhra Pradesh)	5	Yet to be finalized	Expected to be commissioned by 2016
HEECPL-Digha FSRU terminal	Digha region (West Bengal)	4	Yet to be finalized	Expected to be commissioned by 2017
Paradip FSRU Terminal (GAIL)	Odisha	4.0	Yet to be finalized	Expected to be commissioned by 2017
Pipavav port FSRU terminal (SEL)	Gujarat	2.5-5.0	Yet to be finalized	Planned
Mumbai port FSRU (bids invited)	Mumbai, Maharashtra	5	Yet to be finalized	Expected to be completed by 2018
Jaigarh LNG Terminal (H-Energy Gateway Pvt. Ltd)	Jaigarh, Maharashtra	8	Yet to be finalized	Expected to be completed by 2018/19

APGDC – Andhra Pradesh Gas Distribution Corporation; APGIC – Andhra Pradesh Gas Infrastructure Corporation; CPCL – Chennai Petroleum Corporation Ltd; FSRU – floating storage re-gasification unit; GAIL – Gas Authority of India Ltd; GSPC – Gujarat State Petroleum Corporation; HEECPL – H-Energy East Coast Pvt. Ltd; IOCL – Indian Oil Corporation Ltd; LNG – liquefied petroleum gas; MRPL – Mangalore Refinery and Petrochemicals Ltd; MT – million tonnes; MTPA – million tonnes per annum; ONGC – Oil and Natural Gas Corporation; SEL – Swan Energy Ltd

Sources TERI (2014a); The Hindu (2013a, 2013b); Business Standard (2013a, 2013b); Natural Gas Asia (2015); LNG World News (2015); Lok Sabha (2013a, 2013c, 2015a)

France, announced plans to expand the terminal from its current 5 MTPA to 7.5 MTPA by 2016/17 (Natural Gas Asia 2015). The Dabhol LNG terminal operated by Ratnagiri Gas and Power Pvt. Ltd, a JV between GAIL and ONGC, is currently underutilized (at 50% of its capacity) and will operate at full capacity once breakwater facilities are installed by 2016/17. If all the planned capacity additions listed in Table 18 come online, India's import capacity would rise to around 85 MTPA. Table 18 shows the existing and planned LNG terminals across the east and west coast of India. See Map 2.

Natural gas pipelines

India has 15 808 km of natural gas pipelines at present. The government has announced its intention to create a national gas grid by adding another 15 000 km of pipelines (this was announced in the Twelfth Five-year Plan and in the 2014 Union Budget speech). One of the main reasons why the Kochi terminal is underutilized is the delay in the completion of the second phase of the Kochi–Koottanad–Bangalore–Mangalore pipeline (Petronet LNG 2015). Media reports quoting company officials also suggest that plans for the Mangalore terminal have also been postponed owing to the delay in the pipeline connectivity (Natural Gas Asia 2015). On the other hand, the existing pipelines have functioned at a lower capacity owing to the shortage of domestic gas. Since India's imports of natural gas are likely to increase significantly (just the capacity addition of LNG terminals by 2016/17 gives an idea), it is vital that the pipelines connecting these terminals to major demand centres are built simultaneously to ensure minimum idle capacity. Also, scaling-up of the use of piped natural gas (PNG) and compressed natural gas (CNG) (see "Natural gas consumption" section for more details) will require an increased network of pipelines. However, this would also require greater clarity on the pricing of natural gas in India to avoid infructuous investments. Tables 19 and 20 provide details of the pipelines in existence and under construction, respectively.

Natural gas consumption

According to the Gas Utilization Policy introduced by the Government of India, the order of priority of supply of natural gas would be gas-based urea plants, followed by gas-based LPG plants, city gas distribution projects, and gas-based power plants (MoPNG 2008). These guidelines were modified to also include non-priority sectors, including steel plants, refineries, and petrochemical plants. The allocation to the non-core sectors was reduced owing to the falling gas production. However, city gas distribution projects were given priority following an MoPNG notification in February 2014, which stated that "the requirement of natural gas for CNG (transport) and PNG (domestic) shall be met by allocation of domestic gas to the maximum extent possible" (MoPNG 2014a). Table 21 gives the offtake of natural gas since 2010/11 by industry.

Natural gas pricing

One of the key characteristics of the natural gas sector in India is the presence of multiple natural gas prices. Different prices exist even after a new domestic gas pricing policy was approved on 18 October 2014 by the Cabinet Committee on Economic Affairs (CCEA 2014). However, the new gas pricing formula does provide greater clarity than the earlier pricing structure. The new gas pricing formula is as follows:

$$P = (VHH \times PHH + VAC \times PAC + VNBP \times PNBP + VR \times PR)/(VHH + VAC + VNBP + VR)$$

where

- VHH is the total annual volume of natural gas consumed in USA and Mexico;
- VAC is the total annual volume of natural gas consumed in Canada;
- VNBP is the total annual volume of natural gas consumed in EU and FSU, excluding Russia;
- VR is the total annual volume of natural gas consumed in Russia;
- PHH and PNBP are the annual average of daily prices at Henry Hub (HH) and National Balancing Point (NBP), respectively, less the transportation and treatment charges; and
- PAC and PR are the annual average of monthly prices at Alberta Hub and Russia, respectively, less the transportation and treatment charges.

The prices are determined on a half-yearly basis. The data used are of four quarters, with a lag of one quarter, and prices announced are on gross calorific value (GCV), rather than on net calorific value (NCV). The new pricing formula came into effect on 1 November 2014.

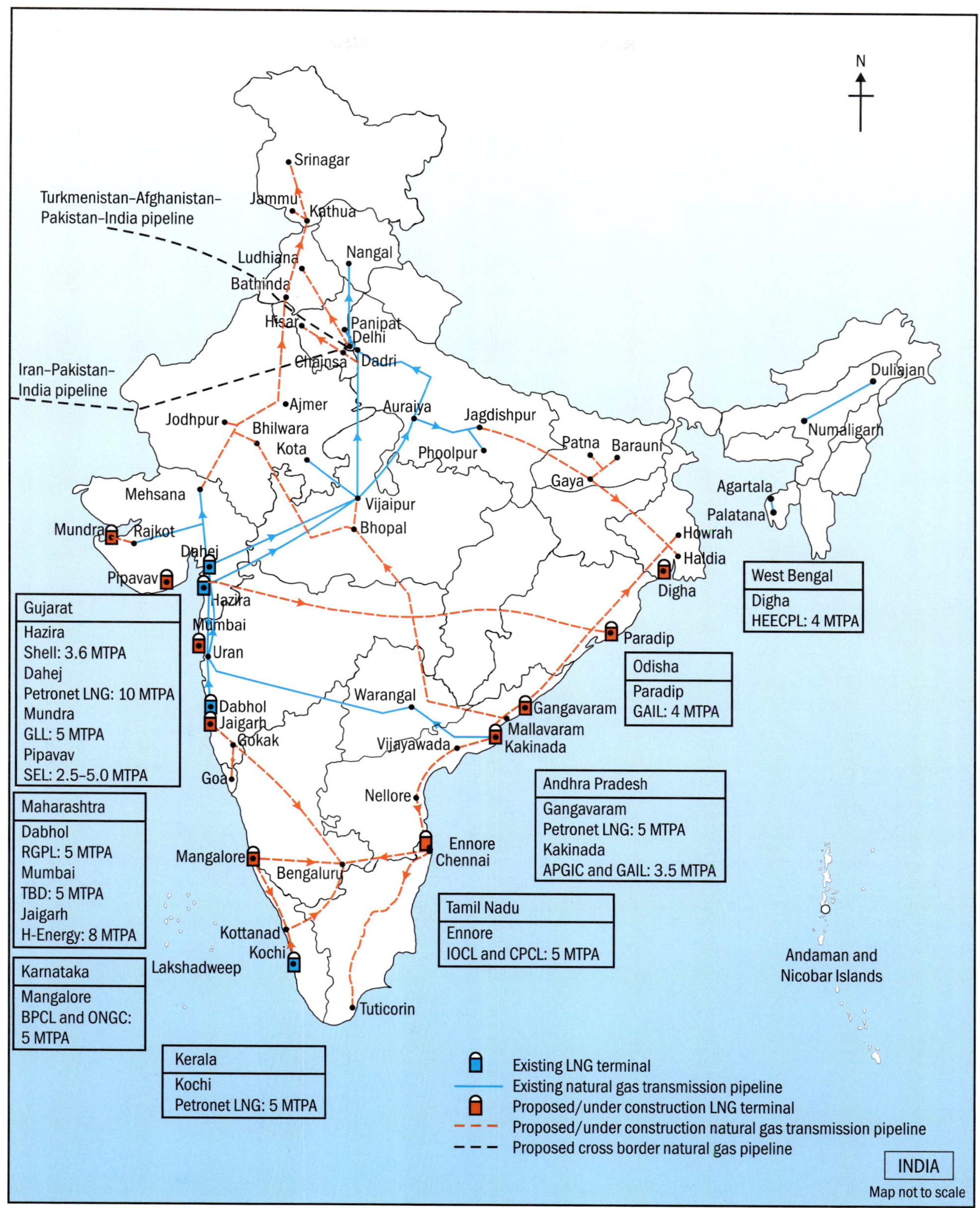

Map 2 LNG terminals and natural gas pipelines in India

Sources MoPNG (2015b); PPAC (2014); TERI (2014a); The Hindu (2013a, 2013b); Business Standard (2013a, 2013b); Natural Gas Asia (2015); LNG World News (2015); Lok Sabha (2013a, 2013c, 2015a)

Table 19 Details of existing natural gas pipelines

Network/Region	Entity	Length (km)	Design capacity (MMSCMD)	Per cent capacity utilization (as on 31 March 2015)
HVJ GREP-DVPL and Spur (Hazira-Vijaipur-Jagdishpur) HVJ/VDPL	GAIL	4 658	53.0	73.13
DVPL-GREP upgradation (DVPL-2 and VDPL)	GAIL	1 119	54	37.63
Chhainsa-Jhajjar-Hisar P/L (CJPL) (including Spur lines) commissioned up to Sultanpur, Jhajjar-Hisar under hold (111 km) Flow of 5 million up to 2011/12	GAIL	265	5	15.8
Dahej-Uran-Panvel (DUPL/DPPL), including Spur lines	GAIL	875	19.9	42.81
Dadri-Bawana-Nangal P/L (DBPL), Dadri-Bawana (106 km), Bawana-Nangal (501 km), Spur line of BNPL (196 km)	GAIL	834.8	31	13.2
Dabhol-Bangalore pipeline (including Spur)	GAIL	1 004	16	5.44
Kochi-Koottanad-Bangalore-Mangalore (Phase 1)	GAIL	41	6	6.50
Assam (Lakwa)	GAIL	8	2.5	20.8
Tripura (Agartala)	GAIL	61	2.3	64.2
Ahmedabad	GAIL	133	2.9	11.0
Rajasthan (Focus Energy)	GAIL	152	2.35	55.74
Bharuch, Badodara (Undera) included RLNG + RIL	GAIL	538	15.4	9.5
Mumbai	GAIL	129	7.0	128.1
KG Basin (included RLNG + RIL)	GAIL	881	16.0	21.6
Cauvery Basin	GAIL	278	8.7	37.52
East-West pipeline (RGTIL)	Reliance	1 469	80.0	26
GSPCL Network, including Spur lines	GSPCL	2 197.7	43.0	53.488
Assam Regional Network	AGCL, DNPL	1 000	6.0	75
Dadri-Panipat	IOCL	140.4	9.5	39.0
Uran Trombay	ONGC	24	6.0	85
Total		15 808	387	

AGCL – Assam Gas Company Ltd; CJPL – Chhainsa-Jhajjar-Hissar Pipeline; DBPL – Dadri-Bawana-Nangal Pipeline; DNPL – Duliajan-Numaligarh Pipeline Ltd; DPPL – Dadri-Panipat Natural Gas Pipeline; DUPL – Dahej-Uran-Panvel Pipeline; DVPL – Dahej-Vijaipur Pipeline; GAIL – Gas Authority of India Ltd; GSPCL – Gujarat State Petroleum Corporation Ltd; HVJ – Hazira-Vijaipur-Jagdishpur; IOCL – Indian Oil Corporation Ltd; MMSCMD – million metric standard cubic metre per day; ONGC – Oil and Natural Gas Corporation; RGTIL – Reliance Gas Transportation Infrastructure Ltd; RIL – Reliance Industries Ltd; RLNG – re-gasified liquefied natural gas; VDPL – Vijaipur-Dadri Pipeline

Source PPAC (2014)

Table 20 Gas pipelines under construction

Network/Region	Entity	Length (km)	Design capacity (MMSCMD)	Status of pipeline laid (as on 31 March 2015)
Kochi-Koottanad-Bangalore-Mangalore	GAIL	1063	16	0
Dabhol-Bangalore (DBPL)	GAIL	410	16	0
Surat-Paradip	GAIL	2112	75	0
Jagdishpur-Haldia	GAIL	1860	32	0
Mallavaram-Bhilwada	GITL	2042	76.25	0
Mehsana-Bathinda	GIGL	2052	77.11	0
Bathinda-Srinagar	GIGL	725	42	0
Kakinada-Srikakulam	APGDC	391	90	0
Shadol-Phulpur	RGPL	312	3.5	0
Ennore-Nellore	KEI-RSOS PEPL	430	36	0
Subtotal		11397		0

APGDC – Andhra Pradesh Gas Distribution Corporation; DBPL – Dadri-Bawana-Nangal Pipeline; GAIL – Gas Authority of India Ltd; GIGL – GSPL India Gasnet Ltd; GITL – GSPL India Transco Ltd; MMSCMD – million metric standard cubic metre per day; RGPL – Reliance Gas Pipeline Ltd

Source PPAC (2014)

Table 21 Offtake of natural gas by industry (in MCM)

	2010/11	2011/12	2012/13	2013/14	2014/15
1. Energy use					
Power	25787	22628	16078	11284	10720
Industry	–	284	269	156	395
Manufacturing	–	30	0	105	138
Road transport	–	0	0	66	0
City gas distribution	3427	5599	5780	5838	5415
Tea plantation	193	175	182	196	188
LPG shrinkage	–	1068	1027	982	1005
International Conference for pipeline system	–	385	387	372	351
Miscellaneous	7546	9064	7976	7479	5934
Total (1)	36953	39233	31699	26477	24146
2. Non-energy use					
Fertilizer	11464	14003	14733	15869	15190
Petrochemical	1309	1858	2486	2405	2890
Refinery	1529	4257	3891	3968	4575
Sponge iron	765	1333	1106	274	154
Total (2)	15066	21451	22216	22517	22809
Grand total (1 + 2)	52019	60684	53915	48994	46955

Source MoPNG (2015b)

This new price applies to all gas produced from "nomination fields" given to ONGC and OIL, all New Exploration Licensing Policy (NELP) blocks, and pre-NELP blocks where PSC provides government approval of gas prices. It also applies to CBM blocks. This price, however, does not apply to (a) small and isolated nomination blocks, (b) blocks where the prices are fixed contractually for a fixed period (although it will apply here after the end of such a contractual period), (c) blocks where the PSC provides a specific formula for natural gas price indexation or fixation, and (d) pre-NELP blocks where no government approval is needed under the PSC.

Further, because of the arbitration on KG-DWN-98-3 block (popularly known as KG-D6), the operator (RIL-BP-NIKO) would continue to pay the earlier price of $4.2/Million Metric British Thermal Unit (MMBTU) and the balance between the new price and this older price would be credited to a gas pool account maintained by GAIL.

However, producers of deep water, ultra-deep water, and high-pressure, high-temperature blocks have requested the government for market-determined prices for gas produced from such blocks owing to higher costs, and higher risks involved. Recognizing these risks, the government had decided that a premium would be given to the gas price determined by the formula. This was also done to incentivize further exploration activities. The Directorate General of Hydrocarbons (DGH) has appointed a consultant to estimate the cost of extraction of gas from such areas to help determine the premium (Lok Sabha 2015b). However, the pricing structure for gas from such wells had not been notified when TEDDY 2015/16 went into publication.

According to the formula, the following are the domestic gas prices in India (PPAC 2015a):

- November 2014 to March 2015: $5.05/MMBTU on GCV basis
- April 2015 to September 2015: $4.66/MMBTU on GCV basis
- October 2015 to March 2016: $3.82/MMBTU on GCV basis

The ministry has also announced pooling of prices for the fertilizer sector, which came into effect on 1 July 2015. The domestic gas will be pooled with re-gasified liquefied natural gas (RLNG) to provide natural gas at a uniform delivered price to all natural-gas-grid-connected urea manufacturing plants for the purpose of manufacturing urea (MoPNG 2015a). To monitor the pool, the Empowered Pool Management Committee (EPMC), comprising senior bureaucrats from MoPNG, Department of Fertilizers, Department of Expenditure, Petroleum Planning and Analysis Cell (PPAC), Fertilizer Industry Coordination Committee (FICC), and GAIL (India) Ltd, is constituted. GAIL has been designated as the pool operator.

Among other responsibilities, the EPMC is mandated to (a) approve plant-wise gas supplies made under the pool mechanism, (b) approve LNG purchase agreements required through medium-term/spot LNG, and (c) monitor both optimum utilization of domestic gas for the pool and operator's activities to ensure transparency.

The operator (GAIL), on the other hand, will (a) collect data regarding anticipated gas supplies to fertilizer units as per the existing contract on a quarterly basis, (b) determine the additional quantity of RLNG required for meeting the demand, (c) determine plant-wise and uniform, weighted-average delivered cost of gas based on the anticipated supply, and (d) procure LNG in a transparent manner.

On the first of every month, the operator is mandated to declare the uniform delivered pool price of gas for the fertilizer (urea) sector for the month.

This pooling, as per the notification of the ministry, will be carried out in two phases. Phase I, from 2015/16 to 2017/18, is for existing fertilizer units along with conversion units, which will be supplied gas when pipeline connectivity is established. Phase II, from 2018/19 onwards, would be for considering requirements of existing units, as well as proposed Brownfield and Greenfield units.

Unconventional fossil fuels

Coalbed methane

Coalbed methane is natural gas extracted from coalbeds instead of conventional reservoirs (Figure 5). In 2014/15, 228 million metric standard cubic metre per day (mmscmd) of CBM was produced in India, up from 162 mmscm in 2013/14

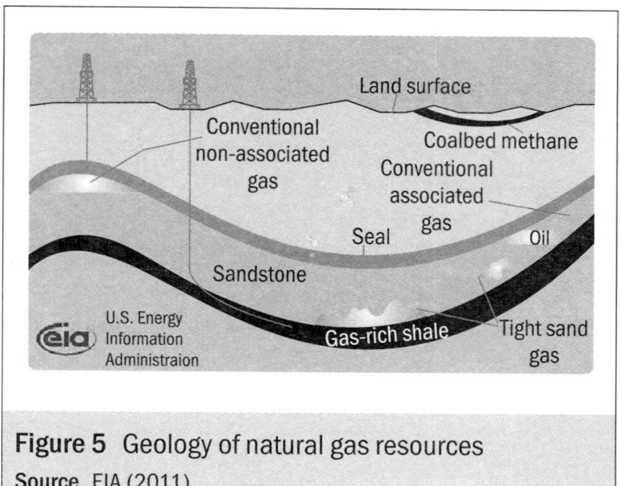

Figure 5 Geology of natural gas resources
Source EIA (2011)

(Table 22). Of this, 224 mmscm was produced in West Bengal's Raniganj South and East, Sohagpur East and West, and Jharia. The companies operating these blocks are Great Eastern Energy Corporation Ltd (GEECL), ESSAR, ONGC, and RIL. The two Raniganj blocks operated by GEECL and ESSAR are the most productive of all CBM blocks. The erstwhile Planning Commission (2011) has projected CBM production to be 1256 mmscm in 2016/17. However, CBM production in the past did not come up to its projected potential. Note that CBM production is included in the data referred in the "Domestic natural gas production" section.

Shale gas

In October 2013, the government announced policy guidelines for the development of shale oil and gas in India. Under this, 55 licenses (50 to ONGC and 4 to OIL) were given to NOCs for the first phase. So far, ONGC has drilled one well, spudded one, and collected the cores from eight other wells. In all, ₹79 crore had been spent on the exploration of shale oil and gas as on 31 December 2014. Exploration efforts of OIL, on the other hand, are at a nascent stage (Rajya Sabha 2015c). The production data from these wells have not yet been generated (Rajya Sabha 2015a).

There are varying estimates of shale gas reserves in India. The Energy Information Administration has reported 584 TCF of gas reserves and 87 billion barrels of shale oil in four basins. Schlumberger estimated 300–2100 TCF shale gas resources, while ONGC has estimated 187.5 (TCF) in five basins (Rajya Sabha 2015b).

Table 22 Coalbed methane production in India (in mmscm)

State	2008/09	2009/10	2010/11	2011/12	2012/13	2013/14	2014/15
West Bengal	20	38	41	79	101	156	224
Madhya Pradesh	0	0	0	2	3	6	2
Jharkhand	0	0	0	4	3	3	2
Total	20	38	41	85	107	162	228

Source MoPNG (2015b)

Renewable energy in fuel retail

The oil and gas industry is not only a key supplier of energy, but also a significant consumer of energy. The various stages of the oil and gas supply chain–from extraction to transformation and retail–are all energy intense activities. Over the past few years, the petroleum industry has been adopting several measures to improve energy efficiency and energy from cleaner sources in order to reduce emissions. One such avenue has been the use of solar energy in retail and marketing operations. This replaces the use of diesel gensets and grid-connected electricity.

In India, till March 2015, there were 53 419 petrol and diesel retail outlets. IOCL is India's largest fuel retailer, with 24 405 retail outlets, followed by HPCL with 13 233 outlets and BPCL with 12 809 outlets. The remaining companies have fewer than 4000 outlets (MoPNG 2015b). To lower greenhouse gas emission from retail operations, IOCL has ensured that more than 2660 of its fuel stations run on solar energy. Their cumulative capacity is about 8.6 MW, with an annual generation capacity of 103 GWh. This has carbon emission reduction potential of 8500 tonnes per year. Further, the company has 69.3 MW of wind capacity for other operations. IOCL intends to add an additional 200 MW of wind and 60 MW of solar power systems (IOCL 2015).

Hindustan Petroleum Corporation Ltd too has been adopting renewable energy to fuel its operations. It has a 75 kW solar installation at its R&D centre. Its overall reported renewable energy consumption in 2013/14 was 134 GJ. The company has also a plan to install renewable energy plants to fuel its retail operations (HPCL 2014b). Finally, BPCL had a renewable energy investment plan of ₹70 crore for setting up 5 MW of solar and 5 MW of wind capacity in 2014/15. It has also been installing solar solar photovoltaic (PV) panels at its R&D centres and refineries. The company has an installed capacity of 5.5 MW of wind and 0.3 MW of solar. In all, 76 of BPCL retail outlets have been solarized (BPCL 2015).

When it comes to natural gas, GAIL (India) Ltd, which is India's largest gas-marketing company, had a total of 118 MW of wind and 5 MW of solar capacity in 2013/14. This generation, however, is grid connected, meaning it does not necessarily fuel its own internal gas -marketing operations. The company is in the process of replacing natural gas-based generators with solar PV installations, which according to its estimates saves 1638 tonnes of CO_2 equivalent per year. The company's main offices and other operations too are increasingly being powered by solar energy. In all, the company generated 464 606 GJ of renewable energy in 2013/14. The company's renewable energy portfolio stood at ₹700 crore in that fiscal year (GAIL 2014).

References

BG Group. 2013. BG Group completes agreement on long-term LNG sales to India. Details available at http://www.bg-group.com/292/where-we-work/india/, last accessed on 10 December 2014

BG Group. 2015. ***Global LNG Market Outlook 2014–15***. UK: BG Group.

BPCL (Bharat Petroleum Corporation Ltd). 2015. ***Sustainable Development Report 2014–15***. Mumbai: BPCL

Business Standard. 2013a. Gail to set up Rs. 3108 cr LNG terminal at Paradip. Details available at http://www.business-standard.com/article/companies/gail-to-set-up-rs-3-108-cr-lng-terminal-at-paradip-113102600351_1.htm, last accessed on 10 December 2014

Business Standard. 2013b. Swan Energy gets green nod for LNG terminal at Pipavav. Details available at http://www.business-standard.com/article/companies/swan-energy-gets-green-nod-for-lng-terminal-at-pipavav-113121600601_1.htm, last accessed on 10 December 2014

CCEA (Cabinet Committee on Economic Affairs). 2014. Revision of domestic gas prices. Details available at http://pib.nic.in/newsite/PrintRelease.aspx?relid=110696, last accessed on 25 November 2014

EIA (Energy Information Administration). 2014a. India. Details available at http://www.eia.gov/countries/cab.cfm?fips=in, last accessed on 30 November 2014

EIA (Energy Information Administration). 2014b. India overview. Details available at http://www.eia.gov/countries/cab.cfm?fips=in, last accessed on 30 November 2014

GAIL (Gas Authority of India Ltd). 2014. ***Sustainability Report 2013–14***. New Delhi: GAIL

HPCL (Hindustan Petroleum Corporation Ltd). 2014b. ***Sustainability Report 2013–14***. Mumbai: HPCL

IGU (International Gas Union). 2015. ***World LNG Report 2015***. Oslo, Norway: IGU

IOCL (Indian Oil Corporation Ltd). 2015. ***Sustainability Report 2014–15***. New Delhi: IOCL

LNG World News. 2015. ***LNG World News: The Industry's LNG News Terminal***. Details available at www.lngworldnews.com/four-companies-bid-to-build-fsru-in-mumbai-port/, last accessed on 11 November 2015

Lok Sabha. 2013a. Unstarred Question No. 2515 (MoPNG). Details available at http://164.100.47.132/lssnew/psearch/QResult15.aspx?qref=144643, last accessed on 28 November 2014

Lok Sabha. 2013c. Starred Question No. 521 (MoPNG). Details available at http://164.100.47.132/LssNew/psearch/QResult15.aspx?qref=140700, last accessed on 2 October 2014

Lok Sabha. 2014. Lok Sabha Questions. Details available at http://164.100.47.132/LssNew/psearch/QResult16.aspx?qref=1963, last accessed on 11 November 2015

Lok Sabha. 2015a. Lok Sabha Questions. Details available at http://164.100.47.132/LssNew/psearch/QResult16.aspx?qref=23886, last accessed on 11 November 2015

Lok Sabha. 2015b. Lok Sabha Questions. Details available at http://164.100.47.132/LssNew/psearch/QResult16.aspx?qref=19822, last accessed on 11 November 2015

MoPNG (Ministry of Petroleum and Natural Gas). 2008. Press note: Decision regarding sale of natural gas by NELP contractors and order of priority to supply gas from RIL's KG-D6 field. Details available at Ministry of Petroleum and Natural Gas, last accessed on 30 November 2014

MoPNG (Ministry of Petroleum and Natural Gas). 2014a. Pricing of CNG (transport) and ONG (domestic) by CGD entities. Details available at http://petroleum.nic.in/docs/gp/Revised%20guidelines%20for%20allocation%20supply%20of%20domestic%20natrural%20gas%20to%20CGD%20entities%20for%20CNG%20(Tpt)%20&%20PNG(Domestic)%20dt14.11.2013.pdf, last accessed on 30 October 2015

MoPNG (Ministry of Petroleum and Natural Gas). 2014c. ***Indian Petroleum and Natural Gas Statistics 2013–14***. New Delhi: MoPNG

MoPNG (Ministry of Petroleum and Natural Gas). 2015a. ***Guidelines for Pooling of Gas in Fertilizer (Urea) Sector***. New Delhi: MoPNG

MoPNG (Ministry of Petroleum and Natural Gas). 2015b. ***Petroleum and Natural Gas Statistics 2014–15***. New Delhi: MoPNG

Natural Gas Asia. 2015. Natural Gas Asia. Details available at http://www.naturalgasasia.com/ongc-bpcl-mitsui-delay-mangalore-lng-re-gasification-terminal-project-16652, last accessed on 11 November 2015

Petronet LNG. 2015. ***Annual Report 2014–15***. New Delhi: Petronet LNG

PIB (Press Information Bureau). 2014a. ***Mechanism for Procurement of Ethanol by Public Sector Oil Marketing Companies (OMCs) to Carry out the Ethanol Blended Petrol Programme***. New Delhi: PIB

PIB (Press Information Bureau). 2015a. ***Allocation of ₹4948 Crore for Gross Budgetary Support to the Indian Strategic Storage Programme for Crude Oil***. New Delhi: PIB

PIB (Press Information Bureau). 2015b. ***Blending of Ethanol: Ministry of Petroleum and Natural Gas***. New Delhi: PIB

PIB (Press Information Bureau). 2015c. ***Building Strategic Reserve of Crude Oil***. New Delhi: PIB

PIB (Press Information Bureau). 2015d. ***From Carbon Subsidy to Carbon Tax – India's Green Actions***. New Delhi: PIB

PIB (Press Information Bureau). 2015e. ***Incentives Given to Sugar Sector to Facilitate Payment of Cane Dues to Farmers: Ministry of Consumer Affairs, Food and Public Distribution***. New Delhi: PIB

PIB (Press Information Bureau). 2015f. ***Installation of Bio-fuel Pumps: Ministry of Road Transport and Highways***. New Delhi: PIB

PIB (Press Information Bureau). 2015g. ***PM at Urja Sangam 2015: Prime Minister's Office***. New Delhi: PIB

PIB (Press Information Bureau). 2015h. ***States urged to Liberalize Restrictions on Ethanol Supply: Ministry of Consumer Affairs, Food and Public Distribution***. New Delhi: PIB

PIB (Press Information Bureau). 2015i. ***Steps for Implementation of Bio-fuel Programme: Ministry of Petroleum and Natural Gas***. New Delhi: PIB

Planning Commission. 2011. ***Report of the Working Group on Petroleum and Natural Gas Sector for the 12th Five-year Plan (2012-17)***. New Delhi: Planning Commission

PPAC (Petroleum Planning and Analysis Cell). 2014. Pipeline and CGD structure. Details available at http://ppac.org.in/content/154_1_PipelineandCGDStructure.aspx, last accessed on 10 December 2014

PPAC (Petroleum Planning and Analysis Cell). 2015a. ***Natural Gas Prices***. New Delhi: PPAC

PPAC (Petroleum Planning and Analysis Cell). 2015b. ***Ready Reckoner: Oil Industry at a Glance***. New Delhi: PPAC

Rajya Sabha. 2015a. ***Starred Question Number 253***. New Delhi: Rajya Sabha

Rajya Sabha. 2015b. ***Unstarred Question Number 1081***. New Delhi: Rajya Sabha

Rajya Sabha. 2015c. ***Unstarred Question Number 1089***. New Delhi: Rajya Sabha

Rajya Sabha. 2015d. ***Unstarred Question Number 1162***. New Delhi: Rajya Sabha

Rajya Sabha. 2015e. ***Unstarred Question Number 1165***. New Delhi: Rajya Sabha

Rajya Sabha. 2015f. ***Unstarred Question Number 273***. New Delhi: Rajya Sabha

Rajya Sabha. 2015g. ***Unstarred Question Number 715***. New Delhi: Rajya Sabha

RBI (Reserve Bank of India). 2015. Table 131: Imports of Principal Commodities – Rupees. Details available at https://rbi.org.in/Scripts/PublicationsView.aspx?id=16572

TERI (The Energy and Resources Institute). 2014a. ***TERI Energy & Environment Data Directory and Yearbook 2013/14***. New Delhi: TERI

TERI (The Energy and Resources Institute). 2014b. ***Evaluation of the Pilot Project on Direct Transfer of Kerosene Subsidies in Kotkasim, Alwar***. New Delhi: Global Subsidies Initiative of the International Institute of Sustainable Development

The Financial Express. 2015. ***Sop on Kerosene to End in 3 years***. New Delhi: The Financial Express

The Hindu. 2013a. LNG terminal to be commissioned by 2016: Saurabh Patel. Details available at http://www.thehindu.com/business/Industry/lng-terminal-to-be-commissioned-by-2016-saurabh-patel/article4912029.ece, last accessed on 20 November 2014

The Hindu. 2013b. Dabhol LNG import terminal gets shipment after six-month gap. Details available at http://www.thehindu.com/business/Industry/dabhol-lng-import-terminal-gets-shipment-after-sixmonth-gap/article5389753.ece, last accessed on 10 December 2014

The Hindu. 2014. ONGC scripts turnaround, plans to ramp up oil output by 23%. Details available at http://www.thehindu.com/business/Industry/ongc-scripts-turnaround-plans-to-ramp-up-oil-output-by-23/article6541342.ece, last accessed on 30 November 2014

Bibliography

BPCL (Bharat Petroleum Corporation Ltd). 2014. ***Annual Report 2013–14***. Mumbai: BPCL

Cairn India. 2008. L&T to build India's first heated pipeline for crude. Details available at https://www.cairnindia.com/sites/default/files/press_releases/LnT21.02.08_0.pdf, last accessed on 10 November 2014

DGH (Directorate General of Hydrocarbons). 2013. ***Annual Report 2012–13***. Noida: DGH

DGH (Directorate General of Hydrocarbons). (n.d.). Gas hydrate. Details available at http://www.dghindia.org/NonConventionalEnergy.aspx?tab=0, last accessed on 20 November 2014

Elvidge, C. D., K. E. Baugh, B. T. Tuttle, A. T. Howard, D. W. Pack, C. Milesi, and E. H. Erwin. 2007. ***A Twelve-year Record of National and Global Gas Flaring Volumes Estimated Using Satellite Data: Final Report to the World Bank***. Washington, DC: World Bank

Forbes. 2014. Key trends impacting global refining margins. Details available at http://www.forbes.com/sites/greatspeculations/2014/03/03/key-trends-impacting-global-refining-margins/, last accessed on 30 November 2014

GE Energy. 2010. Flare gas reduction: recent global trends and policy considerations. Details available at http://www.ge-spark.com/spark/resources/whitepapers/Flare_Gas_Reduction.pdf, last accessed on 20 November 2014

ICRA. 2012. HPCL-Mittal Pipelines Ltd. Details available at http://www.icra.in/Files/Reports/Rationale/HPCL-Mittal%20Pipelines%20Lim-r-07062012.pdf; last accessed on 10 November 2014

IOCL (Indian Oil Corporation Ltd). 2014. ***Annual Report 2013–14***. New Delhi: IOCL

Jayaswal, R. 2013. GAIL India struggles to find buyers for imported LNG. Details available at http://articles.economictimes.indiatimes.com/2013-05-13/news/39228860_1_bc-tripathi-gorgon-lng-d6-gas; last accessed on 25 November 2014

MRPL (Mangalore Refinery and Petrochemicals Ltd). 2014. ***Annual Report 2013–14***. New Delhi: MRPL

Mukherjee, K. and R. Panandiker. 2014. ***LNG—Global Challenges and Opportunities and Imperatives for India***. New Delhi: BCG, ONGC

NRL (Numaligarh Refinery Ltd). 2014. ***Annual Report 2013–14***. Numaligarh: NRL

ONGC (Oil and Natural Gas Corporation). 2013. Sustainability Report 2012-13. Details available at http://www.ongcindia.com/wps/wcm/connect/23a4b0d0-50b6-423f-b48d-8cb0300d1590/SustainabilityReport_13.pdf?MOD=AJPERES&CACHEID=23a4b0d0-50b6-423f-b48d-8cb0300d1590, last accessed on 20 November 2014

Petronet LNG. 2014. ***Annual Report 2013–14***. New Delhi: Petronet LNG

PNGRB (Petroleum and Natural Gas Regulatory Board). 2013. ***"Vision 2030"—Natural Gas Infrastructure in India***. New Delhi: PNGRB

PPAC (Petroleum Planning and Analysis Cell). 2014a. ***Production of Petroleum Products***. New Delhi: PPAC

PPAC (Petroleum Planning and Analysis Cell). 2014b. ***Consumption of Petroleum Products***. New Delhi: PPAC

PPAC (Petroleum Planning and Analysis Cell). 2014c. ***Import/Export of Crude Oil and Petroleum Products***. New Delhi: PPAC

PPAC (Petroleum Planning and Analysis Cell). 2014d. Petroleum subsidies. Details available at http://ppac.org.in/content/150_1_Subsidy.aspx, last accessed on 20 November 2014

PPAC (Petroleum Planning and Analysis Cell). 2014e. ***Natural Gas Imports***. New Delhi: PPAC

PPAC (Petroleum Planning and Analysis Cell). 2014f. Domestic gas price November, 2014–March, 2015. Details available at http://ppac.org.in/WriteReadData/userfiles/file/NG_Price%20Notification%20November14-March2015.pdf, last accessed on 25 November 2014 from PPAC

Rajya Sabha. 2014. ***Starred Question No. 132 Answered on 16-07-2014***. New Delhi: Rajya Sabha

Rao, V. 2014. ***An Overview of Indian Shale Gas Potential and G&G Techniques to Delineate Potential Layers. Infraline Energy's 2nd Annual Conference on "Shale Gas: Critical Technologies for Successful Projects***. New Delhi: Infraline

RBI (Reserve Bank of India). 2014. Table 132: Imports of Principal Commodities – Rupees. Details available at http://www.rbi.org.in/scripts/PublicationsView.aspx?id=15922, last accessed on 5 November 2014

Reuters. 2014. Table: Gross refining margins of Indian state-run, private companies. Details available at http://in.reuters.com/article/2014/09/25/india-diesel-idINL3N0RO2TA20140925, last accessed on 30 November 2014

RIL (Reliance Industries Ltd). 2014. ***Annual Report 2013–14***. Mumbai: RIL

World Bank. 2004. Regulation of associated gas flaring and venting: A global overview and lessons from international experience. Details available at http://www-wds.worldbank.org/external/default/WDSContentServer/WDSP/IB/2004/07/16/000012009_20040716133951/Rendered/PDF/295540Regulati1aring0no10301public1.pdfp, last accessed on 20 November 2014

POWER 5

AT A GLANCE

- The installed generation capacity of power utilities in India as on 31 March 2015 was 271 GW and the gross generation in 2014/15 was 1043 billion units.
- Coal accounted for more than 67% of the total capacity addition during 2014/15.
- During 2014/15, the energy shortage in the country was 3.6%, while the peak deficit was 4.7%.
- The total volume of short-term transactions of electricity as percentage of total electricity generation was around 9% in 2014/15.
- The all-India transmission and distribution losses remained more or less constant, with losses of 23% during 2012/13 compared to 23.7% in 2011/12.

Introduction

The installed capacity in India (utilities) stood at 271 GW in March 2015. Electricity generation during 2014/15 was 1043 billion units (BU). The unrestricted demand during this period was 148 GW (peak) and 1068 BU (energy). The country faced power shortages of varying degrees in different parts. In the meantime, efforts are underway to augment supply, increase the share of renewables, and improve the demand–supply chain with a view to realize the objective of 24/7 supply. A number of technical interventions and policy and regulatory initiatives are also being taken with added focus on improved governance, especially in the distribution sector.

During 2014/15, the power sector in India witnessed a capacity addition of 28 693 MW. The installed capacity in India continues to be dominated by thermal power, which has a share of 61% in the total installed capacity. With a major share in the installed capacity, thermal resources continue to contribute the largest share (84%) in electricity generation (excluding renewables). Coal continues to be the mainstay of total generation, accounting for nearly 87% of the total thermal installed capacity in the country. The electrical energy requirement is anticipated to grow at the rate of 8.5% during the Twelfth Five-year Plan (CEA 2013).

Generation

Installed capacity

During 2014/15, the total installed generating capacity of utilities in India increased from 243 GW (as on 31 March 2014) to 271 GW (as on 31 March 2015), representing an increase of around 11%. At present, the installed capacity comprises 188.8 GW of thermal power (coal, gas, and diesel), 41.2 GW of hydropower, 5.7 GW of nuclear power, and 35.7 GW of renewable power. The fuel mix of installed capacity did not significantly change in 2014/15 compared to 2013/14 (Figure 1).

Of the total installed capacity, continues to contribute the maximum share the private sector (38%), followed by the state sector (35%) and the central sector (27%). The participation of the private sector in power has been increasing rapidly over the past few years. Figure 2 shows the comparison of installed capacity by sector in 2013/14 and 2014/15.

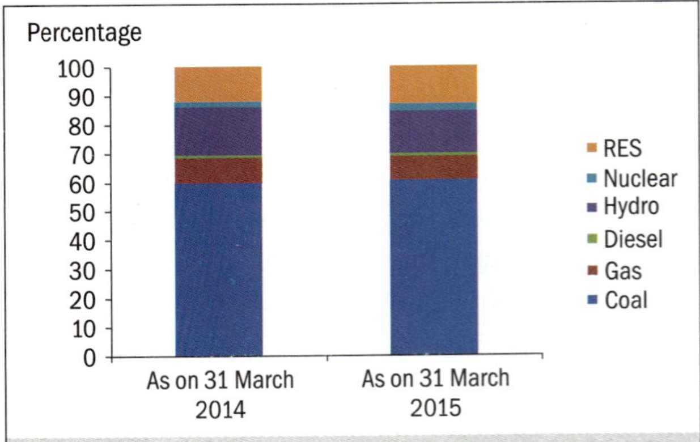

Figure 1 Installed capacity by source (as on 31 March 2015)
RES – renewable energy sources
Source CEA (2015a)

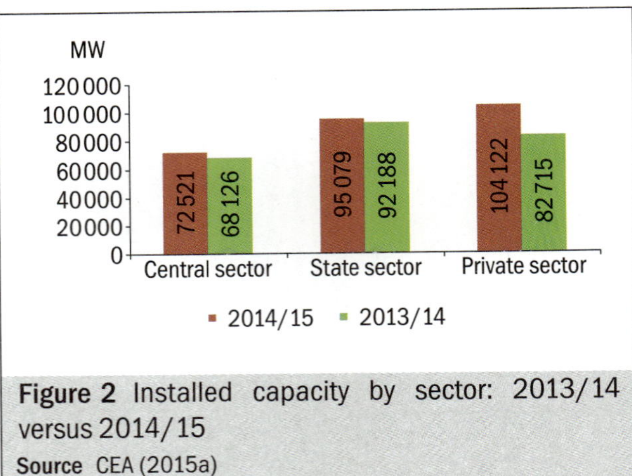

Figure 2 Installed capacity by sector: 2013/14 versus 2014/15
Source CEA (2015a)

Electricity generation

Total electricity generation during 2014/15 was around 1043 BU (excluding renewables). Thermal generation continues to be the mainstay of the total power generation in the country, contributing more than 80% of the total power generation for many years. During 2014/15, thermal power plants contributed nearly 84% (878 BU) of the total power generated in the country. Hydropower (129 BU) and nuclear power (36 BU) constituted the remaining share in electricity generation. Figure 3 shows the trend in electricity generation by source for the last 7 years.

Generating stations in the central sector contributed the maximum share (38%) of the total electricity generation. State, independent power producers (IPPs), and the private sector contributed 35%, 25%, and 2%, respectively (Figure 4).

Captive generation

The installed capacity of captive generation in the country has been gradually increasing over the past few years. The all-India installed capacity of captive plants (CPPs) of 1 MW capacity and above increased to 40 726 MW in 2012/13 from 39 375 MW in 2011/12.

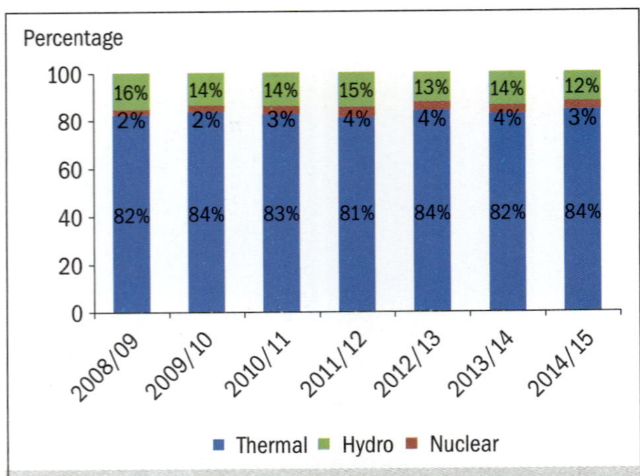

Figure 3 Trend in electricity generation by source (from 2008/09 to 2014/15)
Source CEA (2015b)

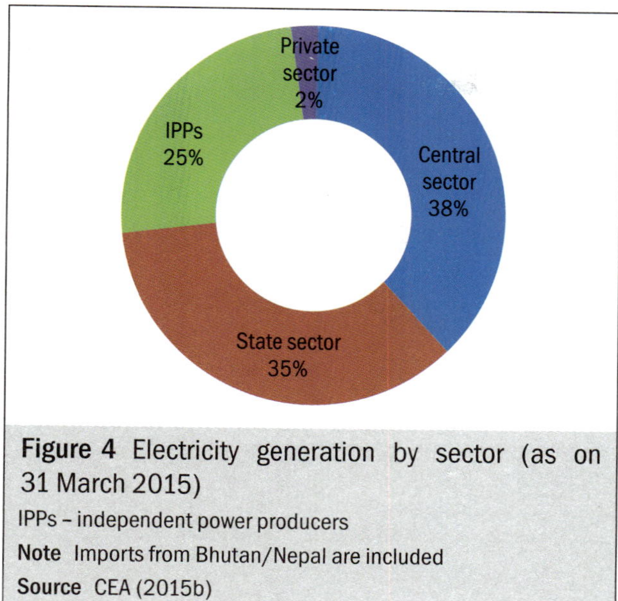

Figure 4 Electricity generation by sector (as on 31 March 2015)

IPPs – independent power producers

Note Imports from Bhutan/Nepal are included

Source CEA (2015b)

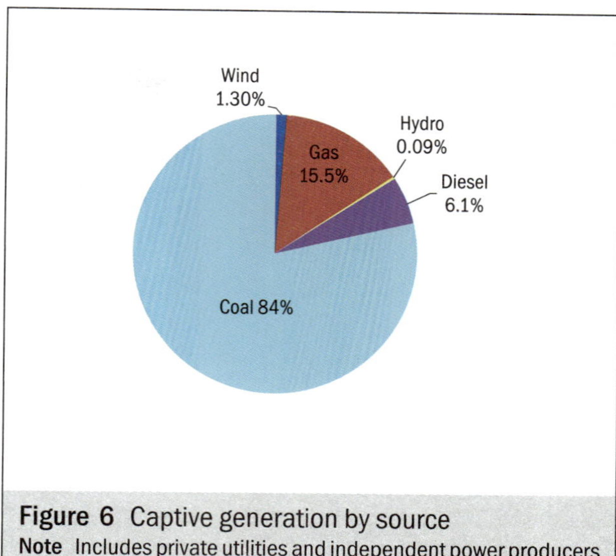

Figure 6 Captive generation by source

Note Includes private utilities and independent power producers

Source CEA (2014a)

Coal and diesel sources continue to dominate the installed capacity in captive generation, accounting for 58.7% and 27.4%, respectively, of the total installed capacity of CPPs in the country. Figure 5 shows the comparison of the installed capacity of captive generation by source during 2011/12 and 2012/13.

The electricity produced from captive generating units during 2012/13 was 144 BU as against 134 BU in 2011/12, representing a growth of around 7%. The prominent source of generation in CPPs continues to be coal, accounting for 84% of the total generation. Figure 6 shows the captive generation by source.

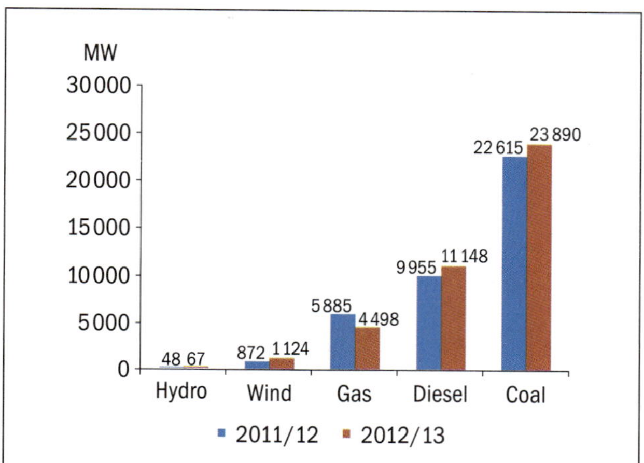

Figure 5 Installed capacity of captive generation by source: 2011/12 versus 2012/13

Source CEA (2014a)

During 2012/13, the dominant industry that used energy from CPPs in the country was iron and steel, which utilized 22% of the total, followed by aluminium (16%).

Plant load factor

The plant load factor (PLF) improved by 1.09% in 2014/15 (65%) compared to 2013/14 (64%). All the sectors have seen an improvement in the PLF of power stations. PLF improved by 0.75%, 2.5%, and 3.5% in the case of central, state, and private sectors, respectively, during 2014/15. Figure 7 shows the trend in annual PLF of the central, state, and private thermal stations.

Transmission and distribution

India's extensive network of transmission lines covers 3 860 110 circuit kilometres (ckt km), operating at various voltage levels. The length of distribution lines up to 500 V is 5 220 446 ckt km. Table 1 compares the length of lines operating at various voltage levels in India from 2007/08 to 2012/13. The growth rate of transmission and distribution lines during 2012/13 was around 4%.

Transmission and distribution losses

The electricity lost in transformation, transmission, and distribution, including unaccounted electricity, was about 23% of the total available electricity in

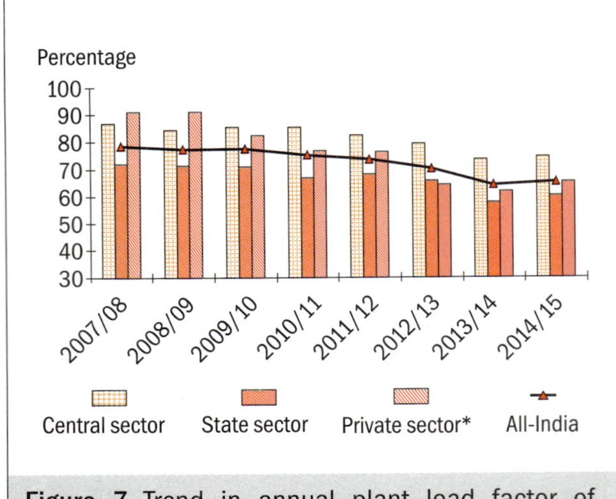

Figure 7 Trend in annual plant load factor of central, state, and private thermal stations from 2007/08 to 2014/15

*Average of PLF in private utilities and private independent power producers

Source CEA (2015c)

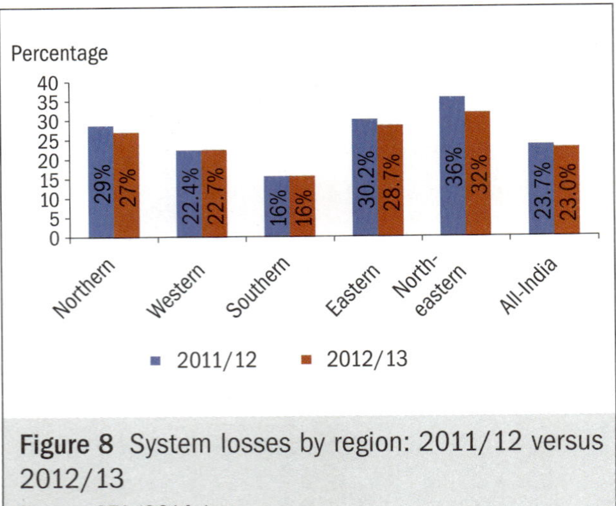

Figure 8 System losses by region: 2011/12 versus 2012/13

Source CEA (2014a)

2012/13 compared to 23.7% in 2011/12. Thus, the percentage loss has remained more or less constant.

The southern region continued to keep the system losses lowest at around 16%, followed by western (22.7%), northern (27%), eastern (28.7%), and north-eastern (32%) regions. The north-eastern region, although it showed the maximum losses as the previous year, witnessed reduction in system losses by around 4%. Figure 8 compares system losses by region during 2011/12 and 2012/13.

Consumption

Power supply position

The estimated peak electricity demand and energy requirement by the end of the Twelfth Five-year Plan (2012–17) have been 199 540 MW and 1355 BU, respectively. For the Twelfth Five-year Plan period, the anticipated growth rate for energy requirements is 8.5% and that for peak demand is 9.5%.

The all-India energy sales during 2012/13 were 708 BU. Industries continue to be the highest consumer, accounting for 35% of the total electricity sold by utilities, followed by domestic (26%) and agriculture (21%) sectors. The sales decreased by

Table 1 Transmission and distribution network (in ckt km)

Voltage level	2007/08	2008/09	2009/10	2010/11	2011/12	2012/13
800 kV	2 057	2 806	3 332	3 341	6 472	6 472
High-voltage direct current	6 428	6 428	6 428	8 008	8 008	8 008
400 kV	73 277	76 239	83 624	90 731	110 408	110 408
230/220 kV	115 452	118 664	123 052	128 267	129 488	138 534
132/110/90 kV	138 137	138 615	142 039	146 790	149 481	155 595
78/66 kV	49 945	51 096	51 024	52 668	53 936	58 403
33/22 kV	335 351	343 625	349 953	372 501	382 745	395 608
15/11 kV	2 231 698	2 326 546	2 438 962	2 676 229	2 785 597	2 954 056
6.6/3.3/2.2 kV	6 519	21 655	22 422	28 195	38 131	33 026
Distribution lines up to 500 V	4 328 549	4 402 303	4 580 262	4 858 571	5 073 564	5 220 446
Total (all-India)	7 287 413	7 487 977	7 801 098	8 365 301	8 737 830	9 080 556

Source CEA (2014a)

10% in 2012/13 compared to 2011/12. The demand is growing in terms of both energy and peak demand. The peak deficit went up to 4.7% in 2014/15 from 4.5% in 2013/14. In absolute terms, the peak deficit has reduced by 903 MW. Figure 9 shows the trend in peak deficit during the past 6 years.

The energy deficit also reduced considerably to 3.6% in 2014/15 compared to 4.2% in the previous year. There was an overall energy shortage of 38 BU during 2013/14 compared to 42 BU in 2013/14. Figure 10 shows the trend in energy deficit during the past 6 years.

Connected load

The total connected load in the country as on 31 March 2013 was 661 GW, up from 587 GW on 31 March 2012, recording an increase of 12.7%. Figure 11 shows the consumer-connected load by category as on 31 March 2013.

During 2012/13, domestic category, which accounted for 38% of the total connected load, emerged as the single largest consumer category in terms of connected load, which was followed by industrial (26%), agricultural (18%), and commercial (11%) categories.

Power and Indian planning

During the Twelfth Five-year Plan, a capacity addition target of 88 537 MW (excluding renewables) was fixed. Of this target, 38 448 MW was achieved by 31 March 2014. Thermal power accounts for 36 889 MW, which is 96% of the total capacity addition. The private sector contributed the maximum (56%), followed by the central (24%) and state (20%) sectors (Table 2).

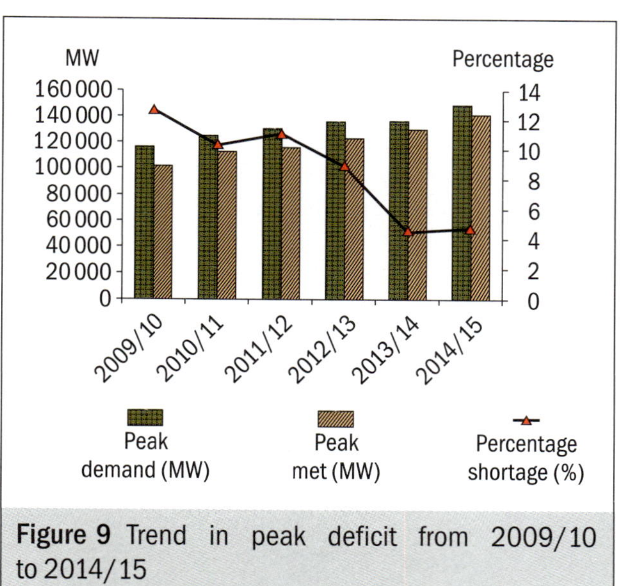

Figure 9 Trend in peak deficit from 2009/10 to 2014/15
Source CEA (2015d)

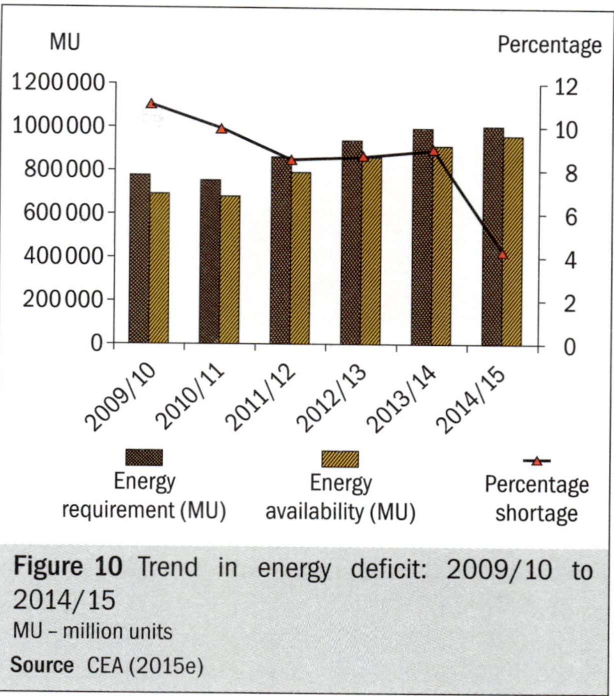

Figure 10 Trend in energy deficit: 2009/10 to 2014/15
MU – million units
Source CEA (2015e)

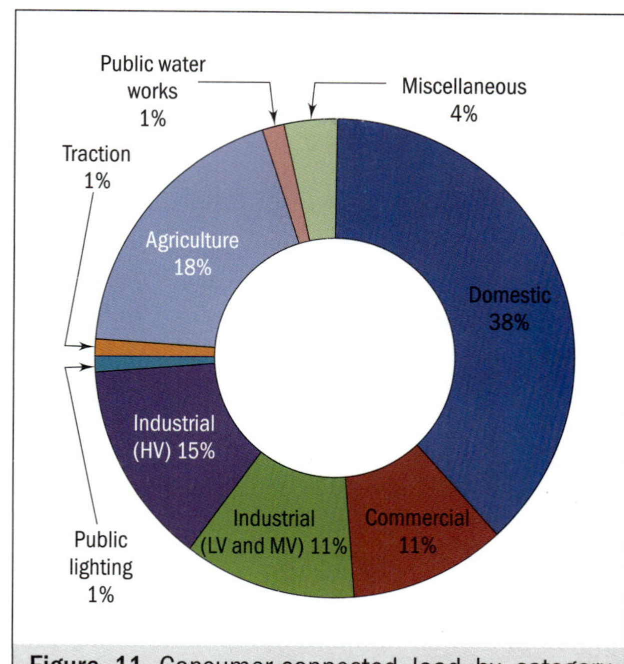

Figure 11 Consumer-connected load by category (as on 31 March 2013)
Source CEA (2014a)

Table 2 Capacity addition target and achievement for the Twelfth Five-year Plan (MW) by sector and source

Source	Thermal		Hydro		Nuclear		Total	
Sector	Target	Achieved till March 2014	Target	Achieved till March 2014	Target	Achieved till March 2014	Target	Achieved till March 2014
Central	14878	6683	6004	1288	5300	0	26182	7971
State	13922	7233	1608	102	0	0	15530	7335
Private	43540	22973	3285	169	0	0	46825	23142
All-India	72340	36889	10897	1559	5300	0	88537	38448

Source CEA (2014b)

Policies, programmes, and regulations

- Central Electricity Authority (Technical Standards for Construction of Electrical Plants and Electric Lines) Amendment Regulations, 2015
- Central Electricity Authority (Installation and Operation of Meters) (Amendment) Regulations, 2014
- Central Electricity Regulatory Commission (Terms and Conditions for Tariff Determination from Renewable Energy Sources) (Fourth Amendment) Regulations, 2015
- Central Electricity Regulatory Commission (Ancillary Services Operations) Regulations, 2015
- Central Electricity Regulatory Commission (Deviation Settlement Mechanism and related matters) (Second Amendment) Regulations, 2015
- Central Electricity Regulatory Commission (Power System Development Fund) (First Amendment) Regulations, 2015
- Central Electricity Regulatory Commission (Sharing of Inter-state Transmission Charges and Losses) (Fourth Amendment) Regulations, 2015
- Central Electricity Regulatory Commission (Grant of Connectivity, Long-term Access and Medium-term Open Access in Inter-state Transmission and Related Matters) (Fifth Amendment) Regulations, 2015
- Central Electricity Regulatory Commission (Open Access in Inter-state Transmission) (Third Amendment) Regulations, 2015
- Central Electricity Regulatory Commission (Sharing of Inter-state Transmission Charges and Losses) (Third Amendment) Regulations, 2015
- Central Electricity Regulatory Commission (Power System Development Fund) Regulations, 2014

Status of power sector reforms

Distribution reforms

State electricity regulatory commissions have been constituted and operationalized in all states. Two joint electricity regulatory commissions have been notified and operationalized: one for the states of Mizoram and Manipur and the other for Goa and the union territories of Andaman and Nicobar Islands, Lakshadweep, Chandigarh, Daman and Diu, Dadra and Nagar Haveli, and Puducherry. Twenty-five states, namely Odisha, Karnataka, Andhra Pradesh, Haryana, Uttar Pradesh, Uttarakhand, Rajasthan, Delhi, Gujarat, Madhya Pradesh, Assam, Chhattisgarh, Maharashtra, West Bengal, Tamil Nadu, Meghalaya, Punjab, Himachal Pradesh, Bihar, Tripura, Jammu and Kashmir, Telangana, Arunachal Pradesh, Nagaland, and Jharkhand, have reorganized their state electricity boards. Further, Tripura and Kerala have created Tripura State Electricity Corporation Ltd and Kerala State Electricity Board Ltd, respectively, as single corporation, to look after generation, transmission and distribution, trading, and state load dispatch centre operations.[1]

The Restructured Accelerated Power Development and Reforms Programme (R-APDRP) was started in 2008 as a revised version of the

[1] Details available at www.cercind.gov.in/serc.html, last accessed on 3 January 2016

Accelerated Power Development Reforms Programme (APDRP launched in 2002/03). The reform programme focussed on reducing aggregate transmission and commercial (AT&C) losses during the Eleventh Five-year Plan. The programme includes determination and verification of baseline AT&C losses, establishment of IT applications for energy accounting and auditing, and strengthening of distribution and transmission systems, distribution strengthening projects, and capacity building.

Under Part-A of R-APDRP,[2] 1412 projects at the cost of ₹5472 crore have been approved for 29 states/union territories and ₹2719 crore has been disbursed till date. Part-A SCADA[3] projects worth ₹1556.23 crore for 72 towns all over India have been sanctioned, and a sum of ₹455.06 crore has been disbursed. Under Part-B of R-APDRP,[4] 1259 projects at the cost of ₹32215.78 crore have been approved for 27 states and ₹4905 crore has been disbursed.

Rural electrification initiatives

The Government of India has launched various programmes for the electrification of rural areas in the country, such as the Pradhan Mantri Gramodaya Yojana in 2001/02, Accelerated Rural Electrification Programme in 2003/04, and Accelerated Electrification of 1 Lakh Villages and 1 Crore Households in 2004/05. Many of these programmes have been merged with the Rajiv Gandhi Grameen Vidyutikaran Yojana (RGGVY), which was introduced in April 2005 for providing electricity to all rural households. The scheme includes setting up of one 33/11 kV (or 66/11 kV) substation in every block, provision of distribution transformers of appropriate capacity in every village/habitation, and decentralized distributed generation and supply system from conventional sources for villages/habitations.

All the states, except Delhi and Goa, have signed agreements under the RGGVY. From April 2005 to October 2015,[5] the cumulative achievement includes electrification of 367201 un-electrified villages, intensive electrification of 264592 partially electrified villages, and release of free electricity connections to 928458 below poverty line households.[6] All the 27 states participating in RGGVY have notified the constitution of district committees to, inter alia, monitor the implementation of the RGGVY, and all states have notified rural areas to take the advantage of exemptions provided in the Act for setting up decentralized distributed generation.

In September 2014, the government introduced the Deendayal Upadhyaya Gram Jyoti Yojana (DDUGJY) to ensure power supply to agricultural consumers for at least 8 h every day and 24/7 uninterrupted electricity supply to all households. The scheme will augment power supply to rural areas and strengthen sub-transmission and distribution systems. The scheme has the following two components: (i) separating agriculture and non-agriculture feeders, facilitating judicious rostering of supply to agricultural and non-agricultural consumers in rural areas and (ii) strengthening and augmenting sub-transmission and distribution infrastructure in rural areas, including 100% metering of distribution transformers/feeders/consumers. The scheme would provide access to rural households and lead to improvement in hours of power supply, reduction in peak load and India's AT&C losses, and improvement in billed energy based on metered consumption. Under the DDUGJY, around ₹40204 crore[7] has been sanctioned for all states and union territories to improve rural connectivity.

Short-term power market

Inter-state trading licensees (traders) have been trading in electricity since 2004. The two power exchanges—Indian Energy Exchange (IEX) and Power Exchange India Ltd (PXIL)—started operations in June 2008 and October 2008, respectively. Till March 2015, there were 47 inter-state trading licensees and two power exchanges.

During 2014/15, the total volume of short-term transactions was 99 BU compared to 105 BU in 2013/14, implying a decrease of 6%. The total volume of short-term transactions of electricity as the percentage of total electricity generation in

[2] Details available at www.apdrp.gov.in/Forms/Summary_Sanctioned_Project_SCADA, last accessed on 3 January 2016
[3] Details available at http://rggvy.gov.in/rggvy/rggvyportal/Report/Details_of_projects_sanctioned_under_DDUGJY.pdf, last accessed on 3 January 2016
[4] Details available at www.apdrp.gov.in/Forms/Summar_Sanctioned_Projects_Part_B, last accessed on 3 January 2016
[5] Details available at www.rggvy.gov.in/rggvy/rggvyportal/dynamic/village_completed.html, last accessed on 3 January 2016
[6] Details available at http://rggvy.gov.in/rggvy/rggvyportal/11p2_plgsheetplan1.jsp, last accessed on 3 January 2016
[7] Details available at http://rggvy.gov.in/rggvy/rggvyportal/Report/Details_of_projects_sanctioned_under_DDUGJY.pdf, last accessed on 3 January 2016

2014/15 was 9%, whereas it was 11% in 2013/14. Figure 12 shows the share of different segments within the total short-term transaction for 2014/15.[8]

The monthly trends in the price[9] of short-term transactions of electricity (unscheduled interchange or UI, bilateral trader transactions, IEX, and PXIL) are shown in Figure 13.[10] It can be observed from the figure that the price of electricity transacted through traders was relatively high compared to the price of electricity transacted through the power exchanges and UI during 2013/14.

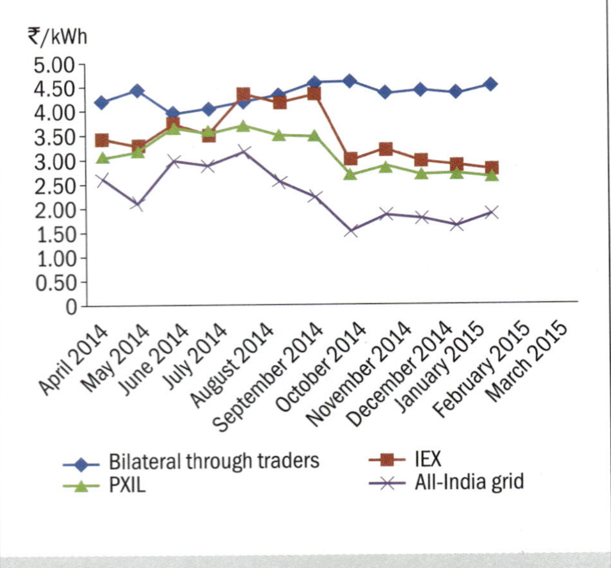

Figure 13 Trends in the price of short-term transactions of electricity (2014/15)
Source CERC (2015)

New developments

Smart grid initiatives in India

During the past couple of years, several programmes and policy initiatives have been undertaken towards developing smart electric grid in India. The Government of India established India Smart Grid Task Force (ISGTF) and India Smart Grid Forum (ISGF) for the development and deployment of smart grid-related activities in India.

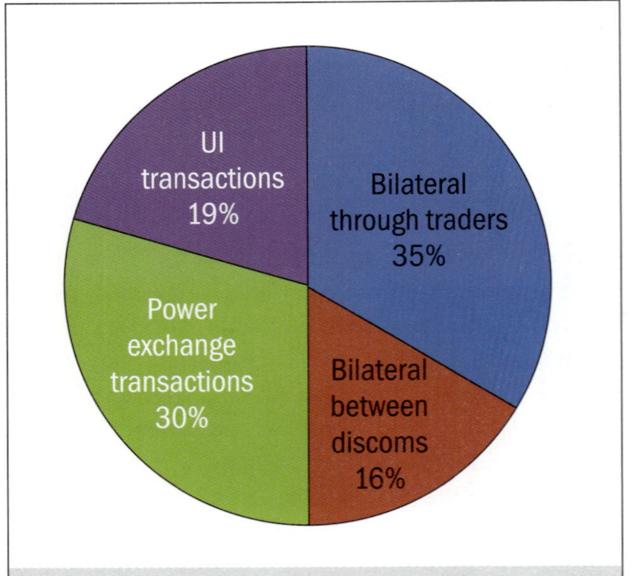

Figure 12 Share of different segments within the total short-term transaction for 2014/15
UI - unscheduled interchange
Source CERC (2015)

Current smart grid projects in the distribution sector of India

In collaboration with the Government of Puducherry, the Power Grid Corporation of India Ltd (PGCIL) is implementing a smart city project. This includes demonstration of smart grid technologies and functionalities such as smart meters, meter data management system, and outage management system in the distribution sector. The Energy and Resources Institute has also developed a first-of-its-kind renewable energy-based Smart Mini-Grid (SMG) system in India, which incorporates optimum utilization of the power generated from renewable energy sources at different operating conditions, for ensuring the supply of reliable and good quality power.

Further, distribution utilities in various states, such as Delhi (Tata Power Delhi Distribution Ltd), West Bengal (West Bengal State Electricity Distribution Company Ltd), Karnataka (Bangalore Electricity Supply Company Ltd and Mangalore Electricity Supply Company), Maharashtra (Maharashtra State Electricity Distribution Company Ltd), Kerala

[8] Details available at http://www.cercind.gov.in/2015/MMC/AnnualReport14-15.pdf, last accessed on 3 January 2016
[9] Average price of UI and weighted average price of other short-term transactions of electricity
[10] Details available at http://www.cercind.gov.in/2015/MMC/AnnualReport14-15.pdf, last accessed on 3 January 2016

(Kerala State Electricity Board), and Haryana (Uttar Haryana Bijli Vitran Nigam), are also implementing smart grid pilot projects with various functionalities and objectives.

In January 2013, the Ministry of Power sanctioned 14 smart grid pilot projects.[11] State-owned distribution utilities in India are implementing the projects in various states to showcase the relevance of smart grid. The current status of these projects is given in Table 3.

Planned smart grid projects in the distribution sector of India

The PGCIL is coming up with three demonstration projects on grid-connected battery energy storage with a capacity of 500 kW. The Government of Odisha has announced a smart grid power network (Disaster Resilient Power Strengthening System) for the cyclone-prone Ganjam district of Odisha, with 60% funding from the Asian Development Bank. Further,

Table 3 Current status of smart grid pilot projects in India

State	Utility	Functionalities	Project status
Assam	Assam Power Distribution Company Ltd	AMI R, AMI I, PLM, OMS and PQM, DG	Project awarded to M/s Phoenix IT Solutions in consortium with Securo, KEPCO, and Nuri Telecom
Chhattisgarh	Chhattisgarh State Power Distribution Company	AMI I, PLM	Price quoted by L1 Bidder has been further revised as the earlier cost was very high
Gujarat	Uttar Gujarat Vij Company Ltd	AMI I, AMI R, and OMS	Board approval expected by the end of October 2015 for placing order on L1 Bidder
Haryana	Uttar Haryana Bijli Vitran Nigam Ltd	AMI I, AMI R, and OMS, SCADA/DMS	Project is executed under a grant from NEDO, Japan. NEDO has completed the feasibility studies and the report has been approved by UHBVN and the Ministry of Power. The project work started on 4 December 2014
Himachal Pradesh	Himachal Pradesh State Electricity Board Ltd	AMI I, PLM, OMS, and PQM	Project awarded to M/S Alstom T&D India for a total cost of ₹ 194.5 million on 28 February 2015
Karnataka	Chamundeshwari Electricity Supply Corporation Ltd, Mysore	AMI R, AMI I, PLM, OMS, and PQM, DG/MG	Project awarded on 30 April 2014 to a consortium led by Enzen Global Solutions Pvt. Ltd
Kerala	Kerala State Electricity Board Ltd	100% AMI industry	Documentation on retendering is completed
Maharashtra	Maharashtra State Electricity Distribution Company Ltd	AMI I, AMI R, and OMS	Project has been cancelled
Puducherry	Electricity Department of the Government of Puducherry	AMI I, AMI R	Five bids received and are under evaluation
Punjab	Punjab State Power Corporation Ltd	AMI, PLM	Contract awarded to M/s Kalkitech. Site survey completed and design documents are being submitted
Rajasthan	Jaipur Vidyut Vitaran Nigam Ltd	AMI R, AMI I, PLM	Project likely to be cancelled
Telangana	Telangana Southern Power Distribution Company Ltd	AMI , PLM, OMS, and PQM	Eight bids received and are under evaluation
Tripura	Tripura State Electricity Corporation Ltd	AMI I, AMI R, PLM	Project cost under revision
West Bengal	West Bengal State Electricity Distribution Company Ltd	AMI I, AMI R, and PLM	Contract awarded to M/s Chemtrols Engineering Ltd on 6 June 2015

AMI R - advanced metering infrastructure Residential consumer; AMI I - advanced metering infrastructure industrial consumer; DG - distributed generation; DG/MG - distributed generation/micro grid; OMS - outage management system; PLM - peak load management; PQM - power quality management; SCADA/DMS - supervisory control and data acquisition/distribution management system

Source Details available at www.indiasmartgrid.org/upload/201510Thu223811.pdf, last accessed on 3 January 2016

[11] Haryana, Mysore, Tripura, Kerala, Puducherry, Gujarat, Andhra Pradesh, Assam, Maharashtra, Chhattisgarh, Himachal Pradesh, Punjab, West Bengal, and Rajasthan

a smart grid pilot project covering around 5800 consumers has also been taken up at Chhatrapur in Odisha with funding from the state government. Tetra Tech and ESTA International are preparing a smart grid roadmap for the Calcutta Electric Supply Corporation Ltd under a US Trade and Development Agency grant.

Smart cities

Although the concept of smart cities is still evolving, the Government of India has embarked on an ambitious programme since 2014 for developing 100 smart cities. The Ministry of Urban Development (MoUD) is made the nodal agency for driving the smart cities programme. The MoUD has conducted several stakeholder consultations and assigned the ISGF with the task of developing a standard framework for infrastructure domains, including a Smart City Maturity Model and model concession agreements for infrastructure services in smart cities.

Demand-side management initiatives

The Government of India, through the Bureau of Energy Efficiency (BEE) and State Designated Agencies (SDAs), has taken several measures to implement demand-side management (DSM) in various consumer categories. Some of the key regulatory and policy initiatives taken by the government are as follows:

- *Framing DSM regulations*: The Forum of Regulators notified the model DSM regulations in 2010. Based on this model, a majority of states such as Maharashtra, Tamil Nadu, Himachal Pradesh, Gujarat, Kerala, Punjab, and Haryana have notified DSM regulations.
- *Standard and labelling*: The BEE introduced the Standard and Labelling (S&L) Programme on 18 May 2006, which is currently invoked for 14 equipment/appliances. The scheme aims to provide the consumer with an informed choice about energy saving and thereby the cost-saving potential of the relevant marketed product.
- *Energy Conservation Building Code*: In 2007, the Government of India launched the Energy Conservation Building Code on a voluntary basis. In line with this, some states such as Kerala and Rajasthan have made mandatory for buildings that have a connected load of 100 kW or more.
- *State energy conservation funds*: States such as Rajasthan, Kerala, and Maharashtra have increased state energy conservation funds under Sections 16 and 56 of the Energy Conservation Act, 2001 to promote energy efficiency activities in the states.
- *Capacity building of DISCOMS*: The Government of India has approved a scheme for the capacity building of DISCOMs in an effort to mainstream DSM in the country. To estimate the feasibility and potential of DSM, carrying out a detailed load research in each utility is essential. Pursuant to this, the Energy Efficiency Services Ltd (EESL) and BEE have entered into a memorandum of understanding (MoU) to facilitate this scheme in 30 selected DISCOMs in the country.
- *Energy service companies*: To develop a viable energy service company (ESCO), the Ministry of Power has set up the EESL, a joint venture of National Thermal Power Corporation Ltd (NTPC), Power Finance Corporation Ltd, Rural Electrification Corporation Ltd, and POWERGRID, to facilitate energy efficiency projects. The EESL will work as an ESCO, as a consultancy organization for Clean Development Mechanism (CDM) and energy efficiency, and as a resource centre for the capacity building of SDAs, utilities, and financial institutions. The EESL will also lead the market-related actions of the National Mission on Enhanced Energy Efficiency (NMEEE).

Various DSM initiatives have been undertaken for different consumer categories in India at the state and central levels. They are as follows:

- *Domestic category*: The BEE has initiated the Bachat Lamp Yojana to promote energy-efficient lighting in India. The key objective of the programme is to replace incandescent bulbs with compact fluorescent lamps (CFLs). Various state utilities in Delhi, Rajasthan, Himachal Pradesh, Uttar Pradesh, and Puducherry are promoting light emitting diode (LED) bulbs among domestic consumers under the Domestic Efficient Lighting Programme (DELP). The number of LEDs sold under DELP scheme as on 26 October 2015 has been depicted in Map 1. Utilities in Maharashtra and Delhi have undertaken replacement programmes to promote energy efficiency in key energy-intensive appliances in households such as air conditioners, refrigerators, and ceiling fans.

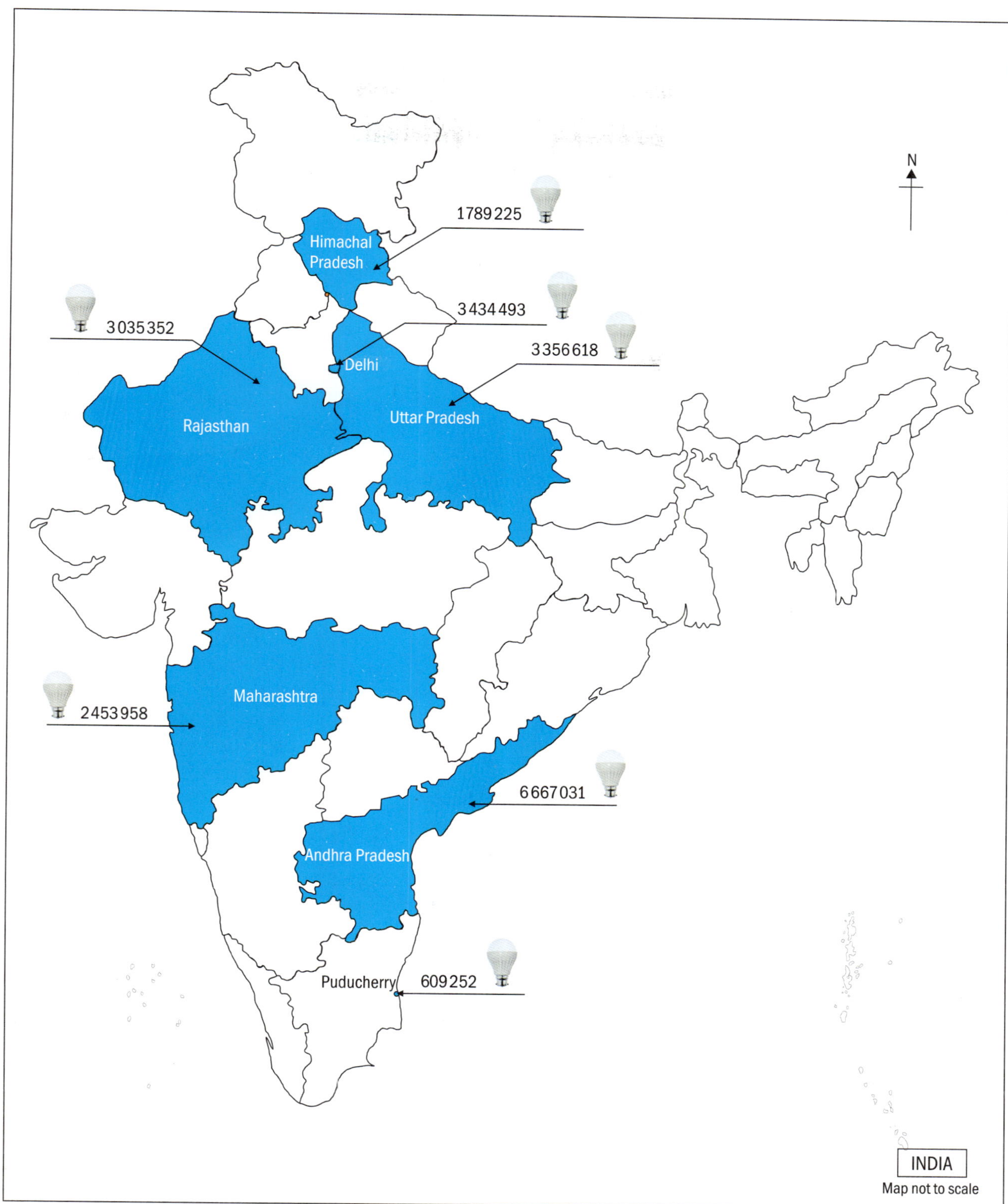

Map 1 Number of LEDs sold till 26 October 2015
Source Details available at http://www.delp.in/

- *Agriculture category*: As part of agricultural DSM, the BEE has come up with pilot projects in Maharashtra, Gujarat, Punjab, Haryana, and Rajasthan to promote energy-efficient pumpsets. Further, some states such as Haryana and Punjab have mandated the use of BEE star-rated pumps (with a minimum of four-star rating), and ISI marked power capacitors, and foot/reflex valves in the agriculture sector. The Government of Gujarat initiated the Jyotigram Yojana in September 2003 to segregate agriculture loads from non-agriculture loads to reduce system losses and improve the reliability of supply. Based on this, the Government of India has come up with a national programme for feeder segregation, which is known DDUGJY.

  The Government of Rajasthan has been promoting solar water pumps through its Department of Horticulture. The beneficiaries of this scheme are provided 86% support, comprising 30% from the Ministry of New and Renewable Energy (MNRE) and 56% from the Rashtriya Krishi Vikas Yojana. States such as Punjab and Tamil Nadu are also promoting solar-based water pumping by giving additional subsidies in addition to the 30% subsidy provided by the MNRE.
- *Industrial category*: The Perform, Achieve, and Trade (PAT) scheme was introduced to improve energy efficiency in energy-intensive large industrial subsectors in the country. Further, energy audits for high tension/low tension industries, which are not included in the PAT scheme, have been made mandatory in states such as Gujarat, Kerala, Maharashtra, and Rajasthan.

  Many states in India have introduced time-of-use charges/exclusive night-time tariff and power factor incentive/rebate tariff structure for better demand management.

  Tata Power in Maharashtra and Delhi and the Jaipur Vidhyut Vitran Nigam Ltd in Rajasthan have implemented pilot demand response programmes in the industrial and commercial sectors for shifting the peak demand to the off-peak period.
- *Municipal DSM initiatives*: Several states in India, such as Rajasthan, Gujarat, and Maharashtra, have taken various initiatives to improve energy efficiency in street lighting and municipal water pumping.

In addition to these initiatives, the BEE and SDAs are undertaking various programmes to create awareness on energy efficiency and conservation, such as Energy Conservation Day, Serve as Volunteer for Energy (SAVE) programme for school children, Energy Clinic Programme, Bal Urja Rakshak Mission, and Urja Samvardhan Upakram.

International cooperation in the power sector

In recent years, active interest has been taken in the enhancement of bilateral cooperation with Bangladesh, Bhutan, Canada, China, Nepal, Sri Lanka, Germany, Japan, United States, and European Union. In September 2014, the cabinet approved the India–Nepal Agreement on electric power trade, cross-border transmission interconnection, and grid connectivity.

India will import about 1400 MW of power from Bhutan. By 2020, Bhutan plans to develop an aggregate capacity of around 11 000 MW as part of the MoU between the Government of India and the Royal Government of Bhutan.

India exports about 50 MW of power to Nepal, for which tariff is decided by the Power Exchange Committee. India also supplies to Nepal about 70 million units of free energy from the Tanakpur Hydroelectric Project (120 MW) under the Mahakali Treaty. A power sale agreement between the Nepal Electricity Authority and the PTC India Ltd has been signed for the import of 150 MW for 25 years.

The transmission line between India and Bangladesh and the high-voltage direct current (HVDC) station in Bheramara, which will enable the supply of 500 MW of power to Bangladesh, have been operationalized, and power flow through the link has started in October 2013. The Bangladesh–India Friendship Power Company, a joint venture between India's NTPC Ltd and the Bangladesh Power Development Board, is developing the 1320 MW Maitri Thermal Power Project.

There has been significant progress in the inter-grid connection between India and Sri Lanka. The feasibility study for the +400 kV, 1000 MW undersea HVDC bipole line from Madurai in India to Anuradhapura in Sri Lanka is under finalization.

Domestic Efficient Lighting Programme

The Domestic Efficient Lighting Programme (DELP) was launched by the Prime Minister of India on 5 January 2015 to provide LED lights to households. The scheme provides LED bulbs at a subsidized rate for replacing traditional or compact florescent lamps.

In comparison to ordinary bulbs, LED bulbs consume around one-tenth of the energy and provide the same or better output. However, the first barrier to accessing LED bulbs is the high cost involved (₹400–500). DELP proposes to overcome this cost barrier by providing on-bill financing. Under this scheme, consumers can purchase the designated number of bulbs on the advance payment of ₹10 per bulb and the balance is recovered from the electricity bill. The scheme also allows consumers to make an upfront payment for an LED bulb. Large-scale and transparent procurement has led to a sharp decline in prices from ₹310 per bulb in January 2014 to ₹73 in June 2015 (PIB 2015a).

The Energy Efficiency Services Ltd (EESL), a public sector body of the Government of India, is implementing the DELP. The EESL has evolved a service model in which it works with electricity distribution companies (DISCOMs) through a benefit-sharing approach. The DELP obviates the need for DISCOMs to invest in the upfront cost of LED bulbs (EESL 2015). This scheme has received support and encouragement from DISCOMS because the EESL makes 100% upfront investment and DISCOMs require no additional budget allocation. The EESL envisions to install around 77 crore domestic LEDs. This will reduce about 20 000 MW of load, save 100 billion kWh of energy, and reduce greenhouse gas emissions by 80 million tonnes per year (PIB 2015b). Under DELP, 186 cities/towns have been enrolled. Till 6 January 2016, the EESL successfully distributed more than 4.47 crore LED bulbs across India (DELP 2016).

The Government of India launched the National DELP Dashboard for continuously monitoring LED distribution under the 100 City Domestic Efficient Lighting Programme managed by the EESL. The continuous monitoring is achieved by aggregating real-time feeds from multiple states where DELP is being implemented (PIB 2015c).

References

CEA (Central Electricity Authority). 2013. ***18th Electric Power Survey (EPS)***. New Delhi: Central Electricity Authority (CEA).

CEA (Central Electricity Authority). 2014a. ***All-India Electricity Statistics, General Review 2014***. New Delhi: CEA

CEA (Central Electricity Authority). 2014b. Executive summary of power sector. Details available at www.cea.nic.in/reports/monthly/executive_rep/mar14.pdf

CEA (Central Electricity Authority). 2015a. All-India installed capacity (in MW) of power stations located in the regions of main land and islands. Details available at www.cea.nic.in/reports/monthly/installedcapacity/2015/installed_capacity-03.pdf

CEA (Central Electricity Authority). 2015b. Energy generation, programme, and plant load factor. Details available at www.cea.nic.in/reports/monthly/generation/2015/March/actual/opm_01.pdf

CEA (Central Electricity Authority). 2015c. Executive Summary Power Sector. New Delhi: CEA

CEA (Central Electricity Authority). 2015d. Peak demand and peak met (revised). Details available at www.cea.nic.in/reports/monthly/powersupply/2015/psp_peak-03.pdf

CEA (Central Electricity Authority). 2015e. Power supply position (revised). Details available at www.cea.nic.in/reports/monthly/powersupply/2015/psp_energy-03.pdf

CERC (Central Electricity Regulatory Commission). 2015. Report of Market Monitoring Cell. New Delhi: CERC

DELP (Domestic Efficient Lighting Programme). 2016. National DELP Dashboard. Details available at www.delp.in, last accessed on 6 January 2016

EESL (Energy Efficiency Services Ltd). 2015. ***Annual Report 2014–15***. New Delhi: EESL

PIB (Press Information Bureau). 2015a. Retail market price of LED bulbs needs correction to make it affordable: Shri Piyush Goyal. EESL distributes over 3 crore LED Bulbs, helps India become energy efficient. Details available at http://pib.nic.in/newsite/PrintRelease.aspx?relid=131891

PIB (Press Information Bureau). 2015b. Power sector lights up; LED bulbs in vogue. Details available at http://pib.nic.in/newsite/mbErel.aspx?relid=126146

PIB (Press Information Bureau). 2015c. Government stands committed to make LED a way of life in India: Shri Piyush Goyal inaugurates National DELP Dashboard for monitoring LED distribution programme. Details available at http://pib.nic.in/newsite/PrintRelease.aspx?relid=123530

TABLE 5.1a Installed capacity (MW) as on 31 March 2015

State/union territory/ central sector		Coal	Gas	Diesel	Total thermal	Nuclear	Hydro	RES	Total
Northern region									
Chandigarh	State	0	0	0	0	0	0	0	0
	Private	0	0	0	0	0	0	4.50	4.50
	Central	32.54	15.32	0	47.86	8.84	56.40	0	113.10
	Subtotal	32.54	15.32	0	47.86	8.84	56.40	4.50	117.60
Delhi	State	135.00	2 050.40	0	2 185.40	0	0	0	2 185.40
	Private	445.50	108.00	0	553.50	0	0	21.47	574.97
	Central	4 421.37	207.61	0	4 628.98	122.08	762.64	0	5 513.70
	Subtotal	5 001.87	2 366.01	0	7 367.88	122.08	762.64	21.47	8 274.07
Haryana	State	3 160.00	25.00	3.92	3 188.92	0	884.51	59.3	4 132.73
	Private	2 165.50	0	0	2 165.50	0	0	77.30	2 242.80
	Central	1 202.03	535.29	0	1 737.32	109.16	531.63	0	2 378.11
	Subtotal	6 527.53	560.29	3.92	7 091.74	109.16	1 416.14	77.30	8 753.64
Himachal Pradesh	State	0	0	0.13	0.13	0	393.60	245.61	639.34
	Private	0	0	0	0	0	1 748.00	478.30	2 226.30
	Central	152.02	61.88	0	213.90	34.08	1 464.90	0	1 712.88
	Subtotal	152.02	61.88	0.13	214.03	34.08	3 606.50	723.91	4 578.52
Jammu and Kashmir	State	0	175.00	8.94	183.94	0	780.00	106.53	1 070.47
	Private	0	0	0	0	0	0	50.00	50.00
	Central	329.32	129.14	0	458.46	77.00	955.45	0	1 490.91
	Subtotal	329.32	304.14	8.94	642.40	77.00	1 735.45	156.53	2 611.38
Punjab	State	2 630.00	25.00	0	2 655.00	0	2 230.23	127.80	5 013.03
	Private	3 154.00	0	0	3 154.00	0	0	365.62	3 519.62
	Central	660.88	263.92	0	924.80	208.04	859.85	0	1 992.69
	Subtotal	6 444.88	288.92	0	6 733.80	208.04	3 090.08	493.42	10 525.34
Rajasthan	State	4 590.00	603.80	0	5 193.80	0	987.96	23.85	6 205.61
	Private	3 196.00	0	0	3 196.00	0	0	4 362.40	7 558.40
	Central	1 014.72	221.23	0	1 235.95	573.00	655.73	0	2 464.68
	Subtotal	8 800.72	825.03	0	9 625.75	573.00	1 643.69	4 386.25	16 228.69
Uttar Pradesh	State	4 923.00	0	0	4 923.00	0	524.10	25.10	5 472.20
	Private	3 345.00	0	0	3 345.00	0	0	964.76	4 309.76
	Central	2 909.95	549.97	0	3 459.92	335.72	1 509.45	0	5 305.09
	Subtotal	11 177.95	549.97	0	11 727.92	335.72	2 033.55	989.86	15 087.05

Contd...

TABLE **5.1a** *Contd...*

State/union territory/ central sector		Coal	Gas	Diesel	Total thermal	Nuclear	Hydro	RES	Total
Uttarakhand	State	0	0	0	0	0	1252.15	62.87	1315.02
	Private	99.00	0	0	99.00	0	400.00	181.45	680.45
	Central	300.50	69.35	0	369.85	22.28	421.44	0	813.57
	Subtotal	399.50	69.35	0	468.85	22.28	2073.59	244.32	2809.04
Central (unallocated)		977.19	290.35	0	1267.54	129.80	648.70	0	2046.04
Total northern region	State	15438.00	2879.20	12.99	18330.19	0	7052.55	651.06	26033.80
	Private	12405.00	108.00	0	12513.00	0	2148.00	6505.80	21166.80
	Central	12000.50	2344.06	0	14344.56	1620.00	7866.23	0	23830.79
	Grand total	39843.50	5331.26	12.99	45187.75	1620.00	17066.78	7156.86	71031.39
Western region									
Chhattisgarh	State	2780.00	0	0	2780.00	0	120.00	11.05	2911.05
	Private	8323.00	0	0	8323.00	0	0	313.45	8636.45
	Central	1490.49	0	0	1490.49	47.52	0	0	1538.01
	Subtotal	12593.49	0	0	12593.49	47.52	120.00	324.50	13085.51
Dadra and Nagar Haveli	State	0	0	0	0	0	0	0	0
	Private	0	0	0	0	0	0	0	0
	Central	44.37	27.10	0	71.47	8.46	0	0	79.93
	Subtotal	44.37	27.10	0	71.47	8.46	0	0	79.93
Daman and Diu	State	0	0	0	0	0	0	0	0
	Private	0	0	0	0	0	0	0	0
	Central	36.71	4.20	0	40.91	7.38	0	0	48.29
	Subtotal	36.71	4.20	0	40.91	7.38	0	0	48.29
Goa	State	0	0	0	0	0	0	0.05	0.05
	Private	0	48.00	0	48.00	0	0	0	48.00
	Central	326.17	0	0	326.17	25.80	0	0	351.97
	Subtotal	326.17	48.00	0	374.17	25.80	0	0.05	400.02
Gujarat	State	4720.00	2321.82	17.28	7059.10	0	772.00	8.00	7839.10
	Private	8620.00	4160.00	0.20	12780.20	0	0	4709.55	17489.75
	Central	2648.27	424.27	0	3072.54	559.32	0	0	3631.86
	Subtotal	15988.27	6906.09	17.48	22911.84	559.32	772.00	4717.55	28960.71
Madhya Pradesh	State	4320.00	0	0	4320.00	0	1703.66	83.96	6107.62
	Private	3950.00	0	0	3950.00	0	0	1480.68	5430.68
	Central	2256.39	257.18	0	2513.57	273.24	1520.00	0	4306.81

Contd...

TABLE 5.1a *Contd...*

State/union territory/ central sector		Coal	Gas	Diesel	Total thermal	Nuclear	Hydro	RES	Total
	Subtotal	10 526.39	257.18	0	10 783.57	273.24	3 223.66	1 564.64	15 845.11
Maharashtra	State	9 560.00	672.00	0	10 232.00	0	2 884.84	208.13	13 324.97
	Private	11 796.00	180.00	0	11 976.00	0	447.00	5 980.17	18 403.17
	Central	3 313.27	2 623.93	0	5 937.20	690.14	0	0	6 627.34
	Subtotal	24 669.27	3 475.93	0	28 145.20	690.14	3 331.84	6 188.30	38 355.48
Central (unallocated)		1 622.35	196.91	0	1 819.26	228.14	0	0	2 047.40
Total western region	State	21 380.00	2 993.82	17.28	24 391.10	0	5 480.50	311.19	30 182.79
	Private	32 689.00	4 388.00	0.20	37 077.20	0	447.00	12 483.85	50 008.05
	Central	11 738.01	3 533.59	0	15 271.60	1 840.00	1 520.00	0	18 631.60
	Subtotal	65 807.01	10 915.41	17.48	76 739.90	1 840.00	7 447.50	12 795.04	98 822.44
Southern region	State	3 085.91	0	0	3 085.91	0	1 721.99	89.50	4 897.40
Andhra Pradesh (new)	Private	1 290.00	1 672.65	16.97	2 979.62	0	0	1 856.50	4 836.12
	Central	1 414.60	0	0	1 414.60	127.16	0	0	1 541.76
	Subtotal	5 790.51	1 672.65	16.97	7 480.13	127.16	1 721.99	1 946.00	11 275.28
Telangana	State	3 606.59	0	0	3 606.59	0	2 012.54	0	5 619.13
	Private	270.00	1 697.75	19.83	1 987.58	0	0	61.25	2 048.83
	Central	1 653.28	0	0	1 653.28	148.62	0	0	1 801.90
	Subtotal	5 529.87	1 697.75	19.83	7 247.45	148.62	2 012.54	61.25	9 469.86
Karnataka	State	2 720.00	0	127.92	2 847.92	0	3 599.80	137.33	6 585.05
	Private	2 060.00	0	106.50	2 166.50	0	0	4 372.90	6 539.40
	Central	1 549.51	0	0	1 549.51	475.86	0	0	2 025.37
	Subtotal	6 329.51	0	234.42	6 563.93	475.86	3 599.80	4 510.23	15 149.82
Kerala	State	0	0	234.60	234.60	0	1 881.50	123.92	2 240.02
	Private	0	174.00	21.84	195.84	0	0	80.03	275.87
	Central	1 002.44	359.58	0	1 362.02	228.60	0	0	1 590.62
	Subtotal	1 002.44	533.58	256.44	1 792.46	228.60	1 881.50	203.95	4 106.51
Tamil Nadu	State	4 770.00	523.20	0	5 293.20	0	2 182.20	122.70	7 598.10
	Private	1 150.00	503.10	411.66	2 064.76	0	0	8 273.04	10 337.80
	Central	3 961.60	0	0	3 961.60	986.50	0	0	4 948.10
	Subtotal	9 881.60	1 026.30	411.66	11 319.56	986.50	2 182.20	8 395.74	22 884.00
Puducherry	State	0	32.50	0	32.50	0	0	0	32.50
	Private	0	0	0	0	0	0	0.03	0.03
	Central	244.57	0	0	244.57	52.78	0	0	297.35
	Subtotal	244.57	32.50	0	277.07	52.78	0	0.03	329.88
Neyveli Lignite Corporation	State	0	0	0	0	0	0	0	0

Contd...

TABLE 5.1a *Contd...*

State/union territory/ central sector		Coal	Gas	Diesel	Total thermal	Nuclear	Hydro	RES	Total
	Private	0	0	0	0	0	0	0	0
	Central	100.17	0	0	100.17	0	0	0	100.17
	Subtotal	100.17	0	0	100.17	0	0	0	100.17
Central (unallocated)		1 463.83	0	0	1 463.83	300.48	0	0	1 764.31
Total southern region	State	14 182.50	555.70	362.52	15 100.72	0	11 398.03	473.45	26 972.20
	Private	4 770.00	4 047.50	576.80	9 394.30	0	0	14 643.75	24 038.05
	Central	11 390.00	359.58	0	11 749.58	2 320.00	0	0	14 069.58
	Grand total	30 342.50	4 962.78	939.32	36 244.60	2 320.00	11 398.03	15 117.20	65 079.83
Eastern region	State	210.00	0	0	210.00	0	0	70.70	280.70
Bihar	Private	0	0	0	0	0	0	43.42	43.42
	Central	2 306.24	0	0	2 306.24	0	129.43	0	2 435.67
	Subtotal	2 516.24	0	0	2 516.24	0	129.43	114.12	2 759.79
Jharkhand	State	1 190.00	0	0	1 190.00	0	130.00	4.05	1 324.05
	Private	900.00	0	0	900.00	0	0	16.00	916.00
	Central	314.93	0	0	314.93	0	70.93	0	385.86
	Subtotal	2 404.93	0	0	2 404.93	0	200.93	20.05	2 625.91
West Bengal	State	5 220.00	100.00	12.06	5 332.06	0	977.00	91.95	6 401.01
	Private	1 941.38	0	0.14	1 941.52	0	0	39.76	1 981.28
	Central	922.45	0	0	922.45	0	271.30	0	1 193.75
	Subtotal	8 083.83	100.00	12.20	8 196.03	0	1 248.30	131.71	9 576.04
Damodar Valley Corporation	State	0	0	0	0	0	0	0	0
	Private	1 050.00	0	0	1 050.00	0	0	0	1 050.00
	Central	6 110.66	90.00	0	6 200.66	0	193.26	0	6 393.92
	Subtotal	7 160.66	90.00	0	7 250.66	0	193.26	0	7 443.92
Odisha	State	420.00	0	0	420.00	0	2 061.92	6.30	2 488.22
	Private	4 650.00	0	0	4 650.00	0	0	110.09	4 760.09
	Central	1 683.04	0	0	1 683.04	0	105.01	0	1 788.05
	Subtotal	6 753.04	0	0	6 753.04	0	2 166.93	116.39	9 036.36
Sikkim	State	0	0	5.00	5.00	0	0	52.11	57.11
	Private	0	0	0	0	0	99.00	0	99.00
	Central	92.10	0	0	92.10	0	75.27	0	167.37
	Subtotal	92.10	0	5.00	97.10	0	174.27	52.11	323.48
Central (unallocated)		1 572.07	0	0	1 572.07	0	0	0	1 572.07

Contd...

TABLE 5.1a *Contd...*

State/union territory/ central sector		Coal	Gas	Diesel	Total thermal	Nuclear	Hydro	RES	Total
Total eastern region	State	7 040.00	100.00	17.06	7 157.06	0	3 168.92	225.11	10 551.09
	Private	8 541.38	0	0.14	8 541.52	0	99.00	209.27	8 849.79
	Central	13 001.49	90.00	0	13 091.49	0	845.20	0	13 936.69
	Subtotal	28 582.87	190.00	17.20	28 790.07	0	4 113.12	434.38	33 337.57
North-eastern region	State	60.00	276.20	20.69	356.89	0	100.00	30.01	486.90
Assam	Private	0	24.50	0	24.50	0	0	4.10	28.60
	Central	0	417.92	0	417.92	0	329.72	0	747.64
	Subtotal	60.00	718.62	20.69	799.31	0	429.72	34.11	1 263.14
Arunachal Pradesh	State	0	0	15.88	15.88	0	0	104.61	120.49
	Private	0	0	0	0	0	0	0.03	0.03
	Central	0	43.06	0	43.06	0	97.57	0	140.63
	Subtotal	0	43.06	15.88	58.94	0	97.57	104.64	261.15
Meghalaya	State	0	0	2.05	2.05	0	282.00	31.03	315.08
	Private	0	0	0	0	0	0	0	0
	Central	0	105.14	0	105.14	0	74.58	0	179.72
	Subtotal	0	105.14	2.05	107.19	0	356.58	31.03	494.80
Tripura	State	0	169.50	4.85	174.35	0	0	16.01	190.36
	Private	0	0	0	0	0	0	5.00	5.00
	Central	0	369.32	0	369.32	0	62.37	0	431.69
	Subtotal	0	538.82	4.85	543.67	0	62.37	21.01	627.05
Manipur	State	0	0	45.41	45.41	0	0	5.45	50.86
	Private	0	0	0	0	0	0	0	0
	Central	0	67.98	0	67.98	0	80.98	0	148.96
	Subtotal	0	67.98	45.41	113.39	0	80.98	5.45	199.82
Nagaland	State	0	0	2.00	2.00	0	0	29.67	31.67
	Private	0	0	0	0	0	0	0	0
	Central	0	46.35	0	46.35	0	53.32	0	99.67
	Subtotal	0	46.35	2.00	48.35	0	53.32	29.67	131.34
Mizoram	State	0	0	51.86	51.86	0	0	36.47	88.33
	Private	0	0	0	0	0	0	0	0
	Central	0	38.29	0	38.29	0	34.31	0	72.60
	Subtotal	0	38.29	51.86	90.15	0	34.31	36.47	160.93
Central (unallocated)		0	104.44	0	104.44	0	127.15	0	231.59
Total north-eastern region	State	60.00	445.70	142.74	648.44	0	382.00	253.25	1 283.69

Contd...

TABLE 5.1a *Contd...*

State/union territory/ central sector		Coal	Gas	Diesel	Total thermal	Nuclear	Hydro	RES	Total
	Private	0	24.50	0	24.50	0	0	9.13	33.63
	Central	0	1 192.50	0	1 192.50	0	860.00	0	2 052.50
	Subtotal	60.00	1 662.70	142.74	1 865.44	0	1 242.00	262.38	3 369.82
Islands	State	0	0	40.05	40.05	0	0	5.25	45.30
Andaman and Nicobar	Private	0	0	20.00	20.00	0	0	5.10	25.10
	Central	0	0	0	0	0	0	0	0
	Subtotal	0	0	60.05	60.05	0	0	10.35	70.40
Lakshadweep	State	0	0	9.97	9.97	0	0	0	9.97
	Private	0	0	0	0	0	0	0.75	0.75
	Central	0	0	0	0	0	0	0	0
	Subtotal	0	0	9.97	9.97	0	0	0.75	10.72
Total (Islands)	State	0	0	50.02	50.02	0	0	5.25	55.27
	Private	0	0	20.00	20.00	0	0	5.85	25.85
	Central	0	0	0	0	0	0	0	0
	Subtotal	0	0	70.02	70.02	0	0	11.10	81.12
All-India	State	58 100.50	6 974.42	602.61	65 677.53	0	27 482.00	1 919.31	95 078.84
	Private	58 405.38	8 568.00	597.14	67 570.52	0	2 694.00	33 857.65	104 122.17
	Central	48 130.00	7 519.73	0	55 649.73	5 780.00	11 091.43	0	72 521.16
	Total	164 635.88	23 062.15	1 199.75	188 897.78	5 780.00	41 267.43	35 776.96	271 722.17

RES – renewable energy sources

Note RES include small hydro projects, biomass gasifier, urban and industrial waste power, and solar.

Source Details available at <http://www.cea.nic.in/reports/monthly/installedcapacity/2015/installed_capacity-03.pdf>, last accessed on 23 November 2015

TABLE 5.1b Growth of installed electrical energy generation capacity (MW), utilities by mode

As on	Hydro	Thermal (including steam, gas, and diesel)	Nuclear	RES	Total
31 December 1950	560	1 153	0	0	1 713
31 March 1956 (end of First Five-year Plan)	1 061	1 825	0	0	2 886
31 March 1961 (end of Second Five-year Plan)	1 917	2 736	0	0	4 653
31 March 1966 (end of Third Five-year Plan)	4 124	4 903	0	0	9 027
31 March 1969 (end of three Annual Plans)	5 907	7 050	0	0	12 957
31 March 1974 (end of Fourth Five-year Plan)	6 966	9 058	640	0	16 664
31 March 1979 (end of Fifth Five-year Plan)	10 833	15 207	640	0	26 680
31 March 1980 (end of Annual Plan)	11 384	16 424	640	0	28 448
31 March 1985 (end of Sixth Five-year Plan)	14 460	27 030	1 095	0	42 585
31 March 1990 (end of Seventh Five-year Plan)	18 307	43 746	1 565	18	63 636
31 March 1992 (end of two Annual Plans)	19 194	48 054	1 785	32	69 065
31 March 1997 (end of Eighth Five-year Plan)	21 658	61 010	2 225	902	85 795
31 March 2002 (end of Ninth Five-year Plan)	26 269	74 429	2 720	1 628	105 046
31 March 2007 (end of Tenth Five-year Plan)	34 654	86 014	3 900	7 761	132 329
31 March 2008 (1st year of Eleventh Five-year Plan)	35 909	91 907	4 120	11 125	143 061
31 March 2009 (2nd year of Eleventh Five-year Plan)	36 878[a]	93 725	4 120	13 242	111 087
31 March 2010 (3rd year of Eleventh Five-year Plan)	36 863	102 454	4 560	15 521	159 398
31 March 2011 (4th year of Eleventh Five-year Plan)	37 567	112 824	4 780	18 455	173 626
31 March 2012 (end of Eleventh Five-year Plan)	38 990	131 603	4 780	24 504	199 877
31 March 2013 (1st year of Twelfth Five-year Plan)	39 491	151 531	4 780	27 542	223 344
31 March 2014 (2nd year of Twelfth Five-year Plan)	40 532	168 255	4 780	31 692	245 259

RES – renewable energy sources

[a] After accounting for deration/upration and reconciliation with utilities, the installed capacity figure of 36 878 MW works out to 36 846 MW.

Note RES include small hydro projects, biomass gasifiers, urban and industrial waste power, and solar power.

Source CEA (Central Electricity Authority). 2014. ***All India Electricity Statistics General Review 2014***. New Delhi: CEA, Government of India

TABLE **5.2** | Annual gross generation in utilities (GWh): 2010/11–2012/13

Type	2010/11	2011/12	2012/13
Diesel	3 181.33	2 648.69	2 448.44
Renewable energy sources	39 244.54	51 226.05	57 448.91
Nuclear	26 266.40	32 286.56	32 866.11
Gas	100 342.35	93 281.34	66 663.91
Hydro	114 415.52	130 511.47	113 720.29
Steam (thermal)	561 298.07	612 497.08	69 1341.21
Total	844 748.21	922 451.19	964 488.87

Source CEA (Central Electricity Authority). 2014. ***All India Electricity Statistics General Review 2014***. New Delhi: CEA, Government of India

TABLE 5.3 Annual plant load factor (%) of thermal/nuclear electricity generation plants, by state/union territory: 2012/13

State/union territory	Thermal			
	Central sector	State sector	Private sector	Nuclear
Northern region				
Haryana	57.86	53.68		
Himachal Pradesh				
Jammu and Kashmir			30.26	
Punjab		78.48		
Rajasthan	58.47	77.54	73.77	85.59
Uttar Pradesh	87.67	53.81	58.57	65.93
Uttarakhand				
Chandigarh				
Delhi	73.77	67.04		
Western region				
Gujarat		62.67	81.22	90.03
Madhya Pradesh	82.06	65.73	34.64	
Chhattisgarh	83.21	77.61	75.08	
Maharashtra		58.10	68.86	80.11
Goa				
Dadra and Nagar Haveli				
Daman and Diu				
Southern region				
Andhra Pradesh	87.78	85.03	65.96	
Karnataka		67.20	87.36	70.60
Kerala				
Tamil Nadu	82.97	77.94	82.56	71.12
Lakshadweep				
Puducherry				
Eastern region				
Bihar	65.58			
Jharkhand		35.12	6.89	
Odisha	83.32	86.47	40.45	
West Bengal	63.51	58.15	77.41	
Damodar Valley Corporation	64.95			
Andaman and Nicobar Islands				
Sikkim				
North-eastern region				
Assam				

Contd...

TABLE **5.3** | *Contd...*

State/union territory	Thermal			
	Central sector	State sector	Private sector	Nuclear
Manipur				
Meghalaya				
Nagaland				
Tripura				
Arunachal Pradesh				
Mizoram				
All-India	79.18	65.54	64.07	78.49

Source CEA (Central Electricity Authority). 2014. ***All India Electricity Statistics General Review 2014***. New Delhi: CEA, Government of India

TABLE **5.4** Installed capacity (MW) and generation (GWh): 1990/91–2012/13 (with respect to non-utilities of installed capacity 1 MW and above)

Year and item	Hydro	Steam	Diesel	Gas	RES	Total
1990/91						
Installed capacity	3.60	4991.06	3031.67	475.25		8501.59
Energy generated	14.89	20005.67	3216.27	1844.98		25081.81
1992/93						
Installed capacity	4.30	5540.74	3726.66	630.81	2.00	9904.51
Energy generated	16.54	24679.89	3851.03	2778.11	2.18	31327.75
1993/94						
Installed capacity	4.30	5792.99	3995.11	773.89	8.40	10574.69
Energy generated	14.89	25407.49	3684.39	3149.50	6.32	32262.59
1994/95						
Installed capacity	4.30	6010.47	4156.12	807.74	34.40	11013.04
Energy generated	14.89	27387.92	4102.69	3407.04	131.32	35043.86
1995/96						
Installed capacity	3.50	6305.30	4199.40	955.41	165.38	11628.98
Energy generated	17.39	28751.69	4334.35	4576.25	462.54	38142.22
1996/97						
Installed capacity	3.50	6151.69	4529.91	1166.38	64.84	11916.31
Energy generated	28.84	29127.94	6573.38	5039.22	45.63	40815.01
1997/98						
Installed capacity	20.75	6628.87	4958.12	1330.44	65.74	13003.92
Energy generated	78.84	30683.70	7463.62	5775.66	48.81	44050.63
1998/99						
Installed capacity	23.69	6799.08	5367.21	1676.06	65.74	13931.77
Energy generated	80.38	31198.19	8154.46	8871.96	48.34	48353.33
1999/2001						
Installed capacity	46.80	7675.66	5737.13	1816.89	59.72	15336.20
Energy generated	98.36	34813.90	10489.42	9932.01	63.47	55397.16
2000/01						
Installed capacity	46.72	7996.79	6217.24	1823.31	72.30	16156.36
Energy generated	99.85	40256.24	11397.99	7782.34	101.54	59637.96
2001/02						
Installed capacity	51.10	8354.03	6521.95	2125.91	92.20	17145.20
Energy generated	116.59	41853.85	10719.66	8835.19	155.53	61680.82

Contd...

TABLE **5.4** *Contd...*

Year and item	Hydro	Steam	Diesel	Gas	RES	Total
2002/03						
Installed capacity	51.10	9 094.16	6 736.10	2 385.15	95.11	18 361.62
Energy generated	90.32	42 229.66	11 155.33	10 198.16	176.43	63 849.90
2003/04						
Installed capacity	61.97	8 585.28	7 195.47	2 745.73	151.85	18 740.31
Energy generated	97.49	39 610.29	13 403.31	14 874.38	187.65	68 173.12
2004/05						
Installed capacity	58.93	8 903.15	7 126.86	2 866.40	147.22	19 102.57
Energy generated	112.77	44 016.99	12 039.86	15 052.05	194.95	71 416.62
2005/06						
Installed capacity	59.98	10 816.28	7 418.14	3 025.16	148.67	21 468.21
Energy generated	236.47	46 264.87	12 253.22	14 664.72	220.42	73 639.70
2006/07						
Installed capacity	60.50	11 397.52	7 723.35	2 976.00	177.67	22 335.04
Energy generated	217.81	56 184.04	9 974.85	15 207.17	215.89	81 799.76
2007/08						
Installed capacity	60.50	11 763.99	8 647.92	4 209.23	304.75	24 986.39
Energy generated	202.21	53568.52	10737.96	25584.57	383.43	90 476.69
2008/09						
Installed capacity	110.79	13 817.53	8 749.67	3 621.05	374.63	26 673.67
Energy generated	146.01	73 626.49	10 082.14	15 306.09	560.43	99 721.16
2009/10						
Installed capacity	55.01	17 183.34	9 456.61	4 368.02	453.89	31 516.87
Energy generated	151.78	77 416.34	8 217.22	19 738.71	609.05	106 133.10
2010/11						
Installed capacity	56.88	19 111.54	9 655.25	5 053.76	566.69	34 444.12
Energy generated	148.79	96 657.13	7 753.53	15 435.17	922.29	120 916.91
2011/12						
Installed capacity	47.59	22 615.39	9 955.23	5 884.95	872.21	39 375.37
Energy generated	131.28	104 862.66	6 244.30	21 971.57	1 178.17	134 387.98
2012/13						
Installed capacity	66.90	23 889.61	11 148.14	4 497.69	1 124.05	40 726.39
Energy generated	118.18	113 166.95	8 205.22	20 768.88	1 750.42	144 009.65

RES – renewable energy sources

Source CEA (Central Electricity Authority). 2014. ***All India Electricity Statistics General Review 2014***. New Delhi: CEA, Government of India

TABLE 5.5 Length of transmission and distribution lines (ckt km) by region and state/union territory as on 31 March 2013

Region/state/union territory	HVDC (OH)	800 kV (OH)	400 kV (OH)	230 kV (OH)	220 kV (OH)	132 kV (OH)	110/90 kV (OH)	78/66 kV (OH)	33 kV (OH)	33 kV (UG)	22/20 kV (OH)	22/20 kV (UG)	15/11 kV (OH)	15/11 kV (UG)	6.6 kV (OH)	6.6 kV (UG)	3.3/2.2 kV (OH)	3.3/2.2 kV (UG)	Distribution lines upto 500 volts	Total
North																				
Haryana	0	0	486	0	3 701	3 994	0	2 793	5 376	0	0	0	112 302	0	0	0	0	0	110 001	238 653
Himachal Pradesh	0	0	278	0	808	1 445	0	730	3 044	0	7 138	0	20 925	0	0	0	46	0	53 809	88 223
Jammu and Kashmir	0	0	134	0	622	1 572	0	221	3653	0	0	0	26 740	0	158	0	0	0	57 182	90 282
Punjab	0	0	1 434	0	8 924	3 219	0	7 712	504	0	0	0	203 415	0	0	0	0	0	153 933	379 141
Rajasthan	0	0	3 019	0	11 806	14 779	0	303	43 633	48	0	0	330 704	43	0	0	0	0	381 974	786 309
Uttar Pradesh	0	411	4 259	0	8 091	13 913	0	3 139	34 972	122	0	0	215 892	300	0	0	0	0	242 700	523 799
Uttarakhand	0	0	328	0	546	1 278	0	41	3 978	0	0	0	33 253	0	0	0	0	0	48 282	87 706
Chandigarh	0	0	0	0	54	0	0	103	28	0	0	0	875	0	0	0	0	0	1 248	2 308
Delhi	0	0	227		660	0	0	1 101	1 413	114	0	0	8 751	1 679	0	0	0	0	34 288	48 233
Central sector	3 210	1 973	21 169	0	4 443	323	0	0	0	0	44	0	12	0	0	0	0	0	23	31 197
Total	3 210	2 384	31 334	0	39 655	40 523	0	16 143	96 601	284	7 182	0	952 869	2 022	158	0	46	0	1 083 440	2 275 851
West																				
Gujarat	0	0	3 717	0	16 352	5 066	0	23 906	75	241	5 905	0	274 099	8 116	0	0	0	0	320 937	658 414
Madhya Pradesh	0	0	2 448	0	11 333	13 983	0	61	41 429	0	0	0	246 464	0	0	0	0	0	377 448	693 166
Chhattisgarh	360	0	1 113	0	2 446	5 051	0	0	16 327	0	0	0	75 168	0	0	0	0	0	133 881	234 346
Maharashtra	1 504	0	7 439	0	15 348	13 031	3 147	3 270	38 803	1 106	28 344	3 154	255 753	6 685	205	1 906	0	0	603 909	983 604
Goa	0	0	0	0	684	0	423	0	1 167	0	0	0	3 179	72	0	0	0	0	9 260	14 785
Dadra and Nagar Haveli	0	0	0	0	49	0	0	131	0	0	0	0	637	0	0	0	0	0	1 786	2 603
Daman and Diu	0	0	0	0	19	0	0	84	0	0	0	0	345	85	0	0	0	0	788	1 321
Central sector	0	3 375	23 375		1 282	0	0	0	3	0	0	0	0	0	0	0	0	0	0	28 035
Total	1 864	3 375	38 092	0	47 513	37 131	3 570	27 452	97 804	1 347	34 249	3 154	855 645	14 958	205	1 906	0	0	1 448 009	2 616 274
South																				
Andhra Pradesh	196	0	4 111	0	14 136	16 264	0	169	42 549	331	0	0	297 728	689	26 107	0	0	0	578 884	981 164
Karnataka	0	0	1 978	0	8 574	0	7 492	8 096	7 545	39	0	0	246 433	3 446	0	0	0	0	544 333	827 936
Kerala	0	0	0	0	2 693	0	4 405	3 007	1 434	0	103	0	53 172	23	0	0	0	0	273 633	338 470
Tamil Nadu	0	0	243	8 920	0	0	15 530	1 186	7 047	909	40 836	24	106 302	3 191	0	0	0	0	588 508	772 696
Lakshadweep	0	0	0	0	0	0	0	0	0	0	0	0	0	163	0	0	0	92	225	480

Contd...

TABLE 5.5 Contd...

Region/state/ union territory	HVDC (OH)	800 kV (OH)	400 kV (OH)	230 kV (OH)	220 kV (OH)	132 kV (OH)	110/90 kV (OH)	78/66 kV (OH)	33 kV (OH)	33 kV (UG)	22/20 kV (OH)	22/20 kV (UG)	15/11 kV (OH)	15/11 kV (UG)	6.6 kV (OH)	6.6 kV (UG)	3.3/2.2 kV (OH)	3.3/2.2 kV (UG)	Distribution lines upto 500 volts	Total
Puducherry	0	0	0	135	0	36	206	0	0	0	760	0	326	0	0	0	0	0	4 541	6 004
Central sector	2 738	0	15 974	0	812	0	0	0	0	0	0	0	0	0	0	0	0	0	0	19 524
Total	2 934	0	22 306	9 055	26 215	16 300	27 633	12 458	58 575	1 279	41 699	24	703 961	7 512	26 107	0	0	92	1 990 124	2 946 274
East																				
Bihar	0	0	75	0	1 445	5 390	0	0	6 495	0	0	0	38 994	0	0	0	0	0	117 654	170 053
Jharkhand	0	0	135	0	712	1 867	0	0	2 517	0	0	0	19 809	0	4 365	0	0	0	19 904	49 309
Odisha	0	0	522	0	5 612	5 313	0	0	12 352	0	0	0	98 369	0	0	0	0	0	114 356	236 524
West Bengal	0	0	1 549	0	2 855	7 011	0	421	12 488	1 654	0	50	151 020	6 588	87	0	22	23	270 292	454 060
Damodar Valley Corporation	0	0	0	0	1 943	3 565	0	0	1 994	0	0	0	0	0	0	0	0	0	0	7 502
Andaman and Nicobar Islands	0	0	0	0	0	0	0	0	425	0	0	0	748	0	0	0	0	0	3 414	4 587
Sikkim	0	0	0	0	0	0	0	273	0	0	0	0	2 109	24	0	0	15	0	2 735	5 156
Central sector	0	713	14 445	0	1 573	419	0	37	0	0	0	0	60	0	0	0	0	0	0	17 247
Total	0	713	16 726	0	14 140	23 565	0	731	36 271	1 654	0	50	311 109	6 612	4 452	0	37	23	528 355	944 438
North-east																				
Assam	0	0	0	0	1 385	2 063	0	727	6 756	0	0	0	58 781	0	0	0	0	0	118 121	187 833
Manipur	0	0	0	0	0	439	0	0	1 133	0	0	0	5 966	0	0	0	0	0	7 834	15 372
Meghalaya	0	0	0	0	0	769	0	0	1 707	0	0	0	9 330	0	0	0	0	0	8 903	20 709
Nagaland	0	0	0	0	0	370	0	392	1 216	0	0	0	5 079	0	0	0	0	0	7 252	14 309
Tripura	0	0	0	0	0	557	0	383	721	0	0	0	10 027	0	0	0	0	0	18 474	30 162
Arunachal Pradesh	0	0	0	0	19	189	0	0	3 074	0	0	0	5 569	0	0	0	0	0	7 681	16 532
Mizoram	0	0	0	0	0	638	0	117	828	0	0	0	4 616	0	0	0	0	0	2 253	8 452
Central sector	0	0	1 950	0	552	1 848	0	0	0	0	0	0	0	0	0	0	0	0	0	4 350
Total	0	0	1 950	0	1 956	6 873	0	1 619	15 435	0	0	0	99 368	0	0	0	0	0	170 518	297 719
All-India	8 008	6 472	110 408	9 055	129 479	124 392	31 203	58 403	304 686	4 564	83 130	3 228	2 922 952	31 104	30 922	1 906	83	115	5 220 446	9 080 556

HVDC – high-voltage direct current; OH – overhead; UG – underground

Source CEA (Central Electricity Authority). 2014. ***All India Electricity Statistics General Review 2014***. New Delhi: CEA, Government of India

TABLE 5.6 Transmission and distribution losses as a percentage of availability in states/union territories: 2001/02–2012/13

State/union territory	2001/02	2002/03	2003/04	2004/05	2005/06	2006/07	2007/08	2008/09	2009/10	2010/11	2011/12	2012/13
State												
Andhra Pradesh	26.8	30.1	27.7	24.0	20.1	18.7	22.4	19.6	18.4	16.6	17.5	19.3
Arunachal Pradesh	53.6	39.0	47.5	43.0	49.7	57.8	67.2	46.9	48.0	47.1	46.3	46.0
Assam	42.8	38.3	39.3	51.8	40.3	33.7	38.6	37.6	32.8	34.2	33.5	30.7
Bihar	51.7	38.0	36.7	38.9	44.0	50.7	48.8	46.4	43.6	50.8	50.9	49.4
Chhattisgarh	33.8	37.9	42.6	28.1	31.1	31.7	29.8	26.4	18.6	15.1	16.5	28.8
Delhi	44.0	45.8	43.7	45.4	42.2	33.0	28.7	22.2	22.1	20.0	19.3	22.1
Goa	25.2	40.3	45.1	36.0	19.7	20.9	21.2	17.1	17.0	15.3	12.4	13.4
Gujarat	26.9	28.5	24.2	30.4	27.9	24.9	26.1	24.1	22.8	19.2	21.8	18.5
Haryana	39.2	37.7	32.1	32.1	30.5	33.4	32.8	30.7	31.0	29.7	28.6	36.0
Himachal Pradesh	25.6	21.2	22.8	28.9	23.6	19.8	17.0	15.5	20.5	22.2	18.6	19.1
Jammu and Kashmir	48.9	45.6	45.5	41.1	44.9	52.0	55.7	58.0	67.4	63.3	61.8	56.6
Jharkhand	26.4	21.2	25.4	19.6	26.8	26.2	23.2	24.3	22.2	17.1	14.3	13.6
Karnataka	33.8	24.6	23.3	26.1	29.8	25.9	18.9	17.0	18.8	17.3	12.7	11.1
Kerala	32.2	27.5	21.6	22.5	23.5	19.1	17.8	13.2	19.6	18.3	17.2	17.7
Madhya Pradesh	44.6	43.3	41.4	41.3	40.1	39.2	35.6	38.5	38.3	37.6	34.5	31.5
Maharashtra	37.3	34.0	34.1	32.4	31.6	31.6	29.8	23.9	25.2	20.7	20.0	21.8
Manipur	62.4	63.7	65.2	70.6	63.1	53.5	63.6	63.4	54.7	50.9	40.5	35.1
Meghalaya	22.7	21.9	16.7	28.4	40.2	35.3	37.6	37.5	39.1	35.8	31.0	23.6
Mizoram	49.8	46.9	55.5	66.1	39.2	38.2	44.6	52.7	53.8	45.6	47.7	37.8
Nagaland	52.3	56.7	55.0	48.3	59.0	54.8	55.6	58.3	56.9	48.2	41.5	40.2
Odisha	47.3	45.4	57.1	44.0	45.6	40.9	39.4	42.7	37.0	42.5	44.6	39.8
Punjab	27.7	24.4	26.0	25.4	27.6	26.6	22.8	23.1	23.4	25.1	23.1	20.3
Rajasthan	43.1	42.6	43.7	44.7	39.9	35.6	34.7	31.5	30.0	27.9	27.9	24.9
Sikkim	31.7	54.9	55.0	50.5	10.7	26.9	36.8	38.8	39.0	33.7	31.1	28.1
Tamil Nadu	16.1	17.3	17.2	19.3	18.7	19.5	18.7	18.1	18.4	13.5	16.3	14.5
Tripura	40.4	40.6	46.4	59.5	41.1	34.8	42.8	35.8	35.6	27.4	39.1	31.7
Uttar Pradesh	37.6	34.2	35.2	34.4	32.6	33.5	28.6	30.9	33.2	34.0	32.4	26.9
Uttarakhand	32.4	25.2	49.2	39.3	36.0	34.5	35.7	41.8	25.3	30.0	28.7	26.9
West Bengal	31.7	25.9	31.0	28.5	24.8	23.6	21.3	16.8	18.3	22.4	23.2	24.1
Union territory												
Andaman and Nicobar Islands	29.2	19.8	26.0	12.6	24.3	23.1	26.9	24.2	19.8	20.7	18.2	18.1
Chandigarh	25.0	24.1	39.1	30.4	31.6	25.1	23.8	22.4	23.2	20.3	23.7	19.3
Dadra and Nagar Haveli	27.2	40.3	15.1	16.0	19.9	11.2	18.1	15.6	11.2	10.1	12.1	9.9

Contd...

TABLE 5.6 *Contd...*

State/union territory	2001/02	2002/03	2003/04	2004/05	2005/06	2006/07	2007/08	2008/09	2009/10	2010/11	2011/12	2012/13
Daman and Diu	7.5	15.0	16.9	15.6	21.6	22.1	23.3	20.1	17.2	16.8	14.5	15.6
Lakshadweep	10.9	11.3	11.9	10.2	11.2	12.9	18.1	24.9	11.6	25.7	22.5	18.6
Puducherry	12.0	21.1	11.6	18.2	18.5	18.8	5.9	12.2	11.8	12.4	14.7	13.5
Damodar Valley Corporation	3.6	3.3	2.7	2.7	1.8	-	-	-	-	-	-	-
All-India	34.0	38.3	32.5	31.3	30.4	28.7	27.2	25.5	25.4	24.0	23.7	23.0

Source CEA (Central Electricity Authority). 2014. ***All India Electricity Statistics General Review 2014***. New Delhi: CEA, Government of India

TABLE 5.7a Aggregate technical and commercial losses (%) for SEBs and PDs selling directly to the consumers: 2011/12–2013/14

Utility	2011/12					2012/13					2013/14				
	Net input energy (MkWh)	Net energy sold (MkWh)	Collection efficiency (%)	Energy realized (MkWh)	AT&C losses (%)	Net input energy (MkWh)	Net energy sold (MkWh)	Collection efficiency (%)	Energy realized (MkWh)	AT&C losses (%)	Net input energy (MkWh)	Net energy sold (MkWh)	Collection efficiency (%)	Energy realized (MkWh)	AT&C losses (%)
BSEB	10 802	5 531	79.60	4 402	59.24	6 952	3 626	77.83	2 822	59.40					
JSEB	9 253	6 063	87.35	5 296	42.76	10 438	6 786	80.77	5 481	47.49	8 276	5 485	5 485	4 786	42.17
Sikkim PD	380	208	76.10	158	58.32	372	227	76.20	173	53.51	391	267	267	112	71.23
Arunachal PD	549	245	77.28	189	65.55	556	238	92.96	221	60.26	573	233	233	182	68.20
Manipur PD	506	325	85.90	279	44.80	548	379	20.98	80	85.49	565	415	415	319	43.55
Mizoram PD	393	252	98.67	249	36.59	405	287	102.42	294	27.55	436	303	303	294	32.53
Nagaland PD	473	293	124.48	365	22.85	501	312	39.69	124	75.30	609	421	421	375	38.37
Tripura PD	840	624	89.12	556	33.76	938	776	90.89	705	24.86	981	712	712	708	27.81
J&K PDD	10 503	4 267	70.99	3 029	71.16	11 129	5 163	84.36	4 355	60.87	11 727	5 754	5 754	5 964	49.14
KSEB	18 026	15 981	99.07	15 833	12.17	19 035	16 838	99.12	16 690	12.32	11 024	9 832	9 832	9 762	11.45
Puducherry PD	2 685	2 322	93.78	2 177	18.91	2 849	2 493	103.86	2 589	9.13	2 727	2 366	2 366	2 286	16.18
Goa PD	3 180	2 753	98.07	2 699	15.12	3 374	2 952	98.14	2 897	14.14	3 366	2 962	2 962	3 005	10.72
Grand total	57 590	38 864	90.66	35 232	38.82	57 097	40 077	90.90	36 431	36.19	40 675	28 750	28 750	27 793	31.67

BSEB – Bihar State Electricity Board; J&K PDD – Jammu and Kashmir Power Development Department; JSEB – Jharkhand State Electricity Board; KSEB – Kerala State Electricity Board; SIKKIM PD – Sikkim Power Department; Goa PD – Goa Power Department

Notes

1. AT&C losses for Sikkim PD (for 2009/10–2011/12) include transmission losses since the figure for transmission losses is not available.
2. Collection efficiency for J & K PDD (for 2008/09–2011/12) has been calculated based on figures for revenue realized available in Resource Plan.

Source PFCL (Power Finance Corporation Ltd). 2014. ***Report on the performance of state power utilities for 2011/12–2013/14***. Details available at <http://www.pfcindia.com>, last accessed on 03 December 2015

TABLE 5.7b Aggregate technical and commercial losses (%) for all discoms of unbundled states: 2011/12–2013/14

State	Utility	2011/12					2012/13					2013/14				
		Net input energy (MkWh)	Energy sold (MkWh)	Collection efficiency (%)	Energy realized (MkWh)	AT&C losses (%)	Net input energy (MkWh)	Energy sold (MkWh)	Collection efficiency (%)	Energy realized (MkWh)	AT&C losses (%)	Net input energy (MkWh)	Energy sold (MkWh)	Collection efficiency (%)	Energy realized (MkWh)	AT&C losses (%)
Bihar	NBPDCL						1506	916	80.84	740	50.85	4553	2777	95.21	2644	41.93
	SBPDCL						3082	1597	104.66	1671	45.77	8465	4481	96.90	4342	48.70
	Total						4588	2513	95.98	2411	47.44	13018	7258	96.25	6986	46.33
Odisha	CESU	7233	4470	87.14	3895	46.15	7402	4663	89.79	4187	43.43	7973	5212	94.11	4905	38.48
	NESCO	5023	3302	92.00	3037	39.54	5045	3283	92.81	3047	39.61	5045	3338	96.03	3205	36.47
	SESCO	2814	1508	88.47	1334	52.60	2930	1661	89.34	1484	49.36	2916	1720	99.69	1715	41.18
	WESCO	6178	3775	92.53	3493	43.46	6391	3945	94.18	3716	41.87	6635	4201	92.81	3899	41.24
	Total	21248	13055	90.08	11759	44.66	21768	13552	91.74	12434	42.88	22569	14471	94.84	13724	39.19
West Bengal	WBSEDCL	29731	21568	92.50	19950	32.90	29542	20534	94.33	19369	34.43	30495	20626	100.47	20722	32.05
	Total	29731	21568	92.50	19950	32.90	29542	20534	94.33	19369	34.43	30495	20626	100.47	20722	32.05
Assam	APDCL	5247	3969	93.24	3701	29.47	5672	4205	91.92	3865	31.85	6277	4763	91.92	4378	30.25
	Total	5247	3969	93.24	3701	29.47	5672	4205	91.92	3865	31.85	6277	4763	91.92	4378	30.25
Meghalaya	MeECL	1509	1075	76.74	825	45.33										
	MePDCL						1469	1042	89.93	937	36.25	1451	1053	89.06	937	35.38
	Total	1509	1075	76.74	825	45.33	1469	1042	89.93	937	36.25	1451	1053	89.06	937	35.38
Delhi	BSES Rajdhani Ltd	10903	9366	97.03	9088	16.65	11231	9827	96.96	9528	15.16	11985	10102	99.44	10045	16.19
	BSES Yamuna Ltd	6204	4844	95.37	4620	25.54	6331	5479	94.82	5195	17.94	6848	5574	103.80	5786	15.51
	TPDDL	7785	6702	97.97	6566	15.67	7897	6968	98.47	6861	13.12	8040	7200	100.77	7256	9.75
	Total	24892	20912	96.95	20274	18.56	25459	22274	96.90	21584	15.22	26873	22876	100.92	23087	14.09
Haryana	DHBVNL	17903	13658	95.00	12974	27.53	18387	14272	92.36	13181	28.31	22056	16838	90.53	15244	30.89

Contd...

TABLE 5.7b Contd...

State	Utility	2011/12					2012/13					2013/14				
		Net input energy (MkWh)	Energy sold (MkWh)	Collection efficiency (%)	Energy realized (MkWh)	AT&C losses (%)	Net input energy (MkWh)	Energy sold (MkWh)	Collection efficiency (%)	Energy realized (MkWh)	AT&C losses (%)	Net input energy (MkWh)	Energy sold (MkWh)	Collection efficiency (%)	Energy realized (MkWh)	AT&C losses (%)
	UHBVNL	16 744	13 203	89.97	11 878	29.06	17 649	12 132	91.69	11 124	36.97	17 720	11 979	90.81	10 878	38.61
	Total	34 647	26 861	92.52	24 852	28.27	36 036	26 404	92.05	24 305	32.55	39 776	28 817	90.65	26 122	34.33
Himachal Pradesh	HPSEBL	7 847	6 918	92.96	6 431	18.04	8 362	7 224	101.99	7 367	11.90	8 549	7 536	96.28	7 255	15.13
	Total	7 847	6 918	92.96	6 431	18.04	8 362	7 224	101.99	7 367	11.90	8 549	7 536	96.28	7 255	15.13
Punjab	PSPCL	41 530	34 600	97.27	33 655	18.96	43 548	36 242	99.11	35 918	17.52	45 238	37 650	98.64	37 138	17.91
	Total	41 530	34 600	97.27	33 655	18.96	43 548	36 242	99.11	35 918	17.52	45 238	37 650	98.64	37 138	17.91
Rajasthan	AVVNL	13 604	10 646	91.86	9 779	28.12	14 950	11 863	100.95	11 975	19.90	15 760	12 500	98.30	12 287	22.04
	JdVVNL	14 821	12 024	93.89	11 290	23.83	16 631	13 615	98.98	13 476	18.97	18 809	14 587	95.82	13 977	25.69
	JVVNL	18 722	15 234	94.40	14 381	23.18	20 619	16 682	97.75	16 308	20.91	21 879	15 784	95.53	15 079	31.08
	Total	47 147	37 904	93.53	35 450	24.81	52 200	42 160	99.05	41 759	20.00	56 448	42 871	96.44	41 343	26.76
Uttar Pradesh	DVVNL	16 052	11 335	84.26	9 551	40.50	17 331	12 577	74.84	9 413	45.69	18 437	13 148	89.09	11 713	36.47
	KESCO	3 089	2 071	103.71	2 148	30.48	3 140	2 154	90.97	1 959	37.61	3 554	2 458	95.02	2 336	34.29
	MVVNL	12 537	9 233	75.48	6 968	44.42	13 147	9 881	72.08	7 122	45.83	14 253	10 711	114.02	12 213	14.32
	Pash VVNL	22 649	16 025	90.53	14 507	35.95	23 674	17 007	92.72	15 770	33.39	24 099	18 537	99.47	18 438	23.49
	Poor VVNL	15 704	11 590	64.54	7 480	52.37	16 034	11 920	64.07	7 637	52.37	16 929	12 743	106.17	13 529	20.09
	Total	70 031	50 254	80.90	40 654	41.95	73 326	53 539	78.26	41 901	42.85	77 272	57 597	101.10	58 229	24.65
Uttarakhand	UtPCL	10 311	8 253	92.65	7 646	25.84	10 789	8 577	96.63	8 288	23.18	11 017	9 065	98.43	8 923	19.01
	Total	10 311	8 253	92.65	7 646	25.84	10 789	8 577	96.63	8 288	23.18	11 017	9 065	98.43	8 923	19.01
Andhra Pradesh	APCPDCL	37 161	31 597	97.71	30 558	17.77	34 779	30 086	97.52	29 339	15.64	36 717	32 378	93.51	30 277	17.54
	APEPDCL	12 595	11 726	96.11	11 269	10.53	12 207	11 402	96.19	10 968	10.15	13 772	13 176	97.65	12 867	6.57
	APNPDCL	11 914	10 244	96.23	9 858	17.26	11 165	9 672	100.33	9 704	13.09	11 914	10 287	91.73	9 436	20.80

Contd...

TABLE 5.7b Contd...

State	Utility	2011/12					2012/13					2013/14				
		Net input energy (MkWh)	Energy sold (MkWh)	Collection efficiency (%)	Energy realized (MkWh)	AT&C losses (%)	Net input energy (MkWh)	Energy sold (MkWh)	Collection efficiency (%)	Energy realized (MkWh)	AT&C losses (%)	Net input energy (MkWh)	Energy sold (MkWh)	Collection efficiency (%)	Energy realized (MkWh)	AT&C losses (%)
	APSPDCL	18 475	16 388	98.99	16 223	12.19	18 418	16 445	97.72	16 071	12.74	20 179	18 024	98.77	17 803	11.77
	Total	80 145	69 955	97.07	67 908	15.27	76 569	67 605	97.75	66 082	13.70	82 582	73 865	95.29	70 383	14.77
Karnataka	BESCOM	24 584	21 030	90.52	19 036	22.57	26 569	22 796	92.72	21 136	20.45	26 786	23 065	94.15	21 716	18.93
	CHESCOM	5 552	4 652	84.74	3 942	28.99	5 860	4 975	81.96	4 077	30.42	5 996	5 112	77.50	3 962	33.92
	GESCOM	6 951	5 442	97.12	5 285	23.96	7 064	5 724	100.86	5 773	18.28	7 268	5 977	84.57	5 055	30.45
	HESCOM	9 593	7 675	95.46	7 327	23.62	10 470	8 389	99.31	8 331	20.44	10 572	8 664	97.10	8 413	20.42
	MESCOM	4 072	3 570	93.61	3 342	17.94	4 280	3 772	96.95	3 657	14.57	4 585	4 038	96.71	3 905	14.83
	Total	50 752	42 369	91.89	38 932	23.29	54 243	45 656	94.13	42 974	20.78	55 207	46 856	91.88	43 051	22.02
Kerala	KSEBL											8 514	7 622	86.26	6 575	22.78
Tamil Nadu	TANGEDCO	66 569	53 938	96.63	52 120	21.70	66 827	52 747	100.46	52 990	20.71	77 315	60 630	99.02	60 034	22.35
Chhattisgarh	CSPDCL	18 238	13 177	98.20	12 939	29.05	20 525	16 090	95.52	15 369	25.12	19 553	14 790	101.58	15 023	23.17
Gujarat	DGVCL	11 764	10 563	96.74	10 219	13.14	12 819	11 337	101.31	11 486	10.40	13 420	12 040	99.39	11 967	10.83
	MGVCL	7 772	6 822	97.52	6 653	14.40	8 074	7 032	97.67	6 868	14.94	8 111	6 955	99.40	6 913	14.77
	PGVCL	21 919	16 332	96.59	15 775	28.03	24 566	17 221	99.27	17 096	30.41	22 740	17 464	98.80	17 255	24.12
	UGVCL	15 670	14 012	96.17	13 475	14.01	17 736	15 321	99.13	15 188	14.37	15 826	14 489	99.29	14 387	9.10
	Total	57 125	47 729	96.63	46 122	19.26	63 195	50 911	99.46	50 638	19.87	60 097	50 948	99.16	50 522	15.93

Contd...

TABLE 5.7b *Contd...*

State	Utility	2011/12					2012/13					2013/14				
		Net input energy (MkWh)	Energy sold (MkWh)	Collection efficiency (%)	Energy realized (MkWh)	AT&C losses (%)	Net input energy (MkWh)	Energy sold (MkWh)	Collection efficiency (%)	Energy realized (MkWh)	AT&C losses (%)	Net input energy (MkWh)	Energy sold (MkWh)	Collection efficiency (%)	Energy realized (MkWh)	AT&C losses (%)
Madhya Pradesh	MP Madhya Kshetra VVCL	13 803	8 577	87.14	7 474	45.85	14 374	9 937	101.30	10 066	29.97	16 440	11 557	100.13	11 573	29.60
	MP Paschim Kshetra VVCL	17 226	12 226	92.39	11 295	34.43	17 824	13 127	97.55	12 805	28.16	18 122	13 978	102.22	14 288	21.15
	MP Purv Kshetra VVCL	11 669	8 178	92.82	7 591	34.94	13 371	9 892	85.98	8 504	36.40	14 526	11 087	85.38	9 466	34.83
	Total	42 698	28 981	90.96	26 360	38.26	45 569	32 956	95.21	31 375	31.15	49 088	36 622	96.46	35 327	28.03
Maharashtra	MSEDCL	95 433	80 132	93.34	74 793	21.63	97 846	83 488	91.48	76 362	21.95	99 575	85 631	99.55	85 245	14.39
Grand total		705 100	561 650	93.36	524 371	25.63	737 533	587 719	94.59	555 928	24.62	790 914	631 547	97.38	615 004	22.24

APCPDCL – Andhra Pradesh Central Power Distribution Company Ltd; APDCL – Assam Power Development Corporation Ltd; APEPDCL – Andhra Pradesh Eastern Power Distribution Company Ltd; APNPDCL – Andhra Pradesh Northern Power Distribution Company Ltd; APSPDCL – Andhra Pradesh Southern Power Distribution Company Ltd; AVVNL – Ajmer Vidyut Vitran Nigam Ltd; BESCOM – Bangalore Electricity Supply Company Ltd; CAEDCL – Central Assam Electricity Distribution Company Ltd; CESCO – Central Electricity Supply Company of Orissa Ltd; CHESCOM – Chamundeshwari Electricity Supply Company Ltd; CSPDCL – Chhattisgarh State Power Distribution Company Ltd; DGVCL – Dakshin Gujarat Vij Company Ltd; DHBVNL – Dakshin Haryana Bijli Vitran Nigam Ltd; DVVNL – Dakshinanchal Vidyut Vitran Nigam Ltd; GESCOM – Gulbarga Electricity Supply Company Ltd; HESCOM – Hubli Electricity Supply Company Ltd; JdVVNL – Jodhpur Vidyut Vitran Nigam Ltd; JVVNL – Jaipur Vidyut Vitran Nigam Ltd; KESCO – Kanpur Electricity Supply Company Ltd; KVVCL – Kshetra Vidyut Vitaran Company Ltd; LAEDCL – Lower Assam Electricity Distribution Company Ltd; MESCOM – Mangalore Electricity Supply Company Ltd; MGVCL – Madhya Gujarat Vij Company Ltd; MP Madhya KVVCL – Madhya Pradesh Madhya Khsetra Vidyut Vidran Company Ltd; MP Paschim KVVCL – Madhya Pradesh Paschim Khsetra Vidyut Vidran Company Ltd; MP Purv KVVCL – Madhya Pradesh Purv Khsetra Vidyut Vidran Company Ltd; MSEDCL – Maharashtra State Electricity Distribution Company Ltd; MVVNL – Madhyanchal Vidyut Vitran Nigam Ltd; NDPL – North Delhi Power Ltd; NESCO – Northern Electricity Supply Company of Orissa Ltd; Pash VVNL – Paschimanchal Vidyut Vitaran Nigam Ltd; PGVCL – Paschim Gujarat Vij Company Ltd; Poorv VVNL – Poorvanchal Vidyut Vitaran Nigam Ltd; SESCO – Southern Electricity Supply Company of Orissa Ltd; WESCO – Western Electricity Supply Company of Orissa Ltd; UAEDCL – Upper Assam Electricity Distribution Company Ltd; UGVCL – Uttar Gujarat Vij Company Ltd; UHBVNL – Uttar Haryana Bijli Vitran Nigam Ltd; UtPCL – Uttarakhand Power Corporation Ltd

Notes

1. AT&C losses (%) = {(units input – units realized)/units input} × 100; Units realized = units billed × collection efficiency; Collection efficiency (%) = (revenue realized/revenue billed) × 100
2. AT&C losses for APSPDCL for 2009/10 to 2011/12 and BSES Rajdhani for 2010/11 include transmission losses since the figures for transmission losses are not available.

Source PPFCL (Power Finance Corporation Ltd). 2014. ***Report on the performance of state power utilities for 2011/12–2013/14***. Details available at <http://www.pfcindia.com>, last accessed on 26 December 2015

TABLE 5.7c Aggregate technical and commercial losses (%) for utilities selling directly to consumers: 2009/10–2011/12

State	Utility	2009/10					2010/11					2011/12				
		Net input energy (MkWh)	Energy sold (MkWh)	Collection efficiency (%)	Energy realized (MkWh)	AT&C losses (%)	Net input energy (MkWh)	Energy sold (MkWh)	Collection efficiency (%)	Energy realized (MkWh)	AT&C losses (%)	Net input energy (MkWh)	Energy sold (MkWh)	Collection efficiency (%)	Energy realized (MkWh)	AT&C losses (%)
Bihar	BSEB	10 802	5 531	79.60	4 402	59.24	6 952	3 626	77.83	2 822	59.40					
	NBPDCL						1 506	916	80.84	740	50.85	4 553	2 777	95.21	2 644	41.93
	SBPDCL						3 082	1 597	104.66	1 671	45.77	8 465	4 481	96.90	4 342	48.70
	Bihar total	10 802	5 531	79.60	4 402	59.24	11 540	6 139	85.26	5 233	54.64	13 018	7 258	96.25	6 986	46.33
Jharkhand	JSEB	9 253	6 063	87.35	5 296	42.76	10 438	6 786	80.77	5 481	47.49	8 276	5 485	87.26	4 786	42.17
Odisha	CESU	7 233	4 470	87.14	3 895	46.15	7 402	4 663	89.79	4 187	43.43	7 973	5 212	94.11	4 905	38.48
	NESCO	5 023	3 302	92.00	3 037	39.54	5 045	3 283	92.81	3 047	39.61	5 045	3 338	96.03	3 205	36.47
	SESCO	2 814	1 508	88.47	1 334	52.60	2 930	1 661	89.34	1 484	49.36	2 916	1 720	99.69	1 715	41.18
	WESCO	6 178	3 775	92.53	3 493	43.46	6 391	3 945	94.18	3 716	41.87	6 635	4 201	92.81	3 899	41.24
	Orissa total	21 248	13 055	90.08	11 759	44.66	21 768	13 552	91.74	12 434	42.88	22 569	14 471	94.84	13 724	39.19
Sikkim	Sikkim PD	380	208	76.10	158	58.32	372	227	76.20	173	53.51	391	267	42.09	112	71.23
West Bengal	WBSEDCL	29 731	21 568	92.50	19 950	32.90	29 542	20 534	94.33	19 369	34.43	30 495	20 626	100.47	20 722	32.05
Eastern total		71 414	46 425	89.54	41 565	41.80	73 660	47 238	90.37	42 690	42.04	74 749	48 107	96.31	46 330	38.02
Arunachal Pradesh	Arunachal Pradesh PD	549	245	77.28	189	65.55	556	238	92.96	221	60.26	573	233	78.11	182	68.20
Assam	APDCL	5 247	3 969	93.24	3 701	29.47	5 672	4 205	91.92	3 865.00	31.85	6 277	4 763	91.92	4 378	30.25
Manipur	Manipur PD	506	325	85.90	279	44.80	548	379	20.98	80	85.49	565	415	76.84	319	43.55
Meghalaya	MeECL/ MePDCL	1 509	1 075	76.74	825	45.33	1 469	1 042	89.93	937	36.25	1 451	1 053	89.06	937	35.38
Mizoram	Mizoram PD	393	252	98.67	249	36.59	405	287	102.42	294	27.55	436	303	97.20	294	32.53
Nagaland	Nagaland PD	473	293	124.48	365	22.85	501	312	39.69	124	75.30	609	421	89.12	375	38.37
Tripura	TSECL	840	624	89.12	556	33.76	938	776	90.89	705	24.86	981	712	99.38	708	27.81
North-eastern total		9 517	6 783	90.87	6 164	35.22	10 089	7 239	86.01	6 226	38.31	10 892	7 900	91.06	7 193	33.94
Delhi	BSES Rajdhani	10 903	9 366	97.03	9 088	16.65	11 231	9 827	96.96	1 959	15.16	11 985	10 102	99.44	10 045	16.19

Contd...

TABLE 5.7c Contd...

State	Utility	2009/10					2010/11					2011/12				
		Net input energy (MkWh)	Energy sold (MkWh)	Collection efficiency (%)	Energy realized (MkWh)	AT&C losses (%)	Net input energy (MkWh)	Energy sold (MkWh)	Collection efficiency (%)	Energy realized (MkWh)	AT&C losses (%)	Net input energy (MkWh)	Energy sold (MkWh)	Collection efficiency (%)	Energy realized (MkWh)	AT&C losses (%)
	BSES Yamuna	6204	4844	95.37	4620	25.54	6331	5479	94.82	7122	17.94	6848	5574	103.80	5786	15.51
	TPDDL	7785	6702	97.97	6566	15.67	7897	6968	98.47	15770	13.12	8040	7200	100.77	7256	9.75
	Delhi total	24892	20912	96.95	20274	18.56	25459	22274	96.90	7637	15.22	26873	22876	100.92	23087	14.09
Haryana	DHBVNL	17903	13658	95.00	12974	27.53	18387	14272	92.36	41901	28.31	22056	16838	90.53	15244	30.89
	UHBVNL	16744	13203	89.97	11878	29.06	17649	12132	91.69	8288	36.97	17720	11979	90.81	10878	38.61
	Haryana total	34647	26861	92.52	24852	28.27	36036	26404	92.05	185477	32.55	39776	28817	90.65	26122	34.33
Himachal Pradesh	HPSEB Ltd	7847	6918	92.86	6431	18.04	8362	7224	101.99	29339	11.90	8549	7536	96.28	7255	15.13
Jammu and Kashmir	J&K PDD	10503	4267	70.99	3029	71.16	11129	5163	84.36	10968	60.87	11727	5754	103.64	5964	49.14
Punjab	PSPCL	41530	34600	97.27	33655	18.96	43548	36242	99.11	9704	17.52	45238	37650	98.64	37138	17.91
Rajasthan	AVVNL	13604	10646	91.86	9779	28.12	14950	11863	100.95	16071	19.90	15760	12500	98.30	12287	22.04
	JDVVNL	14821	12024	93.89	11290	23.83	16631	13615	98.98	66082	18.97	18809	14587	95.82	13977	25.69
	JVVNL	18722	15234	94.40	14381	23.18	20619	16682	97.75	16308	20.91	21879	15784	95.53	15079	31.08
	Rajasthan total	47147	37904	93.53	35450	24.81	52200	42160	99.05	41759	20.00	56448	42871	96.44	41343	26.76
Uttar Pradesh	DVVN	16052	11335	84.26	9551	40.50	17331	12577	74.84	9413	45.69	18437	13148	89.09	11713	36.47
	KESCO	3089	2071	103.71	2148	30.48	3140	2154	90.97	1959	37.61	3554	2458	95.02	2336	34.29
	MVVN	12537	9233	75.48	6968	44.42	13147	9881	72.08	7122	45.83	14253	10711	114.02	12213	14.32
	Pash VVN	22649	16025	90.53	14507	35.95	23674	17007	92.72	15770	33.39	24099	18537	99.47	18438	23.49
	Poorv VVN	15704	11590	64.54	7480	52.37	16034	11920	64.07	7637	52.37	16929	12743	106.17	13529	20.09
	Uttar Pradesh total	70031	50254	80.90	40654	41.95	73326	53539	78.26	41901	42.85	77272	57597	101.10	58229	24.65
Uttarakhand	UtPCL	10311	8253	92.65	7646	25.84	10789	8577	96.63	8288	23.18	11017	9065	98.43	8923	19.01
Northern total		246908	189969	90.54	171991	30.34	260849	201583	92.01	185477	28.89	276900	212166	98.07	208061	24.86
Andhra Pradesh	APCPDCL	37161	31597	96.71	30558	17.77	34779	30086	97.52	29339	15.64	36717	32378	93.51	30277	17.54
	APEPDCL	12595	11726	96.11	11269	10.53	12207	11402	96.19	10968	10.15	13772	13176	97.65	12867	6.57
	APNPDCL	11914	10244	96.23	9858	17.26	11165	9672	100.33	9704	13.09	11914	10287	91.73	9436	20.80

Contd...

TABLE 5.7c *Contd...*

State	Utility	2009/10					2010/11					2011/12				
		Net input energy (MkWh)	Energy sold (MkWh)	Collection efficiency (%)	Energy realized (MkWh)	AT&C losses (%)	Net input energy (MkWh)	Energy sold (MkWh)	Collection efficiency (%)	Energy realized (MkWh)	AT&C losses (%)	Net input energy (MkWh)	Energy sold (MkWh)	Collection efficiency (%)	Energy realized (MkWh)	AT&C losses (%)
	APSPDCL	18 475	16 388	98.88	16 223	12.19	18 418	16 445	97.72	16 071	12.74	20 179	18 024	98.77	17 803	11.77
	Andhra Pradesh total	80 145	69 955	97.07	67 908	15.27	76 569	67 605	97.75	66 082	13.70	82 582	73 865	95.29	70 383	14.77
Karnataka	BESCOM	24 584	21 030	90.52	19 036	22.57	26 569	22 796	92.72	21 136	20.45	26 786	23 065	94.15	21 716	18.93
	CHESCOM	5 552	4 652	84.74	3 942	28.99	5 860	4 975	81.96	4 077	30.42	5 996	5 112	77.50	3 962	33.92
	GESCOM	6 951	5 442	97.12	5 285	23.96	7 064	5 724	100.86	5 773	18.28	7 268	5 977	84.57	5 055	30.45
	HESCOM	9 593	7 675	95.46	7 327	23.62	10 470	8 389	99.31	8 331	20.44	10 572	8 664	97.10	8 413	20.42
	MESCOM	4 072	3 570	93.61	3 342	17.94	4 280	3 772	96.95	3657	14.57	4 585	4 038	96.71	3 905	14.83
	Karnataka total	50 752	42 369	91.89	38 932	23.29	54 243	45 656	94.13	42 974	20.78	55 207	46 856	91.88	43 051	22.02
Kerala	KSEB	18 026	15 981	99.07	15 833	12.17	19 035	16 838	99.12	16 690	12.32	11 024	9 832	99.29	9 762	11.45
	KSEBL											8 514	7 622	86.26	6 575	22.78
	Kerala total	18 026	15 981	99.07	15 833	12.17	19 035	16 838	99.12	16 690	12.32	19 538	17 454	93.60	16 337	16.38
Puducherry	Puducherry PD	2 685	2 322	93.78	2 177	18.91	2 849	2 493	103.86	2 589	9.13	2 727	2 366	96.61	2 286	16.18
Tamil Nadu	TANGEDCO	66 569	53 938	96.63	52 120	21.70	66 827	52 747	100.46	52 990	20.71	77 315	60 630	99.02	60 034	22.35
Southern total		218 177	184 565	95.88	176 970	18.89	219 523	185 339	97.83	181 325	17.40	237 369	201 171	95.49	192 091	19.08
Chhattisgarh	CSPDCL	18 238	13 177	98.20	12 939	29.05	20 525	16 090	95.52	15369	25.12	19 553	14 790	101.58	15 023	23.17
Goa	Goa PD	3 180	2 753	98.07	2 699	15.12	3 374	2 952	98.14	2 897	14.14	3 366	2 962	101.46	3 005	10.72
Gujarat	DGVCL	11 764	10 563	96.74	10 219	13.14	12 819	11 337	101.31	11 486	10.40	13 420	12 040	99.39	11 967	10.83
	MGVCL	7 772	6 822	97.52	6 653	14.40	8 074	7 032	97.67	6 868	14.94	8 111	6 955	99.40	6 913	14.77
	PGVCL	21 919	16 332	96.59	15 775	28.03	24 566	17 221	99.27	17 096	30.41	22 740	17 464	98.80	17 255	24.12
	UGVCL	15 670	14 012	96.17	13 475	14.01	17 736	15 321	99.13	15 188	14.37	15 826	14 489	99.29	14 387	9.10
	Gujarat total	57 125	47 729	96.63	46 122	19.26	63 195	50 911	99.46	50 638	19.87	60 097	50 948	99.16	50 522	15.93

Contd...

TABLE 5.7c Contd...

State	Utility	2009/10					2010/11					2011/12				
		Net input energy (MkWh)	Energy sold (MkWh)	Collection efficiency (%)	Energy realized (MkWh)	AT&C losses (%)	Net input energy (MkWh)	Energy sold (MkWh)	Collection efficiency (%)	Energy realized (MkWh)	AT&C losses (%)	Net input energy (MkWh)	Energy sold (MkWh)	Collection efficiency (%)	Energy realized (MkWh)	AT&C losses (%)
Madhya Pradesh	MP Madhya Kshetra VVCL	13 803	8 577	87.14	7 474	45.85	14 374	9 937	101.30	10 066	29.97	16 440	11 557	100.13	11 573	29.60
	MP Paschim Kshetra VVCL	17 226	12 226	92.39	11 295	34.43	17 824	13 127	97.55	12 805	28.16	18 122	13 978	102.22	14 288	21.15
	MP Purv Kshetra VVCL	11 669	8 178	92.82	7 591	34.94	13 371	9 892	85.98	8 504	36.40	14 526	11 087	85.38	9 466	34.83
	Madhya Pradesh total	42 698	28 981	90.96	26 360	38.26	45 569	32 956	95.21	31 375	31.15	49 088	36 622	96.46	35 327	28.03
Maharashtra	MSEDCL	95 433	80 132	93.34	74 793	21.63	97 846	83 488	91.48	76372	21.95	99 575	85 631	99.55	85 245	14.39
Western total		216 674	172 772	94.29	162 913	24.81	230 509	186 397	94.77	176 651	23.36	231 679	190 953	99.04	189 122	18.37
Grand total		762 690	600 514	93.19	559 603	26.63	794 630	627 796	94.36	592 369	25.45	831 589	660 297	97.35	642 797	22.70

APCPDCL – Andhra Pradesh Central Power Distribution Company Ltd; APDCL – Assam Power Distribution Company Ltd; APEPDCL – Andhra Pradesh Eastern Power Distribution Company Ltd; APNPDCL – Andhra Pradesh Northern Power Distribution Company Ltd; APSPDCL – Andhra Pradesh Southern Power Distribution Company Ltd; AVVNL – Ajmer Vidyut Vitran Nigam Ltd; BESCOM – Bangalore Electricity Supply Company Ltd; BSEB – Bihar State Electricity Board; CESU – Central Electricity Supply Utility of Orissa; CHESCOM – Chamundeshwari Electricity Supply Company Ltd; CSPDCL – Chhattisgarh State Power Distribution Co. Ltd; DGVCL – Dakshin Gujarat Vij. Co. Ltd; DHBVNL – Dakshin Haryana Bijli Vitran Nigam Ltd; DVVNL – Dakshinanchal Vidyut Vitran Nigam Ltd; GESCOM – Gulbarga Electricity Supply Company Ltd; HESCOM – Hubli Electricity Supply Company Ltd; HPSEB Ltd – Himachal Pradesh Electricity Board Ltd; JDVVNL – Jodhpur Vidyut Vitran Nigam Ltd; J&K PDD – Jammu and Kashmir Power Development Department; JSEB – Jharkhand State Electricity Board; JVVNL = Jaipur Vidyut Vitran Nigam Ltd; KESCO – Kanpur Electricity Supply Co. Ltd; KSEB – Kerala State Electricity Board; KSEB – Kerala State Electricity Board Ltd; MP Madhya Kshetra VVCL – MP Madhya Kshetra Vidyut Vitran Co. Ltd; MP Paschim Kshetra VVCL = MP Paschim Kshetra Vidyut Vitran Co. Ltd; MP Purv Kshetra VVCL – MP Purv Kshetra Vidyut Vitran Co. Ltd; MeECL – Meghalaya Energy Corporation Ltd; MESCOM – Mangalore Electricity Supply Company Ltd; MePDCL – Meghalaya Power Distribution Co. Ltd; MGVCL – Madhya Gujarat Vij Company Ltd; MSEDCL – Maharashtra State Electricity Distribution Co. Ltd; MVVNL = Madhyanchal Vidyut Vitran Nigam Ltd; NBPDCL – North Bihar Power Distribution Company Ltd; NESCO – Northern Electricity Supply Company of Orissa Ltd; Pash VVNL – Paschimanchal Vidyut Vitran Nigam Ltd; PGVCL – Paschim Gujarat Vij Co. Ltd; Poorv VVNL – Poorvanchal Vidyut Vitran Nigam Ltd; PPCL – Pragati Power Corporation Limited; SBPDCL – South Bihar Power Distribution Company Ltd; SESCO – Southern Electricity Supply Company of Orissa Ltd; TANGEDCO – Tamil Nadu Generation and Distribution Corporation Ltd; TPDDL – Tata Power Delhi Distribution Ltd; TSECL – Tripura State Electricity Corporation Ltd; UGVCL – Uttar Gujarat Vij Co. Ltd; UHBVNL – Uttar Haryana Bijli Vitran Nigam Ltd; UtPCL – Uttarakhand Power Corporation Ltd; WBSEDCL – West Bengal State Electricity Distribution Co. Ltd; WESCO – Western Electricity Supply Company of Orissa Ltd

Notes

1. AT&C losses for BSES Rajdhani (for 2013/14), BSES Yamuna (for 2013/14), TANGEDCO (for 2012/13 and 2013/14) may include transmission losses since the figures for transmission losses are not separately available.
2. Collection efficiency for J&K PDD for 2009/10–2011/12 has been calculated based on figures for revenue realized available in Resource Plan.

Source PFCL (Power Finance Corporation Ltd) 2014. ***Report on the Performance of State Power Utilities for 2011/12–2013/14***. Details available at <http://www.pfcindia.com>, last accessed on 28 December 2015

TABLE 5.8 Energy sales to ultimate consumer utilities (GWh): 2012/13

State/union territory	Domestic	Commercial	Industrial (low- and medium -voltage)	Industrial (high-voltage)	Public lighting	Traction	Agriculture	Public water works and sewage pumping	Miscellaneous	Total energy sold
Northern region										
Haryana	5 554.81	2 601.18	1 395.42	6 520.42	89.29	596.12	8 279.71	743.99	476.68	26 257.62
Himachal Pradesh	1 616.58	410.98	61.48	4 449.89	13.91	0	74.95	425.58	304.43	7 357.80
Jammu and Kashmir	2 469.39	321.97	955.02		37.67	0	296.74	481.46	600.77	5 163.02
Punjab	9 284.90	2 712.36	2 351.53	9 393.77	145.81	134.83	10 779.03	405.03	617.93	35 825.19
Rajasthan	8 155.85	3 046.82	2 639.05	7 186.35	308.83	404.11	18 324.65	1 479.73	614.83	42 160.22
Uttar Pradesh	19 843.95	5 121.66	3 182.34	11 527.79	754.53	753.40	9 214.73	1 449.82	3 701.33	55 549.55
Uttarakhand	1 799.56	953.94	289.42	4 595.46	69.83	7.83	387.84	302.68	167.55	8 574.11
Chandigarh	586.54	397.55	123.95	137.50	21.98	0	1.40		96.13	1 365.05
Delhi	11 116.35	6 720.97	2797.5	177.51	375.49	436.82	34.22	498.80	986.56	23 144.22
Subtotal	60 427.93	22 287.43	13 795.71	43 988.69	1 817.34	2 333.11	47 393.27	5 787.09	7 566.21	205 396.78
Western region										
Gujarat	10 604.54	4 273.63	8 890.03	22 492.06	289.70	729.91	14 974.84	1 132.16	39.94	63 426.81
Madhya Pradesh	7 771.65	2 069.62	1 074.94	7 443.04	333.57	1 315.54	9 978.28	1 045.70	1 903.27	32 935.61
Chhattisgarh	3 807.42	1 136.63	477.56	5 072.72	92.18	801.82	2 518.08	208.74	0	14 115.15
Maharashtra	22 594.63	12 885.88	6 489.92	30 696.49	1 258.55	2 341.92	22 059.05	1 784.04	174.67	100 285.15
Goa	835.00	216.00	127.03	1 683.18	15.00	0	21.00	101.00	0	2 998.21
Daman and Diu	77.31	38.50	160.22	1 526.38	6.46	0	4.19	1.11	0	1 814.17
Dadra and Nagar Haveli	71.80	27.79	163.31	4 321.30	5.04	0	3.55	0	2.05	4 594.84
Subtotal	45 762.35	20 648.05	17 383.01	73 235.17	2 000.50	5 189.19	49 558.99	4 272.75	2 119.93	220 169.94
Southern region										
Andhra Pradesh	16 049.63	5 979.03	3 916.92	17 113.97	1 476.36	1 772.18	19 935.79	1 135.18	1 060.10	68 439.16
Karnataka	9 191.70	5 917.12	1 811.42	13 784.78	800.63	91.31	17 173.77	2 166.45	502.29	51 439.47

Contd...

TABLE 5.8 *Contd...*

State/union territory	Domestic	Commercial	Industrial (low- and medium -voltage)	Industrial (high-voltage)	Public lighting	Traction	Agriculture	Public water works and sewage pumping	Miscellaneous	Total energy sold
Kerala	8 382.68	3 028.44	1 107.51	3 450.90	316.01	173.78	192.64	55.14	125.45	16 832.55
Tamil Nadu	17 834.00	6 418.00	5 493.13	16 257.78	822.68	736.00	11 248.00	1 618.32	1 649.00	62 076.91
Puducherry	589.81	192.99	197.55	1 387.53	15.84	0	56.75	39.86	12.11	2 492.44
Lakshadweep	27.13	9.29	0.40	0	1.08	0	0	0	0	37.90
Subtotal	52 074.95	21 544.87	12 526.93	51 994.96	3 432.60	2 773.27	48 606.95	5 014.95	3 348.95	201 318.43
Eastern region										
Bihar	2 858.03	730.43	251.12	1 404.76	56.35	448.64	336.53	53.30	581.44	6 720.60
Jharkhand	3 780.00	452.24	181.84	10 709.76	252.00	1 076.52	84.00	150.50	437.50	17 124.36
Odisha	4 299.83	1 171.63	309.57	5 924.22	64.88	930.33	156.63	112.01	582.90	13 552.00
West Bengal	11 070.87	4 904.55	2 010.78	14 375.37	408.15	1 348.93	1 245.56	609.31	346.81	36 320.33
Andaman and Nicobar Islands	100.20	53.14	19.33	0	8.64	0	0.84	0	24.20	206.35
Sikkim	140.58	64.90	147.35	0	5.09	0	0	0	27.53	385.45
Subtotal	22 249.51	7 376.89	2 919.99	32 414.11	795.11	3 804.42	1 823.56	925.12	2 000.38	74 309.09
North-eastern region										
Assam	1 619.00	642.00	74.00	1 295.00	12.00	0	35.00	63.00	465.00	4 205.00
Manipur	203.74	30.81	19.60	14.13	4.85	0	1.58	23.30	81.00	379.01
Meghalaya	403.63	88.46	8.02	600.77	1.29	0	0.48	44.47	111.46	1 258.58
Nagaland	220.50	34.65	13.65	0	5.25	0	0.04	3.15	50.40	327.64
Tripura	403.59	77.17	16.57	37.06	11.32	0	41.96	46.87	86.85	721.39
Arunachal Pradesh	137.65	42.16	33.97	156.60	10.94	0	0.00	7.97	81.64	470.93
Mizoram	197.59	21.10	1.81	1.47	5.36	0	0.08	37.23	21.96	286.60
Subtotal	3 185.70	936.35	167.62	2 105.03	51.01	0	79.14	225.99	898.31	7 649.15
All-India	183 700.44	72 793.59	46 793.26	203 737.96	8 096.56	14 099.99	147 461.91	16 225.90	15 933.78	708 843.39

Source CEA (Central Electricity Authority). 2014. ***All India Electricity Statistics General Review 2014***. New Delhi: CEA, Government of India

TABLE 5.9 Number of consumers and connected load: 2000/01–2012/13

Category	2000/01	2001/02	2002/03	2003/04	2004/05	2005/06	2006/07	2007/08	2008/09	2009/10	2010/11	2011/12	2012/13
Number of consumers (thousand)													
Domestic	88 778.90	93 047.30	97 282.20	99 658.20	106 011.87	110 759.30	119 630.14	133 355.47	148 419.72	164 085.73	183 581.30	206 302.33	216 092.64
Commercial	12 215.70	12 751.30	13 391.80	15 247.90	14 591.45	15 673.30	16 348.94	17 948.32	16 470.46	21 370.84	23 447.63	23 965.39	26 651.50
Industrial (low- and medium-tension)	4 011.60	2 574.10	2 175.30	3 289.50	2 541.04	2 666.08	2 749.69	3 276.15	4 993.54	3 184.62	3 642.99	3 796.78	3 989.77
Industrial (high-tension)	51.90	52.60	51.60	93.90	52.26	67.36	94.74	597.58	490.18	157.82	121.27	245.60	254.05
Public lighting	399.70	448.00	503.40	557.20	516.43	545.18	620.20	659.80	676.27	767.07	646.61	1 159.02	1 242.11
Traction	0.30	0.30	0.40	0.70	0.75	0.33	0.75	0.79	1.89	0.81	1.38	1.86	1.87
Agriculture	12 307.50	12 919.60	12 878.20	12 836.20	13 333.42	13 801.24	14 406.29	15 804.39	15 958.32	17 860.08	21 752.42	23 643.23	24 604.72
Public water works	248.50	280.10	325.40	321.50	434.34	378.68	426.64	452.84	541.50	611.80	682.89	833.02	970.21
Miscellaneous	541.20	395.90	349.50	1 565.90	340.78	253.24	336.08	608.25	692.98	1 180.55	1 107.56	1 581.20	1 975.43
Total	118 555.30	122 469.20	126 957.80	133 571.00	137 822.32	144 144.70	154 613.47	172 703.59	188 244.87	209 219.32	234 984.06	261 528.43	275 782.29
Connected load (MW)													
Domestic	71 027.10	79 055.00	88 114.50	93 625.40	101 429.06	107 864.62	117 543.75	133 932.25	149 462.57	154 755.58	174 535.98	224 235.71	250 629.29
Commercial	18 849.20	20 807.80	21 762.10	24 951.90	27 169.67	29 950.20	33 135.28	38 181.76	37 753.42	45 507.45	55 283.85	62 602.04	76 278.74
Industrial (low- and medium-tension)	30 753.00	30 079.70	30 702.90	31 816.80	36 614.89	33 662.58	35 103.30	40 559.11	55 843.83	44 495.13	52 427.88	66 942.19	70 712.88
Industrial (high-tension)	36 318.11	36 388.60	38 783.80	34 223.10	33 484.18	43 541.95	51 586.16	56 560.83	49 228.11	58 050.90	72 882.07	81 325.64	100 805.15
Public lighting	1 392.40	1 712.90	1 823.50	1 690.70	2 450.19	2 246.42	2 578.12	2 794.50	2 524.81	3 210.22	3 270.29	5 472.21	5 714.00
Traction	2 622.90	3 014.60	3 147.00	3 427.50	3 396.73	3 600.29	3 810.81	3 887.18	4 878.98	5 707.30	6 082.05	5 571.68	5 529.43
Agriculture	48 192.50	51 019.80	53 625.70	51 841.30	56 508.61	55 911.19	62 638.52	66 048.25	71 871.65	80 902.18	94 579.23	111 756.40	118 665.88
Public water works	3179.80	3 857.40	4 176.60	3 650.10	3 870.55	3 745.44	3 787.35	4 287.11	4 996.65	5 252.80	6 307.60	8 320.42	10 043.96
Miscellaneous	3 446.90	3 298.10	2 284.30	3 433.60	2 646.73	1 882.45	2 373.04	3 805.01	5 244.67	9 619.84	11 528.33	21 107.17	23 584.75
Total	215 781.91	229 233.90	244 420.40	248 660.40	267 570.60	282 405.14	312 556.32	350 055.99	381 804.68	407 501.40	476 897.28	587 333.46	661 964.08

Note All figures pertain to 31 March of the respective years.

Source CEA (Central Electricity Authority). 2014. ***All India Electricity Statistics General Review 2014***. New Delhi: CEA, Government of India

TABLE 5.10 Per capita consumption (kWh) of electricity: 2000/01–2012/13

State/union territory	2000/01	2001/02	2002/03[a]	2003/04	2004/05	2005/06	2006/07	2007/08	2008/09	2009/10	2010/11	2011/12	2012/13
Northern region	325	–	550	559	565	603	623	666	688	739	776	833	852
Chandigarh	801	815	648	1 500	887	1 554	1 487	1 557	1 247	1 239	1 283	1 217	1 297
Delhi	672	696	1 426	1 542	1 554	1 767	1 417	1 433	1 374	1 448	1 530	1 587	1 613
Haryana	544	533	997	924	951	1 090	1 208	1 296	1 279	1 491	1 485	1 628	1 722
Himachal Pradesh	343	398	599	794	831	766	872	967	1 016	1 145	1 251	1 289	1 380
Jammu and Kashmir	286	293	592	647	609	711	759	795	894	968	988	1 015	1 043
Punjab	841	836	1 227	1 277	1 245	1 437	1 506	1 614	1 553	1 663	1 736	1 799	1 761
Rajasthan	349	285	566	540	583	572	591	692	747	811	844	927	982
Uttar Pradesh	191	189	316	300	309	312	341	346	372	387	412	450	450
Uttarakhand	–	284	464	825	654	655	707	856	921	930	1 144	1 232	1 297
Western region	532	–	836	837	877	916	969	1 030	1 039	1 061	1 104	1 201	1 284
Chhattisgarh	–	394	676	678	685	686	935	874	1 418	921	1 118	1 320	1 495
Dadra and Nagar Haveli	4 106	3 722	4 135	7 497	8 479	11 568	13 310	14 787	11 094	11 709	13 367	13 767	14 341
Daman and Diu	4 540	4 623	7 376	7 038	7 546	8300	8 950	9 473	6 508	7 202	7 810	7 785	7 927
Goa	810	1 068	1 843	2 178	2 189	1 970	2 098	2 169	2 260	2 005	2 066	2 025	2 045
Gujarat	854	817	1 193	1 175	1 299	1 284	1 331	1 486	1 457	1 559	1 508	1 663	1 796
Madhya Pradesh	295	273	520	475	516	580	582	623	584	618	674	672	753
Maharashtra	551	508	848	879	879	934	975	1 020	969	1 054	1 096	1 204	1 239
Southern region	463	–	660	699	736	758	837	888	904	972	1 010	1 098	1 094
Andhra Pradesh	433	494	673	719	765	723	802	877	928	1 014	1 065	1 157	1 135
Karnataka	412	428	611	642	660	720	806	844	854	873	925	1 081	1 129
Kerala	329	281	377	386	398	424	441	444	444	537	551	594	630
Lakshadweep	248	290	348	342	377	368	402	427	453	429	532	550	592
Puducherry	1 277	1 625	1 761	1 894	2 049	2 509	2 693	2 357	1988	1865	1850	2 125	2 136
Tamil Nadu	599	623	815	866	918	977	1 080	1 145	1 134	1 211	1 233	1 277	1 226

Contd...

TABLE 5.10 *Contd...*

State/union territory	2000/01	2001/02	2002/03[a]	2003/04	2004/05	2005/06	2006/07	2007/08	2008/09	2009/10	2010/11	2011/12	2012/13
Eastern region	201	–	290	352	357	332	358	388	401	446	493	521	552
Andaman and Nicobar Islands	231	253	375	420	346	408	458	515	475	506	499	501	559
Bihar	145	36	82	75	75	86	91	101	107	117	127	134	145
Jharkhand	–	364	468	564	546	549	659	643	696	750	749	790	847
Odisha	343	324	470	695	735	634	665	752	775	838	1 070	1 146	1 209
Sikkim	184	224	247	766	985	430	533	732	806	845	880	886	862
West Bengal	208	218	366	410	414	381	397	439	442	515	538	564	594
North-eastern region	103	–	180	192	219	201	204	229	242	250	268	292	298
Arunachal Pradesh	84	68	132	225	380	298	299	456	447	503	582	683	719
Assam	104	99	160	160	163	171	175	188	199	209	223	250	240
Manipur	69	69	206	218	258	215	195	222	242	207	242	236	353
Meghalaya	170	235	336	416	521	518	547	629	655	613	654	658	690
Mizoram	142	147	300	301	439	250	250	354	378	429	462	507	469
Nagaland	97	57	139	156	181	179	173	199	226	242	265	257	268
Tripura	79	109	227	264	351	191	179	202	204	224	222	254	296
All-India	366	361	567	592	613	631	672	672	734	779	819	884	914

[a] From 2002/03 onwards as per UN methodology, Per capita consumption = (Gross generation + net import)/Population.

Source CEA (Central Electricity Authority). 2014. ***All India Electricity Statistics General Review 2014.*** New Delhi: CEA, Government of India

TABLE **5.11** | Power supply position: peak demand versus peak met (2014/15)

State/system/region	Peak demand (MW)	Peak met (MW)	Surplus/deficit	
			MW	%
Chandigarh	241	241	0	0
Delhi	4 517	4 517	0	0
Haryana	6 370	6 370	0	0
Himachal Pradesh	1 379	1 379	0	0
Jammu and Kashmir	2 191	1 906	-285	-13.0
Punjab	5 778	5 778	0	0
Rajasthan	7 798	7 798	0	0
Uttar Pradesh	12 967	12 347	-620	-4.8
Uttarakhand	1 837	1 762	-75	-4.1
Northern region	41 540	40 003	-1 537	-3.7
Chhattisgarh	3 788	3 677	-111	-2.9
Gujarat	13 207	13 178	-29	-0.2
Madhya Pradesh	6 984	6 984	0	0
Maharashtra	19 933	19 714	-219	-1.1
Daman and Diu	301	301	0	0
Dadra Nagar Haveli	740	740	0	0
Goa	583	552	-31	-5.3
Western region	43 135	42 690	-445	-1.0
Andhra Pradesh	6 794	6 789	-5	-0.1
Telangana	6 813	6 661	-152	-2.2
Karnataka	9 463	9 335	-128	-1.4
Kerala	3 653	3 512	-141	-3.9
Tamil Nadu	13 363	12 380	-983	-7.4
Puducherry	351	350	-1	-0.3
Lakshadweep	8	8	0	0
Southern region	37 801	36 786	-1 015	-2.7
Bihar	3 025	2 945	-80	-2.6
Damodar Valley Corporation	2 562	2 547	-15	-0.6
Jharkhand	1 061	1 043	-18	-1.7
Odisha	3 872	3 850	-22	-0.6
West Bengal	7 532	7 517	-15	-0.2

Contd...

TABLE **5.11** *Contd...*

State/system/region	Peak demand (MW)	Peak met (MW)	Surplus/deficit	
			MW	%
Andaman and Nicobar Islands	40	32	-8	-20.0
Eastern region	17368	17304	-64	-0.4
Arunachal Pradesh	138	114	-24	-17.4
Assam	1395	1282	-113	-8.1
Manipur	150	148	-2	-1.3
Meghalaya	400	312	-88	-22.0
Mizoram	85	80	-5	-5.9
Nagaland	120	112	-8	-6.7
Tripura	265	238	-27	-10.2
North-eastern region	2220	2114	-106	-4.8
All-India	136884	132119	-4765	-3.5

Note Lakshadweep and Andaman and Nicobar Islands have stand-alone systems, power supply position of these does not form part of regional requirement and availability.

Source Details available at <http://www.cea.nic.in/reports/monthly/executivesummary/2015/exe_summary-05.pdf>, last accessed on 11 December 2015

TABLE 5.12 Power supply position: requirement versus availability (April 2014–March 2015)

Region	Requirement (MU)	Availability (MU)	Surplus deficit (-)	
			MU	%
Chandigarh	1 616	1 616	0	0
Delhi	29 231	29 106	- 125	- 0.4
Haryana	46 615	46 432	- 183	- 0.4
Himachal Pradesh	8 807	8 728	- 79	- 0.9
Jammu and Kashmir	16 214	13 119	- 3 095	- 19.1
Punjab	48 629	48 144	- 485	- 1.0
Rajasthan	65 717	65 310	- 407	- 0.6
Uttar Pradesh	103 179	87 062	- 16 117	- 15.6
Uttarakhand	12 445	12 072	- 373	- 3.0
Northern region	332 453	311 589	- 20 864	- 6.3
Chhattisgarh	21 499	21 230	- 269	- 1.3
Gujarat	96 235	96 211	- 24	0.0
Madhya Pradesh	53 374	53 082	- 292	- 0.5
Maharashtra	134 897	133 078	- 1 819	- 1.3
Daman and Diu	2 086	2 086	0	0.0
Dadra and Nagar Haveli	5 307	5 304	- 3	- 0.1
Goa	3 969	3 932	- 37	- 0.9
Western region	317 367	314 923	- 2 444	- 0.8
Andhra Pradesh	59 198	56 313	- 2 885	- 4.9
Telangana	43 337	40 644	- 2 693	- 6.2
Karnataka	62 643	59 926	- 2 717	- 4.3
Kerala	22 459	22 127	- 332	- 1.5
Tamil Nadu	95 758	92 750	- 3 008	- 3.1
Puducherry	2 402	2 376	- 26	- 1.1
Lakshadweep	48	48	0	0
Southern region	285 845	274 184	- 11 661	- 4.1
Bihar	19 294	18 759	- 535	- 2.8
Damodar Valley Corporation	18 222	17 728	- 494	- 2.7
Jharkhand	7 599	7 390	- 209	- 2.8
Odisha	26 482	26 052	- 430	- 1.6
West Bengal	47 086	46 827	- 259	- 0.6
Sikkim	399	399	0	0

Contd...

TABLE **5.12** *Contd...*

Region	Requirement (MU)	Availability (MU)	Surplus deficit (-)	
			MU	%
Andaman and Nicobar Islands	240	180	-60	-25.0
Eastern region	119 322	117 335	-1 987	-1.7
Arunachal Pradesh	677	610	-67	-9.9
Assam	8 527	7 926	-601	-7.0
Manipur	705	678	-27	-3.8
Meghalaya	1 930	1 634	-296	-15.3
Mizoram	455	425	-30	-6.6
Nagaland	688	661	-27	-3.9
Tripura	1 242	1 048	-194	-15.6
North-eastern region	14 224	12 982	-1 242	-8.7
All-India	1 069 211	1 031 013	-38 198	-3.6

Notes
1. Lakshadweep and Andaman and Nicobar Islands have stand-alone systems, power supply position of these does not form part of regional requirement and availability.
2. Cumulative figures for Andhra Pradesh includes the figures of undivided Andhra Pradesh (including Telangana area) for the period April–May 2014. Cumulative figures of Telangana are w.e.f. June 2014 due to bifurcation of Andhra Pradesh.

Source Details available at <http://www.cea.nic.in/reports/monthly/executivesummary/2015/exe_summary-04.pdf>, last accessed on 11 December 2015

TABLE 5.13a Status of consumer metering: 2006/07

State/union territory	Number of consumers (in million)	Metered (in million)	Percentage
Andhra Pradesh	18.34	16.21	88
Arunachal Pradesh	0.11	0.05	46
Assam	1.27	1.21	95
Bihar	1.25	0.62	50
Chhattisgarh	2.47	1.82	74
Delhi	2.71	2.68	99
Goa	0.40	0.39	97
Gujarat (GEB)	7.48	6.96	93
Gujarat (Torrent Power)	1.31	1.31	100
Gujarat (Torrent Power)	0.55	0.55	100
Haryana	4.07	3.66	90
Himachal Pradesh	1.79	1.79	100
Jammu and Kashmir	1.00	0.40	40
Jharkhand	0.65	0.49	75
Karnataka	14.00	12.01	86
Kerala	8.53	8.53	100
Madhya Pradesh	6.49	4.65	72
Maharashtra (BEST)	0.92	0.92	100
Maharashtra (MSEB)	13.53	11.81	87
Maharashtra (REL)	2.55	2.55	100
Manipur	0.17	0.14	82
Meghalaya	0.17	0.08	50
Mizoram	0.13	0.13	99
Nagaland	0.19	0.11	61
Orissa	2.15	1.75	81
Punjab	6.03	5.21	86
Rajasthan	5.85	5.48	94
Sikkim	0.07	0.06	92
Tamil Nadu	3.42	3.03	89
Tripura	0.34	0.30	89
Uttar Pradesh	8.81	8.04	91
Uttarakhand	1.14	1.11	97
West Bengal	6.14	5.95	97

Contd...

TABLE 5.13a *Contd...*

State/union territory	Number of consumers (in million)	Metered (in million)	Percentage
Chandigarh	0.20	0.20	100
Daman and Diu	0.05	0.05	100
Total	124.48	110.46	89

BEST – Brihanmumbai Electric Supply and Transport; GEV – Gujarat Electricity Board; MSEB – Maharashtra State Electricity Board; REL – Reliance Energy Ltd

Source MoP (Ministry of Power). 2007. ***Status Report for March 2007.*** New Delhi: Ministry of Power, Government of India

TABLE 5.13b | Number of consumers (in lakh): 2009/10–2013/14

State/union territory	2009/10	2010/11	2011/12	2012/13	2013/14
Andhra Pradesh	215	228	238	238	238
Arunachal Pradesh	2	2	2	2	2
Assam	17	19	23	23	30
Bihar	32	28	33	42	56
Chhattisgarh	30	37	36	38	39
Goa	5	5	6	6	6
Gujarat	112	115	124	132	139
Haryana	46	48	49	52	54
Himachal Pradesh	19	20	20	21	21
Jammu and Kashmir	12	13	13	13	13
Jharkhand	15	17	20	34	37
Karnataka	168	176	182	190	200
Kerala	97	101	105	108	111
Madhya Pradesh	86	91	111	111	111
Maharashtra	179	194	205	207	207
Manipur	2	2	2	2	2
Meghalaya	3	3	3	3	3
Mizoram	2	2	2	2	2
Nagaland	2	2	2	2	2
Punjab	69	73	76	80	83
Rajasthan	62	91	98	115	112
Sikkim	1	1	1	1	1
Tamil Nadu	216	226	232	239	247
Tripura	4	5	5	6	7
Uttar Pradesh	107	120	131	139	151
Uttarakhand	14	15	16	16	18
West Bengal	76	86	107	117	129
Puducherry	3	3	3	4	4
Total	1596	1723	1845	1943	2025

Source Details available at <http://planningcommission.gov.in/reports/genrep/rep_arpower1305.pdf>, last accessed on 11 December 2015

TABLE 5.14 List of existing inter-regional links

Region	Existing capacity (MW)	
East-West		
Budhipadar-Korba 220 kV 3 ckts	390	
Rourkela-Raipur 400 kV D/C with series comp + TCSC	1 400	
Ranchi-Sipat 400 kV D/C with series comp	1 200	
Rourkela-Raipur 400 kV D/C (2nd) (without series comp)	1 400	
Ranchi-Dharamjayagarh WR Pooling		
Station 765 kV S/C line		2 100
Ranchi-Dharamjayagarh 765 kV 2nd S/C		
Jharsuguda-Dharamjaygarh 765 kV D/C line		4 200
Subtotal	4 390	6 300
West-North		
Auraiya-Malanpur 220 kV D/C	260	
Kota-Ujjain 200 kV D/C	260	
Vindhyachal HVDC back-to-back	500	
Gwalior-Agra 765 kV 2 x S/C	2 200	
Zerda-Kankroli 400 kV D/C	1 000	
Champa Pool-Kurukshetra HVDC Bipole		
Gwalior-Jaipur 765 kV 2 x S/C lines		
RAPP-Sujalpur 400 kV D/C		
Adani (Mundra) - Mahendranagar KVDC Bipole		
Subtotal	4 220	4 500
East-South		
Balimela-Upper Sileru 200 kV S/C	130	
Gazuwaka HVDC back-to-back	1 000	
Talcher-Kolar HVDC bipole	2 000	
Upgradation of Talcher-Kolar HVDC bipole	500	
Angul-Srikakulum		
Subtotal	3 630	
West-South		
Chandrapur HVDC back-to-back	1 000	
Kolhapur-Belgaum 220 kV D/C	260	
Ponda-Nagajhari 200 kV D/C	260	
Raichur-Sholapur 765 kV S/C line (PG)		2 100
Raichur-Sholapur 765 kV S/C line (Pvt. sector)		2 100
Narendra-Kolhapur 765 kV D/C (ch at 400 kV)		

Contd...

TABLE **5.14** *Contd...*

Region	Existing capacity	
Wardha-Hyderabad 765 kV D/C		
Barsur-L. Sileru 200 kV HVDC Monopole (currently not in operation)		200
Subtotal	1 520	4 200
East-North-East		
Birpara-Salakati 200 kV D/C	260	
Maida-Bongaigaon 400 kV D/C	1 000	
Siliguri-Bongaigaon 400 kV D/C (quad) line (private)		1 600
Subtotal	1 260	1 600
North-East-North		
Biswanath Chariali - Agra +/- 900 kV, 3 000 MW HVDC Bipole		
Subtotal		
LILO of Biswanath Chariali-Agra +/- 800 kV, 3 000 MW HVDC Bipole at new pooling station in Alipurduar and addition of second 3 000 MW module		
Various 132 kV inter-regional links (600 MW)	600	
Total cumulative	27 750	46 450

Source MoP (Ministry of Power) 2015. ***Annual Report 2014/15***. New Delhi: Government of India

TABLE 5.15 Details of nuclear power projects as on December 2015

Plant	District and state	Capacity (MWe)	Date of commercial operation
Tarapur Atomic Power Station (unit 1: boiling water reactor)	Thane, Maharashtra	160	28 October 1969
Tarapur Atomic Power Station (unit 2: boiling water reactor)	Thane, Maharashtra	160	28 October 1969
Tarapur Atomic Power Station (unit 3: pressurized heavy water reactor)	Thane, Maharashtra	540	18 August 2006
Tarapur Atomic Power Station (unit 4: pressurized heavy water reactor)	Thane, Maharashtra	540	12 September 2005
Rajasthan Atomic Power Station (unit 1: pressurized heavy water reactor)	Kota, Rajasthan	100	16 December 1973
Rajasthan Atomic Power Station (unit 2: pressurized heavy water reactor)	Kota, Rajasthan	200	1 April 1981
Rajasthan Atomic Power Station (unit 3: pressurized heavy water reactor)	Kota, Rajasthan	220	1 June 2000
Rajasthan Atomic Power Station (unit 4: pressurized heavy water reactor)	Kota, Rajasthan	220	23 December 2000
Rajasthan Atomic Power Station (unit 5: pressurized heavy water reactor)	Kota, Rajasthan	220	4 February 2010
Rajasthan Atomic Power Station (unit 6: pressurized heavy water reactor)	Kota, Rajasthan	220	31 March 2010
Madras Atomic Power Station (unit 1: pressurized heavy water reactor)	Kalpakkam, Tamil Nadu	220	27 January 1984
Madras Atomic Power Station (unit 2: pressurized heavy water reactor)	Kalpakkam, Tamil Nadu	220	21 March 1986
Kaiga Generating Station (unit 1: pressurized heavy water reactor)	Uttar Kannada, Karnataka	220	16 November 2000
Kaiga Generating Station (unit 2: pressurized heavy water reactor)	Uttar Kannada, Karnataka	220	16 March 2000
Kaiga Generating Station (unit 3: pressurized heavy water reactor)	Uttar Kannada, Karnataka	220	6 May 2007
Kaiga Generating Station (unit 4: pressurized heavy water reactor)	Uttar Kannada, Karnataka	220	20 January 2011
Kudankulam Atomic Powr Project (unit 1: pressurized water reactor)	Tirunelveli, Tamil Nadu	1000	31 December 2014
Narora Atomic Power Station (unit 1: pressurized heavy water reactor)	Bulandshahar, Uttar Pradesh	220	1 January 1991
Narora Atomic Power Station (unit 2: pressurized heavy water reactor)	Bulandshahar, Uttar Pradesh	220	1 July 1992
Kakrapar Atomic Power Station (unit 1: pressurized heavy water reactor)	Surat, Gujarat	220	6 May 1993
Kakrapar Atomic Power Station (unit 2: pressurized heavy water reactor)	Surat, Gujarat	220	1 September 1995
Total		5780	
Kudankulam Atomic Power Project (unit 2)	Tirunelveli, Tamil Nadu	1000	Start up process of FAC-Q1 2016-17
Rajasthan Atomic Power Project (unit 7)	Rajasthan	700	June 2016
Rajasthan Atomic Power Project (unit 8)	Rajasthan	700	December 2016
Kakrapar Atomic Power Project (unit 3)	Surat, Gujarat	700	Under review
Kakrapar Atomic Power Project (unit 4)	Surat, Gujarat	700	Under review

MWe – megawatt electrical

Sources

1. Details available at <http://www.npcil.nic.in/main/AllProjectOperationDisplay.aspx>, last accessed on 16 December 2015
2. Details available at <http://www.npcil.nic.in/main/ProjectConstructionStatus.aspx>, last accessed on 16 December 2015

TABLE 5.16 Progress report of village electrification as on 31 October 2015

Region/state/union territory	Total inhabited villages as per Census 2011	Villages electrified as on 31 March 2015		Cumulative achievement as on 31 October 2015	Percentage of villages electrified as on 31 October 2015	Unelectrified villages as on 31 October 2015	Number of unelectrified villages covered under DDUGJY	Remote villages to be electrified through off-grid	Villages to be electrified under state plan
		Numbers	Percentage						
Andhra Pradesh	16 158	16 158	100.0	16 158	100.0	0	0	0	0
Arunachal Pradesh	5 258	3 680	70.0	3 680	70.0	1 578	1 454	124	0
Assam	25 372	22 482	88.6	22 599	89.1	2 773	1 639	1 134	0
Bihar	39 073	36 354	93.0	37 812	96.8	1 261	876	385	0
Chhattisgarh	19 567	18 487	94.5	18 520	94.6	1 047	504	161	382
Goa	320	320	100.0	320	100.0	0	0	0	0
Gujarat	17 843	17 843	100.0	17 843	100.0	0	0	0	0
Haryana	6 642	6 642	100.0	6 642	100.0	0	0	0	0
Himachal Pradesh	17 882	17 846	99.8	17 847	99.8	35	35	0	0
Jammu and Kashmir	6 337	6 205	97.9	6 231	98.3	106	106	0	0
Jharkhand	29 492	26 911	91.2	27 015	91.6	2 477	1 408	455	614
Karnataka	27 397	27 358	99.9	27 358	99.9	39	30	9	0
Kerala	1 017	1 017	100.0	1 017	100.0	0	0	0	0
Madhya Pradesh	51 929	51 457	99.1	51 576	99.3	353	205	148	0
Maharashtra	40 956	40 956	100.0	40 956	100.0	0	0	0	0
Manipur	2 379	2 103	88.4	2 127	89.4	252	252	0	0
Meghalaya	6 459	5 547	85.9	5 548	85.9	911	894	17	0
Mizoram	704	646	91.8	662	94.0	42	42	0	0
Nagaland	1 400	1 318	94.1	1 318	94.1	82	82	0	0
Odisha	47 677	44 227	92.8	44 228	92.8	3 449	3 238	211	0
Punjab	12 168	12 168	100.0	12 168	100.0	0	0	0	0
Rajasthan	43 264	42 769	98.9	42 797	98.9	467	332	0	135
Sikkim	425	425	100.0	425	100.0	0	0	0	0

Contd...

TABLE 5.16 *Contd...*

Region/state/union territory	Total inhabited villages as per Census 2011	Villages electrified as on 31 March 2015		Cumulative achievement as on 31 October 2015	Percentage of villages electrified as on 31 October 2015	Unelectrified villages as on 31 October 2015	Number of unelectrified villages covered under DDUGJY	Remote villages to be electrified through off-grid	Villages to be electrified under state plan
		Numbers	Percentage						
Tamil Nadu	15 049	15 049	100.0	15 049	100.0	0	0	0	0
Tripura	863	837	97.0	840	97.3	23	23	0	0
Telangana	10 128	10 128	100.0	10 128	100.0	0	0	0	0
Uttar Pradesh	97 813	96 285	98.4	96 786	99.0	1 027	1 027	0	0
Uttarakhand	15 745	15 669	99.5	15 669	99.5	76	56	13	7
West Bengal	37 463	37 441	99.9	37 449	100.0	14	14	0	0
Andaman and Nicobar Islands	396	341	86.1	341	86.1	55	0	55	0
Chandigarh	5	5	100.0	5	100.0	0	0	0	0
Dadra and Nagar Haveli	65	65	100.0	65	100.0	0	0	0	0
Daman and Diu	19	19	100.0	19	100.0	0	0	0	0
Delhi	103	103	100.0	103	100.0	0	0	0	0
Lakshadweep	6	6	100.0	6	100.0	0	0	0	0
Puducherry	90	90	100.0	90	100.0	0	0	0	0
Total	597 464	578 957	96.9	581 397	97.3	16 067	12 217	2 712	1 138

Source Details available at <http://www.cea.nic.in/reports/monthly/electrification/2015/village_electrification-10.pdf>, last accessed on 17 December 2015

TABLE 5.17 Electrification of rural households

State/union territory	Total number of rural households	Households electrified	Households using kerosene	Households using solar energy	Households using other oil	Any other	No lighting
Andaman and Nicobar Islands	59 030	46 852	11 393	132	243	68	342
Andhra Pradesh	14 246 309	12 782 453	1 312 644	33 391	36 983	13 624	67 214
Arunachal Pradesh	195 723	108 550	46 175	7 553	767	5 357	27 321
Assam	5 374 553	1 524 221	3 782 653	46 529	7 465	3 464	10 221
Bihar	16 926 958	1 754 673	14 963 756	104 118	50 421	43 027	10 963
Chandigarh	6 785	6 603	162	1	6	6	7
Chhattisgarh	4 384 112	3 070 879	1 235 592	48 935	8 554	6 499	13 653
Dadra and Nagar Haveli	35 408	32 452	2 697	19	8	24	208
Daman and Diu	12 750	12 532	185	4	1	3	25
Delhi	79 115	77 366	1 136	65	133	246	169
Goa	124 674	119 208	4 287	269	80	55	775
Gujarat	6 765 403	5 749 271	869 255	13 874	20 963	19 747	92 293
Haryana	2 966 053	2 585 338	335 860	5 632	9 137	14 591	15 495
Himachal Pradesh	1 310 538	1 265 897	38 895	1 762	1 037	1 158	1 789
Jammu and Kashmir	1 497 920	1 208 527	189 124	20 260	3 969	37 710	38 330
Jharkhand	4 685 965	1 514 050	3 113 279	41 723	11 258	3 574	2 081
Karnataka	7 864 196	6 819 812	965 641	19 187	13 426	5 244	40 886
Kerala	4 095 674	3 772 137	304 225	10 997	4 156	2 447	1 712
Lakshadweep	2 523	2 517	5	0	0	1	0
Madhya Pradesh	11 122 365	6 479 144	4 546 696	36 275	24 901	10 545	24 804
Maharashtra	13 016 652	9 605 299	3 107 049	48 506	32 422	51 490	171 886
Manipur	335 752	205 444	108 164	9 031	1 531	9 296	2 286
Meghalaya	422 197	217 739	193 949	4 224	1 401	1 116	3 768
Mizoram	104 874	72 138	28 159	2 872	644	561	500
Nagaland	284 911	214 319	60 106	1 163	783	4 309	4 231

Contd...

TABLE 5.17 | *Contd...*

State/union territory	Total number of rural households	Households electrified	Households using kerosene	Households using solar energy	Households using other oil	Any other	No lighting
Odisha	8 144 012	2 895 252	5 113 827	31 870	8 464	8 696	85 903
Puducherry	95 133	91 105	3 387	22	113	22	484
Punjab	3 315 632	3 166 394	96 149	4 431	6 832	11 302	30 524
Rajasthan	9 490 363	5 528 360	3 729 431	75 583	36 127	29 467	91 395
Sikkim	92 370	83 277	8 075	293	116	82	527
Tamil Nadu	9 563 899	8 683 426	791 493	8 720	19 254	5 294	55 712
Tripura	607 779	361 573	228 953	13 368	1 470	268	2 147
Uttar Pradesh	25 475 071	6 054 978	19 111 021	141 108	72 589	59 296	36 079
Uttarakhand	1 404 845	1 166 756	204 149	23 789	2 691	2 750	4 710
West Bengal	13 717 186	5 529 496	7 927 731	160 497	29 974	10 168	59 320
Total	167 826 730	92 808 038	72 435 303	916 203	407 919	361 507	897 760

Sources

1. Details available at <http://planningcommission.gov.in/reports/genrep/rep_arpower1305.pdf>, last accessed on 28 December 2015
2. Planning Commission. 2014. ***Annual Report 2013/14 on the Working of State Power Utilities and Electricity Departments.*** New Delhi: Planning Commission, Government of India

TABLE **5.18** Status of hydroelectric potential development, in terms of installed capacity of above 25 MW as on 31 October 2015

Region/state	Identified capacity (IC) as per reassessment study		Capacity under operation		Capacity under construction		Capacity under operation + construction		Capacity yet to be taken up under construction	
	Total (MW)	Above 25 MW	MW	%	MW	%	MW	%	MW	%
Jammu and Kashmir	14 146	13 543	2 969.0	21.92	1 330.0	9.82	4 299.0	31.74	9 244.0	68.26
Himachal Pradesh	18 820	18 540	9 308.0	50.20	2 216.0	11.95	11 524.0	62.16	7 016.0	37.84
Punjab	971	971	1 206.3	100.00	206.0	21.22	1 412.3	100.00	0	0
Haryana	64	64	0	0	0	0	0	0	0	0
Rajasthan	496	483	411.0	85.09	0	0	411.0	100.00	0	0
Uttarakhand	18 175	17 998	3 756.4	20.87	1 430.0	7.95	5 186.4	28.82	12 811.7	71.18
Uttar Pradesh	723	664	501.6	75.54	0	0	501.6	75.54	39.0	5.87
Northern region subtotal	53 395	52 263	18 152.3	34.73	5 182.0	9.92	23 334.3	44.65	28 928.8	55.35
Madhya Pradesh	2 243	1 970	2 395.0	100.00	400.0	20.30	2 795.0	100.00	0	0
Chhattisgarh	2 242	2 202	120.0	5.45	0	0	120.0	5.45	2 082.0	94.55
Gujarat	619	590	550.0	100.00	0	0	550.0	100.00	0	0
Maharashtra	3 769	3 314	2 487.0	75.05	0	0	2 487.0	75.05	827.0	24.95
Goa	55	55	0	0	0	0	0	0	55.0	100.00
Western region subtotal	8 928	8 131	5 552.0	68.28	400.0	4.92	5 952.0	73.20	2 179.0	26.80
Andhra Pradesh	2 366	2 341	1 746.8	74.62	1 010.0	43.14	2 756.8	117.76	0	0
Telangana	2 058	2 019	511.0	25.31	280.0	13.87	791.0	39.18	1 228.0	60.82
Karnataka	6 602	6 459	3 585.4	55.51	0	0	3 585.4	55.51	2 873.6	44.49
Kerala	3 514	3 378	1881.5	55.70	100.0	2.96	1981.5	58.66	1 396.5	41.34
Tamil Nadu	1 918	1 693	1 782.2	100.00	0	0	1782.2	100.00	0	0
Southern region subtotal	16 458	15 890	9 506.9	59.83	1 390.0	8.75	10 896.9	68.58	4 993.2	31.42
Jharkhand	753	582	170.0	29.21	0	0	170.0	29.21	412.0	70.79
Bihar	70	40	0	0	0	0	0	0	0	0
Odisha	2 999	2 981	2 027.5	68.01	0	0	2 027.5	68.01	953.5	31.99
West Bengal	2 841	2829	272.2	9.62	280.0	9.90	552.2	19.52	2 276.8	80.48
Sikkim	4 286	4 248	765.0	18.01	2 526.0	59.46	3 291.0	77.47	957.0	22.53
Eastern region subtotal	10 949	10 680	3 234.7	30.29	2 806.0	26.27	6 040.7	56.56	4 639.3	43.44
Meghalaya	2 394	2 298	282.0	12.27	40.0	1.74	322.0	14.01	1 976.0	85.99
Tripura	15	0	0	0	0	0	0	0	0	0
Manipur	1 784	1 761	105.0	5.96	0	0	105.0	5.96	1 656.0	94.04

Contd...

TABLE 5.18 *Contd...*

Region/state	Identified capacity (IC) as per reassessment study		Capacity under operation		Capacity under construction		Capacity under operation + construction		Capacity yet to be taken up under construction	
	Total (MW)	Above 25 MW	MW	%	MW	%	MW	%	MW	%
Assam	680	650	375.0	57.69	0	0	375.0	57.69	275.0	42.31
Nagaland	1 574	1 452	75.0	5.17	0	0	75.0	5.17	1 377.0	94.83
Arunachal Pradesh	50 328	50 064	405.0	0.81	2 854.0	5.70	3 259.0	6.51	46 805.0	93.49
Mizoram	2 196	2 131	0	0	60.0	2.82	60.0	2.82	2071.0	97.18
North-eastern region subtotal	58 971	58 356	1 242.0	2.13	2 954.0	5.06	4 196.0	7.19	54 160.0	92.81
All-India	148 701	145 320	37 687.9	25.93	12 732.0	8.76	50 419.9	34.70	94 900.3	65.30

Note In addition to above 9 PSS (4785.6 MW) are under operation and 2 PSS (1080 MW) are under construction, DPR of 1 PSS (500 MW) returned to project developed and 1 PSS (1000 MW) is under survey and investigation.

Source Details available at <http://www.cea.nic.in/reports/monthly/hydro/2015/hydro_potential_region-10.pdf>, last accessed on 17 December 2015

TABLE 5.19 Profit and loss of power utilities (₹ million) by state: 2011/12–2013/14

Region/state	2011/12						2012/13						2013/14					
	Profit/loss after tax (accrual basis)	Cash profit (accrual basis excluding depreciation and write offs)	Cash profit (subsidy received basis)	Cash profit (revenue and subsidy on realized basis)	Profit/loss without subsidy	Profit/loss on subsidy received basis	Profit/loss after tax (accrual basis)	Cash profit (accrual basis excluding depreciation and write offs)	Cash profit (subsidy received basis)	Cash profit (revenue and subsidy on realized basis)	Profit/loss without subsidy	Profit/loss on subsidy received basis	Profit/loss after tax (accrual basis)	Cash profit (accrual basis excluding depreciation and write offs)	Cash profit (subsidy received basis)	Cash profit (revenue and subsidy on realized basis)	Profit/loss without subsidy	Profit/loss on subsidy received basis
Bihar	2662	2519	2519	2385	4783	2662	1227	1054	1054	846	4133	1227	367	83	83	213	3023	367
Jharkhand	3211	3094	3094	3362	3961	3211	2668	2532	2532	2577	3768	2668	1511	1394	1394	1157	2477	1511
Odisha	1119	580	580	1110	1119	1119	292	442	442	33	292	292	1643	949	949	978	1643	1643
Sikkim	17	17	17	35	17	17	39	39	39	20	39	39	33	33	33	24	33	33
West Bengal	625	1489	1489	8	625	625	546	1494	1494	1811	546	546	466	1597	1597	224	466	466
Eastern total	6384	4720	4720	6900	9254	6384	3600	1610	1610	67	7607	3600	3021	795	795	2148	6644	3021
Arunachal Pradesh	274	274	274	293	274	274	255	255	255	261	255	255	428	279	279	297	428	428
Assam	525	376	526	645	675	675	305	149	299	356	455	455	621	473	638	728	786	786
Manipur	307	217	217	231	307	307	315	217	217	309	315	315	194	194	194	225	194	194
Meghalaya	204	177	177	265	217	204	291	199	209	669	302	302	274	162	162	196	274	274
Mizoram	127	127	127	129	127	127	200	200	200	198	200	200	192	192	192	195	192	192
Nagaland	214	198	198	172	214	214	212	197	197	265	212	212	191	190	190	92	191	191
Tripura	157	112	112	138	197	157	166	119	119	143	206	166	70	48	48	51	110	70
North-eastern total	1809	1481	1631	1873	2013	1959	1745	1336	1496	2199	1945	1905	1970	1538	1703	1785	2176	2136
Delhi	1710	2319	2319	4147	1710	1710	939	1688	1688	14716	939	939	692	1584	1584	1811	692	692
Haryana	13223	12438	12438	12256	16549	13223	3835	2942	2942	3081	8964	3835	3315	2225	2225	3199	8296	3315
Himachal Pradesh	513	313	313	521	513	513	340	131	131	98	340	340	75	149	149	7	75	75
Jammu and Kashmir	2795	2507	2507	2576	2795	2795	2938	2429	2429	1695	2938	2938	2219	1707	1707	1646	2219	2219
Punjab	559	233	311	5	4663	481	419	1324	1158	1875	4807	253	642	1640	1640	1434	4053	642
Rajasthan	19938	18404	18418	19075	21715	19952	12495	10801	10816	8947	14855	12510	15912	14074	14088	14792	17451	15926
Uttar Pradesh	12006	10928	10928	13988	15602	12006	13154	11924	11924	15700	17844	-13154	17678	16168	16168	15771	22852	17678
Uttarakhand	5	123	123	44	5	5	9	190	190	142	9	9	339	559	559	567	339	339

Contd...

TABLE 5.19 | Contd...

Region/state	2011/12						2012/13						2013/14					
	Profit/loss after tax (accrual basis)	Cash profit (accrual basis excluding depreciation and write offs)	Cash profit (subsidy received basis)	Cash profit (revenue and subsidy on realized basis)	Profit/ loss without subsidy	Profit/loss on subsidy received basis	Profit/ loss after tax (accrual basis)	Cash profit (accrual basis excluding depreciation and write offs)	Cash profit (subsidy received basis)	Cash profit (revenue and subsidy on realized basis)	Profit/loss without subsidy	Profit/ loss on subsidy received basis	Profit/ loss after tax (accrual basis)	Cash profit (accrual basis excluding depreciation and write offs)	Cash profit (subsidy received basis)	Cash profit (revenue and subsidy on realized basis)	Profit/ loss without subsidy	Profit/ loss on subsidy received basis
Northern total	47 330	41 916	41 852	52 515	60 132	47 266	31 395	25 024	25 205	12 590	48 800	31 576	37 525	30 243	30 257	31 604	53 914	37 538
Andhra Pradesh	699	3 614	446	6 609	7 426	3 361	16 654	13 622	13 637	4 766	22 186	16 668	720	2 433	2 433	1 016	7 026	720
Karnataka	159	1 440	1 321	334	1 155	40	791	830	762	458	2 985	859	160	1 313	1 258	278	1 786	216
Kerala	241	707	707	655	241	241	241	750	750	843	241	241	111	627	627	34	111	111
Puducherry	156	128	128	170	158	156	301	271	271	241	303	301	46	22	22	56	51	46
Tamil Nadu	13 321	12 415	12 402	13 067	15 393	13 308	11 443	10 496	10 881	8 975	15 900	11 827	12 677	11 642	11 709	11 999	17 662	12 744
Southern total	12 379	6 783	10 949	19 525	23 891	16 545	28 949	22 810	23 277	12 681	41 133	29 415	13 493	7 290	7 413	10 728	26 414	13 616
Chhattisgarh	1 895	1 612	1 612	1 749	1 897	1 895	501	29	29	228	502	502	1 317	521	521	440	1 317	1 317
Goa	271	268	268	283	271	271	285	276	276	294	285	285	4	1	1	175	4	4
Gujarat	623	2 331	2 331	1 653	477	623	539	2 469	2 469	3 126	571	539	583	2 838	2 838	2 657	517	583
Madhya Pradesh	3 004	2 002	2 002	2 543	4 542	3 004	4 472	3 493	3 495	3 223	6 469	4 474	6 941	5 645	5 651	5 432	9 141	6 947
Maharashtra	37	1 175	1 175	888	39	37	655	3 357	3 357	4 491	653	655	1 534	4 352	4 352	3 523	1 532	1 534
Western total	4 584	375	375	3 809	7 226	4 584	4 064	2 028	2 025	3 872	7 175	4 067	6 145	1 026	1 020	133	9 488	6 152
Grand total	72 486	55 275	59 527	84 622	102 516	76 738	69 753	48 751	49 563	23 531	106 659	70 564	62 154	38 841	39 149	46 131	98 595	62 462

Source PFCL (Power Finance Corporation Ltd). 2014. ***Report on The Performance of State Power Utilities for the Years 2011/12 to 2013/14***. Details available at <http://www.pfcindia.com>, last accessed on 22 December 2015

TABLE 5.20 Status of reforms and restructuring in power sector

State	Constitution of SERC	Number of T/O issued	Year of last T/O	Unbundling/ corporatization (numbers)	11 kV feeder metering (%)	Consumer metering (%)	APDRP sanctions (₹ million)	APDRP investment releases (₹ million)	APDRP utilization (% of sanctions)
Delhi	1999	5	2006	2002 (3)	100	99	2 110	1 050	100
Haryana	1998	6	2007	1999 (2)	100	90	4 320	1 690	71
Himachal Pradesh	2001	5	2007	Ext 07/08	98	100	3 230	3 070	98
Jammu and Kashmir	2002	1	2007		95	40	11 000	7 050	55
Punjab	2001	6	2007	Ext 30/11	100	86	7 160	2 030	60
Rajasthan	1999	4	2007	2000 (3)	100	95	11 930	4 340	74
Uttar Pradesh	1998	6	2008	2000 (4)	100	91	10 690	3 140	93
Uttarakhand	2002	4	2008	2001 (1)	100	97	3 100	2 800	85
Chhattisgarh	2002	2	2006	Ext 07/08	100	74	3 530	1 590	69
Gujarat	1998	4	2007	2005 (4)	100	93	10 830	4 000	91
Goa	2002	*	2002		100	97	2 890	1 130	79
Madhya Pradesh	1998	6	2008	2002 (3)	100	72	6 630	1 790	57
Maharashtra	1999	6	2008	2005 (1)	100	87	16 430	4 360	69
Andhra Pradesh	1999	8	2007	1999 (4)	100	88	11 270	5 670	85
Karnataka	1999	6	2006	1999 (5)	100	86	11 860	4 640	81
Kerala	2002	6	2008	Ext 10/08	100	100	8 590	2 490	50
Tamil Nadu	1999	2	2007	Ext 01/09	100	86	9 480	4 420	76
Bihar	2002	1	2006	Ext 09/08	41	50	8 230	3 130	93
Jharkhand	2002	2	2007	Ext 10/08	86	75	4 240	1 540	54
Orissa	1996	7	2006	1996 (4)	95	81	2 070	740	20
West Bengal	1999	7	2007	2007 (4)	100	97	4 420	930	84
Assam	2002	3	2006	2004 (3)	100	95	6 510	5 800	82
Meghalaya	2005	*	2005	Ext 01/09	100	50	2 270	1 370	59
Arunachal Pradesh	x	*	2000		0	46	830	590	65
Mizoram	JERC notified	*	2005		70	99	1 090	810	81
Manipur	JERC notified	*	2002		21	82	1 420	430	2

Contd...

TABLE 5.20 *Contd...*

State	Constitution of SERC	Number of T/O issued	Year of last T/O	Unbundling/ corporatization (numbers)	11 kV feeder metering (%)	Consumer metering (%)	APDRP sanctions (₹ million)	APDRP investment releases (₹ million)	APDRP utilization (% of sanctions)
Nagaland	x	*	2006		85	61	1220	990	63
Tripura	2003	1	2005	2005 (1)	100	89	1470	980	69
Sikkim	2003	*	2006		100	92	1520	1550	100
Total	25	22		14 (37)	98	88	170340	74120	74

APDRP – Accelerated Power Development and Reforms Programme; JERC – Joint Electricity Regulatory Commission; SERC – State Electricity Regulatory Commission; T/O – tariff order

[x]not constituted

[*]not issued

Notes

1. Memorandum of understanding, memorandum of association, and TPAs have been signed by all states. Distribution in Delhi privatized in 2002 (three companies) and in Orissa in 1994 (four companies). Arunachal Pradesh, Goa, Jammu and Kashmir, Manipur, Mizoram, Nagaland, and Sikkim have state electricity boards. Assam has taken extension up to 15 January 2009 for trading function.
2. Vide Notification No. 1/9/08-EB(PR)196, dated 16 April 2010 Government of Punjab unbundled Punjab State Electricity Board into two companies.
3. The Chhattisgarh State Electricity Board has been reorganized into five companies in accordance with the provisions contained in the Electricity Act 2003 by the Government of Chhattisgarh Vide Notification dated 19 December 2008.

Source Details available at <http://www.powermin.nic.in/distribution/apdrp/projects/pdf/Reforms_and_Restructuring.pdf>, last accessed on 20 December 2011

RENEWABLE ENERGY SOURCES AND TECHNOLOGIES 6

AT A GLANCE

- The Indian solar photovoltaic (SPV) market has witnessed significant growth since the launch of the Jawaharlal Nehru National Solar Mission (JNNSM) in 2010. The total installed capacity has risen from 40 MW in 2010 to 4509 MW till 31 August 2015.[1] An aggregate capacity of 1686 MW was installed in Phase I of JNNSM, and a further capacity of 1000 MW was added in Phase II of the programme.
- A 50 MW solar thermal power plant has been set up at Nokh village in Jaisalmer district of Rajasthan and it has started commercial operations.
- According to the Ministry of New and Renewable Energy (MNRE), about 82 700 family-size biogas plants have been set up till March 2014 against the projected target of 1.06 lakh biogas plants.
- Thirteen states—Delhi, Gujarat, Haryana, Punjab, Uttarakhand, Uttar Pradesh, Rajasthan, West Bengal, Chhattisgarh, Andhra Pradesh, Karnataka, Tamil Nadu, and Kerala—announced their state policies on grid-connected SPV rooftop.
- The current potential for generating energy from surplus agro and forestry residues is estimated to be about 18 000 MW. As on 31 August 2012, the cumulative installed capacity of grid-interactive biomass power generation and bagasse cogeneration was 3313.33 MW and the installed capacity of biomass gasifier for distributed power generation was about 414.648 MW.
- Nearly 80% of manufacturing components in solar thermal technology are domestic. Hence, JNNSM seems to strengthen and complement the "Make in India" initiative.[2]
- The MNRE has identified the potential for 140 million m^2 of collector areas for solar water-heating systems in the country and can save 16 000 GWh of electricity, out of which only 40 million m^2 are potentially techno-economically feasible.[3]
- The state electricity regulatory commissions of 17 states and union territories (UTs) have issued appropriate regulatory orders. These states and UTs include Andhra Pradesh, Chhattisgarh, West Bengal, Delhi, Haryana, Gujarat, Karnataka, Kerala, Tamil Nadu, Uttarakhand, Daman and Diu, Lakshadweep, Chandigarh, Dadra and Nagar Haveli, Andaman and Nicobar Islands, Puducherry, and Goa.

---

[1] Details available at http://mnre.gov.in/mission-and-vision-2/achievements/, last accessed on 20 December 2015

[2] Details available at http://stfi.org.in/images/PPTS/MOM%2028th%20July, 2015.pdf, last accessed on 20 December 2015

[3] Details available at www.cseindia.org/userfiles/State%20of%20CSP%20in%20India%20low.pdf, last accessed on 20 December 2015

Introduction

Renewable energy technologies utilize energy that comes from continually replenished resources such as sunlight, wind, biomass, water, geothermal heat, and ocean energy. The worldwide growth of renewable energy began in the 1990s and accelerated greatly in the 2000s. The total investments in renewable energy in 2013 were $214 billion worldwide, a fall from 2011/12 due to significant cost reductions and also policy uncertainty.[4] India has also announced plans for attracting investments worth $100 billion in the renewable energy sector during the next few years, along with the proposed revision of the National Solar Mission targets to 100 GW of solar power compared to the earlier target of 22 GW by 2022. The government has also announced a target of 60 GW of installed wind power capacity by 2022.

Growth in renewable energy is due to supportive government policies, rising costs of the conventional energy, reduction in costs of renewable energy technology, and economies of scale in manufacturing. Renewable energy applications are also growing in end-use sectors, such as heating/cooling, transport, and power. This chapter summarizes the status of various renewable energy technologies in India, classified into off-grid and on-grid applications.

Status of renewable energy in India

India has around 36.47 GW of installed renewable energy capacity as on August 2015, which is about 13.2% of the total installed power generation capacity.[5] In 2013/14, the total gross grid-integrated renewable energy generation was around 31 692.18 MW as on 31 March 2014, registering a growth of 12.92%.[6] The main renewable energy technologies being developed on grid-connected, large-scale basis are solar photovoltaic (SPV) and concentrated solar power (CSP), onshore wind, biomass power, and small hydropower. So far wind power is the largest renewable energy system in India, reaching 24 088 MW. India is ranked fifth globally for wind installations (REN21 2014). The wind power capacity added to the country in 2013/14 was 2126 MW. This was higher than the year before (2012/13), which saw a drastic dip due to the temporary withdrawal of incentives in the wind energy sector in 2012. Installations fell to 1700 MW in 2012/13 in comparison to 3164 MW in 2011/12.[7] Solar power capacity stands at 4229 MW in the current year.[8] Figure 1 shows the installed capacities of various renewable energy technologies in India.

Power generation applications of renewable energy systems

Off-grid applications

Biomass gasification

Biomass gasification is a process of partial combustion of carbonaceous feed material with oxygen to produce gas which can be effectively used for both electricity generation and thermal applications. Gasifiers can utilize different types of locally available biomass resources, which make them ideal for decentralized rural electrification in areas rich in biomass.

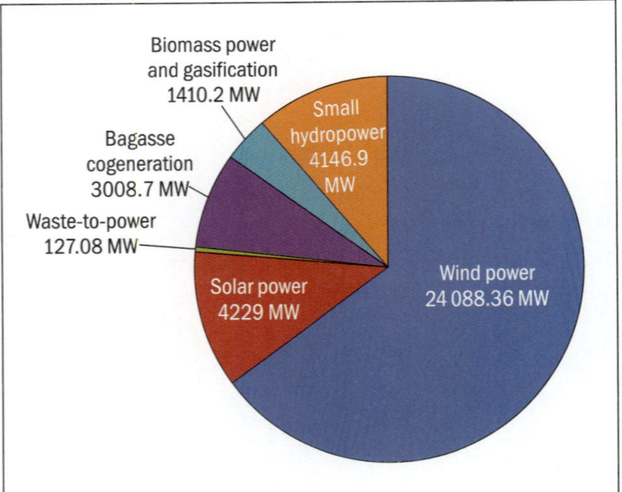

Figure 1 Grid-connected renewable energy installed capacity and power sector in India
Sources CEA (2012) and details available at <http://mnre.gov.in/>

[4] Details available at www.unep.org/pdf/Green_energy_2013-Key_findings.pdf, last accessed on 20 December 2015
[5] Details available at http://powermin.nic.in/power-sector-glance-all-india, last accessed on 20 December 2015
[6] Details available at http://mospi.nic.in/Mospi_New/upload/Energy_stats_2015_26mar15.pdf, last accessed on 20 December 2015
[7] Hindu Business Line (2013)
[8] Details available at http://mnre.gov.in/mission-and-vision-2/achievements/, last accessed on 20 October 2015

India is one of the leaders in the development of biomass gasification technology. Many research organizations, including The Energy and Resources Institute (TERI), Indian Institute of Science (IISc) Bangalore, Ankur Scientific Energy Technologies, Husk Power Systems, Grain Processing Industries (I) Pvt. Ltd, Infinite Energy Pvt. Ltd, GP Green Energy Systems Pvt. Ltd, Rishipooja Energy and Engineering Company, and Radhe Renewable Energy Development Associate are involved in technology development and in providing gasifier-based clean energy solutions. To explore the various advantages of biomass-based gasifier technologies, the Ministry of New and Renewable Energy (MNRE) has set up biomass gasification systems under different schemes: Village Energy Security Programme, Remote Village Electrification Programme, and Biomass Energy and Cogeneration (Non-bagasse) in Industrial and Urban Areas.

In 2014/15, various rice husk gasifier-based projects were successfully installed in 25 villages of Bihar by different entrepreneurs. These projects were based on a sustainable business model, and each project was equipped with 32 kWe of energy production. Additionally, 19 other projects were successfully installed in Rajasthan for electricity supply to the telecom tower, with each project having a capacity of 12 kWe. Table 1 shows the status of bio-power in India till August 2015.

Twenty-six biomass gasifiers were installed in 12 rice mills and 14 other industries in West Bengal, Madhya Pradesh, Tamil Nadu, Uttar Pradesh, and Bihar to meet the demand for electricity and thermal applications. Various firms provide comprehensive services and facilities for installing biomass gasification technology in India. Table 2 gives the list of state-wise biomass gasifiers installed in India.

Biogas

Biogas technology generates an alternative source of energy largely from organic wastes. In 1981/82, the National Biogas and Manure Management Programme (NBMMP) was launched for the promotion of biogas plants based on cattle dung and other organic wastes. The NBMMP mainly caters to setting up of family-size biogas plants that provide energy for cooking and other purposes in rural areas along with making enriched bio-fertilizers for farmers. The availability of clean energy mitigates the drudgery of rural women, reduces pressure on forests, and accentuates social profit. Under the NBMMP, the target for family-size biogas plants for 2015/16 is 1.10 lakh.[9]

The MNRE financially supported the installation of more than 45 000 plants throughout the country in 2014/15. The cumulative achievement of installing family-size biogas plants was 48.23 lakh till 2015.[10]

Table 1 Status of bio-power (grid-connected/off-grid) and family-size biogas plants in India (till August 2015)

Area		Target during 2015/16	Achievement during 2015/16	Cumulative achievement (as on 31 August 2015)
Grid-connected bio-power (capacity in MW)	Biomass power, gasification, and bagasse cogeneration	400	0	4 418.55
	Waste-to-power	10	12	127.08
Total (in MW)		410	12	4 445.63
Off-grid bio-power (capacity in MW_{eq})	Biomass gasifier (rural and industrial)	8	0	170.00
	Biomass (non-bagasse) cogeneration	60	10.50	602.37
	Waste-to-energy	10	0.50	146.51
Total (in MW_{eq})		78	11	918.88
Family-size biogas plants (number in lakh)		1.10	0.05	48.23

Source Details available at <http://mnre.gov.in/mission-and-vision-2/achievements>, last accessed on 20 October 2015

[9] Details available at http://mnre.gov.in/file-manager/offgrid-biogas/Physical-Targets-2015-16-of-NBMMP.pdf, last accessed on 20 October 2015

[10] Details available at http://mnre.gov.in/file-manager/annual-report/2014-2015/EN/Chapter%201/chapter_1.htm, last accessed on 20 October 2015

Table 2 State-wise list of biomass gasifier plants installed in India with their capacity

State	Plant power capacity range (kW)	Number of plants installed	Supported agencies
Arunachal Pradesh	1-20	3	CGPL, OVN
Assam	20-25	2	OVN
Bihar	1-50	3	OVN, NETPRO
Gujarat	1	1	CGPL
Jammu	250-415	2	BETEL
Karnataka	20-415	12	BETEL, CGPL, NETPRO, ENERGREEN
Kerala	75	1	BETEL
Madhya Pradesh	15-100	3	ARUNA, NETPRO
Maharashtra	1	1	CGPL
Manipur	25-35	2	OVN
Mizoram	75	1	CGPL
Nagaland	20	1	OVN
New Delhi	1	1	CGPL
Rajasthan	35	1	NETPRO
Sagar Island	150	1	SYNERGY
Tamil Nadu	7.5-1500	16	ARUNA, ENERGREEN, NETPRO, ARRYA
Tripura	35	4	OVN
Uttar Pradesh	1	1	OVN
West Bengal	1-150	5	CGPL, SYNERGY
Others	5-415	4	ARUNA, BETEL

ARRYA – Arrya Hi-Tech Energy; ARUNA – Aruna Electrical Works (P) Ltd; BETEL – Bioresidue Energy Technology Pvt. Ltd; CGPL – Coastal Gujarat Power Limited; ENERGREEN – Energreen Power Ltd; OVN – OVN Trading Engineers Pvt. Ltd; NETPRO – NETPRO Renewable Energy (India) Pvt Ltd; SYNERGY – Synergy Enviro Engineers (India) Pvt. Ltd

Source Details available at <www.eai.in/ref/ae/bio/csbg/list_of_installation.html>, last accessed on 21 October 2015

During 2014/15, 32 biogas plants were installed with the capacity of 627 kW, and 59 other biogas projects are being implemented with the capacity of 1008 kW.

Solar photovoltaic

Under the Off-grid and Decentralized Solar Application Scheme of the Jawaharlal Nehru National Solar Mission (JNNSM), the MNRE provides a subsidy of 30% of the project cost ranging from ₹21 per W_p to ₹120 per W_p, depending on the capacity of the module and configuration of the SPV system.[11]

The MNRE provides a subsidy from ₹72 to ₹396 for stand-alone SPV power plant installations by the central and state governments in the special category states, including north-eastern states and islands.

For stand-alone installations with battery storage in unelectrified rural areas, central financial assistance (CFA) ranges from of ₹85 per W_p to ₹115 per W_p. CFA for solar water-pumping systems ranges from ₹27 630 per HP to ₹57 600 per HP. Table 3 provides the CFA provided by MNRE for different types of Solar PV systems.

Till December 2014, SPV off-grid systems/power plants of 295.87 MW_p capacity were sanctioned across the country. Some of the major off-grid SPV projects sanctioned during 2014/15 are as follows:

- 250 SPV power plants of 1 kW_p capacity each for Kinnaur, Lahaul, and Spiti areas of Himachal Pradesh
- 977 solar charging stations for 15 districts of Odisha
- 70 micro-grid systems in Uttar Pradesh

[11] Details available at http://mnre.gov.in/file-manager/annual-report/2014-2015/EN/Chapter%204/chapter_4.htm, last accessed on 20 October 2015

Table 3 List of solar photovoltaic systems with different capacity and financed by MNRE

Type of system	Capacity	Central finance assistance by MNRE (₹)
Home lights and small photovoltaic plants	Up to 300 W_p	100 per W_p
LED-based systems	Up to 40 W	160 per W_p
Solar home-lighting systems and small-capacity photovoltaic systems	Between 300 W_p and 1 kW_p	45 per W_p

- Distribution of 10 000 solar lanterns in flood-affected areas of Jammu and Kashmir and another 10 000 in cyclone-affected areas of Andhra Pradesh
- 6000 solar power packs for individuals in Bihar

During 2014/15, solar systems of total capacity 52.77 MW_p, including solar lanterns, home lights, street lights, pumps, and power plants, were installed in various states. Some of the highlights of the completed projects during the current financial year are as follows:

- 24 552 solar street lights installed at various villages in Himachal Pradesh
- Solar power plants having a total capacity of 3500 kW_p installed at various places, including industries in Chhattisgarh
- 18 844 solar street lights installed at Lohiya villages of Uttar Pradesh
- An SPV power plant having a capacity of 100 kW_p installed at the Legislative Assembly of Meghalaya

Solar water-heating systems

Solar water heaters are one of oldest uses of solar energy for thermal applications. Solar water-heating systems are used for meeting hot water requirements in both domestic and commercial sectors. Domestic applications generally include small-scale systems with about 2–10 m^2 area. Commercial applications are large systems with size up to 4000–5000 m^2 area. In these systems, pumps are usually employed for forced circulation of water. According to a conservative estimate, the gross potential in India is 140 million m^2 of the collector area. Of this, the realizable techno-economic potential for deployment is thought to be around 40 million m^2 of the collector area. These solar water-heating systems thus save costly fossil fuels and/or electricity. Industrial solar water heaters of 100 L/day capacity can save around 120 L of furnace oil in a year through preheating. These systems are also effective in demand-side management, saving peak power requirements to a considerable extent. It may be noted that installations of ~2000 m^2 of solar water heaters can typically contribute to a peak load saving of about 1 MW (Krishnamurthy, Shashishekar, and Akashdeep 2010). India has set an ambitious target of installing 20 million m^2 of solar water heater area by 2020, of which 80% share is expected to come from the residential sector. The states with a comparatively good potential for solar resources, namely Andhra Pradesh, Gujarat, Karnataka, Maharashtra, and Tamil Nadu, account for 67% of this potential.

Various financial institutions and banks are implementing the interest subsidy scheme to facilitate installation of solar water-heating systems. A programme on solar water-heating systems is being implemented to accelerate the development and deployment of solar water-heating systems in the country through a combination of financial and promotional incentives. Soft loans, interest subsidy, and capital subsidy are being provided—60% capital subsidy or loan at 5% interest to domestic non-commercial category and 30% capital subsidy or loan at 5% interest to commercial category. Thirty-one banks and financial institutions are participating in the scheme through their branch network in different parts of the country. The cumulative installations of solar water-heating systems in India were 8.90 million m^2 till July 2014.[12]

Solar cooking

Cooking is the major household activity in India that require energy. Fuelwood, kerosene, and liquefied petroleum gas (LPG) are the major energy sources for cooking. There is a gradual shift from fuelwood and kerosene to LPG for cooking, but it is mostly in urban areas. In rural areas, fuelwood and kerosene are still the major energy sources. Solar cooking offers a good alternative for domestic applications. At the

[12] Details available at www.mnre.gov.in/mission-and-vision-2/achievements/, last accessed on 20 October 2015

micro level, solar cookers facilitate financial savings for the consumer; at the macro level, they help in the conservation of LPG and fuelwood. However, installations have remained somewhat limited as these systems are not user friendly. There are many types of solar cookers, such as box type, concentrating (dish), indirect heating, with or without heat storage, and hybrid. Out of these, the box-type solar cookers have reached the commercialization stage. The solar cooker programme was started in India with various promotional efforts, including government subsidies. Soft loans are currently available for the purchase of solar cookers through some designated nationalized banks.

Solar cooking can also be effectively used for institutional or community cooking applications, such as in religious places or hostels. Community cooking systems generally employ steam-based cooking using low pressure steam, which can be easily given using simple solar concentrators operating at moderate temperatures. A parabolic concentrator technology named "Scheffler" was developed to meet this need. This technology is specifically suitable for such applications, owing to its low cost and easy manufacturing and being promoted through government schemes and incentives.

Concentrated solar thermal systems

The industrial sector is the highest consumer of energy among all the sectors. Typical energy demand for the industrial sector is approximately 53% of the total energy consumption. Of this, usually 60%–70% of the energy is used in the form of heat, out of which 70% is used for applications below 250°C.[13] Thus, about 37% of the total energy is consumed in the form of heat by industrial processes,[14] which is a large potential area for the use of solar thermal systems. Numerous industrial processes within this wide range of operating temperatures from 100°C to 250°C can be easily integrated with solar energy. The main application areas are food, dairy, textile, paper, manufacturing, and chemical industries. Various processes in these industries require process heat as the major energy input. Though a wide range of potentially suitable application areas are available, applicability and feasibility in real practice depend on a number of parameters, such as solar system configuration, scale of the system, and the type of conventional fuel in the existing system. The economics of the solar industrial process heat (SIPH) system plays a crucial role in making projects viable for actual installation. The SIPH system needs to be carefully sized to maximize system performance.

Industries have shown increasing interest in this sector, and various indigenously developed technologies are now commercially available. These technologies can operate at different temperature ranges and utilize heat mediums such as pressurized hot water, thermic oil, and low- or high-pressure steam. Some of the available technologies are Scheffler, parabolic troughs, two-axis tracking parabolic dishes, two-axis tracking Fresnelized dishes, linear Fresnel reflectors, evacuated tube, unglazed collector, and so on. The total installed capacity of solar thermal for India stood at 8.63 million m^2, equivalent to 6000 MW (approximately), till 31 December 2014.[15] The major manufacturers of these technologies in India include Clique Solar, Thermax, Forbes Solar, Megawatt Solutions, and KGDS Pvt. Ltd. The MNRE is promoting these systems with various schemes and incentives, such as capital subsidy and soft loans.

Solar air heating

A solar air heater is a type of energy collector in which energy from the sun is captured by an absorbing medium and used for heating air. The main application of solar air heaters is in industry and agriculture which require hot air for drying. Solar air collectors are classified into two categories: unglazed solar collectors and glazed solar collectors. Unglazed solar collectors primarily use ambient air. They are low-cost collectors and require only one penetration into the building. If existing fan inlets are used, then no additional penetrations are necessary. The efficiency of a solar collector is highest when the temperature of the air entering the solar panel is equal to the ambient temperature. This occurs with solar heaters that draw outside air into the solar heater instead of room air. Glazed solar collectors are designed primarily for space heating, and they recirculate building air through a solar air panel where the air

[13] MoSPI (Ministry of Statistics and Programme Implementation). 2015. ***Energy Statistics 2015***. New Delhi: MoSPI

[14] Bhave, A. J. 2012. Industrial process heat applications of solar energy. ***International Journal of Modern Engineering Research*** 2(5):3800–3802

[15] Details available at http://mnre.gov.in/file-manager/annual-report/2014-2015/EN/Chapter%204/chapter_4.htm, last accessed on 20 October 2015

is heated and then directed back into the building. These solar space-heating systems require at least two penetrations into the building and perform only when the air in the solar collector is warmer than the building room temperature. Most glazed collectors are used in the residential sector.

Solar air-heating systems, based on flat plate collectors, are useful in food processing industries for drying various food products. These industries generally require hot air at low temperature as process heat for drying products, such as tea leaves/coffee beans, and also for processing fruits, spices, cereals, mushroom, papad, vegetables, fish, and seafood. Hot air is also required in industries such as leather, textiles, chemicals, rubber, paper, and pharmaceuticals. Major Indian solar air heater manufacturers include ATR Solar, Inter Solar, and Planters Energy Network, and leading international manufacturers include Conserval Engineering, Cansolair, and American Solar.

Grid-connected applications

Biomass-based cogeneration and bio-power

Biomass-based power generation has been increasing rapidly in India with the installation of megawatt-scale plants processing a variety of biomass residues, such as shells, husks, de-oiled cakes, and wood. According to the MNRE, about 32% of the total primary energy use in the country is derived from biomass and more than 70% of the country's population depend on biomass for their energy needs. The MNRE has realized the potential and role of biomass energy in the Indian context and initiated a number of programmes for the promotion of efficient biomass conversion technologies to be used in various sectors of the economy. India has more than 5464 MW biomass-based power plants comprising 4545 MW grid-connected and 918 MW off-grid power plants. Out of the total grid-connected capacity, the major share comes from 3008 MW bagasse cogeneration, followed by 1410 MW biomass combustion-based power plants. While off-grid capacity comprises 602 MW non-bagasse cogeneration, mainly as captive power plants, about 18 MW biomass gasifier systems are being used for meeting electricity needs in rural areas and 152 MW equivalent biomass gasifier systems have been deployed for thermal applications in industries.

Overall, India produces 686 MT of gross crop residue biomass annually, of which 234 MT (34% of gross) is estimated as surplus for bioenergy generation.[16] The potential for power generation from agricultural and agro-industrial residues is estimated at about 18000 MW. With progressive higher steam temperature and pressure, as well as efficient project configuration in new sugar mills and modernization of existing ones, the potential for surplus power generation through bagasse cogeneration in sugar mills is estimated at 7000 MW. The potential for bagasse cogeneration lies mainly in the nine sugar producing states, with the maximum potential of about 1250 MW each in Maharashtra and Uttar Pradesh. Thus, the total estimated biomass power potential is about 25000 MW. About 150 MW equivalent biomass gasifier systems have been set up for grid and off-grid projects. More than 300 rice mills and other industries are using gasifier systems for meeting their captive power and thermal applications. In addition, about 70 biomass gasifier systems are providing electricity to more than 230 villages in the country.

Urban and industrial waste-to-energy

The MNRE has initiated a programme on energy recovery from urban, industrial, and agricultural wastes. The idea is to set up large-size biogas plants to produce biogas or bio-CNG or to generate power. The programme provides CFA for power generation from biogas through biomethanation of urban and agricultural waste/residues, including cattle dung, or for the production of bio-CNG, at ₹2 crore per megawatt, limited to ₹5 crore per project. These projects are set up by state nodal agencies, urban local bodies/municipal corporations, or entrepreneurs.

As per the Niti Ayog Report 2014, biomethanation technology, processing segregated biodegradable wet wastes from households and establishments, with a capacity ranging from 100 kg/day to 10 tonnes/day, has been successfully employed in 172 locations in India. Table 4 provides the details of five biomethanation plants that were successfully operated by municipalities and private operators.

[16] Hiloidhari, Moonmoon, Dhiman Das, and D. C. Baruah. 2014. Bioenergy potential from crop residue biomass in India. ***Renewable and Sustainable Energy Reviews*** 32: 504–512

Table 4 Details of some successfully operated biogas-based power plants with performance

Firm, location, and year of installation	Plant capacity	Installation (month)	Full potential (month)	Quantity/type of waste processed	Number of working days	Quantity of biogas generated	Quantity of manure generated
Katol, Maharashtra, 2010–14	2 MT/day	December 2010	February 2011	800 MT: kitchen, hotels, household	1200 (approximately)	60 000 m^3 (approximately)	50 MT
Bharuch, Ankleshwar, 2006–14	5 MT/day	June 2006	September 2006	4000 MT: kitchen, hotels, household	2500 (approximately)	320 000 m^3 (approximately)	350 MT
Pune Municipal Corporation, 2010–14	5 MT/day	December 2009	March 2010	9000 MT: hotels, kitchen	1200 (approximately)	600 000 m^3	20 MT
Kottayam Medical College, 2009–14	2 MT/day	June 2009	September 2009	2000 MT: hostel kitchen	1400 (approximately)	120 000 m^3	150 MT
Auro Textile, Himachal Pradesh, 2010–14	2 MT/day	June 2010	October 2010	485 MT: kitchen waste and 1995 MT: bio-sludge	1150	63 200 m^3 (meter installed on day 1)	70 MT

MT – million tonnes
Source Niti Ayog (2014)

Based on urban and industrial wastes, nine projects with the capacity of 9.54 MW were initiated to be completed by 31 December 2014. At Fazilka in Punjab, a 1 MW grid-connected power project has been commissioned, which is based on agriculture waste. Table 5 shows the biomass co-generation projects installed in different states during 2014/15.

According to the 2014/15 annual report of the MNRE, 12 projects of about 5.65 MW capacity for producing bio-CNG based on a mix of urban and agricultural wastes were in progress in Pilibhit, Kolkata, Surat, Hanumangarh, Barnala, Amritsar, Kanpur, and Ambala and were expected to be completed in 2015.[17] Around 21 projects based on urban/industrial wastes with an aggregate capacity of about 21 MW were under installation. These projects were based on cattle dung, starch industry wastes, poultry litter, and biogas at distilleries.

Production of bio-compressed natural gas (Bio-CNG) from urban, agricultural, sewage and industrial wastes through biomethanation is also undertaken by the MNRE. A total of 12 projects of Bio-CNG set up under this programme are operational. The total production of Bio-CNG from

Table 5 Biomass co-generation projects installed (MW capacity) in different states during 2014–15

State	Installed capacity (MW) up to 31 December 2014	Projected achievement during 2014/15 (MW)
Haryana	7.25	12.75
Tamil Nadu	7.50	7.50
Uttarakhand	5.00	12.00
Punjab	5.20	5.20
Andhra Pradesh	0.89	8.89
Odisha	5.28	5.06
West Bengal	2.50	2.50
Jharkhand	0.80	3.10
Uttar Pradesh	–	3.00
Total	34.42	60.00

MW – megawatt
Source MNRE (2015)

[17] Details available at http://mnre.gov.in/file-manager/annual-report/2014–2015/EN/Chapter%201/chapter_1.htm, last accessed on 20 October 2015

the 12 projects set up so far is estimated to be about 17 200 kg/day.[18] These projects are set up in Haryana, Karnataka, Maharashtra, Punjab, and Rajasthan.

Programme on energy from urban, industrial, and agricultural wastes/residues including biomass co-generation (non-bagasse) in industry

The industrial sector consumes approximately 35% of the total electricity generated in the country. The ministry is implementing "Programme on Energy from Urban, Industrial and Agricultural Wastes/Residues" including Biomass Co-generation (Non-bagasse) in industries for exploiting the vast potential of biomass power in the form of thermal energy and power for captive use in industry with the following major objectives:

- To promote setting up of projects for the recovery of energy from urban, industrial, and agricultural wastes/residues
- To create conducive conditions and environment, with fiscal and financial regime, to develop, demonstrate, and disseminate utilization of wastes and residues for recovery of energy

According to 2014/15 annual report of the MNRE, 15 biomass cogeneration (non-bagasse) projects with a capacity of 34.42 MW have been completed. In addition, 12 projects with an aggregate capacity of about 51 MW are under implementation.

Onshore wind

India currently had about 23 439.26 MW of installed capacity for wind power till 31 March 2015 and is ranked fifth in the world for total wind installed capacity.[19] The concept of repowering for the coming years would add excellent potential in wind energy. According to a study by Risoe Laboratory, Denmark, the process of repowering will double the capacity.[20] Wind power is a mature and scalable clean energy technology, and India holds a domestic advantage. India has an annual manufacturing capacity of more than 9.5 GW for wind turbines. During 2014/15, India installed a total of 1.33 GW of new wind power capacity.[21] Tamil Nadu, Karnataka, Maharashtra, Rajasthan, and Gujarat are the leading states in wind power generating capacity. On account of new government initiatives on renewables, the Indian wind industry is expected to grow at a capacity of 5 GW per annum for the rest of the decade.[22]

In 2011, the Centre for Wind Energy Technology revised India's wind power potential to 102 778 MW at 80 m height and 2% land availability. This is a significant upward revision from the earlier estimate of approximately 49 130 MW at 50 m height and 2% land availability.[23] During the past year, other research organizations have also estimated India's wind potential using various models for mapping the wind resource. A study by the Lawrence Berkeley National Laboratory, assuming a turbine density of 9 MW/km^2, estimated the total wind potential in India, with a minimum capacity factor of 20%, to be 2006 GW at 80 m hub height and 3121 GW at 120 m hub height.

Offshore wind

India has around 7600 km of coastline with the potential for offshore wind power development. The MNRE's initiative towards the development of offshore wind power includes the approved "National Offshore Wind Energy Policy 2015", effective from 7 October 2015. The National Offshore Wind Energy Policy aims to provide a conducive environment for harnessing offshore wind energy, including setting up of a demonstration offshore wind power project to showcase technology and build investors' confidence.

The National Offshore Wind Energy Authority (NOWA) will be constituted under the MNRE, which will be the nodal agency for offshore wind projects in the country. NOWA will carry out resource assessment and surveys in the exclusive economic zones (EEZs) of the country and simultaneously enter into contract with project developers for the development of offshore wind energy project in the territorial water.[24] As per the National Offshore Wind Energy Policy, preliminary assessments along the coastline have indicated prospects for the development of offshore wind power. Wind resource data collected from the

[18] Details available at www.pib.nic.in/newsite/pmreleases.aspx?mincode=28, last accessed on 20 October 2015
[19] Details available at http://mnre.gov.in/file-manager/UserFiles/wp_installed.htm, last accessed on 20 October 2015
[20] Ahmed (2015)
[21] Details available at http://mnre.gov.in/file-manager/UserFiles/wp_installed.htm, last accessed on 20 October 2015
[22] Details available at www.gwec.net/wp-content/uploads/2015/03/GWEC_Global_Wind_2014_Report_LR.pdf, last accessed on 20 October 2015
[23] Details available at http://niwe.res.in/department_wra_est.php, last accessed on 20 October 2015
[24] Details available at http://pib.nic.in/newsite/erelease.aspx?relid=98283, last accessed on 20 October 2015

coastline of Rameshwaram and Kanyakumari in Tamil Nadu and from the Gujarat coast show a reasonable potential. A preliminary assessment suggests the potential for establishing a 1 GW capacity wind farm each along the coastline of Rameshwaram and Kanyakumari in Tamil Nadu.

In 2014, a memorandum of understanding (MoU) was signed to set up a joint venture company (JVC) for undertaking the First Demonstration Offshore Wind Power Project in the country along the Gujarat coast. The MoU was signed by the MNRE, National Institute of Wind Energy, and a consortium of partners consisting of the National Thermal Power Corporation (NTPC), Power Grid Corporation of India Ltd, Indian Renewable Energy Development Agency, Power Finance Corporation, Power Trading Corporation, and Gujarat Power Corporation Ltd.

The JVC will conduct a detailed feasibility study based on the inputs received from pre-feasibility studies and take necessary steps for implementing the first offshore demonstration wind power project.

The first planned demonstration offshore wind power project along the Gujarat coast will be of about 100 MW capacity. Subsidy has been proposed for setting up evacuation and transmission infrastructure of the offshore wind power project, including financial support for carrying out studies such as wind resource assessment, environment impact assessment, oceanographic survey, and bathymetric studies. The MNRE would assist in obtaining necessary clearances during the implementation of the project. This offshore demonstration project, being the first in the country, will certainly provide enough learning to move into this sector by taking up similar viable projects in the future, enabling India to enter the club of countries already in the business of offshore wind power generation.

Concentrated solar power

Concentrated solar power technologies are used for solar thermal power generation. CSP technologies use mirrors to reflect and concentrate sunlight onto receivers that collect and convert solar energy into heat. This heat/thermal energy can then be used for producing electricity via a steam turbine or heat engine that drives a generator. With the availability of cost-effective storage medium (molten salt), solar thermal power plants can have a higher plant load factor and deliver power even when the sun is down. Under the JNNSM, the Government of India has envisaged cumulative installation of 1000 MW, 4000 MW, and 10 000 MW grid-connected solar power projects by 2013, 2017, and 2022, respectively.

Under Phase I Batch I, 470 MW capacity solar thermal power projects were approved, out of which 420 MW projects were delayed beyond specified time limit with payment of penalties. The government extended the scheduled commercial operation date of the delayed projects by 10 months from May 2013 to March 2014 to allow developers to complete their projects without penalties. The capacity of individual projects varies from 20 MW to 100 MW, and technologies used are compact linear Fresnel and parabolic trough. Out of the seven projects, five projects are in Rajasthan, one in Gujarat, and one in Andhra Pradesh.[25]

The Government of Rajasthan plans to set up 4000 MW of solar thermal capacity by 2018 under the 10 000 ha Bhadla Solar Park Project. A dedicated 400 kV network supported by a strong transmission network of 220 kV and 132 kV in Barmer, Jaisalmer, Jodhpur, and Bikaner has been created as per the Rajasthan Solar Park report. To ensure water access, 58 cusecs of water have been reserved for the CSP plants from the Indira Gandhi Canal Project, about 3 km from the village. Also, a 50 MW solar thermal power plant has been set up at Nokh village in the Jaisalmer district of Rajasthan by the Godawari Green Energy Ltd, a wholly owned subsidiary of the Godawari Power and Ispat Ltd, and it started commercial operations from 19 June 2013. The Rajasthan Renewable Energy Corporation issued the commissioning certificate of the project on 2 July 2013. This 50 MW power plant is India's first concentrated solar thermal power plant, which started commercial operation among the seven solar thermal power projects totalling 470 MW approved for development under Phase I Batch I. Of the 470 MW slated for commissioning by May 2013, the Godawari project is currently the only utility-scale project in commissioning and is the first CSP project in India to produce power.

Solar photovoltaic

The Indian SPV market has seen significant growth since the launch of the JNNSM in 2010. The total installed capacity had risen from 40 MW in 2010 to 4509 MW as on 31 August 2015. An aggregate

[25] Details available at http://nvvn.co.in/Selected%20Projects%20List.pdf, last accessed on 20 October 2015

capacity of 1686 MW was installed in Phase I of JNNSM, and a further capacity of 9000 MW is targeted in Phase II of the programme.[26]

The Twelfth Five-year Plan targets the installation of 10 000 MW of grid solar power, which would include large plants, rooftop plants, and distribution grid plants. Phase II of the JNNSM started in 2013 and will continue until 2017. To meet the targets set for Phase II, the government has decided to take the following measures:

- A target of 750 MW of grid-connected SPV under Phase II, with viability gap funding (VGF) from the National Clean Energy Fund
- Projects set up on a build–own–operate model
- Provision of renewable purchase obligation (RPO) for solar power in the National Tariff Policy
- Grant of subsidy on off-grid applications and generation-based incentive (GBI) facility for bundled power and VGF for grid-connected solar power projects through various interventions announced from time to time
- Setting up ultra-mega solar power projects and solar parks, 1 MW solar parks on the banks of canals, and solar power driven agricultural pump sets for energizing 1 lakh pumps
- Concessional import/excise duty exemption for setting up solar power plants, accelerated depreciation, and tax holiday

Targets under the three phases of JNNSM and the achievements under each target are listed in Table 6. Phase I of the mission has provided several learning experiences, which are listed as follows:

- Reduction in tariff is possible if capacity to be allotted is high
- Experienced companies are interested in large projects
- Transmission and evacuation system is still a major issue
- Assurance of continuous and timely payment is very important from the lenders' point of view
- Resource assessment was another major issue; however, with experience developers are now more confident about it with experience across various parts of the country
- Equipment import is still very high despite domestic content requirements. Domestic manufacturing needs more support
- Financial health of state discoms is hampering growth and increasing the risks for investors

Grid-connected solar rooftop programme

About 10.8 GW of solar capacity is expected to be added in 2016/17, thereby increasing the total operational solar capacity close to 20 GW.[27] TERI has estimated that India has a market potential for 124 GW of rooftop SPV in urban settlements alone.[28] Map 1 shows the installed capacity of grid connected solar power plants across the country. The Rooftop PV and Small Solar Power Generation Programme was launched under Phase II of JNNSM to distinctly focus on deploying an increased number of rooftop SPVs. The objectives of promoting grid-connected solar power during Phase II are as follows:

- To achieve grid parity at the earliest
- To facilitate meeting of solar RPO targets in line with tariff policy

Table 6 JNNSM targets and achievements

Application segment	Target for Phase I (2010-13)	Achievement in Phase I (2010-13)	Cumulative target for Phase II (2013-17)	Achievement in Phase II (2013-17)	Cumulative target for Phase III (2017-22)	Achievement in Phase III (2017-22)
Grid solar power (large plants, rooftop, and distribution grid plants)	1100 MW	1686 MW (including those under state initiative)	9000 MW	4229 MW (as on 31 August 2015)	20 000 MW	To be initiated

Source Details available at <http://mnre.gov.in/; http://pib.nic.in/newsite/pmreleases.aspx?mincode=28>, last accessed on 20 October 2015

[26] Details available at http://mnre.gov.in/file-manager/annual-report/2014-2015/EN/Chapter%204/chapter_4.htm, last accessed on 20 October 2015

[27] Details available at http://cleantechnica.com/2015/10/15/india-likely-add-10-8-gw-solar-power-capacity-2016–17/, last accessed on 20 October 2015

[28] TERI's report on "Accelerating large scale deployment of rooftop solar PV through market transformation and enabling policies" prepared for Shakti Sustainable Energy Foundation

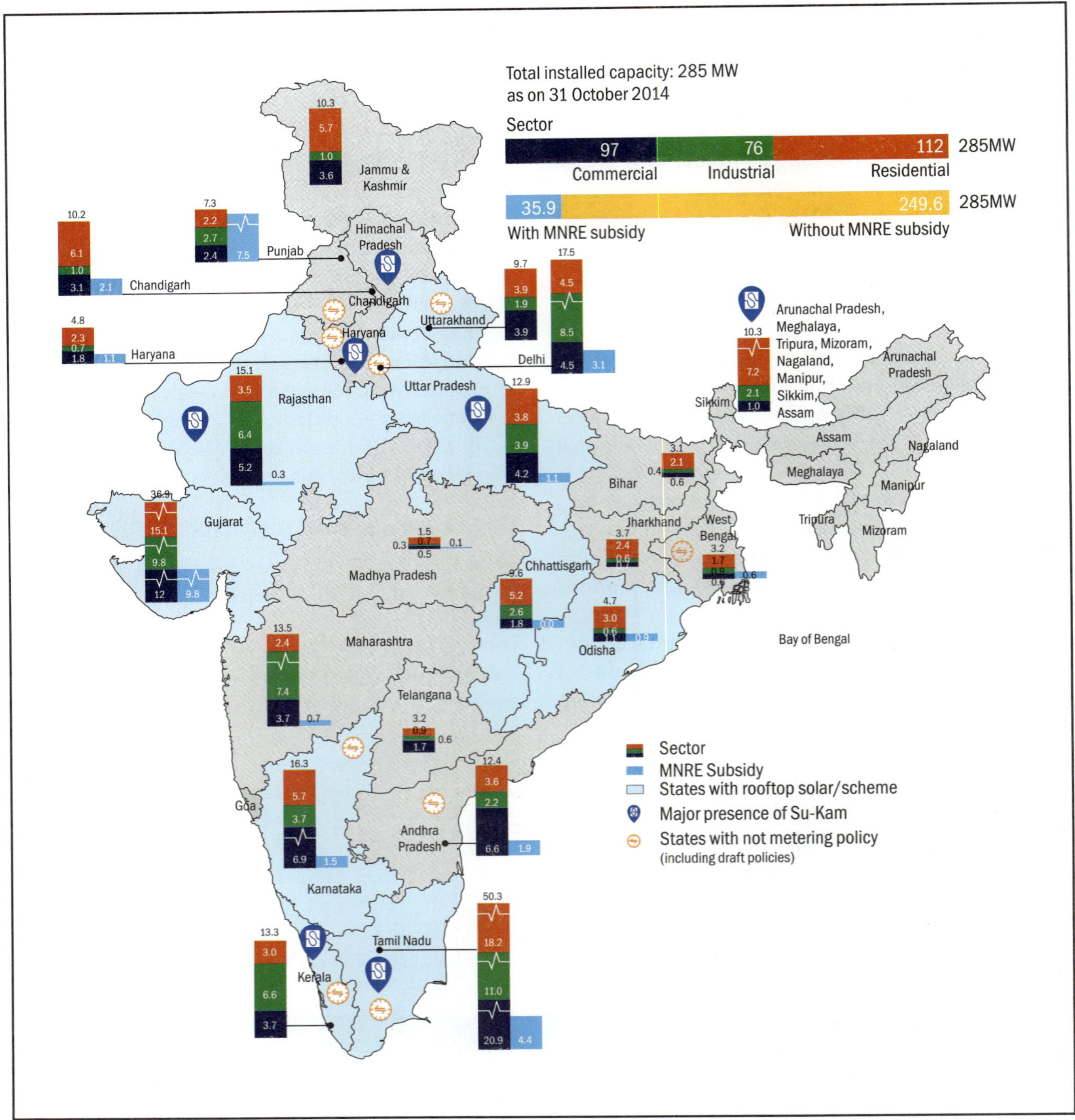

Map 1 Installed capacity of grid-connected solar power plants

Source Details available at http://dev.bridgetoindia.com/wp-content/uploads/2014/11/india-solar-rooftop-map_2014.pdf, last accessed on 20 October 2015

- To encourage wider participation of states in National Solar Mission (NSM)
- To facilitate increased connectivity and grid access for solar projects
- To encourage solar park cluster development
- To promote technology agnostic approach
- To develop domestic technological capability
- To encourage large-scale deployment of grid-connected rooftop solar
- To encourage captive and third-party sale of solar power to meet power shortages in industry
- To encourage local manufacturing
- To support focused research and development (R&D) initiatives
- To develop renewable energy certificate (REC) market
- To develop storage systems

Grid-connected rooftop PV systems shall be connected to the grid, either of 11 kV three-phase line or of 220 V single-phase line, depending on the system installed at the institution/commercial establishment or residential complex. Power generated by these systems would be utilized by industrial and captive loads and excess power is fed back to the grid as long as the grid is available.

Thirteen states—Delhi, Gujarat, Haryana, Punjab, Uttarakhand, Uttar Pradesh, Rajasthan, West Bengal, Chhattisgarh, Andhra Pradesh, Karnataka, Tamil Nadu, and Kerala—have come out with their state policies on grid-connected rooftop SPV.[29] These policies define different targets for installation set by each state along with details of tariff for gross and net metering systems.

Small hydropower plants

India has an estimated potential of about 19 750 MW of small hydropower (SHP) projects (MNRE 2014). Most of the potential is in the Himalayan states as river-based projects and in other states on irrigation canals. The MNRE has created a database of potential sites for small hydro and identified 6474 potential sites with an aggregate capacity of 19 749 MW for projects up to 25 MW capacity (MNRE 2014). So far SHP projects with an aggregate capacity of 3990 MW have been set up; an additional capacity of 747.66 MW is under implementation.[30] Setting up of small hydro projects comes under the purview of state governments. Potential sites are either developed by the state or allotted to private developers. During the Eleventh Five-year Plan, a capacity of 1419 MW was added against 536 MW during the Tenth Five-year Plan. A capacity addition of 1600 MW from SHP projects has been planned during the Twelfth Five-year Plan.

The MNRE provides CFA to set up small/micro hydro projects both in public and in private sectors. Financial support is also given to the state governments for identification of new potential sites, including survey and preparation of detailed project reports, and renovation and modernization of old SHP projects. It also helps the state governments in formulating policies for the development of SHP projects and exploitation of their potential. Under the CFA scheme of the MNRE, capital subsidy is now provided to both private and state projects for renovation and modernization of SHP plants. In addition, technical support is being provided to SHP units through the Alternate Hydro Energy Centre, Indian Institute of Technology Roorkee.

Other technologies

Geothermal energy

India is said to have a geothermal potential of 10 000 MW. Gujarat, Chhattisgarh, Andhra Pradesh, and West Bengal are the first Indian states that plan to build geothermal power plants with capacities ranging from 3 MW to 5 MW.[31] The current projects and potential areas of geothermal resource in India are located in Puga Valley (Jammu and Kashmir), Tatapani (Chhattisgarh), Godavari Basin, Manikaran (Himachal Pradesh), Bakreshwar (West Bengal), Tuwa (Gujarat), and Unai and Jalgaon (Maharashtra).[32] Map 2 shows the potential areas or geothermal hot spots in India.

A number of Indian firms have already signed MoUs for geothermal projects. All these projects involve companies such as Panx Geothermal, LNJ Bhilwara, Tata Power, Thermax, Avni Energy Systems,

---

[29] Details available at http://mnre.gov.in/file-manager/UserFiles/solar-rooftop-states-policy-tariff.htm, last accessed on 20 October 2015

[30] Details available at http://mnre.gov.in/file-manager/annual-report/2014-2015/EN/Chapter%203/chapter_3.htm, last accessed on 20 October 2015

[31] Details available at http://asian-power.com/project/news/gujarat-bids-india%E2%80%99s-first-geothermal-power-plant, last accessed on 20 October 2015

[32] Details available at www.eai.in/ref/ae/geo/geo.html, last accessed on 20 October 2015

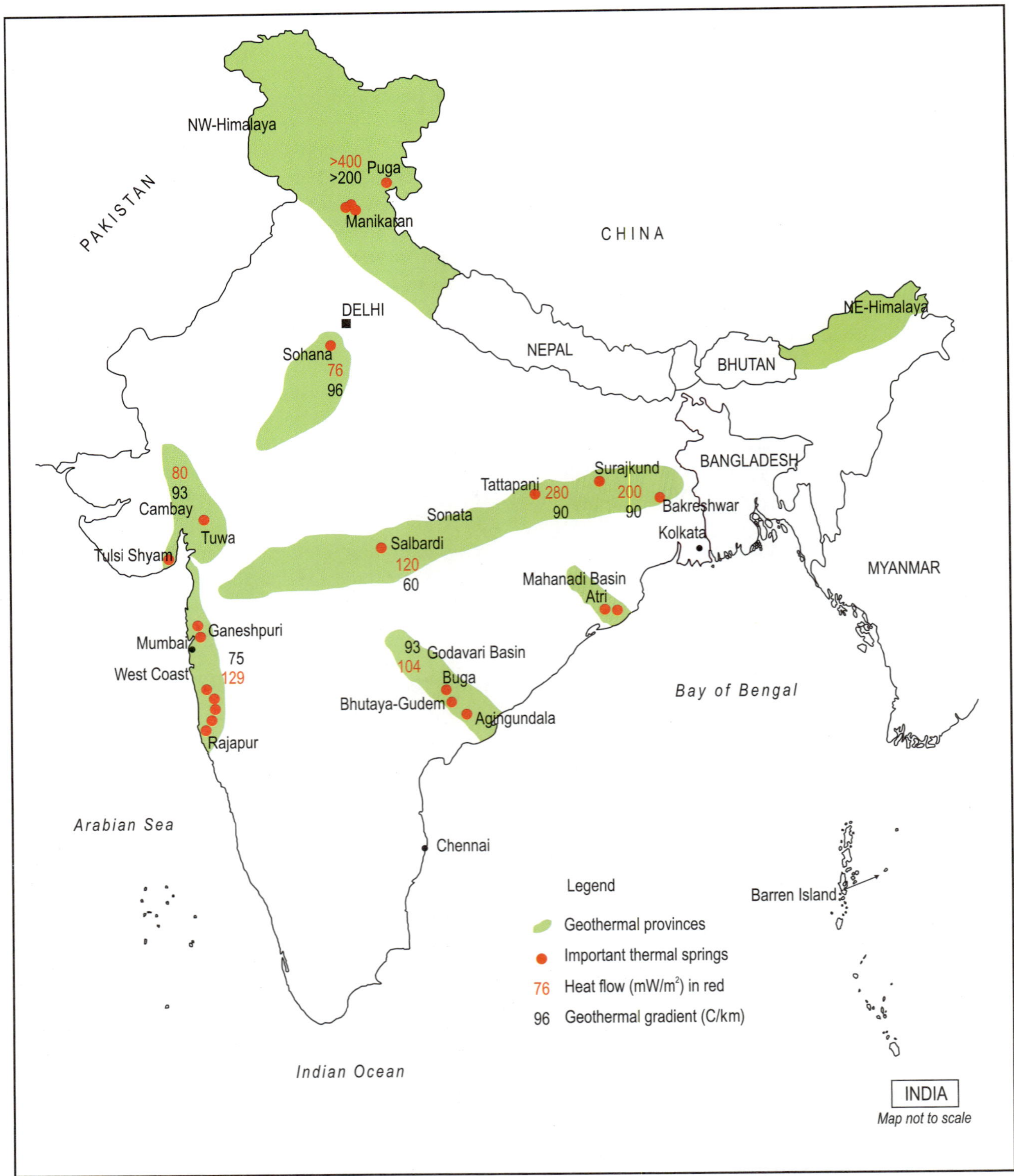

Map 2 Geothermal map of India

Source Details available at www.eai.in/ref/ae/geo/geo.html, last accessed on 20 October 2015

GeoSyndicate Power Pvt. Ltd, ONGC, Mitsubishi, and NTPC.[33] The capital cost of installing 1 MW of geothermal energy works out to be ₹16–20 crore. The government is in the process of forming an agency to take care of exploration of geothermal resources that have been declared as a "national resource".[34]

Hydrogen energy

At present, hydrogen energy is only at the research, development, and demonstration stage. Although hydrogen energy is a niche sector, a lot of development activities are happening in this area. The National Hydrogen Energy Road Map prepared in 2007 for developing the National Hydrogen Energy Programme envisages that 1 million hydrogen-fuelled vehicles would be on Indian roads and 1000 MW aggregate hydrogen-based power-generating capacity would be set up in the country both by 2020.[35]

Environmentally benign processes/technologies for the production of hydrogen have been developed and demonstrated. Materials/alloys/methods have been developed for the storage of hydrogen as metal hydrides. The application of hydrogen in fuel cells for power generation has been demonstrated as a result of initiatives taken by this ministry. Hydrogen-fuelled small power-generating sets and about 15 hydrogen-fuelled motorcycles are being demonstrated in the campus of Banaras Hindu University, Varanasi. Facilities for dispensing hydrogen-blended compressed natural gas fuel have been set up in Faridabad and Delhi, and catalytic combustion systems for residential (cookers) and industrial sectors have also been developed and demonstrated.

The Ministry of Science and Technology, CSIR Laboratories, Ministry of Petroleum and Natural Gas, Defence Research and Development Organization, Indian Space Research Organization, Department of Atomic Energy, oil and gas companies, and private sector automobile companies are also involved in the research, development, and demonstration programmes related to hydrogen.[36]

Fuel cells

The MNRE has launched programme called the "Chemical Sources of Energy Programme" for the development of fuel cell technology. A fuel cell–battery hybrid vehicle with indigenously developed 10 kW proton exchange membrane fuel cell (PEMFC) stack has undergone field performance evaluation. Efforts made are expected to lead to the indigenous production and wider applications of fuel cell systems in the country. Fuel cells are emerging as power sources for automobiles.

The widespread use of fuel cells for power generation, transport, and other applications is expected to reduce dependence on scarce fossil fuels and help in preserving the environment. Some of the achievements made so far are as follows:[37]

- Indigenous base for research and industrial production being established
- R&D projects leading to technology/process/material development
- Prototypes of PEMFCs and phosphoric acid fuel cells developed
- Application of fuel cells demonstrated for decentralized power generation
- A 3 kW capacity uninterrupted power supply (UPS) based on PEMFC developed
- Reformer for a 10 kW PEMFC system developed and tested

Liquid biofuels

To meet the increasing energy needs of the country and provide energy security, the National Policy on Biofuels was announced in December 2009. The major goals of the policy are development and utilization of indigenous non-food feedstocks raised on degraded or wastelands; thrust on R&D on cultivation, processing, and production of biofuels; and a blending mandate of 20% ethanol and biodiesel by 2017.

A variety of fuels can be produced from biomass resources, including liquid fuels such as ethanol, methanol, butanol, biodiesel, and pyrolytic or bio-oil.

[33] Details available at www.eai.in/ref/ae/geo/geo.html, last accessed on 20 October 2015
[34] Details available at www.indianexpress.com/news/state-pitches-for-first-geothermal-energy-plant-in-country/1147784/, last accessed on 20 October 2015
[35] Details available at www.mnre.gov.in/schemes/new-technologies/hydrogen-energy/, last accessed on 20 October 2015
[36] Details available at http://mnre.gov.in/file-manager/UserFiles/faq_hydrogenenergy.htm, last accessed on 20 October 2015
[37] Details available at www.mnre.gov.in/schemes/new-technologies/chemical-energy/, last accessed on 20 October 2015

Biofuels are primarily used to fuel vehicles, but they can also be used in fuel engines or cells for electricity generation. Liquid biofuels are an approach in moving our economy to a more sustainable basis because they move us away from fossil fuels.

Unlike other renewable energy sources, biomass can be directly converted to liquid fuels for our transportation needs. India's first biofuel station would be commissioned in Hassan, Karnataka. According to the Biofuel Task Force and Bharat Petroleum Corporation Ltd, it would be set up at the Biofuel Park campus near Madenur village. Around 0.5 L of biodiesel has to be added to vehicles for every 10 L of conventional diesel.

There are three major technological pathways for making liquid biofuels: biochemical (for example, enzymatic fermentation) to make bio-alcohols; thermochemical (for example, pyrolysis, gasification) to make bio-oil and Fischer–Tropsch fuels; and transesterification to make biodiesel.

Three government institutions—MNRE, Department of Biotechnology (DBT), and Department of Science and Technology (DST)—are directly promoting and supporting different alternative biofuel programmes in India. In the R&D programmes, the MNRE is primarily supporting the thermochemical and chemical routes, such as pyrolysis for bio-oil and transesterification for biodiesel production, whereas DBT is supporting biochemical and other biotechnological routes for bioethanol, biodiesel, and algal biofuel production. The MNRE is formulating a national policy on biofuels, besides supporting research, development, and demonstration of biofuels. In India, pyrolysis through the thermochemical route has not been extensively tapped compared to other thermochemical conversion technologies, such as gasification and combustion for power generation. Research on pyrolysis for liquid fuel and chemicals in India has reached the pilot scale from the bench scale. However, no commercial plant has been built yet. The MNRE's R&D strategy on thermochemical biomass conversion has laid special thrust on basic research on pyrolysis, capability generation, and systematic experimental and modelling work on low and high temperature and pyrolysis is definitely the first step towards this goal.

Ocean energy: ocean thermal energy and tidal energy

India has 159 km^2 of tropical water in the EEZ with a power density of 0.2 MW/km^2. The following sections discuss the developments and advancements of ocean thermal and tidal energy in India would be described.

Ocean thermal energy

Ocean thermal energy conversion (OTEC) can still be considered a nascent technology. However, the potential is significant because oceans capture a large amount of sunlight falling on the earth. In 1998, the erstwhile Ministry of Non-Conventional Energy Sources had sanctioned a project named "Demonstration of 1 MW OTEC Power Plant" to the National Institute of Ocean Technology, Chennai, which was suspended in 2009/10 due to shortage of funds.[38] According to the estimates of MNRE, India has a potential of exploiting 80 000–180 000 MW of OTEC-based power.[39]

Tidal energy

India has a long coastline with estuaries and gulfs where tides are strong enough to move turbines for electrical power generation. The Gulf of Cambay and the Gulf of Kutch in Gujarat on the west coast have the maximum tidal range of 11 m and 8 m with average tidal range of 6.77 m and 5.23 m, respectively. The Ganges Delta in the Sundarbans is approximately 5 m with an average tidal range of 2.97 m. The identified economic power potential is of the order of 8000 MW with about 7000 MW in the Gulf of Cambay, 1200 MW in the Gulf of Kutch in Gujarat, and 100 MW in the Gangetic Delta in the Sundarbans region of West Bengal.[40] To harness adequate power from tidal energy, the height difference between high tide and low tide should be at least five metres. There are only 20 such locations across the globe and India is one of them.

Renewable energy technology integration

The emergence of large amounts of renewable electricity around the country has created new challenges for the transmission and distribution

[38] Details available at www.niot.res.in/index.php, last accessed on 20 October 2015

[39] Details available at www.mnre.gov.in/, last accessed on 20 October 2015

[40] Details available at www.mnre.gov.in/schemes/new-technologies/tidal-energy/, last accessed on 20 October 2015

sectors. Augmentation of transmission capacity enables better management of renewable energy, which is variable in nature. From a long-term perspective, transmission capacity needs to be augmented with conventional sources, energy storage, and adequate reactive power sources in the network. From a short-term perspective, there needs to be better forecasting of renewable energy to help utilities in efficient scheduling and dispatch of renewable energy generation, thereby contributing to grid stability. In the report titled "Green Energy Corridors", the Power Grid Corporation of India has estimated that around ₹43 000 crore would be required to strengthen and develop the evacuation and transmission infrastructure for the planned addition of renewable power capacity in the next 5 years. Currently, renewable energy generated anywhere in the country is absorbed within the state grids.

Challenges in the renewable energy sector

Various challenges and barriers impede the scaling-up of renewable energy technologies in India. The barriers span policy, regulatory, finance, technological, and institutional aspects and can be specific to individual renewable energy sources or can be overarching and hinder renewable energy deployment in general.

Reducing per unit cost and bringing renewables to grid parity pose a challenge. Although costs have fallen drastically over the past few years, there is a need to continuously innovate to increase efficiency and bring down costs further. Innovations can be brought about in various ways. It is possible to harness lower wind speeds with better technology, develop high-efficiency solar cells from carbon nanotube materials, channel the energy of tides and waves to produce electricity, harness the large potential of offshore wind energy, use alternative transport fuels to make travelling less carbon intensive, utilize hydrogen for energy storage, and have a larger grid with lower losses of electricity. There are specific challenges for each renewable fuel source.

For biomass, there is an absence of effective institutional and financing mechanism, reflected by the absence of commercial and service networks (biomass depots for collection, transportation, and delivery of biomass fuels) at the national, regional, and local levels, and limited access to financing and lack of interest on part of the state electricity boards in promoting biomass power generation. Biomass projects have also a long gestation period, and the project development cycle requires several years to be completed. Biomass fuel supply risks are twofold. The first is the physical availability found in the case of all biomass power projects (rainfall, harvesting effectiveness, and productivity). The second is the question of contracted supply encountered when dealing with distributed biomass supplies. The inability to lock up sufficient supplies of biomass from various sources will hinder project finance and implementation.

For wind power, there is a lack of an appropriate regulatory framework to facilitate purchase of renewable energy from outside the host state, inadequate grid connectivity, high wheeling and open access charges in some states, and delays in obtaining land and acquiring statutory clearances, which have hindered faster deployment. The wind sector is also very vulnerable to changes in policies and incentives as observed by drastic drop in installations due to reduction in accelerated depreciation benefit and GBI in the past year.

For solar power, high capital cost and payback period, lower efficiency, need for back-up or storage, insufficient solar radiation data, lack of consumer awareness about technology, lack of trained people and training institutes, absence of financing mechanisms and lack of local infrastructure, political commitment, and R&D are hindering faster deployment (Ansari, Kharb, Luthra, *et al.* 2013).

There are limitations in setting up SHP projects due to difficult and remote terrains in hilly/mountainous regions with severe infrastructural constraints. The locational hurdles also serve to prolong the gestation period and push up the per-megawatt capital costs even as the power evacuation and transmission facilities at the sites remain inadequate. Additionally, there are also delays in acquiring land and obtaining statutory clearances.

A number of factors undermine the current RPO and REC markets in India. The most critical issue is that non-compliance is not penalized, and if it is penalized, the penalty would have to be paid mostly by the indebted state electricity boards. This is compounded by another counterproductive trend: RPO targets are not ambitious enough to create a market pull. Some states such as Rajasthan have reduced their target to match the actual renewable energy production. Very few states have escalating RPO targets. Tamil Nadu has set a target lower than its (considerable) renewable power generation.

The REC market has its own challenges over and above the unpredictable demand from RPOs. Only around 50% of RECs are actually sold, and trading dips drastically during the early months of a financial year. This makes RECs non-bankable. There are a number of further bottlenecks and open questions, for instance, the lack of market aggregators (trading is allowed only through the energy exchanges), the discussion about vintage RECs, or clarifications on whether captive and off-grid renewable energy plants can generate RECs.

International status of renewable energy and India's position

India added more than 4 GW of renewable energy capacity in the previous year. SPV and wind accounted for almost 70% of renewable additions in 2013. Renewables made up approximately 17% of the total additions from all sources during 2013. India was the fourth largest market for wind power in 2013, with more than 1.7 GW of wind power being installed. This was, however, a shrink of 26% over the previous year, primarily due to the devaluation of the rupee against the US dollar and removal of key support policies in 2012. The reinstatement of GBI in late 2013, however, helped revive the market. India leads in the use of concentrating solar thermal systems with at least 145 systems producing steam, primarily for cooking.

India also leads the world in the total capacity of small gasifiers for electricity generation, with a capacity exceeding 155 MW. Additionally, India, Brazil, China, and South Africa are in the lead for the development of large-scale programmes that address the dual challenges of energy access and sustainability.

Offshore wind technology in India

With around 2.5 GW of onshore wind power installation, India has approximately 6% of the world's wind power installed capacity. As the capacity addition for onshore wind power is growing satisfactorily, achieving 60 GW number, which the Indian Government has targeted by 2022 at COP21, is not possible without exploiting substantial offshore resource available across the vast 7517 km coastline of India. The offshore wind power policy, which was under the making since 2013, received the nod from cabinet in September 2015. Preliminary assessment by National Institute of Wind Energy (NIWE) for offshore potential along the coastline of Rameshwaram and Kanyakumari in Tamil Nadu and in Gujarat shows that 60 000 MW and 106 000 MW, respectively, can be installed through the two coastal zones. As a matter of fact, there are two main maritime areas in which structures such as offshore wind farms can be built, namely (i) Indian territorial waters, which extend up to 12 nautical miles (nm) from the baseline and (ii) Exclusive Economic Zone (EEZ), which is beyond the 12 nm limit and up to 200 nm, where under international law India has the right to construct structures such as wind farm installations.

The salient features of the recently approved National Offshore Wind Energy Policy are as follows:

- The objective is to explore, promote, and deploy offshore wind farms in the EEZ, including those under public–private partnership.
- Challenges include resource characterization, subsea cabling, turbine installation including logistics, grid interconnection, transmission infrastructure development, and coastal security during the construction and operation period.
- NIWE has been authorized as nodal agency for carrying out allocation of offshore wind energy blocks. It will also frame guidelines for coordination and allied functions with related ministries and agencies.
- NIWE to call for proposal for establishing offshore wind farm through international competitive bidding, carry out and coordinate resource assessment and surveys, facilitate getting clearances and NOCs from concerned ministries, enter into contract with the developers; sign seabed lease agreement and collect fee from developers, and so on.
- Promote investment in establishing infrastructure including transmission network.
- Promote spatial planning and management of maritime renewable energy resources in the EEZ through suitable incentives.
- Impetus on offshore wind energy for achieving greater energy security goal of the country.
- Reduce country's carbon emissions.
- Encourage indigenization of the offshore wind energy technology.
- Promote research and development in the offshore wind energy sector.
- Create skilled manpower and employment in the offshore wind energy sector.
- Facilitate development of project EPC and operation and maintenance with regard to offshore industry.
- Develop coastal infrastructure and supply chain to support heavy construction and fabrication work and the operation and maintenance activities.
- Government may plan to provide fiscal and financial incentives, and allow foreign direct investment participation, public–private partnership, and international collaborations.
- Government may promote bundling of power generated from offshore wind power projects with conventional power subject to availability of unallocated conventional power for bringing down the cost of power generated.

References

Ahmed, Siraj. 2015. Re-powering the wind power. ***Akshay Urja*** 9(1): 10–15

Ansari, M. F., R. K. Kharb, S. Luthra, S. Shimmi, and S. Chatterji. 2013. Analysis of barriers to implement solar power installations in India. ***Renewable and Sustainable Energy Reviews*** 27: 163–174

CEA (Central Electricity Authority). 2012. Monthly generation report (renewable energy sources). Details available at www.cea.nic.in/reports/articles/god/renewable_energy.pdf

Hindu Business Line. 2013. Wind power capacity addition falls to 1700 MW. Details available at www.thehindubusinessline.com/companies/wind-power-capacity-addition-falls-to-1700-mw/article4610962.ece, last accessed on 20 October 2015

Krishnamurthy, K. N., K. S. Shashishekar, and B. N. Akashdeep. 2010. Performance analysis of vacuum tube collectors for hard water using jacket type heat exchanger. International Journal of Engineering Research and Technology 3(3): 641–651

MNRE (Ministry of New and Renewable Energy). 2014. ***Annual Report 2013–14***. New Delhi: MNRE

MNRE (Ministry of New and Renewable Energy). 2015. ***Annual Report 2014–15***. New Delhi: MNRE

Niti Ayog 2014. ***Report of the Task Force on Waste to Energy (Volume I)***. New Delhi: Niti Ayog, Government of India

REN21. 2014. ***Renewables 2014 Global Status Report: REN21***. Paris: Renewable Energy Policy Network for the 21st Century (REN 21)

Bibliography

MNRE (Ministry of New and Renewable Energy). 2012a. Jawaharlal Nehru National Solar Mission Phase II - Policy Document. Details available at http://mnre.gov.in/file-manager/UserFiles/draft-jnnsmpd-2.pdf

MNRE (Ministry of New and Renewable Energy). 2012b. Press Information Bureau English Releases. Details available at http://pib.nic.in/newsite/erelease. aspx?relid=86860, last accessed on 13 August 2013

MNRE (Ministry of New and Renewable Energy). (n.d.). Ministry of New and Renewable Energy: Small Hydro. Details available at www.mnre.gov.in/schemes/ grid-connected/small-hydro/

NRDC (Natural Resources Defense Council). 2012. Constructing change: accelerating energy efficiency in India's buildings market. Details available at www.nrdc.org/international/files/india-constructing-change-report.pdf, last accessed on 6 September 2013

NREL (National Renewable Energy Laboratory). 2012. Learning about renewable energy. Details available at www.nrel.gov/learning/re_basics. html, last accessed on 19 August 2013

PIB (Press Information Bureau). 2012. Press Information Bureau English Releases. Details available at http://pib.nic.in/newsite/ erelease. aspx?relid=84919, last accessed on 2 August 2013

TABLE 6.1a Cumulative deployment of various renewable energy systems/devices in India as on 31 December 2014

Sector	Achievements during 2014/15 (up to December 2014)	Cumulative achievement as on 31 December 2014
Power from renewables		
Grid-interactive renewable power (MW)		
Wind power	1333.20	22465.03
Small hydropower	187.22	3990.90
Biomass power and gasification	0	1365.20
Bagasse cogeneration	152.00	2800.35
Waste-to-power	1.00	107.58
Solar power (SPV)	430.67	3062.68
Subtotal	2104.09	33791.74
Off-grid/captive power (capacities in MW_{eq})		
Waste-to-energy	8.54	141.27
Biomass (non-bagasse) cogeneration	34.32	561.64
Biomass gasifiers: rural	0.75	18.23
Biomass gasifiers: industrial	6.20	153.40
Aero-generators/hybrid systems (MW)	0.13	2.38
Solar PV systems > 1kW	52.77	227.12
Water mills/micro hydel (numbers)	2.00	15.21
Biogas-based energy system	0.30	4.07
Subtotal (MW_{eq})	105.01	1123.32
Other renewable energy systems		
Family biogas plants (lakh)	0.42	47.95
Solar water heating : collection areas (million m^2)	0.53	8.63

Source MNRE (Ministry of New and Renewable Energy) 2015. ***Annual Report 2014/15***. New Delhi: MNRE, Government of India

TABLE 6.1b State-wise renewable energy potential (MW)

State/union territory	Wind power	Small hydro-power	Biomass power	Bagasse cogeneration	Waste-to-energy	Solar	Total
Andhra Pradesh	14 497	978	578	300	123	38 440	54 916
Arunachal Pradesh	236	1 341	8			8 650	10 235
Assam	112	239	212		8	13 760	14 331
Bihar	144	223	619	300	73	11 200	12 559
Chhattisgarh	314	1 107	236		24	18 270	19 951
Goa		7	26			880	913
Gujarat	35 071	202	1 221	350	112	35 770	72 726
Haryana	93	110	1 333	350	24	4 560	6 470
Himachal Pradesh	64	2 398	142		2	33 840	36 446
Jammu and Kashmir	5 685	1 431	43			111 050	118 209
Jharkhand	91	209	90		10	18 180	18 580
Karnataka	13 593	4 141	1 131	450		24 700	44 015
Kerala	837	704	1 044		36	6 110	8 731
Madhya Pradesh	2 931	820	1 364		78	61 660	66 853
Maharashtra	5 961	794	1 887	1 250	287	64 320	74 499
Manipur	56	109	13		2	10 630	10 810
Meghalaya	82	230	11		2	5 860	6 185
Mizoram		169	1		2	9 090	9 262
Nagaland	16	197	10			7 290	7 513
Odisha	1 384	295	246		22	25 780	27 727
Punjab		441	3 172	300	45	2 810	6 768
Rajasthan	5 050	57	1 039		62	142 310	148 518
Sikkim	98	267	2			4 940	5 307
Tamil Nadu	14 152	660	1 070	450	151	17 670	34 153
Telangana						20 410	20 410
Tripura		47	3		2	2 080	2 132
Uttar Pradesh	1 260	461	1 617	1 250	176	22 830	27 594
Uttarakhand	534	1 708	24		5	16 800	19 071
West Bengal	22	396	396		148	6 260	7 222
Andaman and Nicobar Islands	365	8					373
Chandigarh					6		6
Dadra and Nagar Haveli							0
Daman and Diu	4						4
Delhi					131	2 050	281
Lakshadweep							0
Puducherry	120				3		123
Others					1 022	790	1 812
Total	102 772	19 749	17 538	5 000	2 556	748 990	896 605

Source MNRE (Ministry of New and Renewable Energy). 2015. ***Annual Report 2014/15***. New Delhi: MNRE, Government of India

TABLE 6.2 Wind power cumulative installed capacity (MW) in India, by state: 2006–15

State	2006	2007	2008	2009	2010	2011	2012	2013	2014	2015
Andhra Pradesh	121	123	123	123	236	200	246	448	783	1 038.15
Gujarat	338	637	1 253	1 567	1 864	2 176	2 966	3 175	3 447	3 642.53
Karnataka	585	821	1 011	1 327	1 473	1 730	1 934	2 135	2 324	2 639.45
Kerala	2	2	11	27	28	33	35	35	35	35.1
Madhya Pradesh	40	57	188	213	229	276	376	386	423	876.7
Maharashtra	1 001	1 488	1 756	1 939	2 078	2 311	2 733	3 022	4 065	4 437.9
Rajasthan	358	470	539	738	1 088	1 525	2 071	2 685	2 783	3 308.15
Tamil Nadu	2 895	3 493	3 873	4 305	4 907	5 904	6 988	7 162	7 276	7 456.98
Others	1	1	1	1	4	0	3	4	4	4
Total	5 341	7 091	8 754	10 239	11 907	14 154	17 352	19 051	21 141	23 439

Sources

1. Details available at <http://www.inwea.org/aboutwindenergy.htm>, last accessed on 8 September 2011
2. Details available at <http://www.inwea.org/installedcapacity.htm>, last accessed on 7 January 2016

TABLE 6.3 Wind power installed capacity (MW) in India, by state: 2002/03–2014/15

State	Up to March 2002	2002/03	2003/04	2004/05	2005/06	2006/07	2007/08	2008/09	2009/10	2010/11	2011/12	2012/13	2013/14	2014/15
Andhra Pradesh	93.20	0	6.20	21.80	0.45	0.80	0	0	13.60	55.40	54.10	202.10	335.70	254.80
Gujarat	181.40	6.20	28.90	51.50	84.60	283.95	616.36	313.60	297.10	312.80	789.90	208.30	272.70	195.25
Karnataka	69.30	55.60	84.90	201.50	143.80	265.95	190.30	316.00	145.40	254.10	206.70	201.70	188.70	315.60
Kerala	2.00	0	0	0	0	0	8.50	16.50	0.80	7.40	0	0	0	0
Madhya Pradesh	23.20	0	0	6.30	11.40	16.40	130.39	25.10	16.60	46.50	100.50	9.60	37.40	453.30
Maharashtra	400.30	2.00	6.20	48.80	545.10	485.30	268.15	183.00	138.90	239.10	416.50	288.50	1043.10	372.95
Rajasthan	16.10	44.60	117.80	106.30	73.27	111.75	68.95	199.60	350.00	436.70	545.70	614.00	98.60	524.70
Tamil Nadu	877.00	133.60	371.20	675.50	857.55	577.90	380.67	431.10	602.20	997.40	1083.50	174.6	113.50	181.30
West Bengal	1.10	0	0	0	0	0	0	0	0	0	0	0	0	0
Others	3.20	0	0	0	0	0	0	0	0	0	0	0	0	0
Total	1666.80	242.00	615.20	1111.70	1716.17	1742.05	1663.32	1484.90	1564.60	2349.40	3196.90	1698.80	2089.70	2297.90

Sources

1. Details available at <http://www.cwet.tn.nic.in/Docu/Year_Wise_Installed_Capacity_as_on_28_02_2011.pdf>, last accessed on 8 September 2011
2. Details available at <http://niwe.res.in/information_isw.php>, last accessed on 7 January 2016

TABLE 6.4 Wind power potential in India

State/union territory	Estimated potential (MW)	
	50 m	80 m*
Andaman and Nicobar Islands	2	365
Andhra Pradesh	5394	14497
Arunachal Pradesh*	201	236
Assam*	53	112
Bihar		144
Chhattisgarh*	23	314
Daman and Diu		4
Gujarat	10609	35071
Haryana		93
Himachal Pradesh*	20	64
Jharkhand		91
Jammu and Kashmir*	5311	5685
Karnataka	8591	13593
Kerala	790	837
Lakshadweep	16	16
Madhya Pradesh	920	2931
Maharashtra	5439	5961
Manipur*	7	56
Meghalaya*	44	82
Nagaland*	3	16
Odisha	910	1384
Puducherry		120
Rajasthan	5005	5050
Sikkim*	98	98
Tamil Nadu	5374	14152
Uttarakhand*	161	534
Uttar Pradesh*	137	1260
West Bengal*	22	22
Total	49130	102788

*Wind potential has yet to be validated with measurements

Source MNRE (Ministry of New and Renewable Energy) 2015. ***Annual Report 2014/15***. New Delhi: MNRE, Government of India

TABLE **6.5** Potential, installed, and under implementation projects in small hydropower sector (up to 25 MW), by state/union territory (as on 31 December 2014)

State/union territory	Potential		Projects installed		Projects under implementation	
	Number of sites	Capacity (MW)	Number of sites	Capacity (MW)	Number of sites	Capacity (MW)
Andhra Pradesh	387	978.40	68	221.03	13	32.04
Arunachal Pradesh	677	1 341.38	149	103.91	44	22.23
Assam	119	238.69	6	34.11	3	12.00
Bihar	93	223.05	29	70.70	5	17.70
Chhattisgarh	200	1 107.15	9	52.00	4	115.25
Goa	6	6.50	1	0.05	-	-
Gujarat	292	201.97	6	16.60	-	-
Haryana	33	110.05	7	70.10	2	3.35
Himachal Pradesh	531	2 397.91	165	696.11	26	59.00
Jammu and Kashmir	245	1 430.67	38	149.03	6	16.15
Jharkhand	103	208.95	6	4.05	8	34.85
Karnataka	834	4 141.12	154	1 104.98	16	99.75
Kerala	245	704.10	27	168.92	8	39.25
Madhya Pradesh	299	820.44	11	86.16	3	4.90
Maharashtra	274	794.33	59	335.43	8	35.70
Manipur	114	109.13	8	5.45	3	2.75
Meghalaya	97	230.05	4	31.03	3	1.70
Mizoram	72	168.90	18	36.47	1	0.50
Nagaland	99	196.98	11	29.67	3	3.20
Odisha	222	295.47	10	64.63	4	3.60
Punjab	259	441.38	48	157.40	10	18.25
Rajasthan	66	57.17	10	23.85	-	-
Sikkim	88	266.64	17	52.11	1	0.20
Tamil Nadu	197	659.51	21	123.05	-	-
Tripura	13	46.86	3	16.01	-	-
Uttarakhand	448	1 707.87	101	209.32	44	139.54
Uttar Pradesh	251	460.75	9	25.10	1	10.5
West Bengal	203	396.11	23	98.40	17	84.25
Andaman and Nicobar Islands	7	7.91	1	5.25	-	-
Total	6 474	19 749.44	1 019	3 990.90	233	756.66

Source MNRE (Ministry of New and Renewable Energy). 2015. ***Annual Report 2014/15***. New Delhi: MNRE, Government of India

TABLE 6.6 National Biogas and Manure Management Programme (plants in number)

State/union territory/region	Estimated potential	Cumulative physical achievements as on 31 March 2014	Target (2014/15)	Achievements up to 1 December 2014
Andhra Pradesh	1 065 000	521 756	18 000	5 448
Arunachal Pradesh	7 500	3 472	100	2
Assam	307 000	108 302	6 000	4 927
Bihar	733 000	129 825	0	-
Chhattisgarh	400 000	48 505	4 000	1 107
Delhi	12 900	681	0	-
Goa	8 000	4 093	100	22
Gujarat	554 000	428 676	4 000	522
Haryana	300 000	59 909	1 700	629
Himachal Pradesh	125 000	472 270	350	81
Jammu and Kashmir	128 000	3 044	100	-
Jharkhand	100 000	7 237	100	-
Karnataka	680 000	469 760	10 500	4 526
Kerala	150 000	141 425	4 600	1 002
Madhya Pradesh	1 491 000	345 829	10 400	4 130
Maharashtra	897 000	856 669	13 700	7 357
Manipur	38 000	2 128	0	-
Meghalaya	24 000	9 996	500	-
Mizoram	5 000	4 770	500	250
Nagaland	6 700	7 653	500	194
Odisha	605 000	261 830	7 100	1 308
Punjab	411 000	164 689	9 250	5 132
Puducherry	4 300	578	0	-
Rajasthan	915 000	69 405	1 100	333
Sikkim	7 300	8 735	200	76
Tamil Nadu	615 000	221 704	800	158

Contd...

TABLE 6.6 | Contd...

State/union territory/region	Estimated potential	Cumulative physical achievements as on 31 March 2014	Target (2014/15)	Achievements up to 1 December 2014
Telangana	-	-	0	5 302
Tripura	28 000	3 328	400	30
Uttar Pradesh	1 938 000	437 556	2 300	571
Uttarakhand	83 000	17 541	900	434
West Bengal	695 000	366 316	0	-
Andaman and Nicobar Islands	2 200	137	0	-
Chandigarh	1 400	97	0	-
Dadra and Nagar Haveli	2 000	169	0	-
Khadi and Village Industries Commission	-	-	12 800	1 605
Total	12 339 300	5 178 085	110 000	45 146

Sources

1. MNRE (Ministry of New and Renewable Energy). 2015. ***Annual Report 2014/15***. New Delhi: MNRE, Government of India
2. Details available at <http://www.ireeed.gov.in/policyfiles/432-biogasscheme.pdf>, last accessed on 8 January 2016

TABLE 6.7 State-wise cumulative commissioned biomass power and bagasse cogeneration projects

State	Cumulative capacity (MW)
Andhra Pradesh	380.75
Bihar	43.42
Chhattisgarh	264.90
Gujarat	43.90
Haryana	45.30
Karnataka	603.28
Madhya Pradesh	26.00
Maharashtra	940.40
Odisha	20.00
Punjab	140.50
Rajasthan	101.30
Tamil Nadu	571.30
Uttarakhand	30.00
Uttar Pradesh	776.50
West Bengal	26.00
Total	4013.55

Source MNRE (Ministry of New and Renewable Energy). 2015. ***Annual Report 2014/15***. New Delhi: MNRE, Government of India

TABLE 6.8 Tentative state-wise breakup of renewable power target to be achieved by 2022

State/union territory	Solar power (MW)	Wind power (MW)	Small hydropower (MW)	Biomass power (MW)
Delhi	2 762			
Haryana	4 142		25	209
Himachal Pradesh	776		1 500	
Jammu and Kashmir	1 155		150	
Punjab	4 772		50	244
Rajasthan	5 762	8 600		
Uttar Pradesh	10 697		25	3 499
Uttarakhand	900		700	197
Chandigarh	153			
Northern region	31 119	8 600	2 450	4 149
Goa	358			
Gujarat	8 020	8 800	25	288
Chhattisgarh	1 783		25	
Madhya Pradesh	5 675	6 200	25	118
Maharashtra	11 926	7 600	50	2 469
Dadra and Nagar Haveli	449			
Daman and Diu	199			
Western region	28 410	22 600	125	2 875
Andhra Pradesh	9 834	8100		543
Telangana		2 000		
Karnataka	5 697	6 200	1 500	1 420
Kerala	1 870		100	
Tamil Nadu	8 884	11 900	75	649
Puducherry	246			
Southern region	26 531	28 200	1 675	2 612
Bihar	2 493		25	244
Jharkhand	1 995		10	
Odisha	2377			
West Bengal	5 336		50	
Sikkim	36		50	

Contd...

TABLE **6.8** *Contd...*

State/union territory	Solar power (MW)	Wind power (MW)	Small hydropower (MW)	Biomass power (MW)
Eastern region	12 237	0	135	244
Assam	663		25	
Manipur	105			
Meghalaya	161		50	
Nagaland	61		15	
Tripura	105			
Arunachal Pradesh	39		500	
Mizoram	72		25	
North-eastern region	1 206	0	615	0
Andaman and Nicobar Islands	27			
Lakshadweep	4	600		120
Other new states	99 534	60 000	5 000	10 000

Source Details available at <http://mnre.gov.in/file-manager/UserFiles/Tentative-State-wise-break-up-of-Renewable-Power-by-2022.pdf>, last accessed on 8 January 2016

ENERGY DEMAND

- AGRICULTURE
- INDUSTRY
- TRANSPORT
- HOUSEHOLD ENERGY

ENERGY DEMAND

AGRICULTURE 7

AT A GLANCE

- The agriculture sector provides livelihood and employment to about 54% of India's population and contributes around 13.9% to the gross domestic product, making it a vital element for the inclusive and sustainable growth of the Indian economy.
- About 252.68 million tonnes of foodgrains were produced in 2014/15.
- The agriculture sector is a major consumer of diesel, with about 13% of the total consumption in 2012/13.
- About 159 114 GWh of electricity was consumed in the agriculture sector in 2013/14 (18.03% of the total consumption of electricity in India) compared to 147 462 GWh in 2012/13 (17.89% of the total consumption).
- Electricity has been growing in significance as a preferred input in Indian agriculture, specifically for energizing irrigation pump sets. Around 18 956 850 electric pumps were used in 2013.
- About 696 828 tractors and 56 000 power tillers were sold in 2013/14.
- The fertilizer industry is one of the largest consumers of primary commercial energy. Energy costs comprise 60%–80% of the total manufacturing cost.
- The weighted average energy consumption of ammonia plants in the country was reduced from 52.17 GJ/tonne ammonia in 1987/88 to 36.9 GJ/tonne ammonia in 2011/12, showing an almost 30% improvement in the energy efficiency of ammonia production.
- The agriculture sector in India uses approximately 80% of utilizable water for irrigation, and the average water use efficiency of irrigation projects is assessed to be 30%–35%.
- The total ultimate irrigation potential of the country stands at about 140 Mha, and groundwater contributes more than 79% of the total ultimate potential through minor irrigation.
- In the Solar Pumping Programme for Irrigation and Drinking Water initiated in 2014/15, deployment of 1 million solar pumps has been envisaged for the purpose of irrigation and drinking water by 2021/22.
- Institutional framework for the execution of feeder segregation is recommended since mixed feeder systems impede the disaggregation of power consumption by farm and non-farm use or by losses owing to technical inefficiency or pilferage.

Introduction

The agriculture sector occupies a significant position in the Indian economy as it provides livelihood and employment to more than 54% of the population[1] and contributes around 13.9% to the gross domestic product (GDP) (MoA 2015).[2]

The total geographical area of India is 328.7 million hectares (Mha). Of this, 140.8 Mha is the net sown area and 195.2 Mha is the gross cropped area. The gross and net irrigated areas are 91.5 Mha and 65.3 Mha, respectively, with a cropping intensity of 138.7%. The overall area under foodgrains declined from 124.7 Mha during 2011/12 to 120.7 Mha during 2012/13. As per the fourth advance estimates, the overall area under foodgrains during 2013/14 showed an increase to 126.0 Mha (MoA 2015).

Figure 1 shows the production of foodgrains in India from 1950/51 to 2014/15. India's foodgrain production improved from 51 million tonnes (MT) in the early 1950s to 257.13 MT in 2012/13, with a greater increase in rice and wheat. Foodgrain production estimated at 265.04 MT in 2013/14 witnessed an increase of 7.91 MT compared to the production achieved during 2012/13. According to the fourth advance estimates for 2014/15, the total foodgrain production has been estimated at 252.68 MT. This decline of 12.36 MT compared to previous year's production has occurred on account of the lower production of rice, coarse cereals, and pulses owing to erratic rainfall conditions during the 2014 monsoon season (MoF 2015). As per the fourth advance estimates, the production of rice has shown a decline from 106.65 MT (2013/14) to 104.80 MT (2014/15), wheat from 95.85 MT (2013/14) to 88.94 MT (2014/15), coarse cereals from 43.29 MT (2013/14) to 41.75 MT (2014/15), and pulses from 19.25 MT (2013/14) to 17.20 MT (2014/15).[3]

Energy use in Indian agriculture

Energy use in agriculture at the farm level can be categorized as direct or indirect. Direct energy use in agriculture is in the form of diesel and electricity to operate mobile and stationary equipment to prepare fields, plant and harvest crops, and transport inputs and outputs to and from markets. Indirect energy is consumed off the farm for the manufacture of fertilizers and pesticides. In India, there are two types of diesel fuels: (i) high-speed diesel (HSD), used in automotive applications, and (ii) light diesel oil (LDO), used in stationary applications. LDO is steadily being replaced by HSD, which is primarily a transport fuel. In the agriculture sector, diesel is used in tractors, power tillers, and pump sets. The HSD consumption in agriculture was estimated at 0.429 MT in 2013/14 (Figure 2). This represents around 0.63% of the total HSD consumption in India during 2013/14. In comparison, the other major sectors such as transport, power generation, and

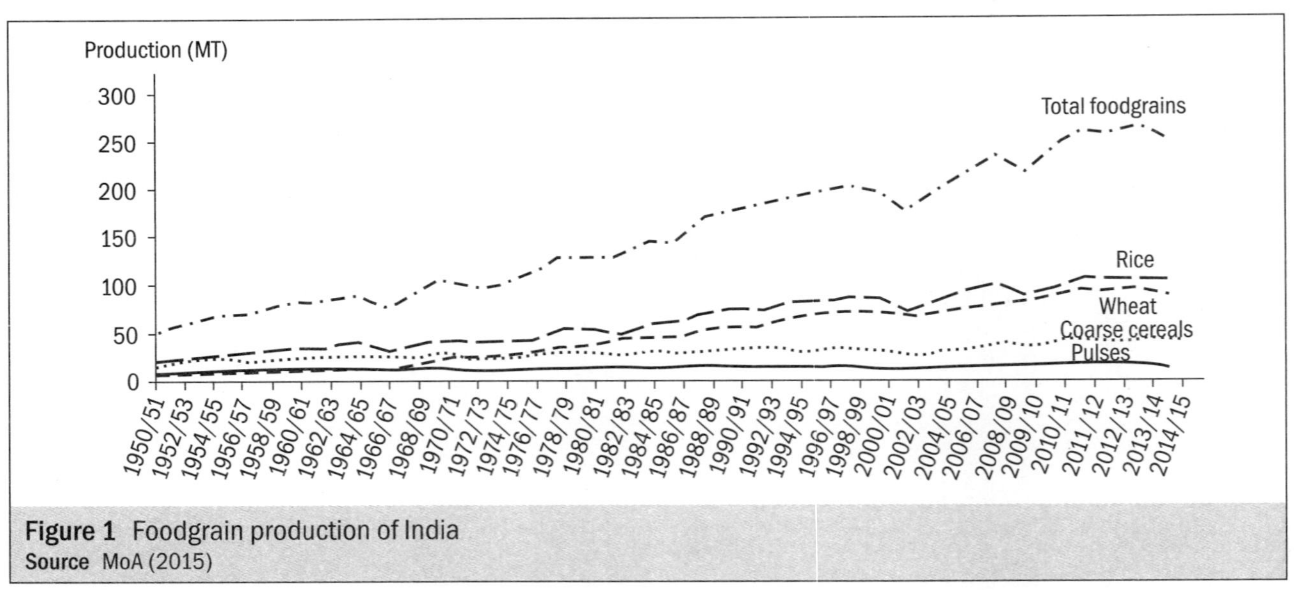

Figure 1 Foodgrain production of India
Source MoA (2015)

[1] Details available at www.censusindia.gov.in

[2] According to the new series of national income released by CSO, at 2011/12 prices the share of agriculture in the total GDP was 18% in 2013/14 (MoF 2015).

[3] Fourth Advance Estimates of Production of Foodgrains for 2014/15. Agricultural Statistics Division, Directorate of Economics and Statistics, Department of Agriculture and Cooperation

industry used 5.159 MT (7.4%), 0.214 MT (0.3%), and 1.628 MT (2.3%), respectively. Miscellaneous service sector with an HSD consumption of 61.414 MT accounts for the largest share (88.78%) of the total HSD consumption.

India is one of the few countries that produces LDO, which is a distillate fuel with a small proportion of residual fuel. It is used in agricultural pump sets by small industries and as start-up fuel for power generators. The consumption of LDO in the agriculture sector has reduced significantly over the years from 28 000 tonnes during 2005/06 to 1300 tonnes in 2013/14. Other fuels that were once used and now almost account for negligible quantities include low sulphur heavy stock and heavy fuel oil, jointly known as fuel oils, which are consumed mainly by fertilizer, chemical, and textiles industries, and mechanical and electrical equipment manufacturers.

An all-India study conducted by M/s Nielsen (India) Pvt. Ltd for the Petroleum Planning and Analysis Cell, Ministry of Petroleum and Natural Gas, Government of India reveals that the agriculture sector is the third major consumer of diesel, with about 13% of the total consumption in 2012/13. Trucks (HCV/LCV), accounting for 28.25% of the diesel consumption, is the major consumer of diesel, followed by private cars and utility vehicles, accounting for 13.15% of the diesel consumption. Within agriculture, the percentages of consumption by tractors, pump sets, and agricultural implements are 7.4%, 2.9%, and 2.7%, respectively. Tractors have a higher consumption, presumably because they are also used for non-agricultural purposes, such as transportation of construction materials (bricks, stones, and mined sand) and people in rural areas. In comparison, the other major sectors such as transport used 70% and other segments including industry used 17%. This implies that agriculture will have major implications for the overall demand and supply of energy.[4]

The total electricity consumption by the agriculture sector increased from 81 673 GWh in 2001/02 (accounting for nearly 25.33% of the total electricity consumption in that year) to 159 144 GWh in 2013/14 (18.03% of the total consumption of electricity during that year) (CEA 2015). Table 1 gives the consumption of electricity from 1955/56 to 2013/14.

Direct uses of energy

In the agriculture sector, tractors and pump sets consume much energy. The use of tractors has risen significantly over the years. In the 1950s, tractors were very limited; however, the number of tractors manufactured gradually rose to 11 000 by 1961.

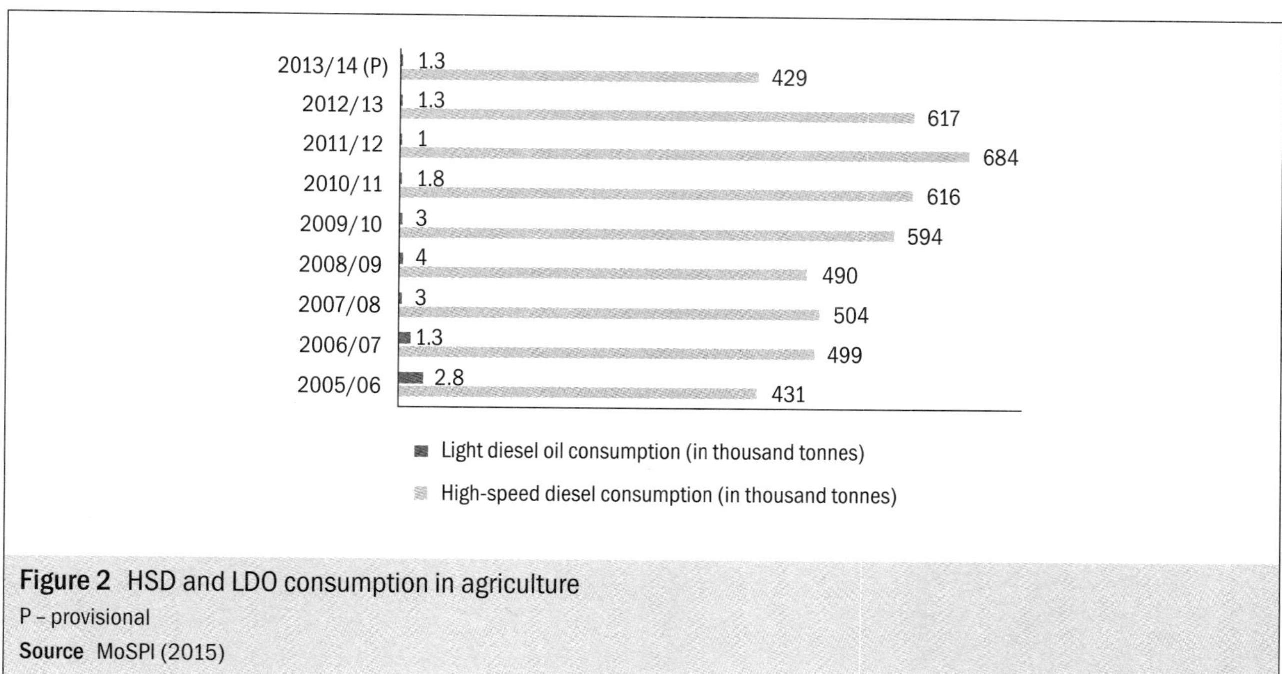

Figure 2 HSD and LDO consumption in agriculture
P – provisional
Source MoSPI (2015)

[4] Press Information Bureau, Government of India, Ministry of Petroleum and Natural Gas, 28 January 2014

Table 1 Electricity consumption in the agriculture sector

Year	Consumption for agricultural purposes (GWh)	Total consumption (GWh)	Share of agricultural consumption (%) in total consumption
1955/56	316	10 150	3.11
1965/66	1 892	30 455	6.21
1978/79	6 310	55 557	11.36
1982/83	17 817	95 589	18.64
1983/84	18 234	102 344	17.82
1984/85	20 960	114 068	18.38
1985/86	23 422	122 999	19.04
1986/87	29 444	135 952	21.66
1987/88	35 267	145 613	24.22
1988/89	38 878	160 196	22.27
1989/90	44 056	175 419	25.11
1990/91	50 321	190 357	26.44
1991/92	58 557	207 645	28.2
1992/93	63 328	220 674	28.7
1993/94	70 699	238 569	29.63
1994/95	79 301	259 630	30.54
1995/96	85 732	277 029	30.95
1996/97	84 019	280 206	29.98
1997/98	91 242	296 749	30.75
1998/99	97 195	309 734	31.38
1999/2000	90 934	312 841	29.07
2000/01	84 729	316 600	26.76
2001/02	81 673	322 459	25.33
2002/03	84 486	339 598	24.88
2003/04	87 089	360 937	24.13
2004/05	88 555	386 134	22.93
2005/06	90 292	411 887	21.92
2006/07	99 023	455 748	21.73
2007/08	104 182	501 977	20.75
2008/09	109 209	553 995	19.79
2009/10	119 492	658 031	18.16
2010/11	129 051	710 673	18.16
2011/12	140 960	785 194	17.95
2012/13	147 462	824 301	17.89
2013/14 (P)	159 144	882 592	18.03

P - provisional
Source CEA (2015)

According to the recent statistics, 254 825 tractors were sold in 2000/01 (82% increase over 1990/91). With the joint effort of the government and the private sector, the sale increased to 696 828 in 2013/14 (173% increase over 2000/01) (Figure 3). The sale of power tillers also increased from 13 563 in 2001/02 to 56 000 in 2013/14 (Figure 4). This reflects the increasing use of HSD in agricultural machinery and implements. Figures 3 and 4 give the sales of tractors and power tillers, respectively, from 1985/86 to 2014/15.

The use of power-driven pump sets in Indian agriculture began in the late 1960s when monsoon failed for consecutive years. Since then there has been a steady increase in the use. This is reflected by a corresponding increase in the consumption of electricity from 17 817 GWh in 1982/83 to 153 116 GWh in 2012/13 (provisional). Thus, there is a clear indication that electricity has been growing in significance as a preferred input in the Indian agriculture, specifically for energizing irrigation pump sets. This increase is also because power supply to the farm sector in India is highly subsidized. This has also led to excessive usage of irrigation pumps and depletion of the water table, resulting in wasteful consumption in the agriculture sector. Table 2 shows the use of diesel and electric pumps in India.

India uses more than 4 billion litres of diesel and around 85 MT of coal per annum to support water pumping for irrigation (KPMG 2014). To meet the agricultural power demand without the need to connect to the grid, the Ministry of New and Renewable Energy (MNRE) launched the Solar Pumping Programme for Irrigation and Drinking Water under the Off-grid and Decentralized Solar Applications Scheme. From 1992 to 2014, 13 964 solar pumps were installed in the country, and the installation of 17 500 solar pumps was sanctioned in 2014/15.

Indirect uses of energy

The consumption of chemical fertilizers during 2013/14 was 24.482 MT, of which 7.434 MT was imported. Consumption of nitrogenous and phosphatic fertilizers decreased from 16.821 MT and 6.653 MT, respectively, in 2012/13 to 16.750 MT and 5.633 MT, respectively, in 2013/14. Consumption of potassic fertilizers witnessed a marginal increase from 2.062 MT in 2012/13 to 2.099 MT in

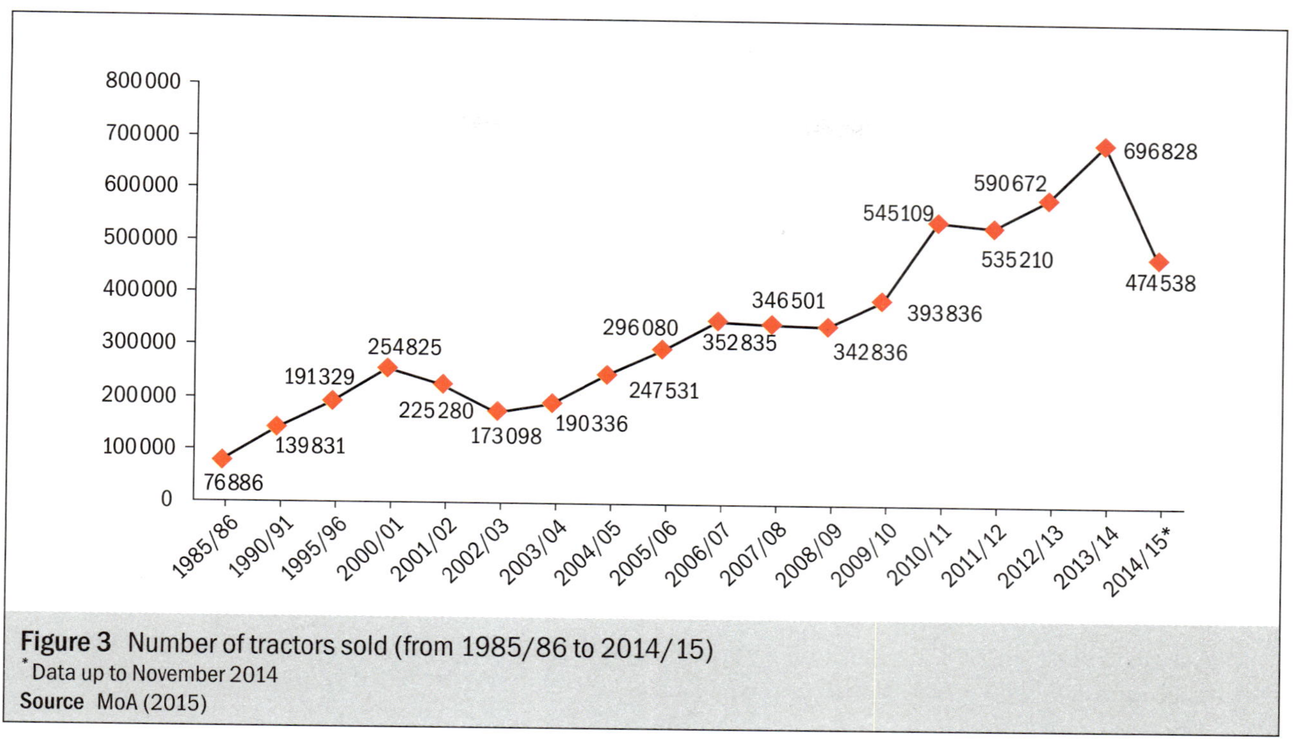

Figure 3 Number of tractors sold (from 1985/86 to 2014/15)
* Data up to November 2014
Source MoA (2015)

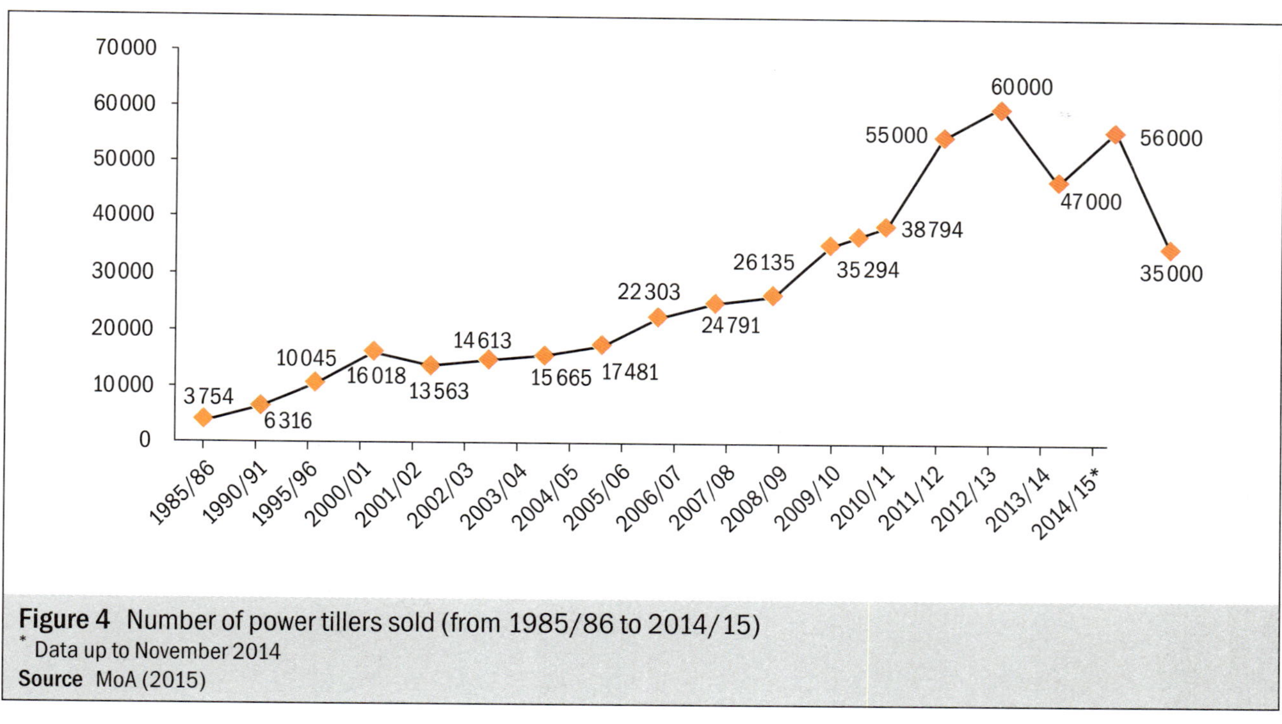

Figure 4 Number of power tillers sold (from 1985/86 to 2014/15)
* Data up to November 2014
Source MoA (2015)

2013/14. A comparison of the production, imports, and consumption of nitrogen, phosphorous, and potassium (NPK) fertilizers between 2009/10 and 2013/14 shows that both availability and consumption have been skewed towards nitrogen or urea since the shift from a product-based subsidy to a nutrient-based subsidy regime in 2010 (Table 3). While urea consumption increased from 59% to 66% of the total consumption in 2012/13 over 2010/11, the per hectare consumption of fertilizers declined from 140 kg to 128 kg over the same period (MoF 2014).

Table 2 Number of diesel and electric pumps used in India

Year	Diesel pumps	Electric pumps
2001	5 856 962	12 823 480
2005	5 877 076	14 445 014
2006	6 026 955	14 803 648
2007	6 186 125	15 350 297
2008	6 328 369	15 674 673
2009	6 479 758	15 963 289
2010	6 627 287	16 693 912
2011	6 783 552	16 760 455
2012	NA	18 423 975
2013*	NA	18 956 850

*Up to August 2013
NA - not available
Source MoP (2013)

Indiscriminate use of NPK has led to imbalanced use of soil nutrients with a substantial distortion in the NPK ratio estimated at 8.2:3.2:1 in 2012/13 which were higher than the recommended national of 4:2:1 NPK balance (MoF 2014).[5]

India meets 80% and 10% of its urea and phosphate requirements, respectively, through domestic production and depends entirely on imports for its potash requirement. The domestic production of urea and complex fertilizers marginally increased in 2013/14 compared to 2012/13, whereas the production of diammonium phosphate (DAP) marginally decreased in 2012/13 compared to 2011/12. The production of urea was estimated to be 22.715 MT in 2013/14, while the productions of DAP and complex fertilizers were 3.611 MT and 6.913 MT, respectively (Figure 5).

The fertilizer industry is one of the largest consumers of primary commercial energy. Energy costs comprise 60%–80% of the total manufacturing cost. Production of nitrogenous fertilizers is highly energy intensive. Efficient production of ammonia, which constitutes the majority composition of nitrogenous fertilizers, has the greatest impact on specific energy consumption as 80% of the energy used for urea production is consumed for ammonia production. The process uses inputs such as air, nitrogen, water, and energy in the form of natural gas, naphtha, fuel

Table 3 Production, imports, and consumption of fertilizers (in thousand tonnes of nutrients)

	1970/71	1980/81	1990/91	2000/01	2007/08	2008/09	2009/10	2010/11	2011/12	2012/13	2013/14	2014/15*
Nitrogenous fertilizers												
Production	830	2 164	6 993	11 004	10 900	10 870	11 900	12 156	12 259	12 194	12 338	9 386
Imports	477	1 510	414	154	3 677	3 844	3 447	4 492	5 240	4 801	3 920	3 579
Consumption	1 487	3 678	7 997	10 920	14 419	15 090	15 580	16 558	17 300	16 821	16 750	NA
Phosphatic fertilizers												
Production	229	842	2 052	3 748	3 807	3 464	4 321	4 222	4 104	3 541	3 714	2 860
Imports	32	452	1 311	396	1 391	2 927	2 756	3 802	4 427	2 797	1 588	1 760
Consumption	462	1 214	3 221	4 215	5 515	6 506	7 274	8 050	7 914	6 653	5 633	NA
Potassic fertilizers												
Imports	120	797	1 328	1 541	2 653	3 380	2 945	4 069	3 335	1 559	1 926	2 076
Consumption	228	624	1 328	1 567	2 636	3 313	3 632	3 514	2 576	2 062	2 099	NA
All fertilizers (NPK)												
Production	1 059	3 006	9 045	14 752	14 707	14 334	16 221	16 378	16 363	15 737	16 092	12 246
Imports	629	2 759	2 758	2 090	7 721	10 151	9 148	12 363	13 002	9 157	7 434	7 415
Consumption	2 177	5 516	12 546	19 702	22 570	24 909	26 486	28 122	27 790	25 536	24 482	NA

*Data for April–December 2014
NA - not available; NPK - nitrogen, phosphorus, and potassium
Source MoF (2015)

[5] Report of the Working Group on Fertilizer Industry for the Twelfth Five-year Plan. Details available at http://planningcommission.gov.in/aboutus/committee/wrkgrp12/wg_fert0203.

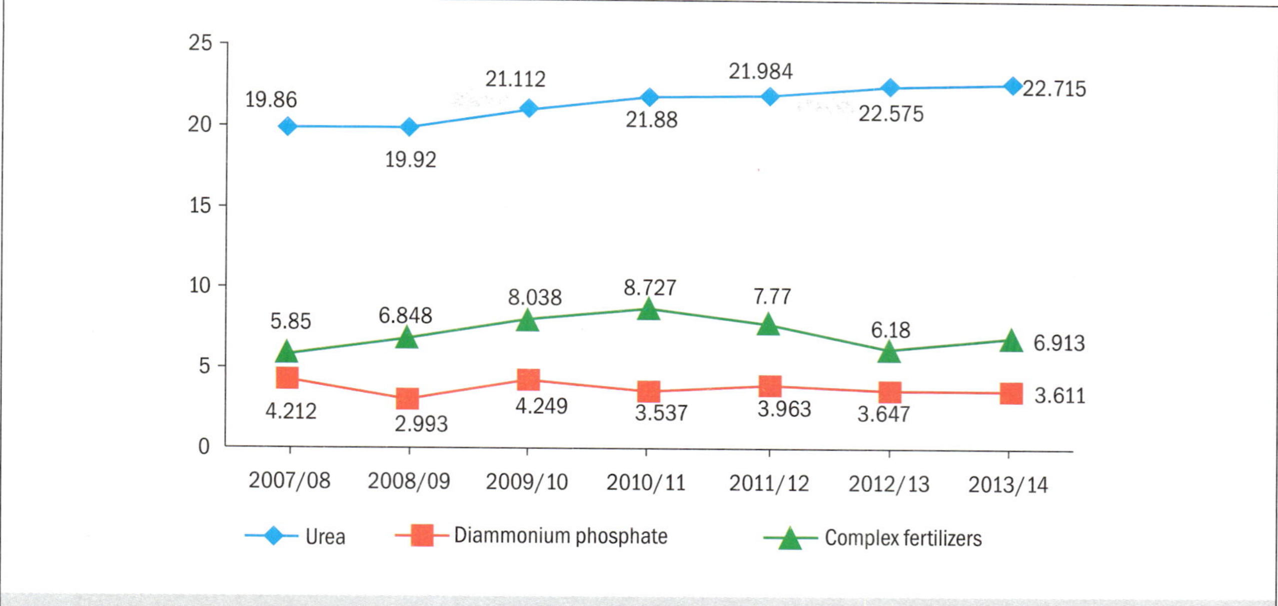

Figure 5 Production of urea, diammonium phosphate, and other complex fertilizers (in MT)
Source MoCF (2014)

oil, coal, and coke oven gas (MoF 2011). Although natural gas is the ideal feedstock for ammonia production, its use in India is restricted owing to its availability. Over the years, ammonia production has become significantly energy efficient owing to better feedstock, advanced process technologies, and increased capacity utilization (MoPNG 2009). As a result of various energy conservation efforts and the addition of capacity through more efficient plants, the weighted average energy consumption of ammonia plants in the country reduced from 52.17 GJ/tonne ammonia in 1987/88 to 36.9 GJ/tonne ammonia in 2011/12, showing an almost 30% improvement in the energy efficiency of ammonia production (FAI 2013). Phosphatic fertilizers are less energy intensive than nitrogenous fertilizers. Phosphate rock and sulphuric acid are the raw materials used in the production of phosphatic fertilizers (MoPNG 2009).

Water use efficiency in agriculture

The irrigation sector is the biggest consumer of water as more than 80% of the available water resources in India are currently utilized for irrigation. The total ultimate irrigation potential of the country stands at about 140 Mha (Figure 6). The share of minor irrigation is higher by 22.96 Mha compared to that of major and medium irrigation. Groundwater contributes more than 79% of the total ultimate potential through minor irrigation (CWC 2013). State-wise irrigation potential in India is depicted in Map 1.

The requirement of freshwater for irrigation is growing continuously. As per the estimates of the National Commission for Integrated Water Resources Development, the total withdrawal/utilization in 2010 for all types of uses was estimated at 710 billion cubic metres (BCM) for the high projection scenario, out of which irrigation accounted for nearly 78%. For 2050, 1180 BCM has been estimated for the high projection scenario, out of which irrigation will be accounted for

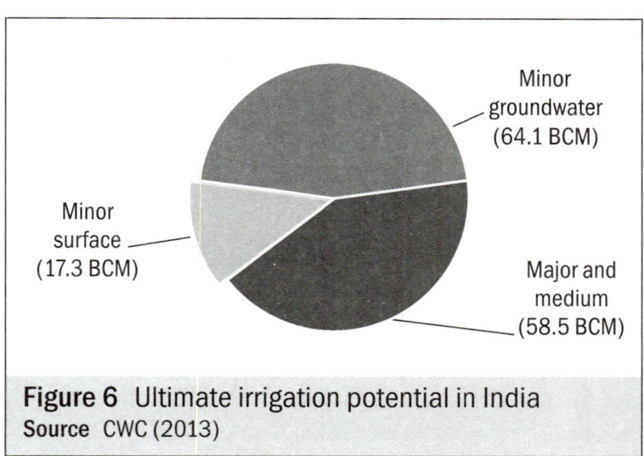

Figure 6 Ultimate irrigation potential in India
Source CWC (2013)

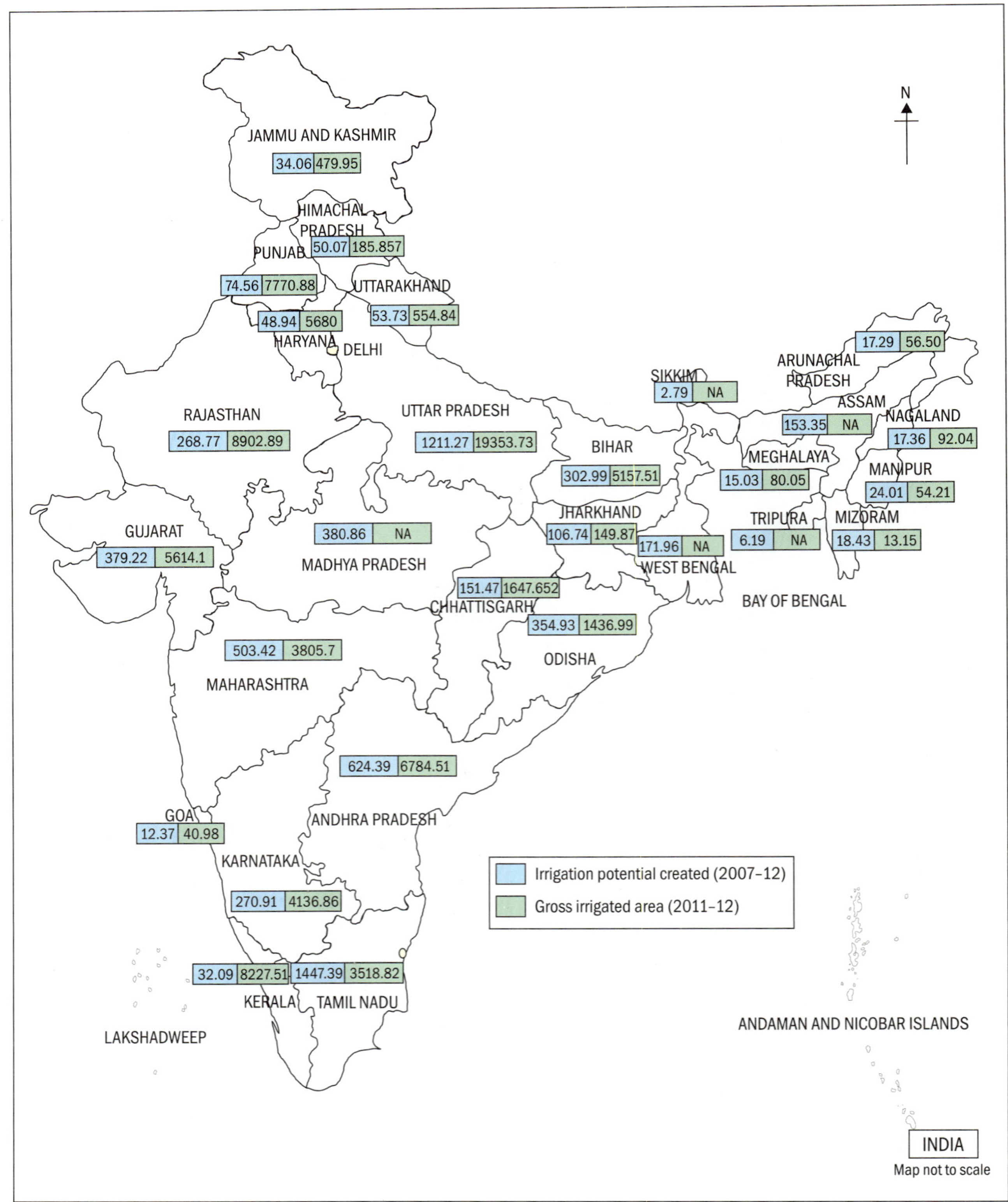

Map 1 State-wise irrigation potential in India ('000 hectares)

NA - not available

Source Central Water Commission, P&P Directorate and Ministry of Water Resources (Minor Irrigation Division)

Disclaimer The state of Andhra Pradesh has been bifurcated into two states, Telangana and the residuary state of Andhra Pradesh (Seemandhra), with effect from 02 July 2014. The map represents Andhra Pradesh as one state based on data and information available.

Table 4 Water requirement for irrigation in India (in BCM)

Water use	1997/98	2010			2025			2050		
		Low	High	%	Low	High	%	Low	High	%
Surface water irrigation	318	330	339	48	325	366	43	375	463	39
Groundwater irrigation	206	213	218	31	236	245	29	253	344	29
Total water use (irrigation)	524	543	557	78	561	611	72	628	817	68
Total water use	629	694	710	100	784	843	100	973	1 180	100

Source CWC (2013)

nearly 68% (Table 4). With the given constraint of the total utilizable water resources of 1123 BCM (MoUD 2010), it is apparent that measures are needed to improve the efficiency of irrigation water use.

The average efficiency of water use for irrigation projects in India has been assessed to be only 30%–35% because of huge conveyance and distribution losses, indicating a vast potential for saving water in irrigation (Table 5). Water use efficiency has an impact on energy savings because less power is utilized in lifting water from wells for irrigating a given area under crops. Improvement in water use efficiency in agriculture is essential because irrigation sources are declining, energy costs are making irrigation more expensive, global demand for food, feed, and fibre is increasing, and production is being stretched to more arid environments. Efficiency gains are often possible through suitable crop selection, proper irrigation scheduling, effective irrigation techniques, and the use of alternative water sources.

Table 5 Efficiencies of various facilities for utilization of irrigation water

Water use and methods	Efficiency (%)
Irrigation efficiencies	
Conveyance	
Through unlined canal for surface water	55-60*
Through lined canal for surface water	70-75*
Application for both surface and groundwater	
Flood irrigation	65
Furrow irrigation	80
Sprinkler	85
Drip	90
Overall efficiency for surface water system	30-65
Overall efficiency for groundwater system	65-75

*Conveyance efficiency of the canal depends on many factors, such as the length of the canal, type of soil, material used for lining, and so on.

Source CWC (2014)

Currently, 63 Mha or 45% of the net cropped area is irrigated (MoF 2014). One of the main reasons for this low coverage is the predominant use of flood irrigation, where water use efficiency is very low. Available estimates indicate that the water use efficiency under flood irrigation is only about 35%–40% because of high distribution losses.

In traditional surface irrigation methods, the losses in water conveyance and application are large. These losses can be considerably reduced by adopting micro-irrigation methods. Among all the irrigation methods, micro-irrigation in the form of drip/sprinkler irrigation is the most efficient one and can be practised in a large variety of crops, especially in vegetables, orchard crops, flowers, and plantation crops. Besides saving water, micro-irrigation also results in productivity gains (in the range of 20%–50%), reduces the growth of weeds, controls soil erosion, and is less labour intensive (Table 6).

In India, there has been a growth in the area under micro-irrigation during the past 15 years. At present, around 3.87 Mha is under micro-irrigation, while it was only 40 ha in 1960. Of the 3.87 Mha under micro-irrigation, 1.42 Mha is under drip irrigation and 2.44 Mha is under sprinkler irrigation. There is a huge potential for the expansion of micro-irrigation in India. As per some projections, 42.23 Mha in the country has the potential for micro-irrigation (Palanisami, Mohan, Kakumanu, *et al.* 2011). The National Committee on Plasticulture Applications in Horticulture, Ministry of Agriculture, Government of India has estimated that 27 Mha of area in the country has the potential for drip irrigation. The potential and actual area under drip and sprinkler irrigation across different states in India is given in Table 7.

Table 6 Comparison of water use in traditional and drip irrigation systems for different crops

Crop	Traditional method		Drip irrigation method		Saving in water (%)	Increase in yield (%)	Water use efficiency (yield/ha)/(mm/ha)	
	Water (mm/ha)	Yield (tonnes/ha)	Water (mm/ha)	Yield (tonnes/ha)			Traditional method	Drip irrigation method
Banana	1 760	57.5	970	87.5	45	52	0.033	0.090
Grapes	532	26.4	278	21.5	48	23	0.050	0.117
Citrus	1 660	100	640	150	61	50	0.060	0.234
Tomato	300	32	180	48	39	50	0.107	0.267
Brinjal	90	28	42	21	53	14	0.311	0.500
Chilli	100	4.2	42	6.1	62	44	0.042	0.145
Sugarcane	2 150	128	940	170	65	33	0.060	0.181
Cotton	90	2.3	42	2.9	53	27	0.026	0.069

Source CIAE (2008)

Table 7 State-wise potential and actual area under micro-irrigation in India

State	Drip			Sprinkler			Total		
	Potential area ('000 ha)	Actual area ('000 ha)	%	Potential area ('000 ha)	Actual area ('000 ha)	%	Potential area ('000 ha)	Actual area ('000 ha)	%
Andhra Pradesh	730	363.07	49.74	387	200.95	51.93	1 117	564.02	50.49
Bihar	142	0.16	0.11	1 708	0.21	0.01	1 850	0.37	0.02
Chhattisgarh	22	3.65	16.58	189	59.27	31.36	211	62.92	29.82
Goa	10	0.76	7.62	1	0.33	33.2	11	1.09	9.95
Gujarat	1 599	169.69	10.61	1 679	136.28	8.12	3 278	305.97	9.33
Haryana	398	7.14	1.79	1 992	518.37	26.02	2 390	525.5	21.99
Himachal Pradesh	14	0.12	0.83	101	0.58	0.58	115	0.7	0.61
Jharkhand	43	0.13	0.31	114	0.37	0.32	157	0.5	0.32
Karnataka	745	177.33	23.8	697	228.62	32.8	1 442	405.95	28.15
Kerala	179	14.12	7.89	35	2.52	7.19	214	16.64	7.77
Madhya Pradesh	1 376	20.43	1.48	5 015	117.69	2.35	6 391	138.12	2.16
Maharashtra	1 116	482.34	43.22	1 598	214.67	13.43	2 714	697.02	25.68
Nagaland	11	0	0	42	3.96	9.43	53	3.96	7.48
Odisha	157	3.63	2.31	62	23.47	37.85	219	27.2	12.37
Punjab	559	11.73	2.1	2 819	10.51	0.37	3 378	22.24	0.66
Rajasthan	727	17	2.34	4 931	706.81	14.33	5 658	723.82	12.79
Tamil Nadu	544	131.34	24.14	158	27.19	17.21	702	158.52	22.58
Uttar Pradesh	2 207	10.68	0.48	8 582	10.59	0.12	10 789	21.26	0.2
West Bengal	952	0.15	0.02	280	150.03	53.58	1 232	150.18	12.19
Others	128	15	11.72	188	30	15.96	316	45	14.24
Total	11 659	1 428.46	12.25	30 578	2 442.41	7.99	42 237	3 870.86	9.16

Source Palanisami, Mohan, Kakumanu, *et al.* (2011)

The Government of India launched National Mission on Micro Irrigation in 2010 to boost convergence of micro-irrigation activities under major government programmes such as National Food Security Mission, Integrated Scheme of Oilseeds, Pulses, Oil Palm and Maize (ISOPOM), and Technology Mission on Cotton for increasing water use efficiency, crop productivity, and farmers' income. To achieve convergence of investments in irrigation at the field level and ensure access to some means of protective irrigation to all agricultural farms in the country, the Pradhan Mantri Krishi Sinchayee Yojana was formulated with an outlay of ₹50 000 crore for a period of 5 years (2015/16 to 2019/20).

Agriculture and Indian planning

Having recognized the criticality of the energy–water–food nexus and the need to minimize energy wastage and power losses in the agriculture sector, the Indian Government has devised policies and programmes to highlight and address this issue.

Tariff Policy

The Tariff Policy, 2006, notified in continuation of the National Electricity Policy, 2005, highlights the undesirability of the provision of free electricity and suggests reasonable tariffs for agriculture use keeping in mind groundwater sustainability and cost of supply. It advocates differential tariffs based on the condition of the groundwater level. It, however, proposes higher electricity subsidies for poor farmers residing in areas with adverse groundwater levels as long as sustainable use of groundwater can be managed. The policy also suggests that metering of supply to agricultural and rural users can be made consumer-friendly through the involvement of panchayat institutions and cooperative societies in managing local distribution run by franchisees through commercial arrangements. It recommends the use of cost-effective options, such as "self-closing load limiters" for metering in the cases of "limited-use consumers" who are eligible for subsidized electricity (MoP 2006).

Twelfth Five-year Plan

The Twelfth Five-year Plan emphasizes the criticality of the deterioration of the groundwater table in India due to power subsidies in the agriculture sector. It points to potential measures such as physical segregation of power feeders for rural lighting and agriculture, provision of rationed but reliable supply to farmers at a flat tariff, development of high-voltage distribution system (HVDS), and use of energy-efficient pumps to reduce energy losses and conserve groundwater. The need for substantial investments to take forward these initiatives is highlighted (Planning Commission 2013).

Agriculture Demand-side Management

Initiated by the Bureau of Energy Efficiency under the Ministry of Power, the Agriculture Demand-side Management (AgDSM) Programme commenced in 2009 as part of the Eleventh Five-year Plan. The programme seeks to establish an institutional environment for encouraging market-driven interventions in the agriculture pumping sector to facilitate energy conservation, improve groundwater extraction efficiencies, and reduce subsidy expenditures. Using a public–private partnership approach, the replacement of existing pump sets with energy-efficient ones is being undertaken in specific rural pump set feeder areas (BEE 2009). Pilot projects have been undertaken in eight states, including Maharashtra, Gujarat, Punjab, Haryana, and Andhra Pradesh (MoP 2012). The project involves selection of appropriate location and feeders, awareness generation and capacity building of farmers and utility employees, development of detailed project reports and business model for the intended project followed by choosing of energy service organization for project implementation, and effective monitoring and verification procedures (BEE 2009).

Solar Pumping Programme for Irrigation and Drinking Water under Off-grid and Decentralized Solar Applications: Jawaharlal Nehru National Solar Mission

A programme on solar pumping was assumed by the MNRE in 1992 (MNRE 2014a). However, to establish an enabling environment for the penetration of solar technology, a larger programme, the Jawaharlal Nehru National Solar Mission (JNNSM), was undertaken in 2010/11 as part of the eight missions under the National Action Plan on Climate Change. The JNNSM was to be undertaken in three phases—Phase I (2010–13), Phase II (2013–17), and Phase III (2017–22)—as a part of which a capacity of 20 GW of grid-connected and 2 GW of off-grid supply was targeted by 2022 (MNRE 2010, 2012). Owing to the thrust on off-grid solar energy applications, a sub-scheme for Off-grid and Decentralized Solar Applications was rolled out, which also included promotion of solar water pumps. Financial support in the form of both capital subsidy (30%) and interest subsidy (5% interest bearing loan) was provided (MNRE 2010). Between 1992 and 2014, 13 964 solar pumps were installed (MNRE 2014a).

In 2014/15, a new Solar Pumping Programme for Irrigation and Drinking Water was initiated under the Off-grid and Decentralized Solar Applications Scheme. With a proposed duration of 5 years, the deployment of 1 million solar pumps was envisaged for the purpose of irrigation and drinking water by 2021/22. The objective of the scheme is to improve access to energy, develop models to foster solar pumping technologies in rural areas in a scalable fashion, and explore the potential of solar pump programmes to support rural development. The MNRE is to implement the programme in India in coordination with the Ministry of Agriculture through state nodal agencies, National Bank for Agriculture and Rural Development (NABARD), and the Ministry of Drinking Water and Sanitation. State governments will be involved in identifying targets for pump replacements based on various parameters, including water availability, cropping patterns, affordability for beneficiaries, and availability of state funds and others. State nodal agencies would be responsible for implementation, monitoring and evaluation, as well as awareness generation and training (MNRE 2014a). Initially, the installation of 1 lakh solar photovoltaic (SPV) water pumping systems was proposed for 2014/15 with a financial assistance of ₹400 crore. However, a reallocation of targets was done and the revised targets include the installation of 50 000 solar pumps for irrigation and 20 000 solar pumps for drinking water based on the response from various states (MNRE 2014a, 2014b).

Capital Subsidy Scheme of Government of India for Promoting Solar Photovoltaic Water Pumping Systems for Irrigation

In mid-2014, NABARD undertook a scheme that promoted the deployment of 10 000 SPV water pumping systems for irrigation. However, the MNRE also launched a new scheme in November 2014 to support the installation of 30 000 solar pumping units for irrigation. Therefore, the previous scheme was merged with the latter MNRE scheme. As the scheme provides credit-linked, back-ended capital subsidy, NABARD will be responsible for channelizing the subsidy. However, in addition to NABARD (under direct lending), projects financed by other commercial banks, regional rural banks, and State Cooperative Agricultural Rural Development Banks will also be eligible for subsidy (NABARD 2014).

State-level Initiatives

Andhra Pradesh, Gujarat, Haryana, Punjab, Karnataka, Maharashtra, Madhya Pradesh, and Rajasthan have undertaken rural load segregation so that electricity supply for agriculture and irrigation can be distinct from that for non-farm rural consumers. This initiative is driven by the objective to ensure a regular supply of quality power to rural consumers and to spur socio-economic development in rural areas. Through the separation of feeders, the aim of utilities is also to limit and measure the power provided free for irrigation, besides attempting to impose a stricter schedule while maintaining its lower price (Planning Commission 2013).

For instance, Gujarat segregated feeders as part of the Jyotigram Yojna in 2003/04 and also installed meters. The agriculture feeders provided farmers with quality power for a fixed duration on a pre-announced schedule. Feeder segregation was undertaken in more than 18 000 villages as part of the scheme that ended in 2006. Rajasthan initiated feeder segregation in 2005 through its Feeder Renovation Programme, which covered 8126 feeders and also included other system improvements such as HVDS on agricultural feeders and DT (distribution transformer) meters. The programme is still ongoing. Haryana undertook rural load segregation , which was initiated in 2005/06 and completed in 2010, and installed 1226 feeders with bulk meters.

Highlights of the World Bank study on lighting rural India: load segregation experience in selected states

Heavily subsidized power in the agriculture sector and the absence of metering have resulted in the wastage of power and overexploitation of groundwater. In this scenario, states have attempted feeder segregation since mixed feeder systems impede the disaggregation of power consumption by farm and non-farm use or by losses due to technical inefficiency or pilferage. Focusing on Gujarat, Rajasthan, Andhra Pradesh, and Haryana, the World Bank study[6] revealed that the drivers for feeder segregation are to (i) ensure rural development through reliable power supply, (ii) improve energy accounting and auditing to facilitate a restatement of power consumption and losses in the agriculture sector, and (iii) enhance load management by rotating supply for farmers.

Overall the study highlighted that the technical approach followed by states was based on their unique vision, policies, and status of the power sector. The findings[7] in Rajasthan and Gujarat revealed that rural feeder segregation ushered in transparency in agricultural consumption with a significant improvement in the quality of supply, besides facilitating accurate estimations of distribution losses. However, since project scopes were distinct for each state and costs varied on the approach and architecture of the system, direct cost comparisons were not feasible.

In the Vinchhiya Rural Subdivision (Gujarat) study site, although distribution losses increased despite decreased sales to agriculture, feeder segregation helped improve peak demand management and load rotation on agriculture-dominant feeders. In contrast, in the Bassi Subdivision (Rajasthan), power sale to agriculture increased, while distribution losses decreased. However, the financial implications of the initiatives appeared to be different in states. Gujarat has controlled subsidy and financial losses, whereas in Rajasthan, higher agriculture sales to both metered and unmetered connections alongside dwindling subsidy disbursement led to continuing financial losses. Owing to the lack of data at the feeder level, the impact of load segregation on subsidy transparency could not be assessed.

The study also revealed that most of the states under study (with the exception of Rajasthan) showed the absence of a specific institutional framework for the execution of feeder segregation. The schemes also failed to include remote meter reading or advanced metering infrastructure for the capture of online metering data required to measure, analyse, and monitor agricultural power consumption. This prevented an evaluation of the direct benefits of load segregation. Furthermore, there was insignificant monitoring and evaluation of project execution or outcomes.[8]

The study also pointed out that while feeder segregation is amenable in situations where metering of agriculture loads is absent, it should not be envisioned as a substitute to metering as it can facilitate transparency in agricultural consumption and mechanisms for direct subsidy delivery. Furthermore, load segregation must be accompanied by institutional and governance reforms at the level of utilities catering to IT-based metering, data collection and analysis, preservation of segregated feeders, and monitoring and accountability systems, besides the development of necessary infrastructure and human resources.

The study cautions that although feeder segregation has delivered good results in improving power supply to farmers and non-agriculture consumers, it may not be appropriate for all states. It makes the following recommendations: (i) setting rules of engagement and principles for the design of sustainable and feasible rural power initiatives through a common strategic framework, (ii) establishing a centralized knowledge hub that supports states in assuming programmes for improving rural power supply, and (iii) disseminating the experiences and outcomes of demonstration projects. It also provides a guidance note to enable state governments and utilities to adopt the approach to rural power supply that best fits their context and proposes an institutional framework for rural load segregation.

Source ASTAE (2014)

---

[6] The study mainly assessed the interventions in Gujarat and Rajasthan. For various reasons, detailed evaluation for Andhra Pradesh and Haryana was not possible.

[7] The findings were at the subdivision level. These were aggregated at the level of distribution utilities and then at the state level.

[8] Nevertheless, a third-party evaluation post-scheme was undertaken by Gujarat and Haryana for designing a monitoring and evaluation framework to estimate power consumption by agricultural consumers post-segregation and bring transparency in devising subsidies.

References

ASTAE (Asia Sustainable and Alternative Energy Program). 2014. ***Lighting Rural India: Load Segregation Experience in Selected States***. South Asia Energy Studies, Washington, DC: World Bank

BEE (Bureau of Energy Efficiency). 2009. ***Launch of Agricultural Demand-Side Management (AgDSM) Pilot Project***. New Delhi: Bureau of Energy Efficiency, Ministry of Power, Government of India

CEA (Central Electricity Authority). 2015. ***All-India Electricity Statistics: General Review 2015***. New Delhi: CEA, Government of India

CIAE (Central Institute of Agricultural Engineering). 2008. ***Agricultural Engineering Data Book***. Bhopal: CIAE

CWC (Central Water Commission). 2013. ***Water and Related Statistics***. New Delhi: CWC, Government of India

CWC (Central Water Commission). 2014. ***Guidelines for Improving Water Use Efficiency in Irrigation, Domestic and Industrial Sectors***. New Delhi: CWC, Government of India

FAI (Fertilizer Association of India). 2013. Performance improvements in ammonia plants in India. Details available at http://ietd.iipnetwork.org/content/performance-improvements-ammonia-plants-india

KPMG. 2014. ***Feasibility Analysis for Solar Agricultural Water Pumps in India***. New Delhi: KPMG Advisory Services Pvt. Ltd

MNRE (Ministry of New and Renewable Energy). 2010. ***Jawaharlal Nehru National Solar Mission, Building Solar India, Guidelines for Off-Grid and Decentralized Solar Applications and Rooftop and Other Small Solar Power Plants***. New Delhi: Ministry of New and Renewable Energy, Government of India

MNRE (Ministry of New and Renewable Energy). 2012. ***Jawaharlal Nehru National Solar Mission. Phase II: Policy Document***. New Delhi: MNRE, Government of India

MNRE (Ministry of New and Renewable Energy). 2014a. ***Supplementary Guidelines for Implementation of Solar Pumping Programme for Irrigation and Drinking Water under Off-Grid and Decentralized Solar Applications Scheme***. New Delhi: MNRE, Government of India

MNRE (Ministry of New and Renewable Energy). 2014b. ***Reallocation of Targets under 100 000 of Solar Pumping Programme***. New Delhi: MNRE, Government of India

MoA (Ministry of Agriculture). 2015. ***Annual Report 2014/2015***. New Delhi: MoA, Government of India

MoCF (Ministry of Chemicals and Fertilizers). 2014. ***Indian Fertilizer Scenario 2014***. New Delhi: MoCF, Government of India

MoF (Ministry of Finance). 2011. Agriculture and food management. In ***Economic Survey 2010/11***, pp. 187–216. New Delhi: MoF, Government of India

MoF (Ministry of Finance). 2014. Agriculture and food management. In ***Economic Survey 2013/14***, pp. 137–159. New Delhi: MoF, Government of India

MoF (Ministry of Finance). 2015. Prices, agriculture and food management. In ***Economic Survey 2014/15***, pp. 69–89. New Delhi: MoF, Government of India

MoP (Ministry of Power). 2006. ***Tariff Policy, Resolution No. 23/2/2005-R&R (Vol. III).*** The Gazette of India, Extraordinary, Part I, Section 1. New Delhi: MoP, Government of India

MoP (Ministry of Power). 2012. ***Report of the Working Group on Power for Twelfth Plan (2012–17)***. New Delhi: MoP, Government of India

MoP (Ministry of Power). 2013. ***Annual Report 2012/13***. New Delhi: MoP, Government of India

MoPNG (Ministry of Petroleum and Natural Gas). 2009. ***Basic Statistics on Indian Petroleum and Natural Gas 2009/10***. New Delhi: MoPNG, Government of India

MoSPI (Ministry of Statistics and Programme Implementation). 2015. ***Energy Statistics 2015***. New Delhi: MoSPI, Government of India

MoUD (Ministry of Urban Development). 2010. ***Report for National Mission on Sustainable Habitat***. New Delhi: MoUD, Government of India

NABARD (National Bank for Agriculture and Rural Development). 2014. ***The Scheme of Ministry of New and Renewable Energy for Promoting Solar Photovoltaic Water Pumping Systems for Irrigation Purpose***. New Delhi: NABARD

Palanisami, K., K. Mohan, K. R. Kakumanu, and S. Raman. 2011. Spread and economics of micro-irrigation in India: Evidence from nine states. ***Economic and Political Weekly***, Vol. XLVI, No. 26 and 27, 25 June 2011

Planning Commission. 2013. ***Twelfth Five-Year Plan (2012–2017): Faster, More Inclusive and Sustainable Growth,*** Volume I. New Delhi: Planning Commission, Government of India

ENERGY DEMAND

INDUSTRY 8

AT A GLANCE

- Industrial output increased by 1.9% during April–October 2014, compared to 0.2% during the same period last year.
- The total installed capacity of aluminium in the country during 2012/13 was 1.9 million tonnes per annum (MTPA). Aluminium production capacity in India at the end of the Twelfth Five-year Plan (that is, 2016/17) is expected to be about 4.7 million tonnes (MT).
- The installed production capacity of crude steel during 2013/14 was 101.02 MTPA as against 97.02 MTPA in 2012/13. Crude steel production grew at a compound annual growth rate of 7% during the past 5 years ending 2013/14.
- The country has achieved 80% self-sufficiency in the production capacity of urea. Hence, India can substantially manage its requirement of nitrogenous fertilizers through indigenous industry and imports. Similarly, 50% of indigenous capacity has been developed to meet the domestic requirements of phosphatic fertilizers.
- The textile industry accounts for 14% of industrial production, which is 4% of gross domestic product, employs 45 million people, and accounts for 13% of the country's total export basket. During 2014/15, about 60% of the total cloth production was from the power loom sector.
- The cement industry comprises 190 large-cement plants with an installed capacity of 350 MT and more than 350 mini-cement plants with an estimated capacity of 11.10 MTPA. Large cement plants contribute about 97% to the total installed capacity.
- During 2013/14, alkali chemicals together accounted for about 67% of the total production of major chemicals (alkali, inorganic and organic chemicals, pesticides, dyes and dyestuffs). The production of alkali chemicals in 2014/15 (up to September 2014) was 3.279 MT compared to 3.207 MT during the same period in 2013/14, indicating a growth of 2.2%.
- There are 750 pulp and paper mills in India, producing different varieties of paper, such as writing and printing paper, packaging paper, coated paper, and specialty grade paper. The paper, paperboard, and newsprint production in the country is 10.11 MT against the installed capacity of 12.70 MT. The annual consumption of paper, paperboard, and newsprint in the country is around 11.15 MT.

Introduction

The Indian economy registered a growth of 5.5% in the first half of the current financial year. This is distinctly higher than the growth recorded in the first half and the second half of 2013/14. This is also in tandem with the projection of the full-year growth of 5.4%–5.9% made in the Economic Survey 2013/14, as well as in line with the initial expectations of the government of a gradual pickup in growth. The index of industrial production (IIP) witnessed positive growth in the first half of 2014/15, with the first-quarter growth better than the second quarter. Industrial output increased by 1.9% during April–October 2014 compared to 0.2% growth registered during the same period last year.

The industrial performance measured in terms of IIP has shown a fluctuating trend over the last few years. Industrial growth peaked at 15.5% in 2007/08 (manufacturing growth rate was 18.4%) and then declined to 2.5% in 2008/09. It, however, recovered to 5.3% in 2009/10 and further improved to 8.2% in 2010/11. Subsequently, the industrial growth decelerated to 2.9% in 2011/12 and 1.1% in 2012/13. Industrial production again moderated, recording a negative growth of −0.1% in 2013/14. The growth in IIP during April–December 2014/15 improved to 2.1% along with improvement in the growth of all sectors. The details are given in Table 1.

Aluminium industry

Aluminium is the second most widely used metal after steel. The popularity of the metal is due to its excellent properties and versatility. The electrical sector in the country is the largest consumer of aluminium (48%), followed by transport (15%), construction (13%), consumer durables (7%), machinery and equipment (7%), packaging (4%), and others (6%) (MoM 2013a). Per capita consumption of aluminium in India is among the lowest in the world, which is only 1.4 kg compared to the world average of about 8 kg. Four major primary producers—National Aluminium Company Ltd (NALCO), Hindalco Industries Ltd (HIL), Bharat Aluminium Company Ltd (BALCO), and Vedanta Aluminium Ltd (VAL)—are at the forefront of aluminium production in the country. Aluminium is endlessly recyclable, and there is no loss of properties or quality during the recycling process. Aluminium recycling is primarily concentrated in the small and medium enterprise sector, with only one large-scale aluminium recycling unit of HIL located at Taloja, which has a capacity of 25 000 tonnes per annum.

Bauxite ore is used for producing aluminium. The basic production processes are Bayer process for alumina refining and Hall–Heroult process for aluminium smelting. About 1 tonne of alumina is produced from 3 tonnes of bauxite, and about 1 tonne of aluminium is produced from 2 tonnes of alumina. The estimated bauxite reserves of the country (proven and probable) stand at 593 million tonnes (MT). India occupies the seventh position in terms of world's bauxite reserve. Odisha and Andhra Pradesh account for more than 90% of country's metallurgical-grade resources.

Table 1 Annual growth rate of industrial production (%)

Period	Mining and quarrying	Manufacturing	Electricity	Overall
Weight	14.2	75.5	10.3	100
2008/09	2.6	2.5	2.7	2.5
2009/10	7.9	4.8	6.1	5.3
2010/11	5.2	8.9	5.5	8.2
2011/12	-2.0	3.0	8.2	2.9
2012/13	-2.3	1.3	4.0	1.1
2013/14	-0.6	-0.8	6.1	-0.1
2014/15 (April–December)	1.7	1.2	10.0	2.1

Source MoCI (2015)

Production

The total installed capacity of aluminium in the country during 2012/13 was 1.9 MT per annum or MTPA (MoM 2013a). It is expected that India's aluminium production capacity at the end of the Twelfth Five-year Plan period (2016/17) would be about 4.7 MT. The production of aluminium by primary aluminium producers from 2011/12 to 2014/15 (upto December 2014) is given in Table 2.

Energy consumption

Aluminium smelter and alumina refinery are the two major energy-consuming centres in primary aluminium manufacturing. Aluminium is highly recyclable, and more than 70% of the total metal produced till date is still in use. Aluminium recycling

Table 2 Production of aluminium by primary aluminium producers				
Name of the company	Aluminium production (in tonnes)			
	2011/12	2012/13	2013/14	2014/15
NALCO	413 089	403 384	316 492	243 572
HIL	581 985	547 416	618 286	589 606
Vedanta Group	676 380	773 946	795 355	648 766
Total	1 671 454	1 724 746	1 730 133	1 481 944

HIL – Hindalco Industries Ltd; NALCO – National Aluminium Company Ltd
Source MoM (2015)

is less capital intensive than primary metal production as it requires only 5% of energy; between 13 000 and 15 000 units of power are required for producing 1 tonne of aluminium through the primary route. Besides, it keeps the greenhouse gas emission levels to about 5% of the actual emission experienced during primary production.

A threshold limit of energy consumption above 7500 tonnes of oil equivalent (TOE) has been defined for the aluminium sector for classification of units as designated consumers under Perform, Achieve, and Trade (PAT) scheme. At present, there are 10 designated consumers in the sector. The specific energy consumption (SEC) for 2014 for different countries is given in Table 3.

Energy conservation measures

- Increase in anode length to optimize current density.
- Increase in anode ring bus size to reduce voltage drop.

Table 3 Specific energy consumption for 2014		
Location	SEC (kWhAC/ tonne)	Deviation from world average (%)
North America	14 870	+4%
South America	15 038	+5%
Europe	15 513	+9%
Asia (excluding China)	14 714	+3%
China	13 596	-5%
Gulf Cooperation Council	14 889	+4%
Africa	14 569	+2%
Oceania	14 770	+3%
World average	14 289	

SEC – specific energy consumption
Source Details available at <www.world-aluminium.org/statistics/primary-aluminium-smelting-energy-intensity/>

- Installation of the state-of-the-art cathode block pre-heater to reduce resistance by eliminating uneven heating before putting into new pots.
- Installation of pregnant liquor flashing unit in the precipitation area to reduce steam consumption.
- Automatic voltage control can be replaced by electronic pot controller to reduce DC power consumption in Soderberg alumina electrolysis.

Iron and steel industry

Consumption of steel is regarded as an indicator of economic development. The steel sector contributes nearly 2% to the country's gross domestic product (GDP) and employs more than 0.6 million people. India is the fourth largest producer of crude steel and the largest producer of direct reduced iron (DRI) or sponge iron in the world. In the 5 years corresponding to the Twelfth Five-year Plan (2012–17), the domestic demand of total finished steel is likely to grow at an annual average growth of more than 10% compared to the average annual growth of 8% achieved between 1991/92 and 2010/11.

Crude steel is primarily made through two routes: the large integrated plants use blast furnace and basic oxygen furnace (BOF), while smaller plants use DRI and/or scrap steel and electric arc furnace (EAF) and/or induction furnace.

Capacity and production

Steel production in India gained momentum with the setting up of three steel plants in the public sector, Steel Authority of India Limited or SAIL (erstwhile Hindustan Steel Limited), in the late 1950s and the fourth in the early 1970s. These plants, along with the Indian Iron and Steel Company (IISCO) (now a part of SAIL), Visvesvaraya Iron and Steel Plant (VISL), and Tata Iron and Steel Company or

TISCO (now Tata Steel), were the only integrated steel producers till the 1980s. Vizag Steel Plant and Rashtriya Ispat Nigam Ltd (RINL) came into production in the early 1990s.

Secondary producers consist of suppliers of processed inputs for steelmaking, primary steelmakers, and independent re-rollers. The 1970s saw the emergence of the secondary sector: small-scale steel producers in the private sector who opted for production of pig iron and sponge iron in the form of DRI and set up DRI-based EAF/induction furnace routes to meet primary local demands. India has been the world's largest producer of sponge iron since 2003. Over the years, the coal-based route has emerged as the key contributor and accounted for 88% of the total sponge iron production in the country in 2013/14 and 89% during April–December 2014/15 (provisional). India is also an important producer of pig iron. Post-liberalization, with the setting up of several units in the private sector, not only the imports have drastically reduced but India has also become a net exporter of pig iron. The private sector accounted for 93% of the total production for sale of pig iron in the country during 2013/14.

The installed production capacity of crude steel during 2013/14 was 101.02 MTPA as against 97.02 MTPA in 2012/13. The crude steel production grew at a compounded annual growth rate of 7% during the past 5 years ending 2013/14. The installed capacity and production of crude steel in recent years are given in Table 4.

The share of different processes adopted in crude steel production in the country is given in Table 5. The share of the electric route of steelmaking, particularly the induction furnace route, which accounted for 34% of the total crude steel production in the country during 2013/14, has emerged as the key driver of crude steel production.

Table 4 Crude steel capacity and production in India (in MTPA)

Year	Capacity	Production	Capacity utilization (%)
2010/11	80.36	70.67	88
2011/12	90.87	74.29	82
2012/13	97.02	78.42	81
2013/14	101.02	81.69	81
2014/15 (April–December)*	75.76**	62.39	82

*Provisional
**Pro-rata, based on annual data
Source MoS (2015)

Table 5 Percentage share of different process routes in crude steel production

Process route	Share (%)		
	2009/10	2013/14	2014/15 (April–December)*
Basic oxygen furnace	45	43	44
Electric arc furnace	25	23	22
Induction furnace	30	34	34
Total	100	100	100

*Provisional
Source MoS (2015)

Energy consumption

A large variation in SEC is found among steel plants because of differences in size, technology, raw materials, and products. The SEC of Indian steel plants ranges from 6.09 Gcal per tonne of crude steel (tcs) (25.3 GJ/tcs) to 8.17 Gcal/tcs (34.2 GJ/tcs). On an average, SEC is 7.18 Gcal/tcs (30 GJ/tcs) in India, which is almost twice that of world's leading plants. Although the SEC of Indian steel plants has declined significantly since 1990, mainly due to the entry of large private sector players, there is still a significant scope to improve the overall energy efficiency of the sector.

A threshold limit of energy consumption above 30 000 TOE has been defined for the iron and steel sector for classification of units as designated consumers under PAT scheme. At present, there are 67 designated consumers in the sector.

Energy conservation measures

- Improved charging system in sintering process.
- Adoption of sinter cooler waste heat recovery process.
- Use of multi-silt burners in sintering furnace.
- Adoption of top gas recovery turbine in blast furnace.
- Optimization of the efficiency of coke oven plants.
- Use of coke dry quenching process.
- Use of coke oven gas as a supplementary fuel in other areas.
- Use of hot stove waste heat recovery in existing/new blast furnace.

- Use of iron-smelting technology "Corex" (for green field projects).
- Adoption of pulverized coal injection in existing/new blast furnace.
- Use of ultra-high power transformers with EAF.
- Increase in hot blast temperature and ensure proper coking coal blend.
- Use of energy-efficient regenerative burners in reheating furnaces.
- Installation of vapour absorption refrigeration system for utilizing waste heat in furnaces.
- Reuse of sludge after treatment by removing moisture.
- Reduction in sulphur content of coke oven gas before reuse.
- Adoption of continuous casting in mini-steel plants.

Fertilizer industry

In the success of India in the agriculture sector, in terms of not only meeting the total requirement of food grains but also generating exportable surpluses, the significant role played by chemical fertilizers is well recognized and established. As of now, the country has achieved 80% self-sufficiency in the production capacity of urea. Thus, India can substantially manage its requirement of nitrogenous fertilizers through the indigenous industry and imports. Similarly, 50% of the indigenous capacity has been developed to meet the domestic requirements of phosphatic fertilizers; however, raw materials and intermediates for these fertilizers are largely imported. The entire requirement of potash (K) is met through imports, as there are no viable sources/reserves in the country.

Capacity and production

At present, there are 30 large urea plants in the country, 21 units of diammonium phosphate (DAP) and complex fertilizers, and 2 units of ammonium sulphate as a by-product. Besides, there are 97 medium- and small-scale units in operation, producing single superphosphate (SSP). The installed capacity reached 13.258 MT of nitrogen and 7.06 MT of phosphatic nutrient in 2014/15, making India the third largest fertilizer producer in the world. The production of fertilizers during 2013/14 was 12.378 MT of nitrogen and 3.714 MT of phosphates. The fertilizer production estimated for 2014/15 is 12.683 MT of nitrogen and 4.017 MT of phosphate, representing a growth rate of 2.46% in nitrogen and 8.16% in phosphate. The installed capacity of different types of fertilizers in India and the installed capacity of various fertilizer plants (urea) are given in Tables 6 and 7, respectively. The production of urea, DAP, and complexes for the last few years is provided in Table 8 (MoCF 2015).

Table 6 Installed capacity of various fertilizers (in MTPA)

Product	Number of units	Total installed capacity
Urea	30	21.597
DAP	12	8.332
Complexes	21	6.071
SSP	97	10.121

DAP – diammonium phosphate; SSP – single superphosphate

Energy consumption

Fertilizer production consumes approximately 1.2% of world's total energy on an annual basis. Ammonia production accounts for the bulk or approximately 90% of industry's total, with approximately 166 MT produced annually. The synthesis of ammonia through Haber–Bosch process is the source of nearly all nitrogen fertilizers. (IFA 2014)

Indian fertilizer industry has kept pace with technological development with upgrade and utilization of better feedstock. It is comparable to the best in the world in terms of efficiency, capacity utilization, energy consumption, and utilities such as power and water. The Indian fertilizer industry is benchmarked as one of the best in the world in terms of operational efficiency, energy consumption, maintenance of safety, and environmental standards. Figure 1 shows the considerable improvements in energy consumption in urea plants between 1987/88 and 2013/14.

Ammonia accounts for 80% of the energy required for the production of urea, the rest being steam and power required for the conversion of ammonia and CO_2 to urea. The specific energy consumed to produce 1 tonne of urea is the major indicator of the operating efficiency of the plant since the cost of energy constitutes 60%–70% of the total cost of production. Over the years, the fertilizer industry has been consistently making efforts to improve efficiency by taking suitable measures for reduction in energy consumption. Energy consumption is expected to reduce further when all

Table 7 Installed capacity of fertilizer plants (urea) in India

Name of the plant	Total installed capacity ('000 tonnes)	Production (2013/14) ('000 tonnes)
NFL: Nangal-II	478.5	394.5
NFL: Bathinda	511.5	560.4
NFL: Panipat	511.5	511.2
NFL: Vijaipur-I	864.6	1006.3
NFL: Vijaipur-II	864.6	1162.4
BVFCL: Namrup-II	240.0	70.6
BVFCL: Namrup-III	315.0	235.4
RCF: Thal	1706.8	1993.4
RCF:-Trombay-V	330.0	352.6
MFL: Chennai	486.8	486.7
IFFCO: Kalol	544.5	600.4
IFFCO: Phulpur	551.1	651.8
IFFCO: Phulpur Expn	864.6	950.9
IFFCO: Aonla-I	864.6	1103.0
IFFCO: Aonla-II	864.6	1074.2
KRIBHCO: Hazira	1729.2	2209.9
GSFC: Vadodara	370.6	322.1
SFC: Kota	379	403.2
DIL: Kanpur/KFCL	722.0	313.2
ZIL: Goa	399.3	376.2
SPIC: Tuticorin	620	286.2
MCF: Mangalore	380	379.1
GNFC: Bharuch	636	696.4
IGFL: Jagdishpur	864.6	1035.5
NFCL: Kakinada-I	597.3	646.8
NFCL: Kakinada-II	597.3	780.0
CFCL: Gadepan-I	864.6	990.6
CFCL: Gadepan-II	864.6	950.6
TCL: Babrala	864.6	1136.5
KSFL: Shahjahanpur	864.6	1035.3
Total	20752.4	22715.5

BVFCL - Brahmaputra Valley Fertilizers Corporation Ltd; CFCL - Chambal Fertilizers and Chemicals Ltd; DIL - Duncans Industries Limited; GNFC - Gujarat Narmada Valley Fertilizers and Chemicals Limited; GSFC - Gujarat State Fertilizers and Chemicals Ltd; IFFCO - Indian Farmers Fertiliser Cooperative Ltd; IGFL - Indo Gulf Fertilizers Ltd; KFCL - Kanpur Fertilizers & Cement Ltd; KRIBHCO - Krishak Bharti Cooperative Ltd; KSFL - Kribhco Shyam Fertilizers Ltd; MCF - Mangalore Chemicals and Fertilizers Ltd; MFL - Madras Fertilizers Ltd; NFCL - Nagarjuna Fertilizers and Chemicals Ltd; NFL - National Fertilizers Ltd; RCF - Rashtriya Fertilizers and Chemicals Ltd; SFC - Shriram Fertilizers and Chemicals; SPIC - Southern Petrochemical Industries Corporation Ltd; TCL - Tata Chemicals Ltd; ZIL - Zuari Industries Ltd

Source MoCF (2014)

the fuel oil-based plants and naphtha-based plants will switch over to gas as feedstock. A threshold limit of energy consumption above 30000 TOE has been defined for the fertilizer sector for classification of units as designated consumers under PAT scheme. Presently, there are 29 designated consumers in the sector

Energy conservation measures

- Installation of low temperature shift guard with additional heat recovery to preheat boiler feed water.
- Conversion of a single-stage flash vessel system in regeneration section to a multi-stage flash vessel system with ejectors.
- Improved burner design for higher combustion efficiency, with saving in fuel and reduction in NOx.
- Use of new materials in construction (micro-alloy) for primary reformer tubes for improved efficiency and capacity.
- Installation of vortex mixer at the bottom of the reactor in urea plants to improve conversion efficiency and reduce steam consumption.

Textile industry

Indian textile industry has an overwhelming presence in the economic life of the country. Apart from providing one of the basic necessities of life, the textile industry also plays a pivotal role through its contribution to industrial output, employment generation, and export earnings of the country. The textile industry accounts for 14% of industrial production, which is 4% of GDP, employs 45 million people, and accounts for 13% share of the country's total export basket. The textile sector is one of the largest providers of employment along with agriculture. The Indian textile industry represents a widely diverse spectrum of activities with the handspun and handwoven sector at the one end and the capital-intensive, sophisticated mill sector at the other end. The decentralized power looms, hosiery, and knitting sectors form the largest section of the textile industry. The close linkage of the Indian textile industry to agriculture and the ancient culture and traditions of the country makes the textile sector unique compared to the textile industry of other countries.

Table 8 Production of urea, diammonium phosphate, and complexes (in MT)

Year	2009/10	2010/11	2011/12	2012/13	2013/14
Urea	21.112	21.881	21.984	22.575	22.715
DAP	4.246	3.537	3.963	3.647	3.611
Complex fertilizers	8.038	8.727	7.770	6.180	6.913

DAP – diammonium phosphate
Source Indian Fertilizer Scenario (2014)

Capacity, production, and technological status

The power loom sector is a decentralized and unorganized sector and forms one of the most important segments of the textile industry in terms of fabric production and employment generation. It provides employment to 60.86 lakh people and contributes 58.26% of the total cloth production in the country. The decentralized power loom/ hosiery and knitting sectors form the largest section of the textile sector. During 2014/15, about 60% of the total cloth production had come from the power loom sector. The production of cloth in the mill sector has shown a steady growth in the last decade and the highest growth rate of 13% was recorded during 2010/11. The production of cloth from the mill sector during 2014/15 (April–March) was 2491 million m^2, which was about 2% less than that of the corresponding period of 2013/14.[1] The break-up of production of spun yarn in the country is given in Table 9.

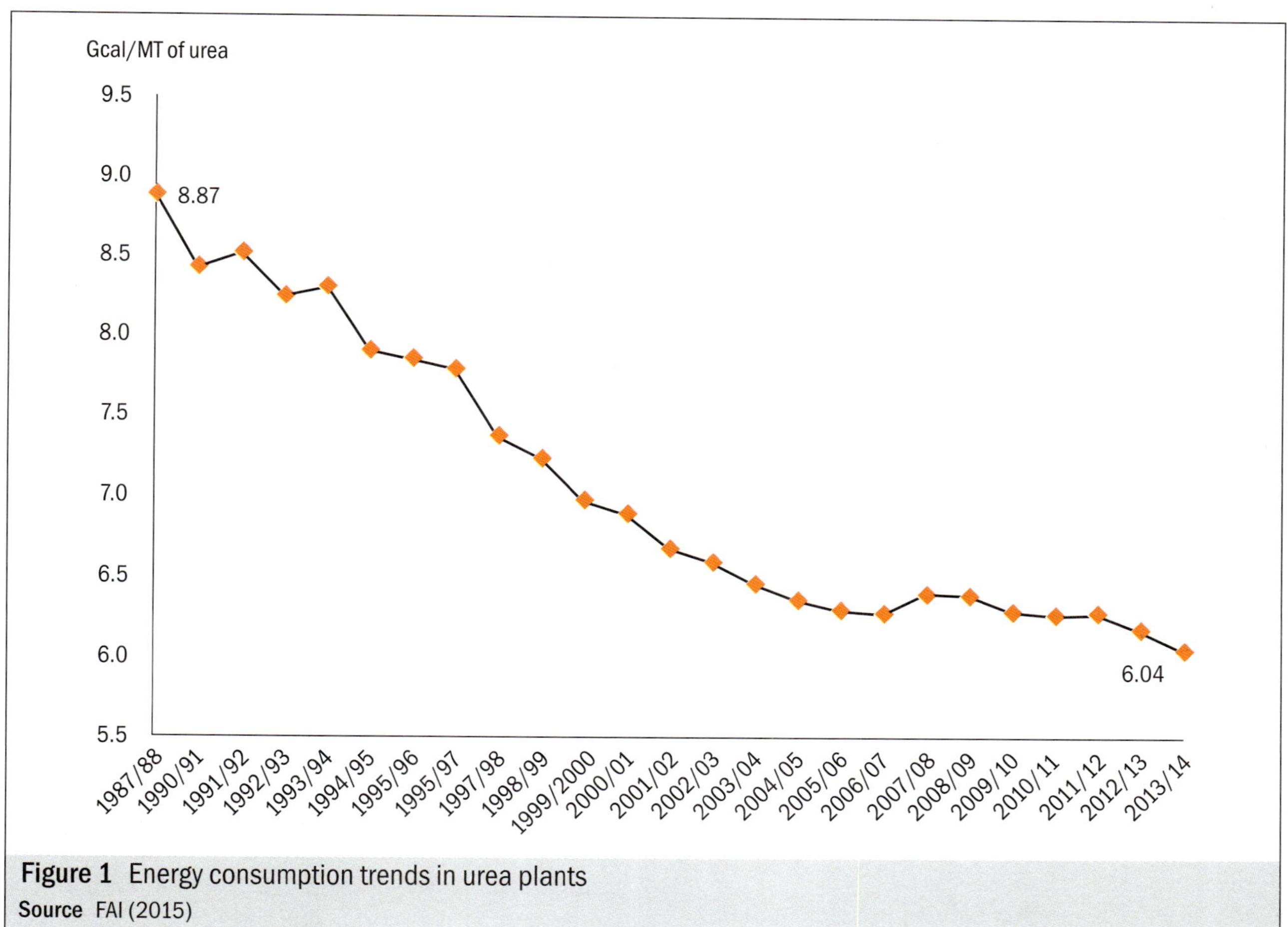

Figure 1 Energy consumption trends in urea plants
Source FAI (2015)

[1] Details available at http://texmin.nic.in/ermiu/pdata/prod_sec_cloth.asp, last accessed on 29 October 2015

Table 9 Production of spun yarn (in million kg)				
Year	Cotton	Blended/ mixed	100% non-cotton	Total
2007/08	2948	677	378	4003
2008/09	2899	655	361	3915
2009/10	3079	707	407	4193
2010/11	3491	796	427	4713
2011/12	3127	789	457	4373
2012/13	3583	828	457	4868
2013/14	3932	897	484	5313
2014/15	4057	915	513	5485

Source Details available at <http://texmin.nic.in/ermiu/pdata/prod_var_yarn.asp>, last accessed on 23 October 2015

Energy consumption

Textile industry is highly energy intensive. Wet processing or dyeing operation consumes almost 50% of the energy in a composite mill. Thermal energy (steam and hot water) is primarily used to process, dye, print, and dry the cloth during wet processing. There is a large scope to save energy in boilers, steam distribution, and drying operation in a textile mill.

Electricity is the major energy form used in spinning plants. Typically, about 75% of electricity is used in spinning machines, 15% in dehumidification, and the rest in lighting and air compressors. A threshold limit of energy consumption above 3000 TOE has been defined for the textile sector for classification of units as designated consumers under PAT scheme. At present, there are 90 designated consumers in the sector.

Energy conservation measures

- Energy efficiency improvement in the humidification plant.
- Installation of energy-efficient fans in ring frame machines.
- Installation of energy-efficient motors and variable frequency drives in machines and pumps.
- Conversion of thermic fluid heating system to direct gas firing system in stenters and dryers.
- Temperature control system in processing machines.
- Recovery of condensate in wet processing plants.
- Energy efficiency improvement in cylinder dryer.
- Waste heat recovery in stenters, merceriser machines, and bleaching system.
- Replacing electric heating with thermic fluid heating in polymerizer machine.
- Installation of photocells for speed frames.
- Installation of flat belts for spinning ring frame machines.

Cement industry

Cement industry occupies an important place in the Indian economy because of its strong linkages to other sectors such as construction, infrastructure, coal, and power. India, with a share of about 7%, is the second largest cement producer in the world after China. Cement is one of the most technologically advanced industries in the country. The modern Indian cement plants are state-of-the-art plants and comparable to the best in the world. The cement industry comprises 190 large cement plants with an installed capacity of 350 MT and more than 350 mini-cement plants with an estimated capacity of 11.1 MTPA (DIPP 2015). A few large cement plants are owned by the central and state governments. Large cement plants contribute about 97% to the total installed capacity. Since the basic raw material for cement production is limestone, the industry is concentrated around limestone reserves in the country. India has abundant reserves of limestone. The private housing sector is the major consumer of cement (53%), followed by the government infrastructure sector (Planning Commission 2011a).

Production

India produces different types of cement, such as ordinary Portland cement, Portland Pozzolana Cement, Portland blast furnace slag cement, and others such as white and specialized cement. The estimated production of cement was 235.11 MT in 2012/13 and 256.04 MT in 2013/14. The installed capacity of major cement plants in India is given in Table 10.

Energy consumption

Cement is one of the most technologically advanced industries in the country. The modern Indian cement units are the state-of-the-art plants and comparable to the best in the world. The induction of advanced technology and the state-of-the-art plants has helped

Table 10 Capacity share of major cement plants (2012/13)

Company	Capacity (in MTPA)
UltraTech Cement	53.90
ACC Ltd	28.80
Ambuja Cements Ltd	27.95
Jaypee Cement Ltd	32.55
India Cements Ltd	15.55
Madras Cements Ltd	12.72
J K Group	7.47
J K Lakshmi Cement	5.3
Shree Cement Ltd	13.50
Dalmia Cement	11.8
Chettinad Cement	11.0
Century Textiles and Industries Ltd	7.80
Kesoram Industries	7.25
Lafarge India Pvt. Ltd	7.75
Others	75.60
Total	318.94

Source MoM (2013b)

the cement industry immensely in improving efficiency by conserving energy and fuel and addressing environmental concerns. The cement industry is highly energy intensive. Energy costs account for 35%–45% of the production cost in a cement plant. Production of clinker is the most energy-intensive stage of cement production. The average thermal energy consumption is estimated to be about 725 kcal/kg of clinker. Electricity is used in the mills, fans, conveyors, and other areas of the plant. The average electricity consumption is about 80 kWh per tonne of cement. In the Indian cement industry, there is a wide variation in the SEC from plant to plant owing to large variations in vintage, capacity, product mix, technology, and the level of technology upgrade undertaken by plants. The general trend of SEC is given in Table 11.

The best thermal and electrical energy consumptions currently achieved by the Indian cement industry are about 667 kcal/kg clinker and 67 kWh/tonne cement, respectively, which are comparable to the best reported figures of 660 kcal/kg clinker and 65 kWh/tonne cement in a developed country such as Japan. It is expected that the industry's average thermal energy consumption by the end of the Twelfth Five-year Plan (2012–17) will be about 710 kcal/kg clinker and the average electrical energy consumption will come down to 78 kWh/tonne cement.

A threshold limit of energy consumption above 30 000 TOE has been defined for the cement sector for classification of units as designated consumers under PAT scheme. At present, there are 85 designated consumers in the sector.

Energy conservation measures

- Modification of pre-heater top-stage cyclones for higher efficiency and low pressure drop.
- Optimization of coal fineness for kiln and calciner firing.
- Installation of waste heat recovery systems for power production from pre-heater and cooler exhaust gases.
- Modifications of grate cooler and cooler fans with static pre-grate.

Table 11 General trend of SEC in Indian cement plants

Process	Specific heat consumption (kcal/kg clinker)	Specific electrical energy consumption (kWh/tonne cement)
Wet process	1300	115
Semi-dry process	900	110
Dry process		
Four-stage pre-heater pre-calcinator	800	95
Five-stage pre-heater pre-calcinator	750	88
Six-stage pre-heater pre-calcinator	726	78

Source NCCBM (2015)

- Reducing the feed size of limestone by installing high-efficiency crusher as a pre-grinder before the cement mill.
- Replacement of pneumatic with mechanical conveying system for cement mills along with weigh feeders.
- Installation of roller press.
- Increase in blended cement production.
- Arresting air leakage to decrease load on pre-heater fan and kiln/raw mill electro-static precipitator (ESP) fan.
- Use of VFD in different drives: boiler feed pumps, raw mill, coal mill, cooler drives.
- Use of renewable energy resources/alternative materials in kiln and power plant.

Chlor-alkali industry

The chlor-alkali industry mainly produces three key products: caustic soda, soda ash, and chlorine. During 2013/14, the alkali chemicals together comprised about 67% of the total production of major chemicals (alkali chemicals, inorganic chemicals, organic chemicals, pesticides, dyes, and dye stuffs). The production of alkali chemicals in 2014/15 (up to September 2014) was 3.279 MT, compared to 3.207 MT during the same period in 2013/14, implying a growth of 2.2%. Caustic soda (sodium hydroxide) is produced by the electrolysis of brine by the diaphragm or membrane cell process. The membrane cell process is a new approach and it accounts for more than 95% of the caustic soda production in India. Electrolysis also produces chlorine and hydrogen as the by-products. Power accounts for about 65% of the production cost of caustic soda.

Table 13 Average operating SEC of the chlor-alkali sector

Financial year	Average SEC kWh/tonne of caustic soda produced
1990/91	3351
1994/95	3130
1999/2000	2977
2009/10	2350

Source SSEF (2013a)

Production

The production of caustic soda and chlorine grew at the rate of 5.4% during the Eleventh Five-year Plan period. The break-up of production details of alkali chemicals during the last few years is given in Table 12.

Energy consumption

The SEC is about 2400–2500 kWh/tonne of caustic soda production in the membrane cell process and about 3100–3400 kWh/tonne of caustic soda production in the diaphragm cell process. Table 13 gives the trend of SEC in the Indian chlor-alkali industry.

A threshold limit of energy consumption above 12 000 TOE has been defined for the chlor-alkali sector for classification of units as designated consumers under PAT scheme. Currently, there are 22 designated consumers in the sector.

Energy conservation measures

Caustic soda manufacturing is a highly energy-intensive process and consumes about 2.5 MW/tonne of caustic soda. In the chlor-alkali

Table 12 Product-wise installed capacity and production of alkali chemicals

Product category	Installed capacity			Production				
	2011/12	2012/13	2013/14	2009/10	2010/11	2011/12	2012/13	2013/14
Soda ash	2.951	2.951	2.951	2.058	2.299	2.411	2.438	2.392
Caustic soda	3.003	3.052	3.052	2.237	2.334	2.409	2.376	2.392
Liquid chlorine	2.160	2.125	2.125	1.534	1.637	1.658	1.673	1.697
Total	8.114	8.128	8.128	5.829	6.270	6.478	6.487	6.481

industry, the electrolysis is the most energy-intensive phase and power is the main form of energy used in this sector. Some of the energy-saving options for this sector are as follows:

- Conversion of existing mercury cell-based plants to membrane cell plants.
- Use of high-performance membranes for reducing energy consumption in the electrolysis process.
- Use of plate evaporators.
- Optimization of the operating hours of the agitator.
- Regular and timely replacement of membranes and recoating of electrodes to optimize power consumption.
- Use of advanced technologies such as oxygen-reducing electrodes (oxygen-depolarized cathodes).
- Installation of advanced control systems to control brine feed temperature.
- Replacement of inefficient chlorine compressors with high energy efficiency chlorine compressors.
- Utilization of hydrogen generated from electrolysis in the vapour absorption machine.

Pulp and paper industry

The paper industry in India is one of the oldest industries having a bearing on the socio-economic development of the country. As per the 2012 statistics, 750 pulp and paper mills are operating in India, producing different varieties of paper, such as writing and printing paper, packaging paper, coated paper, and specialty grade paper. The paper, paperboard, and newsprint production in the country is 10.11 MT against an installed capacity of 12.70 MT. The annual consumption of paper, paperboard, and newsprint in the country is around 11.15 MTPA (Tandon, Negi, and Mathur 2013).

Production

The industry is highly fragmented and consists of large, medium, and small paper mills having capacities that range from 5 tonnes to 800 tonnes per day. The paper production from small and medium mills accounts for 60%, while 40% comes from large mills. Scandinavian countries, USA, the Russian Federation, China, Indonesia, and Japan are the major players in the pulp and paper sector. These countries have some of the best raw materials available for paper production, possess cutting-edge technologies, and control the global trade.

With the increase in domestic demand, the paper industry in India has become more promising. The majority of the pulp and paper units are small in size. About 70% of the total installed capacity of paper production in India is accounted by Gujarat, West Bengal, Odisha, Andhra Pradesh, Karnataka, and Maharashtra. The states of Uttar Pradesh, Tamil Nadu, Haryana, Kerala, Bihar, and Assam together account for about 25% of the total paper production in India. Under the Newsprint Control Order, the total installed capacity of newsprint in the country was about 2.5 MTPA in 2013/14.

Production of paper from different raw materials is given in Table 14. The installed capacity and production of some of the large pulp and paper mills in India during 2010/11 are given in Table 15.

Energy consumption

The pulp and paper industry is highly energy intensive. Energy cost accounts for about 16%–25% of the cost of production of paper. The SEC of integrated wood-based plants varies between 23 GJ/tonne and 37 GJ/tonne compared to 18–22 GJ/tonne of paper internationally.

A threshold limit of energy consumption above 30 000 TOE has been defined for the pulp and paper sector for classification of units as designated consumers under PAT scheme. Currently, there are 31 designated consumers in the sector.

Energy conservation measures

- Adoption of extended delignification process in the pulping process.
- Adoption of pressure screens instead of centrifugal screens in the pulping section.

Table 14 Production of paper from different raw materials in 2010/11 (in MT)

Variety of paper	Wood based	Agro based	Recycled fibre/waste paper based	Total
Writing/printing	2.36	0.73	0.81	3.90
Packaging	0.77	1.50	3.15	5.42
Newsprint	0.03	Nil	0.76	0.79
Grand total	3.16	2.23	4.72	10.11

Source Planning Commission (2011b)

Table 15 Installed capacity and production of selected pulp and paper mills for 2010/11		
Company	Capacity (tonnes)	Production (tonnes)
ABC Papers	105 000	60 553
Abhishek Industries	175 000	123 639
Andhra Paper Mills	175 000	176 452
BIMT	800 000	796 718
Emami Paper Mills Ltd	145 000	142 494
Hindustan Newsprint Ltd	100 000	100 546
ITC-PSPD	550 000	547 931
J K Paper Ltd	240 000	273 755
N R Agarwal Industries Ltd	170 000	158 031
Rainbow Papers Ltd	159 500	109 054.9
Rama Paper Mills Ltd	61 000	48 627.9
Star Paper Mills	75 000	68 592
TNPL	275 575	245 008
Westcoast Paper	320 000	173 638
Yash Paper Mills	39 960	26 598

BILT - Ballarpur Industries Ltd; ITC-PSPD - ITC Paperboards and Specialty Papers Division; TNPL - Tamil Nadu Newsprint and Papers Ltd
Source SSEF (2013b)

- Installation of low-intensity refiners for wood fibres.
- Installation of multi-port cylinder dryer in the paper machine.
- Optimization of steam consumption in the paper mill dryer.
- Conversion of existing boiler to fluidized bed combustion boiler.
- Use of long-tube falling-film evaporators for concentration of black liquor.
- Installation of advanced condensate recovery and reuse system.
- Installation of high-pressure cogeneration system.
- Installation of shoe press machine instead of roller press machine in paper mills.
- Gasification of black liquor.
- Biomethanation of wastewater in agro-based mills.

Micro, small, and medium enterprises

The micro, small, and medium enterprise (MSME) sector has emerged as a highly vibrant and dynamic sector of the Indian economy during the past five decades. MSMEs not only play a crucial role in providing large employment opportunities at comparatively lower capital cost than large industries but also help in the industrialization of rural and backward areas, thereby reducing regional imbalances and ensuring more equitable distribution of national income and wealth. MSMEs are complementary to large industries as ancillary units, and this sector contributes enormously to the socio-economic development of the country. In addition to providing a wide range of services, the MSME sector produces more than 6000 products, ranging from traditional to high-tech items. As per the Fourth All-India Census of MSMEs, the total number of enterprises (registered and unregistered) in the MSME sector was about 361.76 lakh with a total employment of 805.24 lakh. Rural areas accounted for 55.34% of the total working enterprises in the MSME sector. The classification of the total number of enterprises by nature of activity indicates that 31.79% of the enterprises in the MSME sector were engaged in manufacturing, whereas 68.21% of the enterprises were engaged in providing services. The distribution of the total number of MSME units according to the source of energy is given in Table 16.

Policy initiatives

Industry plays an important role in the Indian economy. Through the 2015/16 Union Budget, the government has taken several steps to bring structural changes in the Indian economy, with specific focus

Table 16 Distribution of enterprises by main source of energy	
Source of energy	Number of enterprises (in lakh)
No energy required	198.18 (55%)
Electricity	117.01 (32%)
Oil	14.39 (4%)
Others	10.44 (3%)
Coal	6.48 (2%)
Traditional energy/firewood	7.39 (2%)
LPG/CNG	4.04 (1%)
Non-conventional energy	0.88 (0.2%)
Not recorded	2.95 (1%)

CNG - compressed natural gas; LPG - liquefied petroleum gas
Source MMSME (2015)

on promoting industrial productivity. One of the key initiatives, "Make in India", focuses on increasing the manufacturing activity in the country. Some of the recent policy initiatives related to the industry sector are summarized as follows:

- The Union Budget 2015/16 has targeted to double the infrastructure spending to ₹1080 billion. This is expected to result in sustained recovery in the demand for cement and other construction materials.
- Duties and tariffs directly levied on cement have been increased marginally from 12.4% + ₹120 per tonne to 12.5% + ₹125 per tonne.
- To help raise debt by finance companies, a ₹200 billion National Investment and Infrastructure Fund has been planned. The budget also plans to issue tax-free bonds for roads, railways, and irrigation projects and aims to rationalize the tax regime for Infrastructure Investment Trusts.
- In the Union Budget 2015/16, the clean energy cess on coal (domestic and imported) has been increased to ₹200 per tonne from ₹100 per tonne.
- The Union Budget 2015/16 has increased the rail freight rate for cement by 2.7% and for coal by 6.3%.
- The Union Budget 2015/16 has increased the basic excise duty on metals from 12.36% to 12.50%.
- The basic customs duty on metallurgical coal has been increased from 2.5% to 5%, and the special additional duty on iron and steel scrap has been reduced from 4% to 2%.
- To boost investments, the Union Budget 2015/16 has granted an additional depreciation of 20% to new plant and machinery installed by a manufacturing unit or a unit engaged in the generation and distribution of power.

Budgetary allocation under the Technology Upgradation Funds Scheme has been reduced to ₹15.2 billion for 2015/16 from ₹18.6 billion in 2014/15 (CRISIL Research 2015; MoF 2014).

Indian iron and steel industry: use of latest technologies

Steel is one of the most important products of the modern world and of strategic importance for the development of any nation. Consumption of steel is regarded as an indicator of economic development. Global crude steel production has witnessed a shift from Western Hemisphere to Eastern Hemisphere since the beginning of 2000. Asia is the largest producer of steel and it contributed about 68% to the global steel production in 2013.

India is the fourth largest producer of crude steel and the largest producer of direct reduced iron (DRI) or sponge iron in the world. The Indian steel sector contributes nearly 2% to India's GDP and employs more than 0.6 million people. The Indian steel industry exists in different sizes with varying degrees of vertical and horizontal integration. After the liberalization of economic policies in the 1990s, coupled with deregulation, the Indian steel industry has undergone a major structural change. During the pre-reform period, it was concentrated with public sector units and TISCO, but reforms in the early 1990s have brought a number of private players in a big way.

The iron and steel industry is the largest consumer of energy among all the industrial sub-sectors in the country. The average specific energy consumption of integrated steel plants based on the BF-BOF route is in the range of 6–8 Gcal/tcs compared to 4 Gcal/tcs for the most efficient steel plant and the world average of 4.5–5.0 Gcal/tonne.

With the introduction of energy-efficient technologies, process modernization, expansion of existing steel plants, and installation of green field units with the latest technology, it is expected that the average specific energy consumption per tonne of crude steel by Indian steel plants will be gradually reduced from the current level of 6.5 Gcal/tcs to 6 Gcal/tcs by 2020/21, 5.5 Gcal/tcs by 2025/26, and 5 Gcal/tonne by 2030/31.

Some of the futuristic technologies that can be used by the Indian steel sector to improve energy efficiency and utilize the lean grade and abundantly available raw materials include the following:

- **Injection of plastic waste in blast furnace:** Plastic waste has a higher calorific value than coal and its injection allows a reduction in coke production. The theoretical maximum limit for plastic injection at the tuyère level is about 0.070 tonne/tonne of hot metal and is set according to the thermo-chemical and kinetic conditions of the blast furnace. The steel industry in Japan effectively reuses both industrial and municipal plastic waste, and about 1% of plastic waste is added to the coal charge in the country's steel industry.
- **Finex technology:** The Finex process was developed jointly by Posco and Siemens VAI. In the Finex process, iron ore fines are first reduced to DRI in a series of fluidized-bed reactors by means of a coal-based reduction gas. The DRI fines are then hot compacted and charged into the melter gasifier for melting. The tapped product, liquid hot metal, is equivalent in quality to the hot metal produced in a blast furnace. Finex export gas is also a valuable by-product that can be used for a variety of industrial applications, such as power generation, heating applications, or for additional DRI/HBI (hot briquetted iron) production. A Finex plant with the production capacity of 600 000 tonnes per year was first commissioned in 2003 at Posco Pohang Works. On the basis of its excellent operational results, Posco subsequently commissioned two more plants in 2007 and 2014.
- **ITmk3® Technology:** Kobe Steel has developed the ITmk3® Process, which produces high-purity granular iron by utilizing iron ore concentrate and non-coking coal. In the ITmk3® Process, iron ore concentrate and non-coking coal are mixed together and agglomerated into green pellets, which are fed into a rotary hearth furnace in which reduction and melting process occurs simultaneously to make granular iron (iron nuggets equivalent to pig iron). The ITmk3® Process can complete the reduction reaction process in approximately 10–12 min compared to 8 h in the BF-BOF process and 6 h in the DR process. The first commercial plant using this technology was commissioned in 2010 in cooperation with Steel Dynamics Inc., Minnesota, USA, with a production capacity of 500 000 tonnes per year.

Some of these technologies have attracted Indian steelmakers owing to their inherent benefits, and feasibility studies have been undertaken to use these technologies in green field projects. It is expected that a plant of 3 MTPA capacity using the Finex process and another plant of 1.5 MTPA capacity based on the ITmk3® Process will be commissioned in the country by 2025.

Source Bectron Technology Pvt. Ltd (2015)

References

Bectron Technology Pvt. Ltd. 2015. ***Status Report on Indian Iron and Steel Sector***. Mumbai: Bectron Technology Pvt. Ltd

CRISIL Research. 2015. ***CRISIL Budget Analysis 2015***. Details avai-lable at http://www.moneycontrol.com/news_html_files/news_attachment/2015/CRISILBudget.pdf, last accessed on 7 December 2015

DIPP (Department of Industrial Policy and Promotion). 2015. ***Annual Report 2014/15***. New Delhi: DIPP, Government of India

FAI (Fertiliser Association of India). 2015. ***Sectoral Report on Fertilisers-Ammonia: Urea Plants***. Mumbai: FAI

IFA (International Fertilizer Industry Association). 2014. ***Ammonia Production: Moving Towards Maximum Efficiency and Lower GHG Emissions.*** IFA, Fertilizer Facts. Details available at http://www.fertilizer.org/imis20/images/Library_Downloads/2014_ifa_ff_ammonia_emissions_july.pdf?WebsiteKey=411e9724-4bda-422f-abfc-8152ed74f306&=404%3bhttp%3a%2f%2fwww.fertilizer.org%3a80%2fen%2fimages%2fLibrary_Downloads%2f2014_ifa_ff_ammonia_emissions_july.pdf

MMSME (Ministry of Micro, Small and Medium Enterprises). 2015. ***Annual Report 2014–15***. New Delhi: MMSME

MoCF (Ministry of Chemicals and Fertilizers). 2014. ***Indian Fertilizer Scenario 2014***. New Delhi: Department of Fertilizers, MoCF

MoCF (Ministry of Chemicals and Fertilizers). 2015. ***Annual Report 2014–15***. New Delhi: Department of Chemicals and Petrochemicals, MoCF

MoCI (Ministry of Commerce and Industry). 2015. ***Annual Report 2014–15***. New Delhi: Department of Industrial Policy and Promotion, MoF

MoF (Ministry of Finance). 2014. ***Mid-year Economic Analysis 2014–15***. Details available at http://finmin.nic.in/reports/MYR201415English.pdf

MoM (Ministry of Mines). 2013a. ***Indian Minerals. Yearbook 2013 (Part II: Metals and Alloys)***. New Delhi: MoM

MoM (Ministry of Mines). 2013b. ***Indian Minerals. Yearbook 2013 (Part III: Mineral Reviews)***. New Delhi: MoM

MoM (Ministry of Mines). 2015. ***Annual Report 2014–15***. New Delhi: MoM

MoS (Ministry of Steel). 2015. ***Annual Report 2014–15***. New Delhi: MoS

NCCBM (National Council for Cement and Building Materials). 2015. ***Current Technology Status, Energy Consumption and Possible Future Trends in Cement Industry Sector***. Faridabad: NCCBM

Planning Commission. 2011a. ***Report of the Working Group on Cement Industry for XII Five-year Plan***. New Delhi: Planning

Planning Commission. 2011b. ***Report of the Working Group on Pulp and Paper for XII Five-year Plan***. New Delhi: Planning Commission

SSEF (Shakti Sustainable Energy Foundation). 2013a. ***Technology Compendium on Energy Saving Opportunities: Chlor-Alkali Sector***. Hyderabad: Confederation of Indian Industry

SSEF (Shakti Sustainable Energy Foundation). 2013b. ***Technology Compendium on Energy Saving Opportunities: Pulp and Paper Sector***. Hyderabad: Confederation of Indian Industry

Tandon, R., S. D. Negi, and R. M. Mathur. 2013. Waste paper collection mechanism in India: current status and future requirements. ***IPPTA*** 25(3): 37–40

Bibliography

MoT (Ministry of Textiles). 2015. ***Annual Report 2014–15.*** New Delhi: MoT

SMERA's union budget analysis (FY 2015–16). Details available at https://www.smera.in/documents/SMERA_Union_Budget_Analysis_(FY-2015-16).pdf

SSEF (Shakti Sustainable Energy Foundation). 2013. ***Technology Compendium on Energy Saving Opportunities in Iron and Steel Sector.*** Hyderabad: Confederation of Indian Industry

TABLE 8.1 Installed capacity ('000 tonnes), production ('000 tonnes), and capacity utilization (%) in the major steel industry: 2007/08–2014/15

Crude/liquid steel												
Producers	2007/08			2008/09			2009/10			2010/11		
	Working capacity	Production	% of utilization	Working capacity	Production	% of utilization	Working capacity	Production	% of utilization	Working capacity	Production	% of utilization
Public sector												
Bhilai Steel Plant	3925	5055	129	3925	5183	132	3925	5108	130	3925	5329	136
Durgapur Steel Plant	1802	1914	106	1802	1886	105	1802	1966	109	1802	1961	109
Rourkela Steel Plant	1900	2093	110	1900	2083	110	1900	2128	112	1900	2160	114
Bokaro Steel Plant	4360	4127	95	4360	3577	82	4360	3599	83	4360	3592	82
Ispat Steel Plant	500	458	92	500	417	83	500	400	80	500	411	82
Alloy Steel Plants	234	157	134	234	168	72	234	205	88	234	200	85
Salem Steel Plant												
Visvesvaraya Iron and Steel Plant	118	158	134	118	95	81	118	103	87	118	108	92
Total (SAIL)	12839	13962	109	12839	13409	104	12839	13509	105	12839	13761	107
RINL	2910	3129	108	2910	2963	102	2910	3205	110	2910	3235	111
Total (public sector)	15749	17091	109	15749	16372	104	15749	16714	106	15749	16996	108
Private sector												
Tata steel Ltd	5000	5013	100	5000	5646	113	6800	6563	97	6800	6856	101
Major steel producers[1]	11400	9538	84	14800	10218	69	18233	14381	79	18433	14549	79
Other EAF units/Corex-BOF/MBF-EOF	6831	5282	77	8614	8147	95	8419	8357	99	9140	9332	102
Induction furnace units	20865	16933	81	22180	18054	81	25800	19824	77	30241	22939	76
Total (private)	44096	36766	83	50594	42065	83	59252	49125	83	64614	53676	83
Grand total	59875	53857	90	66343	58437	88	75001	65839	88	80363	70672	88

Crude/liquid steel												
Producers	2011/12			2012/13			2013/14			2014/15 (Apr-Dec)		
	Working capacity	Production	% of utilization	Working capacity	Production	% of utilization	Working capacity	Production	% of utilization	Working capacity	Production	% of utilization
Public sector												
Bhilai Steel Plant	3925	4901	125	3925	5008	128	3925	5136	131	2944	3526	120
Durgapur Steel Plant	1802	1914	106	1802	2034	113	1802	2019	112	1352	1516	112
Rourkela Steel Plant	1900	2170	114	1900	2209	116	1900	2291	121	1425	2047	144
Bokaro Steel Plant	4360	3647	84	4360	3757	86	4360	3776	87	3270	2887	88
Ispat Steel Plant	500	330	66	500	135	27	500	127	25	375	13	3
Alloy Steel Plants	234	200	85	234	131	56	234	122	52	176	78	44
Salem Steel Plant	180	96	53	180	73	41	180	91	51	135	89	66
Visvesvaraya Iron and Steel Plant	118	91	77	118	64	54	118	13	11	89	29	33
Total (SAIL)	13019	13349	103	13019	13411	103	13019	13575	104	9764	10185	104
RINL	2910	3128	107	2910	3071	106	2910	3202	110	2183	2396	110
Total (public sector)	15929	16477	103	15929	16482	103	15929	16777	105	11974	12581	105
Private sector												
Tata steel Ltd	6800	7128	105	9600	8130	85	9600	9155	95	7200	6892	96
Major steel producers[1]	25540	17015	67	25540	18424	72	25540	18308	72	19155	14204	74
Other EAF units/Corex-BOF/MBF-EOF	11580	9735	84	12010	9695	81	13458	9874	73	10094	7696	76
Induction furnace units	31017	23936	77	33945	25685	76	36494	27579	76	27371	21019	77
Total (private)	74937	57814	77	81095	61934	76	85092	64916	76	63819	49811	78
Grand total	90866	74291	82	97024	78416	81	101021	81693	81	75766	62392	82

[1] Major steel producers include Essar, JSW Ispat, JSWL, and JSPL.

EOF – energy optimizing furnace; MBF – mini blast furnace; RNL – Rashtriya Ispat Nigam Ltd; SAIL – Steel Authority of India Ltd

Source MoS (Ministry of Steel). 2015. ***Annual Report 2014/15.*** New Delhi: MoS, Government of India

TABLE 8.2 Installed capacity (lakh tonnes), production (lakh tonnes), and capacity utilization (%) in the fertilizer industry: 2000/01–2014/15

Year	2000/01	2001/02	2002/03	2003/04	2004/05	2005/06	2006/07	2007/08	2008/09	2009/10	2010/11	2011/12	2012/13	2013/14	2014/15
Nitrogenous															
Installed capacity	119.8	121.6	122.4	121.66	122.08	122.88	122.9	122.9	122.9	120.61	120.4	120.45	120.45	127.66	132.58
Production	109.42	106.90	105.08	105.57	113.05	113.33	115.25	109.03	109.00	119.02	121.6	125.8	123.99	123.77	126.83
Capacity utilization	92.00	87.90	86.40	86.40	93.40	94.50	95.60	95.20	95.20	98.8	100.9	104.4	102.9	97	95.7
Consumption	109.20	113.10	104.74	110.76	117.14	127.23	137.74	144.19	150.90	155.80	165.58	173	183.36	165.25	NA
Phosphatic															
Installed capacity	49.87	51.12	53.33	54.01	54.80	54.60	57.36	58.75	58.92	56.59	56.28	56.28	56.28	67.84	70.6
Production	37.34	38.35	39.07	36.26	40.38	42.03	44.40	37.14	34.17	43.22	42.23	44.32	38.76	39.58	40.17
Capacity utilization	77.60	75.50	74.10	67.80	75.50	78.50	78.50	64.70	58.50	76.8	75	78.7	68.9	58.3	56.9
Consumption	42.15	43.82	40.18	41.24	46.24	52.04	55.43	55.15	65.06	72.74	80.5	79.14	59.55	54.58	NA
Potash															
Consumption	15.67	16.67	16.01	15.98	20.6	24.13	23.34	26.36	33.12	36.32	35.14	25.76	18.13	19.76	NA

NA – not available

Source MoCF (Ministry of Chemicals and Fertilizers). 2015. ***Annual Report 2014/15***. New Delhi: MoCF, Government of India

TABLE 8.3 Import (lakh tonnes) of fertilizers: 2000/01–2014/15

Year	2000/01	2001/02	2002/03	2003/04	2004/05	2005/06	2006/07	2007/08	2008/09	2009/10	2010/11	2011/12	2012/13	2013/14	2014/15
Nitrogenous	1.54	2.69	0.66	1.32	4.09	13.85	26.88	36.77	38.44	34.47	44.92	52.40	46.90	38.08	93.86
Phosphatic	3.96	4.29	1.70	3.38	2.96	11.21	13.23	12.53	29.27	27.56	38.02	44.27	27.78	15.90	17.60
Potash	15.41	17.01	14.38	15.48	20.45	27.47	20.69	26.53	33.8	29.44	40.69	33.35	12.3	13.33	20.76
Total	20.91	23.99	16.74	20.18	27.50	52.53	60.80	75.83	101.51	91.47	123.63	130.02	86.98	67.31	132.22

Source MoCF (Ministry of Chemicals and Fertilizers). 2015. ***Annual Report 2014/15***. New Delhi: MoCF, Government of India

TABLE 8.4 Production of aluminium (in tonnes): 2006/07–2014/15

Company	2006/07	2007/08	2008/09	2009/10	2010/11	2011/12	2012/13	2013/14	2014/15 (Apr-Dec)
NALCO	358 734	359 213	361 262	431 488	443 600	413 089	403 384	316 492	243 572
HINDALCO	442 686	477 723	523 453	555 404	543 670	581 985	547 416	618 286	589 606
MALCO	37 652	37 635	23 224						
BALCO	313 189	358 671	356 781	268 425		676 980	773 946	795 355	648 766
VAL			82 031	269 083	642 249				
Total	1 152 261	1 233 242	1 346 751	1 524 751	1 629 519	1 671 454	1 724 746	1 730 133	1 481 944

BALCO – Bharat Aluminium Company Ltd; HINDALCO – Hindustan Aluminium Company Ltd; MALCO – Madras Aluminium Company Ltd; NALCO – National Aluminium Company Ltd; VAL – Vedanta Aluminium Ltd

Source MoM (Ministry of Mines). 2014. ***Annual Report 2014/15***. New Delhi: MoM, Government of India

TABLE **8.5** Number of working, medium, small, and micro enterprises/industries: 1992/93–2013/14

Year	Number of MSME (in lakh)
1992/93	73.51
1993/94	76.49
1994/95	79.60
1995/96	82.84
1996/97	86.21
1997/98	89.71
1998/99	93.36
1999/2000	97.15
2000/01	101.10
2001/02	105.21
2002/03	109.49
2003/04	113.95
2004/05	118.59
2005/06	123.42
2006/07	361.76
2007/08**	377.36
2008/09**	393.7
2009/10**	410.8
2010/11**	428.73
2011/12**	447.64
2012/13**	447.54
2013/14**	488.46

** Projected
MSME – micro, small, and medium enterprises
Note Including activities of wholesale/retail trade, legal,education and social services, hotels and restaurants, transport and storage, and warehousing (except cold storage).
Source MMSME (Ministry of Medium, Small, and Micro Enterprises). 2015. ***Annual Report 2014/15***. New Delhi: MMSME, Government of India

TRANSPORT 9

AT A GLANCE

Roads

- The share of road sector in the total transport by road and rail was estimated to be 64.5% of freight and 85.9% of passenger movement in 2011/12, as compared to 60.7% of freight and 83.1% of passenger movement in 2001/02.
- The total number of registered vehicles in India increased from 58.9 million in 2002 to 159.5 million in 2012.

Railways

- Railways registered a 0.28% decrease in passenger traffic from 8421 million passengers in 2012/13 to 8397 million passengers in 2013/14.
- Passenger kilometres increased by 5.5% from 1098.1 billion in 2012/13 to 1158.74 billion in 2013/14.
- Revenue earning freight traffic handled by the railways during 2013/14 was 1051.64 million tonnes (MT), registering a 4.31% increase, compared to 1008.09 MT in 2012/13.
- Revenue earning freight traffic measured in terms of net-tonne kilometres increased by 2.5% from 649.6 billion in 2012/13 to 665.8 billion in 2013/14.

Port and shipping

- Total traffic handled at all ports increased from 980.49 MT in 2013/14 to 1052.5 MT in 2014/15, registering a growth of 7.3%.
- The major ports handled a total traffic of 581.33 MT during 2014/15 compared to 555.50 MT during 2013/14, representing an increase of about 4.65%. The non-major ports handled about 416.97 MT in 2013/14, which increased to 471.19 MT in 2014/15, representing an increase of about 13%. The total shipping tonnage decreased from 10.41 million gross tonnage (GT) in 2013 to 10.31 million GT in 2014.

Aviation

- The number of total domestic passengers travelling in 2013/14 was 61.7 million against 58.7 million in 2012/13, registering a growth of 5.1%.
- The total domestic freight carried in 2013/14 was 0.39 MT against 0.36 million tonnes carried in 2012/13, registering a growth of 8.3%.

Transport energy consumption

- The road sector consumed 36.24 MT of high-speed diesel (HSD), 14.2 MT of motor spirit, and 0.223 MT of liquefied petroleum gas (LPG) in 2010/11 as compared to 33.7 MT of HSD, 12.8 MT of motor spirit, and 0.225 MT of LPG in 2009/10.
- The consumption of HSD for locomotive services increased from 2699.6 million litres in 2011/12 to 2780.9 million litres in 2013/14. For purposes other than locomotive services, consumption of HSD in the railways was 30.74 million litres in 2012/13, which increased to 33.38 million litres in 2013/14. Locomotives used 14410.7 million kWh of electricity for traction purposes in 2013/14.
- The shipping sector consumed 0.54 MT of HSD, 0.002 MT of light diesel oil (LDO), and 0.37 MT of furnace oil in 2011/12 as compared to 0.56 MT of HSD, 0.0038 MT of LDO, and 0.77 MT of furnace oil in 2010/11.
- The consumption of aviation turbine fuel was 5.33 MT in 2014 as compared to 5.26 MT in 2013, registering an increase of 1.4%.

Introduction

India's transport system is one of the largest systems in the world and serves a land area of 3.3 million km^2 and a population of more than 1.21 billion (MoHF 2011). The rapid growth of economic activity in India has led to increased trade and market participation in both urban and rural areas of the country. As a result, demands for mobility and related energy used for transport within the country have increased. The transport sector has become pivotal for ensuring continued economic development of the nation. The contribution of the transport sector to the country's gross domestic product (GDP) increased from 5.7% in 1999/2000 to 6.5% in 2011/12.[1] This is also evident from the growth rates of both freight and passenger traffic achieved in the road and rail sectors. However, the growth of traffic, both freight and passenger, was much steeper in the road sector than the growth rate achieved by the railways.

Between 1986/87 and 2007/08, freight moved by road increased from 224 million tonnes (MT) to 1559 MT (596%). During the same period, freight traffic carried by the railways increased from 255.4 MT to 768.72 MT (201%) (RITES 2012). The primary reason for the less growth in the railways was the deliberate change in policy that preferred bulk movement where the unit of movement gradually changed from a wagon to a train. Non-bulk commodities that cannot be moved in large volumes at a time shifted to road transport.

The network of national highways increased markedly from 76 818 km in 2011/12 to 79 116 km in 2012/13, achieving a 3% growth (MoRTH 2012b). Compared to this, India's railway network grew by only 0.5% during the same period.

Between 2009 and 2012, the number of vehicles in the country grew at the compound annual growth rate (CAGR) of 10.8%. There has been a steep rise in the number of private vehicles (both two- and four-wheelers) in the last decade, mainly in cities.

Key trends in the consumption of energy in the transport sector

The transport sector accounted for 34.24% of the total volume of petroleum products consumed in India in 2009/10 (MoPNG 2012). This includes the total consumption by road transport, railways, air transport, and water transport. Of all the products consumed within the sector, high-speed diesel (HSD) consumption was the maximum. Of the total consumption of 56.32 MT of HSD in 2009/10, the transport sector accounted for 59.1% (33.74 MT) (MoPNG 2012). This energy consumption pattern clearly reveals the heavy dependence of the transport sector on petroleum products. Road transport has been the major consumer of HSD over the years, as shown in Figure 1, followed by railways and shipping.

The use of alternative fuels in the transport sector remains relatively modest. Of the total energy consumed in the sector, 98.5% is met through petroleum products and the remaining 1.5% by electricity (TERI 2008).

Road

Brief overview

India has one of the largest road network of 5.23 million km in the world, comprising national highways, expressways, state highways, major district roads, other district roads, and village roads. Road transport was responsible for 4.8% of India's GDP in 2011/12. Road traffic has been growing rapidly in the past few decades. The share of road traffic in the total traffic movement by roads and

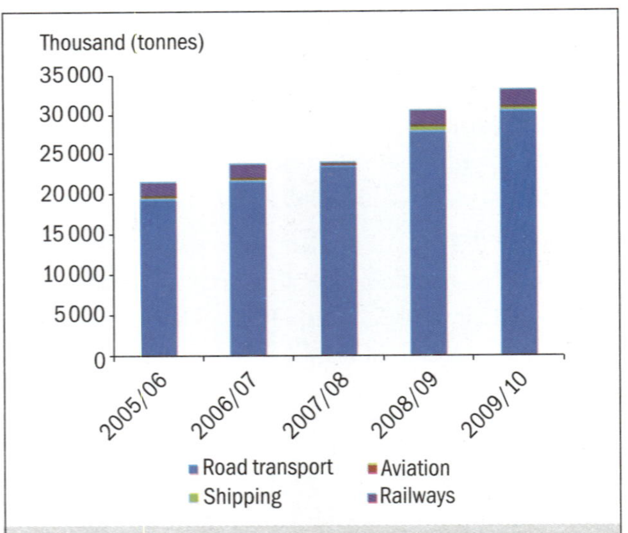

Figure 1 Consumption of high-speed diesel in different transport modes (in '000 tonnes)
Source MoPNG (2015)

[1] Details available at http://indiabudget.nic.in/es2011–12/echap-10.pdf

railways was around 64.5% of freight movement and 85.9% of passenger movement in 2011/12, as compared to 61.3% of freight movement and 38.7% of passenger movement in 1950/51 (MoRTH 2014, Planning Commission 2007). Table 1 gives the type and length of different road categories in India.

The Government of India has entrusted the National Highways Authority of India with the responsibility of implementing the National Highways Development Project (NHDP). The plan is a flagship programme of the government and is spread over 7 phases with an estimated expenditure of about ₹600 000 crore. Phases I and II of NHDP included the development of national highways to four/six lane standards of the Golden Quadrilateral connecting four major metropolitan cities, North–South and East–West Corridors, and road connectivity of major ports of the country. Under Phases I and II, 12 639 km of road out of the total 13 390 km was completed. The Golden Quadrilateral has been completed. Phase III involves four laning of 12 109 km of high-density traffic corridors not included in Phases I and II, connectivity of state capitals under NHDP (Phases I and II), and connectivity of centres of tourism and places of economic importance, out of which 6634 km has been completed. Phase IV envisages upgradation of about 13 203 km of national highways to two-lane paved shoulder, and Phase V plans six laning of 6500 km of the existing four-lane national highways. So far 1585 km under Phase IV and 2264 km under Phase V have been completed. Phase VI envisages the development of 1000 km of expressways under the public–private partnership (PPP) model, which is yet to be started. In Phase VII, ring roads, bypasses, grade separators, flyovers, elevated roads, tunnels, road overbridges, underpasses, and service roads will be constructed under the built–operate–transfer model. Under this phase, 22 km of the total 700 km has been completed.[2,3] Details of the various NHDP phases are given in Table 2.

During the Eleventh Five-year Plan, 9044 km of national highways was completed under NHDP. The target for the Twelfth Five-year Plan is to add 85 000 km by the end of the plan period. It has also been

Table 1 Type and length of different road categories in India (as on 31 March 2013)

Road type	Length of road (km)
National highways	79 116
State highways	169 227
Other public works departments roads	1 066 747
Rural roads	3 159 639
Urban roads	446 238
Project roads	310 955
Total	5 231 922

Source MoRTH (2015)

Table 2 Status of NHDP phases

NHDP project	Total length (km)	Completed (km)	Under implementation (km)	Contracts under implementation (number)	Balance length for award (km)
Golden Quadrilateral	5 846	5 846	0	0	–
North–South, East–West (Phases I and II)	7 142	6 414	461	39	267
Port connectivity	402	379	23	3	–
NHDP Phase III	12 109	6 634	3 308	79	2 167
NHDP Phase IV	13 203	1 585	4 668	57	6 950
NHDP Phase V	6 500	2 264	1 401	22	2 835
NHDP Phase VI	1 000	–	–	–	1 000
NHDP Phase VII	700	22	19	1	659
Total	46 902	23 144	9 880	201	13 878

Source Details available at <www.nhai.org/whatitis.asp>, last accessed on 5 January 2016

[2] Details available at www.nhai.org/whatitis.asp; last accessed on 5 January 2016

[3] Details available at http://pmgsy.nic.in/pmg31.asp#2, last accessed on 5 January 2016

planned to develop 1000 km of expressways under NHDP Phase IV and another 500 km of expressways comprising 200 km through budgetary allotments and 300 km through PPP mode (MoRTH 2012a).

Considering the significance of an adequate and well-defined rural road network and its impact on economic development, the Government of India launched the Pradhan Mantri Gram Sadak Yojana (PMGSY) on 25 December 2000 as a 100% centrally sponsored scheme. The PMGSY is a nationwide plan to provide good all-weather road connectivity to unconnected villages. The PMGSY aims to provide road connectivity to all

1. villages with a population of 1000 persons and above by 2003,
2. villages in hilly states with a population of 500 persons and above by 2007, and
3. tribal and desert area villages with a population of 500 persons and above by 2003 or with a population of 250 persons and above by 2007.[4]

The government is trying to set high and uniform technical and management standards and also facilitating policy development and planning at the state level to ensure sustainable management of rural roads.[5] By 2015[6], more than 197 210 habitations were connected and nearly 0.425 million km of road length was completed, with a total expenditure of ₹11.6 million. The scheme has completed 100 482 new connectivity works and 38 708 upgrade works.

The central government has announced the "Bharat Mala" project, which is envisaged across 13 states on a 5300 km stretch, starting from Gujarat, passing through Rajasthan, Punjab, Jammu and Kashmir, Himachal Pradesh, Uttarakhand, Uttar Pradesh, Bihar, Sikkim, Assam, Arunachal Pradesh, and ending across the Indo-Myanmar border of Manipur and Mizoram. The road will play strategic and economic roles in providing accessibility to border regions and improving border trade. The government also plans to connect the Bharat Mala with the "Sagar Mala" project, which is an initiative to modernize India's ports and develop the coastline. The Sagar Mala project aims to develop existing ports, industrial clusters, and hinterlands and provide connectivity through roads, rails, and inland and coastal waterways.

India had 159.5 million registered motor vehicles till 31 March 2012, implying a CAGR of 10.5% for 2002–12 (MoRTH 2014). Table 3 shows the total number of registered motor vehicles in India during 2009–12.

The break-up of motor vehicle population in Table 3 reflects the dominance of two-wheelers, with a share of 72% in the total vehicle population, followed by passenger cars at 13.5%, buses at 1.05%, goods vehicles at 4.8%, and other vehicles (a heterogeneous category that includes three-wheelers, trailers, and tractors) at about 8.25% in 2012 (as on 31 March).

Personalized modes constituting mainly two-wheelers and cars accounted for about 86% of the total number of motor vehicles in the country. In contrast to personal vehicles, the share of buses in the total number of registered vehicles has declined from 1.2% in 2001 (3% in 1981) to 1% in 2012, indicating slow growth in public transport modes (MoRTH 2013). Even though the share of public transport buses has been decreasing compared to other personal modes, the number of buses has been increasing steadily.

Table 3 Total number of registered motor vehicles in India (2009–12) (in million)

Year (as on 31 March)	All vehicles	Two-wheelers	Cars, jeeps, and taxis	Buses*	Goods vehicles	Others**
2009	114.951	82.402	15.313	1.486	6.041	9.710
2010	127.746	91.598	17.109	1.527	6.432	11.080
2011	141.866	101.865	19.231	1.604	7.064	12.102
2012	159.491	115.419	21.568	1.677	7.658	13.169

*Includes omni buses

**Others include tractors, trailers, three-wheelers (passenger vehicles)/LMV, and other miscellaneous vehicles that are not separately classified.

Source MoRTH (2014)

[4] Details available at http://pmgsy.nic.in/pmg31.asp#2, last accessed on 5 January 2016

[5] Details available at http://omms.nic.in/, last accessed on 5 January 2016

[6] Details available at http://omms.nic.in/, last accessed on 30 October 2015

The rise in the number of buses in recent years can be attributed to the Jawaharlal Nehru National Urban Renewal Mission (JNNURM), a massive city modernization scheme launched by the government in 2005 to improve the quality of life and infrastructure in cities. Under JNNURM, cities increased the number of public buses in their city transport services. A number of cities in the country have also taken up bus rapid transit systems (BRTS) to promote sustainable and world-class public transport systems.

The number of registered vehicles in five metropolitan cities—Delhi, Bengaluru, Chennai, Hyderabad, and Pune—accounted for 49.3% of the total registered vehicles among the 35 million-plus cities (MoRTH 2014). Pune recorded the highest CAGR of 12.94% for vehicles during 2001–11, followed by Hyderabad and Chennai, which recorded CAGRs of 12.3% and 10.64%, respectively. Figure 2 shows the rise in the registered vehicles during 1951–2012.

Key trends in the consumption of petroleum products

Within the transport sector, the road sector is the largest consumer of energy generated from petroleum products. As roads have emerged as the dominant means of transportation for both passenger and freight, the sector now accounts for the largest share (72%) of HSD consumption in India (MoPNG 2012). The consumption of HSD in the road transport sector increased from 33.74 MT in 2009/10 to 36.24 MT in 2010/11, registering an increase of nearly 7.4%. The consumption of motor spirit in 2010/11 was 14.2 MT compared to 12.82 MT in 2009/10 (MoPNG 2011). A recent phenomenon has been the increased consumption of liquefied petroleum gas (LPG), owing to the increased use of cleaner fuel in three-wheelers. The consumption of LPG by the transport sector during 2009/10 was 0.225 MT, constituting 1.7% of the overall LPG consumption in the country. The consumption remained almost the same during 2010/11 (MoPNG 2012). The road sector is highly energy intensive compared to railways and water transport. Additionally, freight movement by road is more energy intensive than passenger movement. Figure 3 shows the share of different petroleum products consumed in the road transport sector during 2011/12.

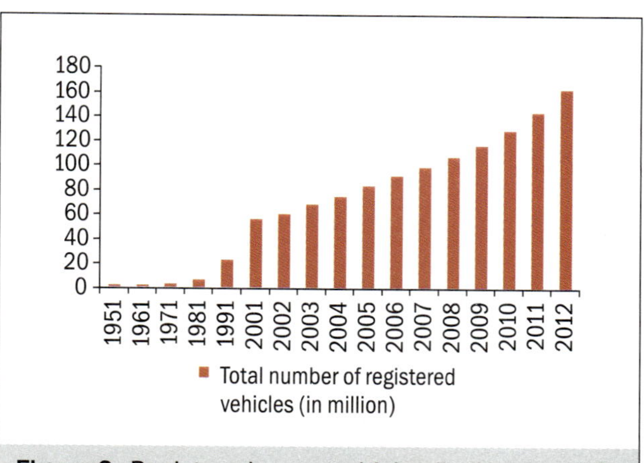

Figure 2 Registered motor vehicles (in thousand) for 1951–2012

Source MoRTH (2013)

Urban transport

National Urban Transport Policy

The National Urban Transport Policy (NUTP), formulated by the Ministry of Urban Development (MoUD) in 2006, aims to guide state-level action plans for meeting the ever-increasing mobility needs of the urban population. The NUTP envisages to ensure safe, affordable, quick, comfortable, reliable, and sustainable access to the growing number of

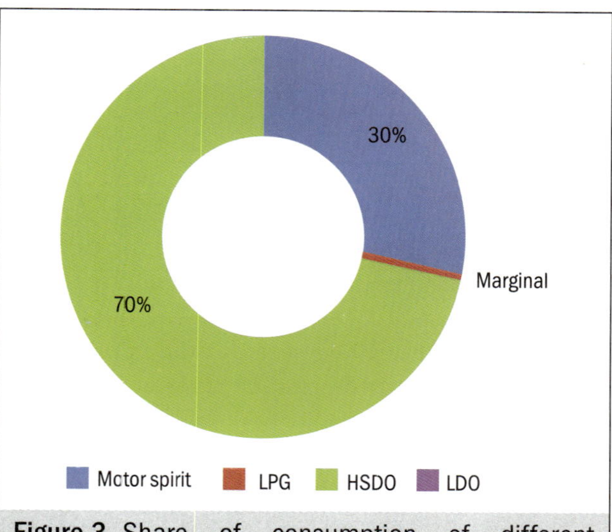

Figure 3 Share of consumption of different petroleum products in the road transport sector

HSDO – high-speed diesel oil

Source MoPNG (2011, 2012)

city residents. The policy document emphasizes on the need for integrated land use and transportation planning, increased priority to public transport, non-motorized transport, and improved parking facilities. The policy envisages the setting up of Unified Metropolitan Transport Authority in cities with a million-plus population.

Jawaharlal Nehru National Urban Renewal Mission

The Jawaharlal Nehru National Urban Renewal Mission was launched in 2005 for a 7-year period (up to March 2012) as a reform-based programme to fast-track planned infrastructure development and, hence, improve the quality of life in the "mission cities". For the implementation of the programme, 67 cities/urban agglomerations (UAs) were selected. Cities/UAs with 4 million-plus population as per the 2001 census were classified as Class A cities, while cities/UAs with 1 million-plus but less than 4 million population as per the 2001 census were classified as Class B cities. Under Class C, selected cities/UAs (state capitals and other cities/UAs of religious/historic and touristic importance) were included. The development of urban transport infrastructure was one of the important aspects of the mission.

Under the JNNURM, 515 projects across 31 states/union territories at the cost of ₹580.38 billion were sanctioned till December 2009, of which 85 were roads/flyovers and 34 were urban transport projects. In addition, 44 more projects were approved, bringing the total number of sanctioned projects to 559 since the inception of JNNURM (GoI 2012). Under the JNNURM, 21 BRTS in 10 cities across six states have been taken up at the cost of ₹52.04 billion. In addition, 20 transport infrastructure projects, such as parking facilities, information technology services, traffic infrastructure, and transit management centres, have also been sanctioned under this scheme at the total cost of ₹16.66 billion. The mission has sanctioned another 97 projects for the improvement of road infrastructure across 23 cities at the cost of ₹81.22 billion (MoUD 2011). Till July 2013, 56 projects related to roads/flyovers/road overbridges, 5 mass rapid transport projects, and 10 other urban transport projects were completed under the JNNURM.

The current government has launched the Atal Mission for Rejuvenation and Urban Transformation (AMRUT), a new initiative for the development of infrastructure in the country, under which 500 cities will be selected. Urban transport features prominently under the larger umbrella of the project and would focus on ferry vessels for inland waterways (excluding port/bay infrastructure), buses, footpaths/walkways, sidewalks, foot overbridges, facilities for non-motorized transport (for example, bicycles), multi-level parking, and BRTS. AMRUT has a budget of ₹50 000 crore for 5 years from 2015/16 to 2019/20 and will be operated as a centrally sponsored scheme. The project fund would be 90% of the annual budgetary allocation for 2015/16, and incentives for reforms will be offered only from 2016/17 onwards. Another flagship programme of the current government is Smart Cities. The prospective smart cities would be chosen through a "City Challenge Competition", which would evaluate the preparedness and capacity of cities sponsored by state governments. Twenty cities would be taken up for a Union grant of ₹100 crore per year for initially 5 years.

Railways

Brief overview

The Indian Railways is one of the largest rail networks in the world. Till 31 March 2014, it had a network of 65 808 route kilometres with a running track length of 89 919 km (MoR 2015). The rail network consists of largely broad gauge (86.6%), followed by metre gauge (9.8%) and narrow gauge (3.6%). The metre gauge system and a part of the narrow gauge system are gradually converted to broad gauge under the unigauge programme. The Indian Railways has a total electrified route of 21 614 km and an electrified running track of 39 661 km (MoR 2015).

The Indian Railways has a fleet of 66 392 passenger service vehicles, 245 267 wagons, and about 10 499 locomotives, including 43 steam, 5633 diesel, and 4823 electric locomotives, carrying in 2013/14 about 8397 million originating passengers and 1058.81 MT of originating freight through 7112 stations (MoR 2015). Table 4 gives details of track length (in km) by gauge.

The total freight tonnage carried by the Indian Railways was 1058.81 MT in 2013/14, of which the revenue earning freight traffic was 1051.64 MT. The Indian Railways carried 666.728 billion net tonne kilometres of total freight,

Table 4 Length of traction (in km) by gauge (as on 31 March 2014)

Gauge	Route (km)	Running track (km)	Total track (km)
Broad gauge (1 676 mm)	58 177	81 914	107 513
Metre gauge (1 000 mm)	5 334	5 708	6 688
Narrow gauge (762 mm and 610 mm)	2 297	2 297	2 579
Total	65 808	89 919	116 765

Source MoR (2015)

including 665.810 billion of revenue earning traffic in 2013/14. The freight earnings amounted to ₹915.7 billion in 2013/14 (MoR 2015).

During 2013/14, the railways carried 8397 million passengers as against 8421 million passengers in 2012/13, thus registering a volume decrease of about only 0.28%. Passenger kilometres grew by 5.5% from 1 098.1 billion in 2012/13 to 1 158.74 billion in 2013/14 (MoR 2014). Figure 4 shows the trend of originating passenger and passenger kilometre growth in railways from 1980/81 to 2013/14. Figure 5 shows the trend of originating freight and freight tonne kilometre growth in Indian Railways.

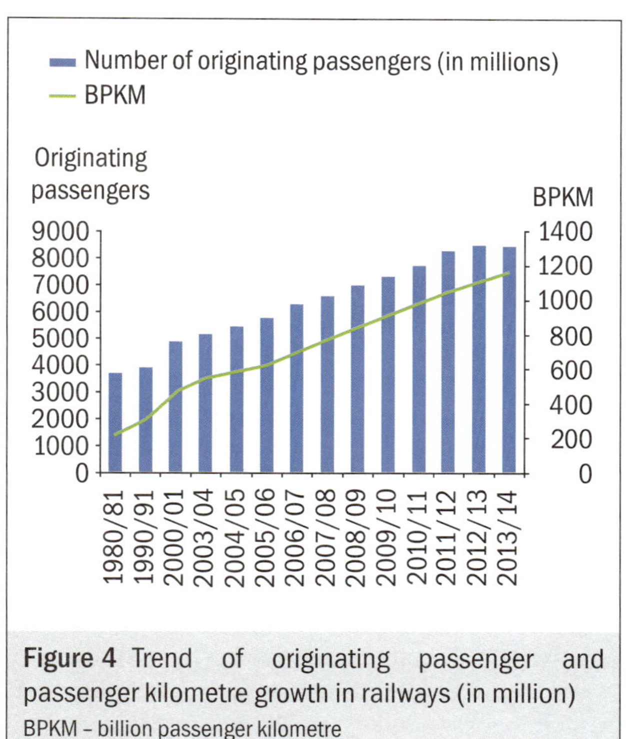

Figure 4 Trend of originating passenger and passenger kilometre growth in railways (in million)
BPKM – billion passenger kilometre
Source MoR (2015)

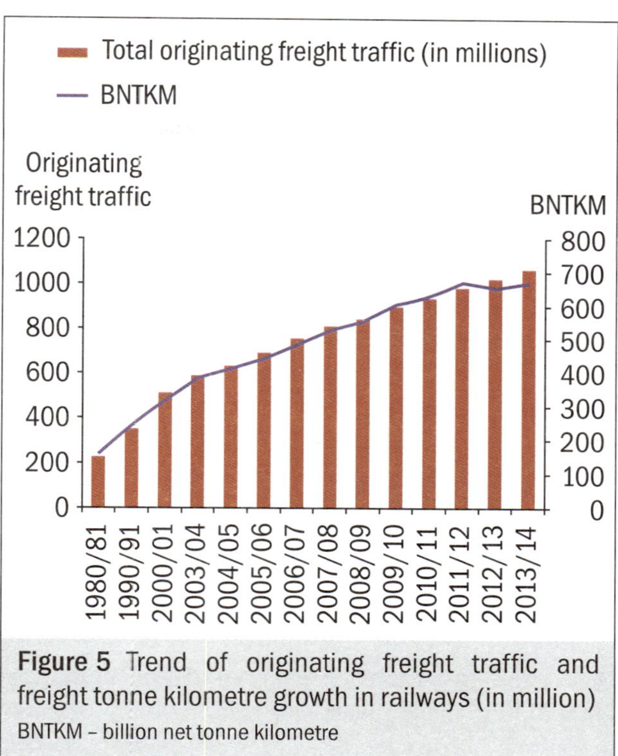

Figure 5 Trend of originating freight traffic and freight tonne kilometre growth in railways (in million)
BNTKM – billion net tonne kilometre
Source MoR (2015)

The market share of railways has been gradually decreasing because of the capacity constraints faced by the Indian Railways. The lack of investments has aggravated the situation further. High freight rates have been the other main reason for the declining share of freight traffic as they are driving away customers. High freight rates have been imposed to make up for the low passenger fares. The Indian Railways posted a loss of ₹3206.71 million in passenger business in 2013/14 (MoR 2015).

The total investment in the railways for the Eleventh Five-year Plan stood at ₹2033 billion, which was ₹299 billion less than the planned outlay. The Twelfth Five-year Plan has projected an investment of ₹5192 billion in the railways between 2012 and 2017.

Key trends in energy consumption

Rail transportation is six times more energy efficient and four times more economical than road transportation.[7] The social cost in terms of

[7] Details available at http://pib.nic.in/feature/fe0199/f1101991.html

environmental damage or degradation is significantly low in rail transportation. The consumption of electricity for locomotive services increased from 13 853 million kWh in 2012/13 to 14411 million kWh in 2013/14. The consumption of HSD for locomotive services increased from 2699.6 million litres in 2011/12 to 2780.9 million litres in 2013/14 (MoR 2015). For purposes other than locomotive services, consumption of HSD was 30.74 million litres in 2012/13 and increased to 33.38 million litres in 2013/14 (MoR 2015). Direct consumption of coal by the railways is almost negligible due to the phasing out of steam locomotives. The consumption of coal for locomotives services was 0.001 MT in 2013/14 (MoR 2015). Figure 6 shows the growth in the consumption of electricity and HSD by the railways for locomotive services from 1990/91 to 2013/14.

The Indian Railways has been striving to take environment-friendly initiatives and save fuel over the years. As a result of these initiatives, locomotives have been upgraded with various innovative technologies such as microprocessor control, roller suspension bearing, auxiliary power unit, and low idle features to reduce the fuel consumption. Other technology initiatives that helped in reducing fuel consumption and carbon emission include the common rail direct injection system introduced in diesel engines, electronic fuel injection system, REMMLOT systems, and guidance optimized locomotive driving systems for better control. The Indian Railways is harnessing renewable energy wherever possible and it has harnessed about 10 MWp of solar energy through different applications till date. Railways aims to source at least 10% of its total energy demand from wind and solar energy (MoR 2015).

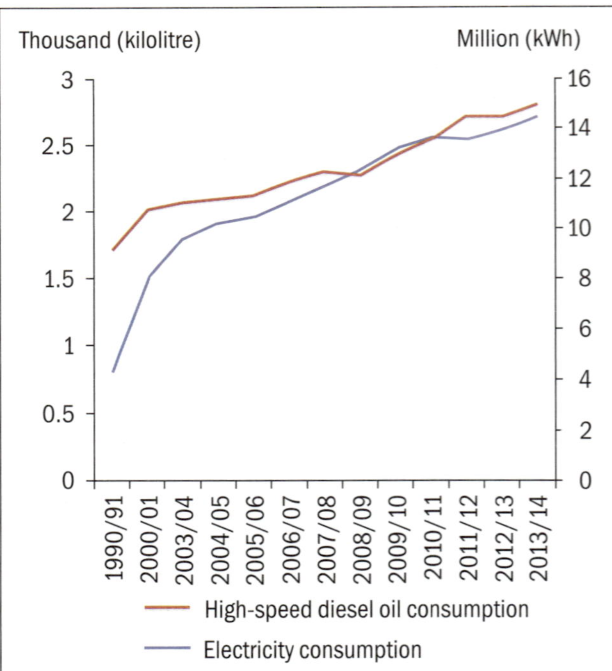

Figure 6 Growth in the consumption of electricity and high-speed diesel by the railways for locomotive services
Source MoR (2015)

Ports, shipping, and inland water transport

Brief overview

Maritime transport handles more than 90% by volume and 70% by value of India's trade. There are 12 major ports and 187 minor/intermediate ports along the 7517 km long coastline of the country (MoS 2014a). The major and non-major[8] ports together handled 1052.52 MT of cargo in 2014/15 as compared to 980.49 MT in 2013/14, registering a growth of 7.3%. The growth rates of cargo handled at major and non-major ports in 2014/15 were 4.6% and 12.13%, respectively, compared to 1.8% and 7.5% in 2013/14 (MoS 2015). The major ports handled 581.33 MT of traffic during 2014/15 compared to 555.50 MT during 2013/14 (MoS 2014a, 2015). The non-major ports handled about 416.97 MT in 2013/14, which increased to 471.19 MT in 2014/15 (MoS 2014b). These ports were responsible for about 45% of the country's total maritime freight traffic in 2014/15. The growth in cargo handled by non-major ports in 2014/15 was 13.0% compared to 7.5% in 2013/14.

The major commodities handled at Indian ports include crude oil and petroleum products, dry cargo bulks (coal, iron ore, and steel), dry cargo minor bulk (metals and minerals, forest and steel products, sugar), and containers. POL (petroleum, oil, and lubricants) was the largest commodity group handled at major ports in terms of composition, with a share of 32.5%. It was followed by other cargo (20.5%), container traffic (20.5%), coal (20.3%), iron ore (2.9%), fertilizer and fertilizer raw materials (FRM) (2.8%), and food grain

[8] Major ports are ports under Central Government jurisdiction and minor ports are ports under state government jurisdiction.

(0.5%) (MoS 2015). The Indian ports handled 13.88 million passengers during 2010/11 compared to 13.98 million passengers in 2009/10. In 2010/11, the major ports handled about 344000 passengers, while the minor ports handled about 13.53 million passengers.

The Eleventh Five-year Plan had an outlay of ₹303 billion for the ports sector, but the total expenditure was only ₹63 billion. The outlay for the Twelfth Five-year Plan for the ports sector is ₹158 billion. For now, the government is not building any new ports, and the private sector is mainly engaged in the construction of new ports.

Shipping

India is among the 20 leading countries in the world owning merchant fleet. It had a fleet strength of 1204 vessels with gross registered tonnage (GRT) of 10.31 million in 2014 (MoRTH 2014) (as on 31st December 2014) compared to 1199 vessels and 10.38 million GRT in 2013. This reflects an addition of five vessels with a decrease of 0.07 million GRT (MoRTH 2014). Of the total fleet, 358 ships with 9.09 million GRT cater to India's overseas trade and the rest to coastal trade (MoS 2015).

The shipping sector had an outlay of ₹150 billion in the Eleventh Five-year Plan and against this the expenditure was ₹98 billion. The Twelfth Five-year Plan has an outlay of ₹101 billion for the shipping sector.

Inland waterways

India has about 14500 km of navigable waterways, which comprise rivers, canals, backwaters, and creeks. Although inland water transport (IWT) is the most cost-effective, fuel-efficient, and environment-friendly mode of transportation for bulk cargo, its modal share of freight traffic is only 0.28% (TERI 2009). The total cargo movement in India's waterways (comprising the three national waterways and waterways in Goa and Maharashtra, which carry most of the cargo) decreased from 412.52 lakh tonnes in 2012/13 to 322.63 lakh tonnes in 2013/14 (MoRTH 2014). The decline in the total cargo movement in 2013/14 over 2012/13 is mainly attributed to the decline in cargo movement in Goa's waterways due to a ban on iron ore mining/export imposed by the Supreme Court of India from 5 October 2012. Subsequently, there was a 21.04% decrease in tonne kilometres through inland waterways in 2013/14 (24183 tonne km in lakh) as compared to 2012/13 (30629 tonne km in lakh).

The Sagar Mala project, announced in 2003 by the then Prime Minister, has been revived by the present government. Under this project, 10 coastal economic regions will be developed along the coastline to create industrial clusters supported by inward linkages through intermodal (rail, land, and inland waterways) freight options for the smooth evacuation of cargo to and from ports. The development of port-based industrial parks, captive industries, and ancillary facilities such as ship repair, shipbuilding, ship recycling, logistics parks, warehousing, maritime zones/services, offshore storage, drilling platforms, bunkering, container freight stations, and industries requiring significant import of raw materials and industries with large export potential will also be carried out.

Among the new developments in IWT, the "Jal Marg Vikas" project was announced in the 2014 budget. This project will undertake a capacity augmentation of the National Waterway 1 from Haldia to Allahabad ensuring navigation of 1500 dead weight tonnes (DWT) vessels from Haldia to Allahabad through waterways. An integrated waterways transportation grid with the aim to provide road, rail, and port connectivity to all existing and proposed national waterways has also been proposed. The five national waterways and Indo-Bangladesh Protocol Route will also be developed under the same proposal. For this project, an investment of ₹22763 crore has been estimated. Table 5 gives the details of various national waterways in India.

Key trends in energy consumption

Water transport is one of the most environment-friendly, cost-effective, and efficient modes of transportation. During 2010/11, the shipping sector consumed 0.56 MT of HSD, 0.004 MT of light diesel oil (LDO), and 0.78 MT of furnace oil (MoPNG 2012). During 2011/12, the shipping sector consumed 0.54 MT of HSD, 0.002 MT of LDO, and 0.37 MT of furnace oil. Figure 7 shows the consumption of petroleum products by the shipping sector during 2011/12.

Aviation

Brief overview

The aviation sector in India has seen phenomenal growth in the recent years even though there were signs of slowdown in passenger traffic in 2008 due to

Table 5 Details of National Waterways in India

National waterways	Waterway stretch	Length (km)	Year of declaration
NW 1	The Ganga-Bhagirathi-Hooghly river system from Haldia to Allahabad	1 620	1986
NW 2	The Brahmaputra River from Dhubri to Sadiya	891	1988
NW 3	The West Coast Canal from Kottapuram to Kollam along with Udyogamandal and Champakara canals	205	1993
NW 4	The Kakinada-Puducherry stretch of canals and the Kaluvelly Tank, Bhadrachalam-Rajahmundry stretch of River Godavari and Wazirabad-Vijayawada stretch of River Krishna	1 078	2008
NW 5	The Talcher-Dhamra stretch of the Brahmani-Kharsua-Dhamra river system, Geonkhali-Charbatia stretch of the East Coast Canal, Charbatia-Dhamra stretch of River Matai, and Mangalgadi-Paradip stretch of Mahanadi delta rivers	588	2008

NW – national waterway
Source MoS (2014a)

the steep rise in fuel costs coupled with the global economic recession. The scheduled domestic passenger traffic volume declined from 44.4 million in 2007/08 to 39.5 million in 2008/09, and the total domestic cargo traffic volume declined from 368 415 tonnes to 341 763 tonnes during the same period (DGCA 2012). Signs of recovery were visible in the second half of 2009. The total number of domestic passengers travelling by air in 2013/14 increased to 61.7 million from 58.7 million in 2012/13 and recorded 59.139 billion passenger kilometres on Indian carriers. The total domestic freight carried in 2013/14 was 0.39 MT as against 0.36 MT in 2012/13 (DGCA 2015). Figure 8 shows the trend in the growth of domestic passengers and cargo in the aviation sector from 2004/05 to 2013/14.

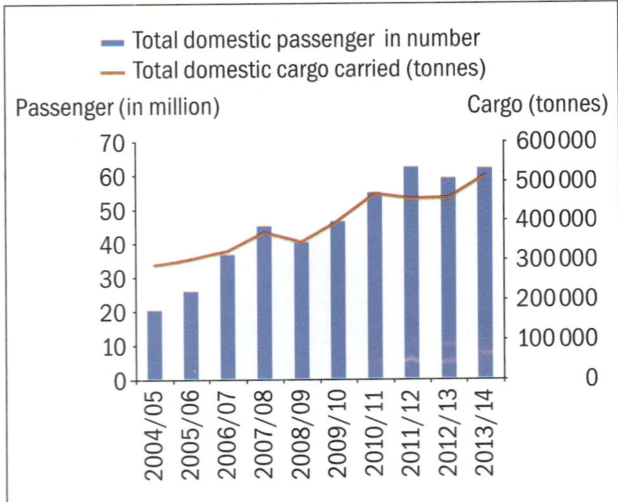

Figure 8 Trend in the growth of domestic passengers and cargo in the aviation sector
Source DGCA (2015)

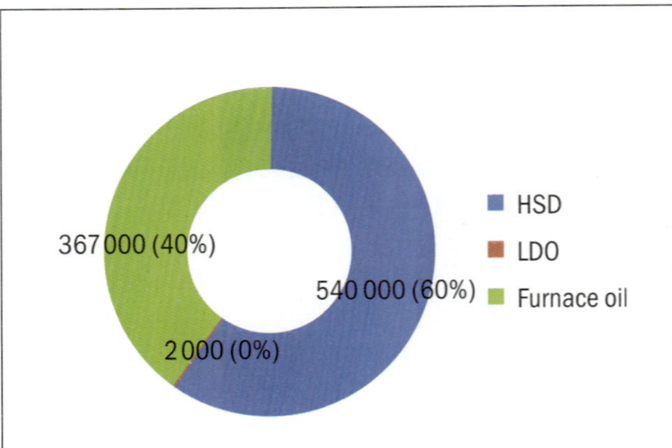

Figure 7 Consumption of petroleum products in shipping (tonnes, %) during 2011/12
Source MoPNG (2012)

In 2011/12, there were nine domestic scheduled airlines and 76 international airlines operating in India. In 2012, 147 companies were holding non-scheduled operator's permit (as on 31 December) compared to 134 companies in 2011 (MoCA 2013). The contribution of civil aviation and its ancillary sectors to India's GDP was about 1.5% in 2009. The sector is expected to generate about 9.95 million jobs within the country (MoCA 2012).

Airports

The Airports Authority of India (AAI) manages 125 airports, which include 68 operational airports (11 international airports, 8 customs, and 49 domestic airports), 26 civil enclaves[9] (3 international, 4 customs, and 19 domestic), and 31 non-operational domestic airports. The AAI also provides Air Traffic Management Services over Indian air space and adjoining oceanic areas with the help of ground installations at all airports and 25 other locations to ensure safe aircraft operations. The total number of passengers handled by Indian airports from 2005/06 to 2010/11 grew at the average rate of 11.5% and cargo grew at the average annual rate of 11%. The passenger handling capacity grew threefold from 72 million in 2005/06 to 233 million in 2010/11. Four international airports were successfully completed through the PPP mode, including greenfield development of Hyderabad and Bengaluru international airports and the modernization of Delhi and Mumbai international airports.

The investment in the Eleventh Five-year Plan was projected at ₹3613.8 million as against the original estimate of ₹3096.8 million.[10] The Eleventh Five-year Plan included the development of three greenfield airports and six existing airports in the North-East, as well as upgradation of airports in Jammu, Dehradun, Agatti, and Port Blair, under the scheme for modernization of 35 non-metro airports. During the Eleventh Five-year Plan, the cargo handling capacity increased from 0.5 MT in 2005/06 to 3.3 MT in 2010/11. During the plan period, four international airports (Hyderabad, Bengaluru, Delhi, and Mumbai) were completed through the PPP mode. Also during the same period, the Airport Economic Regulatory Authority (AERA) was established to safeguard the interests of users and service providers at Indian airports.[11] In the Twelfth Five-year Plan, the focus will be on more airports, higher capacity, supporting infrastructure, finance, and human resources to facilitate growth in the sector. The total investment in the airport sector in the Twelfth Five-year Plan is estimated to be around ₹8771.4 million. The AAI has identified 15 operational airports for taking up operation and maintenance, and five greenfield airports, including Navi Mumbai, Goa, Kannur, Chandigarh, and Kota, have been identified for development through the PPP mode in the Twelfth Five-year Plan. The Draft National Civil Aviation Policy, 2015 also proposes development of no-frills airports based on the demand by reviving the existing airstrips at the cost of ₹50 crore.

Key trends in energy consumption

The consumption of aviation turbine fuel (ATF) has increased rapidly since 2000/01 when the sector experienced phenomenal growth in passenger traffic (MoPNG 2012). The use of ATF increased from 4.63 MT in 2009/10 to 5.32 MT in 2011/12. The overall consumption of ATF by the aviation sector in 2015 was 5.3 MT,[12] recording a 1.4% growth over the 2013 values (MoPNG 2012, 2015).

Key legislations and regulations

Roads

- Road Transport Corporation Act, 1950
- National Highways Act, 1956
- National Highways Rules, 1957
- Motor Vehicles Act, 1988 (under amendment)
- Central Motor Vehicle Rules, 1989
- National Highways Act, 1995
- National Highways Authority of India Act, 1998
- Central Road Fund Act, 2000
- Control of National Highways (Land and Traffic) Act, 2002 (1)
- National Highways Tribunal Rules, 2003 (1)
- Carriage by Road Act, 2007
- National Highways Fee (Determination of Rates and Collection) Rules, 2008
- Carriage by Road Rules, 2011 (1)
- Motor Vehicles (Amendment) Bill, 2012

Railways

- Railway Claims Tribunal Act, 1987
- The Railways Act, 1989
- Railways (Amendment) Act, 2008
- The Metro Railways (Construction of Works) Act, 1978
- Delhi Metro Railway (Operation and Maintenance) Act, 2002
- The Indian Railway Board Act, 1905

[9] Civil enclave is a part of airbase used for civil aviation purposes, instead of building a new civilian airport.
[10] Details available at www.infrastructure.gov.in/pdf/inv-infra.pdf
[11] Details available at http://planningcommission.gov.in/aboutus/committee/wrkgrp12/transport/report/wg_civil.pdf
[12] Provisional figures

Ports, shipping, and inland water transport[13]

- Indian Ports Act, 1908
- The Inland Vessels Act, 1917
- Merchant Shipping Act, 1958
- Major Ports Trust Act, 1963
- Inland Waterways Authority of India Act, 1985
- Multimodal Transportation of Goods Act, 1993
- Amendment of Multimodal Transportation of Goods Act, 1993
- Indian Maritime University Act, 2008
- Major Ports Regulatory Authority Bill, 2009
- Shipping Trade Practices (STP) Bill, 2010

Civil Aviation

- Aircraft Act, 1934
- Aircraft Rules, 1937
- Airports Authority of India Act, 1994, as amended by the Amendment Act, 2003
- Air Corporations (Transfer of Undertakings and Repeal) Act, 1994
- The Airports Economic Regulatory Authority of India Act, 2008
- AERA, Appellate Tribunal (Forms for Application and Appeal and Fee) Rules, 2010
- Aircraft (Security) Rules, 2011
- Airports Authority of India (Major Airports) Development Fees Rules, 2011

[13] Details available at http://shipping.nic.in/index1.p?linkid=164&langid=1

Delhi Metro

The Delhi Metro Rail Corporation (DMRC) started its operations on 25 December 2002 and it is the world's 13th largest metro system in terms of length. Planned and built in different phases, the metro is still expanding its network. Phases 1 and 2 have been completed, and phase 3 is now under way. Phase I covering of 65 km of lines was completed in 2005, and phase 2 covering 125 km was completed in the next 4.5 years. The network currently consists of six operational lines and 141 stations along 193 operational kilometres. It has additional six stations on the Airport Express link, which provides rapid metro connectivity between New Delhi and Indira Gandhi International Airport. The DMRC is expanding its network to better connect Delhi with the neighbouring cities of Noida, Ghaziabad, Faridabad, Bahadurgarh, and Gurgaon. Delhi Metro has been successful in achieving very high ridership levels. In 2014/15, the DMRC carried 870.6 million passengers. It has been largely successful in raising the standard of rail-based mass transit system in the country, and its success story has inspired other cities in India to invest in metro rail infrastructure.

Delhi Metro has been instrumental in changing the opinion of people about public transport systems and has thereby contributed significantly in reducing vehicular traffic and carbon emissions in the city. A 2011 study by the Central Road Research Institute reveals that Delhi Metro has been responsible for taking off about 1.17 lakh vehicles from Delhi roads. The United Nations has certified Delhi Metro as the first metro rail and rail-based system in the world to get carbon credits. The system has been responsible for reducing global warming by lowering considerable carbon emissions by facilitating modal shift to a sustainable model. In 2014/15, the DMRC was responsible for reducing about 0.16 million tonnes of CO_2. Delhi Metro is also the first railway project in the world to claim carbon credits for regenerative braking. The regenerative braking, modal shift, and energy efficiency projects of the DMRC are expected to reduce approximately 5.7 lakh tonnes of CO_2 emissions annually. These projects have been registered by the DMRC under the Clean Development Mechanism (CDM). In 2008, the DMRC became the first metro or railway project in the world to be registered by the United Nations under CDM, which enabled it to claim carbon credits. The carbon mitigation efforts adopted by the DMRC and their results have led the Gold Standard Foundation to certify Delhi Metro. The emission reduction per passenger kilometre achieved by Delhi Metro is mainly due to its highly improved resource efficiency compared to other modes. The improvements in environment attributed to Delhi Metro were possible through the establishment of an efficient public transportation system and the use of electricity as the energy source.

Sources Details available at www.delhimetrorail.com; DMRC (2014); Details available at available at https://cdm.unfccc.int/Projects/Validation/DB/B0HZXF5XRHA0LZGP3DHW7QHPDSSA61/view.html;DMRC (2015)

References

DMRC (Delhi Metro Rail Corporation Ltd). 2014. ***Annual Report 2013-14***. New Delhi: DMRC

DMRC (Delhi Metro Rail Corporation Ltd). 2015. Environmental Policy: ***A Way Forward 2015***. New Delhi: DMRC

DGCA (Directorate General of Civil Aviation). 2012. ***Time Series Data of Domestic Traffic Carried by Indian Operators in the Last 25 years***. New Delhi: DGCA

Directorate General of Civil Aviation (DGCA). 2015. Details available at http://dgca.nic.in/reports/stat-ind.htm, last accessed on 29 October 2015

GoI (Government of India). 2012. ***UPA Report to People 2011-2012***. New Delhi: Ministry of Information and Broadcasting, Government of India

MoCA (Ministry of Civil Aviation). 2012. ***Annual Report 2011/12***. New Delhi: MoCA, Government of India

MoHF (Ministry of Home Affairs). 2011. ***Census 2011***. New Delhi: MoHF

MoCA (Ministry of Civil Aviation). 2013. ***Annual Report 2012-13***. New Delhi: MoCA, Government of India

MoPNG (Ministry of Petroleum and Natural Gas). 2011. ***Basic Statistics on Indian Petroleum & Natural Gas***. New Delhi: MoPNG, Government of India

MoPNG (Ministry of Petroleum and Natural Gas). 2012. ***Basic Statistics on Indian Petroleum & Natural Gas***. New Delhi: MoPNG, Government of India

MoPNG (Ministry of Petroleum and Natural Gas). 2015. ***Indian Petroleum and Natural Gas Statistics 2014-15***. New Delhi: MoPNG, Government of India.

MoR (Ministry of Railways). 2014. ***Indian Railways Annual Statistical Statements 2012-13***. New Delhi: MoR, Government of India

MoR (Ministry of Railways). 2015. ***Indian Railways Annual Statistical Statements 2013-14***. New Delhi: MoR, Government of India

MoRTH (Ministry of Road Transport and Highways). 2012a. ***Annual Report 2011-12***. New Delhi: MoRTH, Government of India

MoRTH (Ministry of Road Transport and Highways). 2012b. ***Report of the Working Group on the Central Roads Sector***. New Delhi: MoRTH, Government of India

MoRTH (Ministry of Road Transport and Highways). 2013. ***Annual Report 2012/13***. New Delhi: MoRTH, Government of India

MoRTH (Ministry of Road Transport and Highways). 2014. ***Basic Road Transport Statistics***. New Delhi: MoRTH, Government of India

MoRTH. 2015. ***Basic Road Statistics of India 2013-14***. New Delhi: MoRTH, Government of India

MoS (Ministry of Shipping). 2014a. ***Annual Report 2013-2014***. New Delhi: MoS, Government of India

MoS (Ministry of Shipping). 2014b. ***Update on Indian Ports (31 March 2014)***. New Delhi: MoS, Government of India

MoS (Ministry of Shipping). 2015. ***Update on Indian Ports (31 March 2015)***. New Delhi: Ministry of Shipping, Government of India

MoUD (Ministry of Urban Development). 2011. ***Jawaharlal Nehru Urban Renewal Mission Annual Update***. New Delhi: MoUD, Government of India

Planning Commission. 2007. ***Report of the Working group on Roads for the Eleventh Five-year Plan***. New Delhi: Planning Commission, Government of India

RITES. 2012. ***Total Transport System Study on Traffic Flows and Modal Costs (Highways, Railways, Airways and Coastal Shipping)***. New Delhi: Planning Commission, Government of India

TERI (The Energy and Resources Institute). 2008. ***Pathway to a Low Carbon Economy***. New Delhi: TERI

TERI (The Energy and Resources Institute). 2009. ***The Energy Data Directory and Yearbook 2009***. New Delhi: TERI

Bibliography

CORE (Central Organization for Railway Electrification). 2012. Details available at www.core.indianrailways.gov.in/, last accessed on 30 October 2012

DGCA (Directorate General of Civil Aviation). 2009. ***India Air Transport Statistics 2008/09***. New Delhi: DGCA

EXIM-Bank of India. 2010. ***Indian Shipping Industry: A Catalyst for Growth***. New Delhi: Export-Import Bank of India

IWAI (Inland Waterways Authority of India). 2009. ***Annual Report 2008/09***. New Delhi: IWAI, Government of India

MoCA (Ministry of Civil Aviation). 2010. ***Annual Report 2009/10***. New Delhi: MoCA, Government of India

MoCA (Ministry of Civil Aviation). 2011. ***Report of the Working Group on Civil Aviation for the 12th Five Year Plan (2012-17)***. New Delhi: Planning Commission

MoF (Ministry of Finance). 2010. ***Economic Survey 2009/10***. New Delhi: MoF, Government of India

MoF (Ministry of Finance). Various Years. ***Economic Survey***. New Delhi: Economic Division, MoF, Government of India

MoR (Ministry of Railways). 2010a. ***Annual Report and Accounts 2008/09***. New Delhi: MoR, Government of India

MoR (Ministry of Railways). 2010b. ***White Paper on India Railways***. New Delhi: MoR, Government of India

MoR (Ministry of Railways). 2012a. ***Indian Railways Annual Report and Accounts 2010-11***. New Delhi: MoR, Government of India

MoR (Ministry of Railways). 2012b. ***Indian Railways Year Book 2010-11. New Delhi***: MoR, Government of India

MoR (Ministry of Railways). 2012c. ***Working Group Report for the 12th Plan-Railway Sector***. New Delhi: Planning Commission

MoR (Ministry of Railways). 2013. ***Indian Railways Annual Statistical Statements 2011-12***. New Delhi: MoR, Government of India

MoR (Ministry of Railways). 2014. ***Indian Railways Year Book 2013-14***. New Delhi: MoR, Government of India

MoR (Ministry of Railways). Various Years. ***Indian Railways Year Book***. New Delhi: MoR, Government of India

MoRD (Ministry of Rural Development). 2011. ***Working Group on Rural Roads 12th Five-year Plan***. New Delhi: Planning Commission

MoRTH (Ministry of Road Transport and Highways). 2012a. ***Basic Road Statistics of India 2008–09, 2009–10 and 2010–11***. New Delhi: MoRTH, Government of India

MoRTH (Ministry of Road Transport and Highways). 2012b. ***Indian Shipping Statistics***. New Delhi: MoRTH, Government of India

MoRTH (Ministry of Road Transport and Highways). 2012c. ***Road Transport Year Book (2009–10 and 2010–11)***. New Delhi: MoRTH, Government of India

MoRTH (Ministry of Road Transport and Highways). 2012d. ***Statistics of Inland Water Transport***. New Delhi: MoRTH, Government of India

MoRTH (Ministry of Road Transport and Highways). 2013. ***Road Transport Year Book 2011–12***. New Delhi: MoRTH, Government of India

MoRTH (Ministry of Road Transport and Highways). 2014a. ***Indian Shipping Statistics***. New Delhi: MoRTH, Government of India

MoRTH (Ministry of Road Transport and Highways). 2014b. ***Statistics of Inland Water Transport***. New Delhi: MoRTH, Government of India

MoRTH (Ministry of Road Transport and Highways). Various Years. ***Basic Road Transport Statistics***. New Delhi: MoRTH, Government of India

MoS (Ministry of Shipping). 2009a. ***Indian Shipping Statistics 2007***. New Delhi: MoS, Government of India

MoS (Ministry of Shipping). 2009b. ***Basic Port Statistics of India, 2006/07***. New Delhi: MoS, Government of India

MoS (Ministry of Shipping). 2010. ***Annual Report 2009/10***. New Delhi: MoS, Government of India

MoS (Ministry of Shipping). 2012a. Ministry of Shipping: Ports Wing. Details available at http://shipping.gov.in/index1.php?lang=1&level=0&linkid=16&lid=64, last accessed on 31 October 2012

MoS (Ministry of Shipping). 2012b. ***Planning Commission Working Group Report for Ports for the 12th Five Year Plan***. New Delhi: Planning Commission

MoS (Ministry of Shipping). 2015. ***Annual Report 2014–15***. New Delhi: Government of India.

MoSPI (Ministry of Statistics and Programme Implementation). 2009. ***Energy Statistics 2008***. New Delhi: MoSPI, Government of India

MoSPI (Ministry of Statistics and Programme Implementation). Various Years. ***National Accounts Statistics***. New Delhi: MoSPI, Government of India

MoSPI (Ministry of Statistics and Programme Implementation). 2009. ***Energy Statistics 2008***. New Delhi: MoSPI, Government of India

MoSPI (Ministry of Statistics and Programme Implementation). Various Years. ***National Accounts Statistics***. New Delhi: MoSPI, Government of India

NHAI (National Highways Authority of India). 2014. Details available at www.nhai.org/roadnetwork.htm, last accessed on 30 September 2014

Planning Commission. n.d. ***The Working Group Report on Road Transport for the Eleventh Five Year Plan***. New Delhi: Government of India

TERI (The Energy and Resources Institute). 2010. ***The Energy Data Directory and Yearbook 2009***. New Delhi: TERI

TABLE 9.1 Key data on transport: 2000/01–2013/14

Parameter	Unit	2000/01	2001/02	2002/03	2003/04	2004/05	2005/06	2006/07	2007/08	2008/09	2009/10	2010/11	2011/12	2012/13	2013/14
Railways															
Gross revenue receipts	Million rupee	360 110	393 578	427 415	449 106	490 466	563 155	647 858	732 766	816 590	892 293	966 810	1062 453	1 261 804	1 432 139
Route/length (total)	km	63 028	63 140	63 122	63 221	63 465	63 332	63 327	63 273	64 015	63 974	64 460	64 600	65 436	65 808
Route/length (electrified)	km	14 856	15 994	16 272	16 776	17 495	17 907	17 786	18 274	18 559	18 927	19 607	20 275	20 884	21 614
Station	Number	6 843	6 931	6 906	7 031	7 133	6 974	6 909	7 025	7 030	7 083	7 133	7 146	7 172	7 112
Locomotive (total)	Number	7 566	7 739	7 681	7 818	7 916	8 025	8 153	8 330	8 592	8 889	9 213	9 549	9 956	10 499
Steam	Number	54	53	52	45	44	44	43	44	43	42	43	43	43	43
Diesel	Number	4 702	4 815	4 699	4 769	4 807	4 793	4 816	4 843	4 963	5 022	5 137	5 197	5 345	5 633
Electric	Number	2 810	2 871	2 930	3 004	3 065	3 188	3 294	3 443	3 586	3 825	4 033	4 309	4 568	4 823
Passenger coaches	Number	42 677	44 063	44 757	46 488	48 313	49 705	51 255	53 555	55 086	57 527	59 635	61 899	63 878	66 392
Electric multiple-unit coaches	Number	4 668	4 879	4 957	5 278	5 388	5 894	6 454	6 641	6 984	7 487	8 053	8 617	9 184	9 371
Wagons	Number	222 193	216 717	214 760	227 752	222 409	207 983	207 723	204 034	212 835	220 549	229 997	239 321	244 818	246 267
Number of employees	Thousand	1 545	1 511	1 472	1 442	1 424	1 412	1 398	1 394	1 386	1 362	1 332	1 306	1 307	1 334
Average wage per employee	Rupee	121 281	126 650	133 967	145 040	158 140	169 770	173 799	186 067	290 784	382 472	394 112	456 357	527 295	588 523
Number of passengers originating	Million	4 833	5 093	4 971	5 112	5 378	5 725	6 219	6 524	6 920	7 246	7 651	8 224	8 421	8 397
Suburban passengers	Million	2 861	2 999	2 934	2 986	3 178	3 329	3 514	3 689	3 802	3 875	4 061	4 377	4 477	4 552
Non-suburban passengers	Million	1 972	2 094	2 037	2 126	2 200	2 396	2 705	2 835	3 118	3 370	3 590	3 847	3 944	3 845
Passenger-kilometre	Billion	457	491	515	541	576	616	695	770	838	903	979	1 047	1 098	1 159
Tonnes originating revenue-earning freight traffic	Million	474	493	519	557	602	667	728	794	833	888	922	969	1 008	1 052
Net tonne km (total traffic)	Billion	316	336	356	384	411	442	483	523	552	601	625	669	651	667

Contd...

TABLE 9.1 | Contd...

Parameter	Unit	2000/01	2001/02	2002/03	2003/04	2004/05	2005/06	2006/07	2007/08	2008/09	2009/10	2010/11	2011/12	2012/13	2013/14
Net tonne km (revenue-earning traffic)	Billion	312	333	353	381	407	440	481	521	551	601	626	668	650	666
Road															
Road length	'000 km	3374	3427	3529	3622	3929	4004	4141	4110	4472	4582	4677	4865	5232	–
Surfaced	'000 km	1602	1662	1701	1761	1847	1911	1997	2036	2325	2433	2525	2699	2988	–
Unsurfaced	'000 km	1772	1765	1827	1861	2083	2093	2143	2074	2147	2150	2152	2167	2244	–
Motor vehicles registered	Thousand	54991	58924	67007	72718	81502	89618	96707	105353	114951	127746	141866	159491	182445	–
Ports															
Major ports	Number	12	12	12	12	12	13	13	13	13	13	13	13	13	13
Cargo handled at major ports	Million tonnes	281	288	314	345	384	424	464	519	531	561	570	560	546	555
Inland water transport															
Length of navigable waterways	km	–	–	–	14500	14500	14500	–	–	15198	–	–	–	14426	14500

MT – million tonnes

Sources

1. MoR (Ministry of Railways). 2014. ***Indian Railways Annual Report and Accounts 2012/13***. New Delhi: MoR, Government of India
2. MoS (Ministry of Shipping). 2014. ***Statistics of Inland Water Transport 2013/14***. New Delhi: MoS, Government of India
3. MoRTH (Ministry of Road Transport and Highways). 2012. ***Road Transport Year Book 2009/10 and 2010/11***. New Delhi: Transport Research Wing, MoRTH, Government of India
4. MoRTH (Ministry of Road Transport and Highways). 2013. ***Basic Road Statistics of India 2011/12***. New Delhi: Transport Research Wing, MoRTH, Government of India
5. MoRTH (Ministry of Road Transport and Highways). 2015. ***Road Transport Year Book 2012/13***. New Delhi: Transport Research Wing, MoRTH, Government of India
6. Details available at <http://morth.nic.in/writereaddata/sublinkimages/table-12458822488.htm>, last accessed on 2 November 2008
7. Details available at <http://ipa.nic.in/oper.htm>, last accessed on 8 January 2015
8. Details available at <http://indianrailways.gov.in/railwayboard/uploads/directorate/stat_econ/IRSP_2013-14/pdf/Statistical_Summary/Summary%20Sheet_Eng.pdf>, last accessed on 11 January 2016
9. Details available at <http://india-wris.nrsc.gov.in/wrpinfo/?title=Inland_Waterways>, last accessed on 11 January 2016

TABLE 9.2 Gross domestic product in the transport sector (million rupee): 1999/2000–2011/12

Item	1999/2000	2000/01	2001/02	2002/03	2003/04	2004/05	2004/05	2005/06	2006/07	2007/08	2008/09	2009/10	2010/11	2011/12
	(1999/2000 prices)						(2004/05 prices)							
At current prices														
Railways	211 220	214 430	224 080	246 780	269 190	296 450	291 620	307 710	374 290	436 080	474 780	555 710	556 800	604 760
Road transport	681 390	768 360	834 120	962 030	1 129 190	1 326 870	1 440 970	1 634 360	1 895 210	2 145 280	2 472 980	2 785 080	3 332 370	3 940 550
Water transport	37 280	46 720	46 940	44 760	66 670	87 120	64 340	72 890	81 280	92 410	104 210	114 510	131 630	137 210
Air transport	34 510	35 720	36 180	36 910	50 300	55 900	55 900	63 090	92 990	107 580	107 440	117 710	155 140	173 020
Total	964 400	1 065 230	1 141 320	1 290 480	1 515 350	1 766 340	1 852 830	2 078 050	2 443 770	2 781 350	3 159 410	3 573 010	4 175 940	4 855 540
At constant prices														
Railways	211 220	219 960	235 350	248 640	263 360	282 570	291 620	313 390	348 320	382 350	411 610	447 630	474 040	509 450
Road transport	681 390	734 550	762 290	843 050	954 210	1 072 570	1 440 970	1 567 250	1 701 580	1 839 740	1 964 900	2 101 440	2 289 980	2 499 520
Water transport	37 280	39 440	41 390	45 000	49 040	53 780	64 340	70 880	80 150	89 660	91 930	105 030	109 400	112 650
Air transport	34 510	36 000	33 980	37 780	41 040	49 250	55 900	69 600	87 940	103 780	99 030	112 470	129 990	141 360
Total	964 400	1 029 950	1 073 010	1 174 470	1 307 650	1 458 170	1 852 830	2 021 120	2 217 990	24 15 530	2 567 470	2 766 570	3 003 410	3 262 980

Source CSO (Central Statistical Organization). 2013. ***National Accounts Statistics***. New Delhi: CSO, Department of Statistics, Ministry of Planning and Programme Implementation, Government of India

TABLE 9.3 Consumption ('000 tonnes) of petroleum, oil, and lubricants in the transport sector: 2007/08–2014/15

Sector	2007/08	2008/09	2009/10	2010/11	2011/12	2012/13	2013/14	2014/15(P)
High-speed diesel oil								
Road transport	2345.42	2379.49	2434.27	2483.41	2558.39	2207.17	205.85	1579.84
Aviation	2.85	1.75	1.71	1.01	1.37	1.36	1.23	1.23
Shipping	618.80	745.21	667.91	561.34	539.75	413.09	365.21	343.50
Railways	2035.76	2166.17	2260.68	2370.88	2429.26	2538.31	2630.80	2692.69
Light diesel oil (regular)								
Road transport	0	0	0	0	0.01	0.02	0.65	2.21
Shipping	33.19	12.97	4.49	3.79	1.70	2.15	2.43	2.09
Railways	2.35	2.13	1.78	1.12	0.83	0.90	0.76	0.65
Furnace oil								
Road transport	54.85	33.53	0.22	0.09	0.67	1.35	6.15	10.17
Shipping	260.17	431.06	556.60	776.15	366.92	272.70	306.74	334.48
Railways	0	3.95	3.12	3.89	3.38	2.60	2.45	1.80

P – provisional

Source MoPNG (Ministry of Petroleum and Natural Gas). 2015. ***Indian Petroleum and Natural Gas Statistics 2014/15***. New Delhi: MoPNG, Government of India

TABLE 9.4 Total length (kilometre) of national highways (as on 31 March): 2000–13

State/union territory	2000	2001	2002	2003	2004	2005	2006	2007	2008	2009	2010	2011	2012	2013
State														
Andhra Pradesh	3 910	4 038	4 038	4 038	4 472	4 472	4 472	4 472	4 472	4 537	4 537	4 537	4 537	5 022
Arunachal Pradesh	352	392	392	392	392	392	392	392	392	1 992	1 992	1 992	2 027	2 027
Assam	2 706	2 836	2 836	2 836	2 836	2 836	2 836	2 836	2 836	2 836	2 836	2 836	2 940	2 940
Bihar	3 539	4 915	3 502	3 502	3 537	3 537	3 642	3 642	3 642	3 642	3 642	3 642	4 105	4 168
Chhattisgarh	–	–	1 774	1 774	2 184	2 184	2 184	2 184	2 184	2 184	2 184	2 184	2 289	2 289
Goa	269	269	269	269	269	269	269	269	269	269	269	269	269	269
Gujarat	2 241	2 461	2 461	2 461	2 871	2 871	3 245	3 245	3 245	3 245	3 245	3 245	4 032	3 828
Haryana	1 361	1 361	1 361	1 361	1 468	1 468	1 512	1 512	1 512	1 512	1 518	1 518	1 633	1 633
Himachal Pradesh	1 188	1 188	1 188	1 188	1 208	1 208	1 208	1 208	1 208	1 409	1 409	1 409	1 506	1 506
Jammu and Kashmir	738	823	823	823	823	823	1 245	1 245	1 245	1 245	1 245	1 245	1 245	1 695
Jharkhand	–	–	1 413	1 413	1 805	1 805	1 805	1 805	1 805	1 805	1 805	1 805	2 170	2 374
Karnataka	3 394	3 570	3 570	3 570	3 843	3 843	3 843	3 843	3 843	4 396	4 396	4 396	4 396	4 642
Kerala	1 230	1 440	1 440	1 440	1 440	1 440	1 440	1 440	1 457	1 457	1 457	1 457	1 457	1 457
Madhya Pradesh	5 174	6 438	4 664	4 664	5 200	5 200	4 670	4 670	4 670	4 670	5 027	5 027	5 064	5 116
Maharashtra	3 626	3 626	3 626	3 626	4 176	4 176	4 176	4 176	4 176	4 176	4 191	4 191	4 257	4 498
Manipur	954	954	954	954	959	959	959	959	959	959	959	959	1 317	1 317
Meghalaya	717	717	717	717	810	810	810	810	810	810	810	810	1 171	1 171
Mizoram	857	927	927	927	927	927	927	927	927	927	927	927	1 027	1 027
Nagaland	369	369	369	369	494	494	494	494	494	494	494	494	494	494
Odisha	2 863	3 301	3 301	3 301	3 704	3 704	3 704	3 704	3 704	3 704	3 704	3 704	3 704	4 416
Punjab	1 328	1 553	1 553	1 553	1 557	1 557	1 557	1 557	1 557	1 557	1 557	1 557	1 557	1 557
Rajasthan	4 381	4 481	4 597	4 597	5 585	5 585	5 585	5 585	5 585	5 585	5 585	5 585	7 130	7 180
Sikkim	62	62	62	62	62	62	62	62	62	62	62	62	149	149
Tamil Nadu	3 681	3 758	3 758	3 758	4 183	4 183	4 462	4 462	4 462	4 832	4 832	4 832	4 943	4 943

Contd...

TABLE 9.4 | *Contd...*

State/union territory	2000	2001	2002	2003	2004	2005	2006	2007	2008	2009	2010	2011	2012	2013
Tripura	400	400	400	400	400	400	400	400	400	400	400	400	400	400
Uttarakhand	–	–	1075	1075	1991	1991	1991	1991	1991	2042	2042	2042	2042	2042
Uttar Pradesh	4570	5758	4942	4942	5599	5599	5874	5874	5874	6774	6774	6774	7818	7818
West Bengal	1951	1951	1951	1951	2325	2325	2377	2377	2524	2578	2578	2578	2681	2681
Union territory														
Andaman and Nicobar Islands	–	–	–	–	300	300	300	300	300	300	300	300	300	300
Chandigarh	24	24	24	24	24	24	24	24	24	24	24	24	24	24
Delhi	72	72	72	72	72	72	72	72	72	72	80	80	80	80
Puducherry	53	53	53	53	53	53	53	53	53	53	53	53	53	53
All-India	52010	57737	58112	58112	65569	65569	66590	66590	66754	70548	70934	70934	76817	79116

Note Separate figures for Chhattisgarh, Jharkhand, and Uttarakhand for 2000 and 2001 are not available and are included in Madhya Pradesh, Bihar, and Uttar Pradesh, respectively.

Sources

1. MoRTH (Ministry of Road Transport and Highways). Various years. ***Basic Road Transport Statistics***. New Delhi: Transport Research Wing, MoRTH, Government of India
2. MoRTH (Ministry of Road Transport and Highways). 2013. ***Basic Road Statistics of India 2011/12***. New Delhi: Transport Research Wing, MoRTH, Government of India
3. MoRTH (Ministry of Road Transport and Highways). 2015. ***Basic Road Statistics of India 2012/13***. New Delhi: Transport Research Wing, MoRTH, Government of India
4. MoSRTH (Ministry of Shipping, Road Transport and Highways). 2007. ***Road Transport Year Book 2004/05***. New Delhi: Transport Research Wing, MoSRTH, Government of India

TABLE 9.5 Registered motor vehicles (thousand): 1951–2013

Year	Two-wheelers	Cars, jeeps, taxis	Buses	Goods vehicles	Others	All vehicles
1951	27	159	34	82	4	306
1956	41	203	47	119	16	426
1961	88	310	57	168	42	665
1966	226	456	73	259	85	1099
1971	576	682	94	343	170	1865
1976	1057	779	115	351	398	2700
1981	2618	1160	162	554	897	5391
1986	6245	1780	227	863	1462	10577
1991	14200	2954	331	1356	2533	21374
1996	23252	4204	449	2031	3850	33786
2000	34118	6143	562	2715	5319	48857
2001	38556	7058	634	2948	5795	54991
2002	41581	7613	635	2974	6121	58924
2003	47519	8599	721	3492	6676	67007
2004	51922	9451	768	3749	6828	72718
2005	58799	10320	892	4031	7457	81499
2006	64743	11526	992	4436	7921	89618
2007	69129	12649	1350	5119	8460	96707
2008	75336	13950	1427	5601	9039	105353
2009	82402	15313	1486	6041	9710	114952
2010	91598	17109	1527	6432	11080	127746
2011	101865	19231	1604	7064	12102	141866
2012	115419	21568	1677	7658	13169	159491
2013	132550	24853	1894	8597	14551	182445

Note Buses include omni buses since 2001. Others include tractors, trailers, three-wheelers (passenger vehicles), light motor vehicles, and other miscellaneous vehicles that are not separately classified.

Source MoRTH (Ministry of Road Transport and Highways). 2015. ***Road Transport Year Book 2012/13***. New Delhi: Transport Research Wing, MoRTH, Government of India

TABLE 9.6 Registered motor vehicles (number) in India: 31 March 2013

State/union territory	Multi-axled/ articulated vehicles, trucks and lorries	Light motor vehicles (goods)	Buses	Taxis	Light motor vehicles (passengers-auto)	Total transport	Two-wheelers	Cars	Jeeps	Omni buses	Tractors	Trailers	Others	Total non-transport	All vehicles
State															
Andhra Pradesh	253 471	301 439	73 787	125 962	546 813	1 301 472	9 592 270	1 077 849	25 703	58 310	188 439	129 867	313 793	11 386 231	12 687 703
Arunachal Pradesh															151 279
Assam	2 175	67 167	18 982	42 827	71 633	202 784	1 230 895	383 729	18 621	1 229	21 955	18 309	0	1 674 738	1 877 522
Bihar	89 754	13 457	25 992	66 564	141 174	336 941	2 592 137	184 792	94 764	a	249 987	137 472	21 134	3 280 286	3 617 227
Chhattisgarh	97 515	58 466	13 071	15 247	21 890	206 189	2 768 727	175 606	18 202	36 613	129 116	91 324	11 465	3 231 053	3 437 242
Goa	42 759	13 486	9 956	15 427	3 955	106 035[a]	639 594	184 631	c	0	3 277	e	4 835	832 337	938 372
Gujarat	319 207	499 277	70 615	92 455	605 194	1 586 748	11 500 292	1 602 129	173 373	a	525 567	333 458	50 886	14 185 705	15 772 453
Haryana	307 509	137 511	42 800	27 946	114 465	630 231	4 146 906	1 134 616	130 503	656	545 192	0	11 490	5 969 363	6 599 594
Himachal Pradesh	71 843	56 174	13 510	33 043	5 557	180 127	433 890	187 734	14 011	2 427	25 656	212	32 004	695 934	876 061
Jammu and Kashmir	40 751	56 230	26 888	31 285	16 407	171 561	530 594	265 024	25 001	2 687	18 052	3 318	4 549	849 225	1 020 786
Jharkhand	209 790	202 638	14 329	252 268	227 414	930 461[a]	2 041 835	239 655	71 790	16 392	53 674	36 173	26 797	2 486 316	3 416 777
Karnataka	247 639	258 701	69 718	162 894	320 482	1 180 736[b]	8 575 104	1 420 767	43 263	116 987	386 376	267 436	73 017	10 882 950	12 063 686
Kerala	78 796	364 323	404 153	121 825	666 808	1 751 514[ab]	4 464 849	1 331 253	137 547	3 830	12 287	3 656	152 886	6 106 308	7 857 822
Madhya Pradesh	131 098	111 713	40 633	140 260	96 817	520 521	6 877 681	493 412	52 784	a	577 703	214 897	23 041	8 239 518	8 760 039
Maharashtra	402 366	739 725	110 121	184 066	655 299	2 091 627[a]	15 457 123	2 592 565	453 380	19 414	461 932	337 877	74 234	19 396 525	21 488 152
Manipur	10 732	6 445	3 732	2 956	14 030	37 895	212 917	24 417	14 432	1 624	3 852	790	0	258 032	295 927
Meghalaya	26 827	8 873	4 945	18 133	7 674	66 453[a]	73 092	56 209	16 913	3	786	2 833	3 732	153 568	220 021
Mizoram	4 630	6 880	1 187	9 225	3 554	25 476	70 449	15 561	4 510	0	286	102	4 952	95 860	121 336
Nagaland	90 225	19 569	5 750	7 437	15 328	138 309	66 353	55 749	29 034	539	2 598	1 094	2 199	157 566	311 231[e]
Odisha	134 990	132 625	23 004	68 641	78 131	437 391	3 311 553	214 047	46 148	3 668	91 944	81 082	29 707	3 778 149	4 215 540
Punjab	125 898	75 860	30 160	18 539	66 734	317 191	4 729 594	616 549	63 527	a	517 743	1 172	17 163	5 945 748	6 262 939
Rajasthan	401 983	76 396	88 616	103 690	145 422	816 107	7 465 863	659 542	288 056	a	768 645	73 732	a	9 255 838	10 071 945
Sikkim	3 051	1 353	633	9 378	a	14 415	3 808	11 879	6 004	48	123		a	21 862	36 277
Tamil Nadu	492 741	404 244	156 470	307 079	319 074	1 679 608	15 336 823	1 668 913	59 319	19 957	204 927	74 142	188 372	17 552 453	19 232 061

Contd...

TABLE 9.6 Contd...

State/union territory	Multi-axled/ articulated vehicles, trucks and lorries	Light motor vehicles (goods)	Buses	Taxis	Light motor vehicles (passengers-auto)	Total transport	Two-wheelers	Cars	Jeeps	Omni buses	Tractors	Trailers	Others	Total non-transport	All vehicles
Telangana	120 442	168 851	38 245	65 226	236 625	629 389	4 701 734	717 431	14 889	39 658	85 441	50 330	137 605	5 747 088	6 376 477
Tripura	9 674	12 338	2 291	2 286	23 599	50 188	158 533	22 033	11 685	a	783	551	1 149	194 734	244 922
Uttarakhand	25 174	26 924	8 997	28 536	14 077	103 708	1 064 349	227 596	11 300	1 734	44 838	4 508	2 103	1 356 428	1 460 136
Uttar Pradesh	186 404	213 657	40 501	89 393	195 194	725 149	13 724 495	1 205 374	228 836	23 646	1 088 058	14 055	38 571	16 323 035	17 048 184
West Bengal	193 943	242 896	51 660	105 542	78 452	672 493	4 257 670	877 588	8 851	122 050	102 048	19 700	50 334	5 438 241	6 110 734
Total	4 121 387	4 277 218	1 390 746	2 148 130	4 691 802	16 910 719	126 029 130	17 646 650	2 062 446	471 472	6 111 285	1 898 090	1 276 018	155 495 091	172 572 445
Union territory															
Andaman and Nicobar Islands	2 519		958	489	4 107	8 073	60 354	16 474	c	c	c	c	541	77 369	85 442
Chandigarh	2 794	24 283	2 265	3 519	2 664	35 525	760 327	310 273	c		224	d		1 070 824	1 106 349
Dadra and Nagar Haveli	9 355	3 809	352	154	647	14 317	54 244	24 027	581	7	62	208	0	79 129	93 446
Daman and Diu	3 946	3 629	543	55	1 245	9 457[a]	56 893	23 810	575	51	369	221	169	82 088	91 545
Delhi	5 176	126 539	19 912	63 082	11 407	317 963[b]	4 980 507	2 416 974	67 821	5	1 630	a	383	7 467 320	7 785 283
Lakshadweep	0	879	0	262	479	1 620	8 561	165	149	0	111	0	499	9 485	11 105
Puducherry	4 593	10 635	3 987	762	6 321	26 298	600 278	65 016	1 321	3 691	2 305	705	0	673 316	699 614
Total	28 383	169 774	28 017	68 323	26 870	413 253	6 521 164	2 856 739	70 447	3 754	4 701	1 134	1 592	9 459 531	9 872 784
Grand total	4 149 770	4 446 992	1 418 763	2 216 453	4 718 672	17 323 972	132 550 294	20 503 389	2 132 893	475 226	6 115 986	1 899 224	1 277 610	164 954 622	182 445 229

[a] includes motorcycles on hire

[b] includes other vehicles not covered in transport vehicles

[c] included in cars

[d] included in tractors

[e] includes 15 356 government vehicles for which category-wise break-up is not available

Source MoRTH (Ministry of Road Transport and Highways). 2015. ***Road Transport Year Book 2012/13***. New Delhi: Transport Research Wing, MoRTH, Government of India

TABLE 9.7 Indian shipping and gross registered tonnage (million): 2000–14

Item	2000	2001	2002	2003	2004	2005	2006	2007	2008	2009	2010	2011	2012	2013	2014
Number of ships															
Coastal	312	329	424	429	454	485	526	573	616	662	700	750	804	835	846
Overseas	237	228	193	196	215	236	250	277	296	312	340	372	350	364	358
Total	549	557	617	625	669	721	776	850	912	974	1 040	1 122	1 154	1 199	1 204
Shipping Corporation of India	108	97	90	89	84	84	81	79	82	83	94	104	86	85	77
Gross registered tonnage (million)															
Coastal	0.71	0.73	0.80	0.81	0.81	0.82	0.84	0.89	0.96	0.98	1.01	1.05	1.09	1.14	1.22
Overseas	6.24	6.24	5.40	5.82	6.89	7.43	7.58	8.14	8.35	8.50	9.15	10.01	9.33	9.25	9.09
Total	6.95	6.97	6.20	6.62	7.70	8.24	8.42	9.03	9.31	9.48	10.17	11.06	10.42	10.38	10.31
Shipping Corporation of India	2.80	2.66	2.55	2.73	2.58	2.84	2.76	2.73	2.98	3.16	3.65	4.00	3.40	3.46	3.32

Note Figures pertain to the end of the year or quarter.

Source MoS (Ministry of Shipping). 2015. ***Indian Shipping Statistics 2014***. New Delhi: MoS, Transport Research Wing, Government of India

TABLE 9.8 Capacity available (million tonnes) at major ports, by commodity: 2000–14

Commodity	2000	2001	2002	2003	2004	2005	2006	2007	2010	2011	2012	2013	2014
Petroleum, oil, and lubricants	118.50	128.50	135.35	149.55	154.85	157.35	162.35	179.10	211.65	220.76	228.70	278.90	291.90
Iron ore	43.00	43.50	43.60	43.60	51.00	51.00	55.80	57.50	66.80	79.50	79.50	72.00	72.00
Coal	16.80	32.80	44.20	44.20	44.20	44.20	46.25	46.25	46.25	54.25	65.95	65.95	65.95
Fertilizer	3.45	4.70	5.60	5.60	4.10	4.10	7.60	9.10	9.10	9.30	9.30	9.30	11.30
Container	24.70	28.90	37.00	39.15	49.55	48.30	62.10	88.08	110.02	137.53	137.88	139.76	140.21
General cargo	51.60	53.05	78.20	80.65	85.80	92.55	122.20	136.12	155.55	168.79	175.20	179.00	219.16
Total	258.05	291.45	343.95	362.75	389.50	397.50	456.30	516.15	599.37	670.13	696.53	744.91	800.52

Note Data as on 31 March of every year.

Sources

1. GoI (Government of India). 1997. ***Annual Plan 1996/97***. New Delhi: Planning Commission, GoI
2. MoSRTH (Ministry of Shipping, Road Transport and Highways). 2009. ***Basic Port Statistics of India 2006/07***. New Delhi: Transport Research Wing, MoSRTH, Government of India
3. Details available at <http://ipa.nic.in/mon_cap.xls>, last accessed on 12 January 2015

TABLE 9.9 Cargo traffic handled (million tonnes) at major ports, by commodity: 2005/06–2014/15

Commodity	2005/06	2006/07	2007/08	2008/09	2009/10	2010/11	2011/12	2012/13	2013/14	2014/15
Petroleum, oil, and lubricants	142.09	154.34	168.94	176.14	175.08	179.17	179.10	185.98	187.25	188.89
Iron ore	79.17	80.59	91.97	94.04	100.33	87.06	60.40	28.47	26.22	16.54
Fertilizers	12.19	14.13	16.68	18.23	17.72	19.99	20.39	14.74	13.70	16.38
Coal	58.76	59.98	65.19	70.40	71.71	72.73	78.79	86.66	104.12	118.73
Containers	61.98	73.44	92.05	93.14	101.24	114.11	120.10	119.82	114.67	119.38
Others	69.38	81.31	84.33	78.59	95.01	96.97	101.36	110.12	109.52	121.43
Total	423.57	463.78	519.16	530.53	561.09	570.03	560.14	545.79	555.49	581.34

Source Details available at <http://ipa.nic.in/oper.htm>, last accessed on 12 January 2015

TABLE 9.10 Performance indicators of selected ports: 2009/10–2013/14

Name of port	Average turnaround time (days)					Average pre-berthing detention (days)					Average output per ship berth day (tonnes)				
	2009/10	2010/11	2011/11	2012/13	2013/14	2009/10	2010/11	2011/12	2012/13	2013/14	2009/10	2010/11	2011/12	2012/13	2013/14
Kolkata	6.80	6.21	5.45	4.72	4.51	1.31	1.23	0.77	0.61	0.56	1 917	2 253	2 503	2 762	2 963
Haldia	5.01	4.45	3.62	3.95	3.77	4.39	3.73	2.54	2.29	2.21	6 243	6 563	6 728	6 078	6 130
Paradip	9.04	7.73	6.33	4.39	4.62	6.30	5.04	3.69	1.65	1.94	13 853	14 243	15 995	16 625	18 179
Visakhapatnam	4.78	5.84	5.68	5.39	4.73	1.90	2.81	2.84	2.50	1.84	10 484	10 334	10 704	10 641	10 925
Ennore	2.11	2.78	2.17	2.95	4.24	0.37	0.65	0.76	1.33	2.38	21 665	17 699	27 505	27 741	22 357
Chennai	4.04	4.36	3.91	3.24	2.46	1.35	1.61	1.16	0.80	0.41	11 428	10 984	10 352	12 046	14 268
Tuticorin	3.90	4.00	4.94	4.31	3.92	1.36	1.29	1.91	1.31	1.19	6 934	7 035	6 733	7 452	9 633
Cochin	2.08	2.20	1.82	1.58	1.76	0.85	1.03	1.05	1.09	0.97	11 089	11 752	15 784	15 878	15 881
New Mangalore	3.06	2.70	2.95	3.29	3.18	0.81	0.59	0.79	1.04	0.81	13 896	14 211	13 957	15 921	16 314
Mormugao	8.91	10.43	7.68	5.06	4.50	3.46	4.07	2.94	1.62	1.47	5 002	4 409	10 530	11 484	10 018
Jawaharlal Nehru Port	2.01	2.64	1.94	2.48	2.26	0.98	1.51	1.13	1.31	1.08	21 563	20 393	19 227	23 319	23 014
Mumbai	4.61	4.96	5.22	5.58	4.25	1.06	1.23	1.37	1.62	1.18	6 122	6 042	6 476	8 709	7 057
Kandla	5.03	5.90	6.42	6.33	5.66	2.60	3.32	3.74	3.58	2.72	13 549	14 137	14 272	15 728	15 729
All ports	4.63	5.29	4.56	4.29	3.84	2.16	2.32	2.05	1.79	1.48	9 215	9 140	10 575	11 812	12 179

Source MoS (Ministry of Shipping). 2015. ***Basic Port Statistics of India 2013/14***. New Delhi: Ministry of Shipping, Government of India

TABLE 9.11 Performance of the railways: 2003/04–2013/14

Traffic	2003/04	2004/05	2005/06	2006/07	2007/08	2008/09	2009/10	2010/11	2011/12	2012/13	2013/14
Total revenue earning freight traffic (million tonnes)	557.39	602.10	666.51	727.75	793.89	833.39	887.79	921.73	969.05	1 008.09	1 051.64
Coal (million tonnes)	251.75	271.40	294.25	313.33	336.83	369.63	396.15	420.37	455.81	496.42	508.06
Raw material for steel plants (million tonnes)	43.65	44.26	51.35	53.22	11.18	10.85	11.60	13.30	14.51	15.60	17.33
Pig iron and finished steel (million tonnes)	14.66	15.24	17.74	21.04	25.79	28.58	31.85	32.82	35.15	35.31	38.95
Iron ore (million tonnes)	26.64	36.41	41.24	38.84	136.69	130.58	132.74	118.46	104.70	111.41	124.27
Cement (million tonnes)	49.25	53.77	61.19	73.13	78.99	86.24	93.15	99.08	107.66	105.87	109.80
Foodgrains (million tonnes)	44.32	46.52	41.64	41.84	38.23	35.51	38.69	43.45	46.40	49.03	55.09
Fertilizers (million tonnes)	23.73	28.75	32.65	34.26	35.83	41.35	43.68	48.22	52.70	46.21	44.70
POL (million tonnes)	32.02	32.00	33.45	31.69	35.88	38.08	38.88	39.29	39.77	40.61	41.16
Other goods (million tonnes)	71.37	73.75	93.00	120.40	94.47	92.57	101.05	106.74	112.35	107.63	112.28
Net tonne km per wagon per day (broad gauge)	2570.00	2677.00	2960.00	3238.00	3539.00	8687.00	9022.00	9247.00	9261.00	8453.00	8471.00
Passenger traffic originating (million)	5112.00	5378.00	5725.00	6219.00	6524.00	6920.00	7246.00	7651.00	8224.00	8421.00	8397.00
Passenger km (billion)	541.21	575.70	615.61	694.76	769.96	838.03	903.47	978.51	1046.52	1098.10	1158.74

POL – petroleum, oil, and lubricants

Source MoR (Ministry of Railways). 2015. ***Indian Railways Year Book 2013/14***. New Delhi: Railway Board, MoR, Government of India

TABLE 9.12 Railway traffic: 2000/01–2013/14

Traffic	2000/01	2001/02	2002/03	2003/04	2004/05	2005/06	2006/07	2007/08	2008/09	2009/10	2010/11	2011/12	2012/13	2013/14
Passengers (million)														
Suburban (million)	2861	2999	2934	2986	3178	3329	3514	3689	3802	3876	4061	4377	4477	4552
Non-suburban (million)	1972	2094	2037	2126	2200	2396	2705	2835	3118	3370	3590	3847	3944	3845
Total (million)	4833	5093	4971	5112	5378	5725	6219	6524	6920	7246	7651	8224	8421	8397
Revenue earning goods traffic (MT)	473.5	492.5	519	557	602	667	728	794	833	888	922	969	1008	1052
Gross earning (million rupee)														
Passenger traffic	104832	111623	125408	132598	140725	150808	171760	197833	218665	234144	257056	282464	313228	363523
Goods traffic	230454	245868	262315	274032	304892	355347	410732	464255	517493	569115	606871	677436	834788	915709

MT – million tonnes

Source MoR (Ministry of Railways). 2015. ***Indian Railways Year Book 2013/14***. New Delhi: Railway Board, MoR, Government of India

TABLE 9.13 Electrification of railway routes and tracks (kilometre): 2000/01–2013/14

Route/track	2000/01	2001/02	2002/03	2003/04	2004/05	2005/06	2006/07	2007/08	2008/09	2009/10	2010/11	2011/12	2012/13	2013/14
Routes														
Electrified	14856	15994	16272	16776	17495	17907	17786	18274	18559	18927	19607	20275	20884	21614
Total	63028	63140	63122	63221	63465	63332	63327	63273	64015	63974	64460	64600	65436	65808
Running tracks														
Electrified	27937	29567	29974	30589	32686	33540	33623	34700	35471	35811	36007	38669	38524	39661
Total	81865	82354	82492	83859	84260	84370	85389	85158	86937	87087	87114	89801	89236	89919
Number of stations	6853	6856	6906	7031	7133	6974	6909	7025	7030	7083	7133	7146	7172	7112

Source MoR (Ministry of Railways). 2015. ***Indian Railways Year Book 2013/14***. New Delhi: Railway Board, MoR, Government of India

TABLE 9.14 Energy intensity in railways (consumption per thousand gross tonne-km): 2010/11–2013/14

Fuel	2010/11			2011/12			2012/13			2013/14		
	Broad gauge	Metre gauge	Total including narrow gauge	Broad gauge	Metre gauge	Total including narrow gauge	Broad gauge	Metre gauge	Total including narrow gauge	Broad gauge	Metre gauge	Total including narrow gauge
Passenger and proportion of mixed												
Coal (kg)	–	427.9	568.2	–	360.2	579.0	–	759.7	1335.9	–	875	1509.9
Diesel oil (litre)	4.0	5.7	4.1	4.0	5.3	4.1	3.9	5.6	4.0	3.83	5.35	3.91
Electricity (kWh)	19.5	–	19.5	19.9	–	19.9	19.4	–	19.4	19.0	–	19.0
Goods and proportion of mixed												
Coal (kg)	–	–	–	–	–	–	–	–	–	–	–	–
Diesel oil (litre)	2.3	7.1	2.3	2.2	7.4	2.2	2.2	7.7	2.2	2.21	8.67	2.22
Electricity (kWh)	6.8	–	6.8	7.1	–	7.1	6.8	–	6.8	6.43	–	6.43

Source MoR (Ministry of Railways). 2015. ***Indian Railways Annual Report and Accounts 2013/14***. New Delhi: Railway Board, MoR, Government of India

TABLE 9.15 Trends in relative rail and road traffic: 2000/01–2012/13

Mode	2000/01	2001/02	2002/03	2003/04	2004/05	2005/06	2006/07	2007/08	2008/09	2009/10	2010/11	2011/12	2012/13
Freight traffic													
Road (billion tonne-km)	494.0	515.0	545.0	595.0	646.0	658.9	766.2	851.7	920.2	1015.1	1128.4	1212.4	1285.1
Rail (billion tonne-km)	312.4	333.2	353.2	381.2	411.3	441.8	481.0	521.3	551.4	600.5	625.7	667.6	649.6
Road (%)	61.3	60.7	60.7	61.0	61.1	59.9	61.4	62.0	62.5	62.8	64.3	64.5	66.4
Rail (%)	38.7	39.3	39.3	39.0	38.9	40.1	38.6	38.0	37.5	37.2	35.7	35.5	33.6
Total (billion tonne-km)	806.4	848.2	898.2	976.2	1057.3	1100.7	1247.2	1373.0	1471.6	1615.6	1754.1	1880.0	1934.7
Passenger traffic													
Road (passenger-km)	2075.5	2413.1	2814.7	3070.2	3469.3	4251.7	4545.8	4860.3	5196.5	5555.9	5940.3	6351.2	6790.5
Rail (passenger-km)	457.0	490.9	515.0	541.2	575.7	615.6	694.8	770.0	838.0	903.4	978.5	1046.5	1098.1
Road (%)	82.0	83.1	84.5	85.0	85.8	87.4	86.7	86.3	86.1	86.0	85.9	85.9	85.1
Rail (%)	18.0	16.9	15.5	15.0	14.2	12.6	13.3	13.7	13.9	14.0	14.1	14.1	14.9
Total (billion tonne-km)	2532.5	2904.0	3329.7	3611.4	4045.0	4867.3	5240.6	5630.3	6034.5	6459.3	6918.8	7397.7	7888.6

Note Figures for road transport from 2006/07 to 2011/12 have been estimated by the Transport Research Wing, Ministry of Road Transport and Highways on the basis of actual GDP growth rate and elasticity of road transport with respect to GDP.

Source MoSRTH (Ministry of Shipping, Road Transport and Highways). 2015. ***Road Transport Year Book 2012/13***. New Delhi: MOSRTH, Government of India

TABLE 9.16 | Consumption of fuel/energy in Indian Railways: 2009/10–2013/14

Type	For traction					For other than traction purposes (including manufacturing units)				
	2009/10	2010/11	2011/12	2012/13	2013/14	2009/10	2010/11	2011/12	2012/13	2013/14
Electricity (million kWh)	13 087.31	13 571.53	13 449.98	13 853.44	14 410.70	2 506.17	2 483.96	2 458.34	2 478.37	2 492.08
HSD oil (million litre)	2 400.47	2 516.04	2 705.08	2 699.62	2 780.90	45.29	44.02	40.21	46.72	58.33
Coal (million tonne)	0.002	0.001	0.001	0.001	0.001	0.001	0.001	0.001	0.001	0.001

HSD - high-speed diesel

Source MoR (Ministry of Railways). 2015. ***Indian Railways Year Book 2013/14***. New Delhi: Railway Board, MoR, Government of India

TABLE 9.17 Traffic trends in terms of aircraft, passenger, and cargo movements: 2002/03–2014/15

	2002/03	2003/04	2004/05	2005/06	2006/07	2007/08	2008/09	2009/10	2010/11	2011/12	2012/13	2013/14	2014/15(P)
Aircraft movements (number)													
International airports	132 790	132 934	163 300	190 880	215 570	248 540	270 340	282 200	300 190	309 290	313 910	335 970	345 980
Domestic airports	504 157	506 042	554 300	647 340	862 020	1 059 090	1 036 190	1 048 690	1 093 570	1 233 600	1 164 900	1 200 650	1 260 260
Total	636 947	638 976	717 600	838 220	1 077 590	1 307 630	1 306 530	1 330 890	1 393 760	1 542 890	1 478 810	1 536 620	1 606 240
Passengers (million)													
International airports	16.65	16.63	19.42	22.37	25.78	29.82	31.584	34.368	37.907	40.800	43.030	46.620	50.690
Domestic airports	32.04	32.07	39.86	50.97	70.62	87.07	77.295	89.387	105.523	121.510	116.370	122.300	157.980
Total	48.69	48.70	59.28	73.34	96.40	116.89	108.879	123.755	143.430	162.310	159.400	168.920	208.670
Cargo (tonne)													
International airports	693 362	693 173	823 600	920 150	1 021 260	1 146 750	1 149 920	1 270 710	1 496 240	1 467 900	1 406 330	1 443 060	1 539 720
Domestic airports	375 080	375 029	456 700	477 150	529 640	568 230	552 060	690 900	852 200	812 090	784 220	836 090	991 220
Total	1 068 442	1 068 202	1 280 300	1 397 300	1 550 900	1 714 980	1 701 980	1 961 610	2 348 440	2 279 990	2 190 550	2 279 150	2 530 940

P – provisional

Source MoCA (Ministry of Civil Aviation). 2015. ***Annual Report 2014/15***. New Delhi: MoCA, Government of India

TABLE 9.18 Growth of civil aviation: 2000/01–2014/15

Indicator	2000/01	2001/02	2002/03	2003/04	2004/05	2005/06	2006/07	2007/08	2008/09	2009/10	2010/11	2011/12	2012/13	2013/14	2014/15*
Fleet strength (number)															
Air India	26	29	30	35	36	34	36	–	–	–	–	–	–	–	–
Indian Airlines	52	53	53	56	61	64	70	–	–	–	–	–	–	–	–
Air India Ltd								122	108	113	98	91	95	95	99
Revenue tonne-km (million rupee)															
Air India	1501.40	1399.00	1561.00	1773.00	2218.00	2364.00	2219.60	–	–	–	–	–	–	–	–
Indian Airlines	777.30	755.50	845.10	877.50	1017.30	1140.90	1236.30	–	–	–	–	–	–	–	–
Air India Ltd								3729.00	3284.00	3533.00	3677.00	3603.00	3346.00	3910.00	4188.16
Number of passengers carried (million)															
Air India	3.30	3.13	3.46	3.84	4.44	4.44	4.30	–	–	–	–	–	–	–	–
Indian Airlines	5.99	5.53	5.65	5.90	7.13	7.86	8.57	–	–	–	–	–	–	–	–
Air India Ltd								13.32	11.78	11.75	12.78	13.43	14.18	15.41	16.95
Passengers handled (million)															
AAI airports	42.03	39.98	43.72	48.78	59.28	73.35	93.84	63.71	44.25	50.87	59.64	68.40	68.39	71.65	57.46
Joint venture international airports	–	–	–	–	–	–	2.56	53.18	64.62	72.88	83.79	93.90	91.01	97.27	80.62
Total at Indian airports	–	–	–	–	–	–	96.40	116.89	108.87	123.76	143.43	162.30	159.40	168.92	138.09
Cargo handled (thousand tonnes)															
AAI airports	842.15	854.28	979.36	1068.66	1278.47	1397.30	1529.52	723.46	561.42	592.95	726.52	703.43	650.41	637.72	517.84
Joint venture international airports	–	–	–	–	–	–	21.39	991.52	1140.57	1366.76	1621.92	1576.56	1540.14	1641.43	1391.32
Total at Indian airports	–	–	–	–	–	–	1550.91	1714.98	1701.99	1959.71	2348.44	2279.99	2190.55	2279.15	1909.16

AAI – Airports Authority of India

*Up to December 2014

Source MoF (Ministry of Finance). 2015. ***Economic Survey 2014/15***. New Delhi: MoF, Government of India

TABLE 9.19 Physical performance of Air India Limited: 2011/12–2014/15

Particulars	2011/12	2012/13	2013/14	2014/15(Apr–Dec 14) (P)
Available seat km (million)	45 015	39 909	45 428	45 026
Passenger km (million)	30 577	28 879	33 279	33 153
Passenger load factor (%)	67.9	72.4	73.3	73.6

P – provisional

Source MoCA (Ministry of Civil Aviation). 2014. ***Annual Report 2013/14***. New Delhi: MoCA, Government of India

TABLE 9.20 Share of different modes of transport in gross domestic product at basic and constant prices: 2011/12–2013/14

Sector	2011/12	2012/13	2013/14
Total transport	4.7	4.9	4.8
Railways	0.8	0.8	0.9
Road transport*	3.1	3.2	3.1
Water transport*	0.1	0.1	0.1
Air transport*	0.1	0.1	0.1
Services*	0.8	0.8	0.8

*Unadjusted Financial Intermediation Services Indirectly Measured (FISIM)

Source MoRTH (Ministry of Road Transport and Highways). 2015. ***Road Transport Year Book 2012/13***. New Delhi: MoRTH, Transport Research Wing, Government of India

TABLE 9.21 Production of motor vehicles (number) in India: 2003/04–2013/14

Category	2003/04	2004/05	2005/06	2006/07	2007/08	2008/09	2009/10	2010/11	2011/12	2012/13	2013/14
M&HCVs	166 123	214 807	219 295	294 258	294 957	192 283	250 133	344 542	384 801	280 677	221 626
LCVs	108 917	138 896	171 788	225 724	254 049	224 587	317 423	408 193	544 335	551 972	477 238
Total CVs	275 040	353 703	391 083	519 982	549 006	416 870	567 556	752 735	929 136	832 649	698 864
Cars	782 562	960 487	1 046 133	1 238 021	1 426 212	1 516 967	1 932 620	2 453 113	2 775 124	2 662 520	2 508 665
MUVs	206 998	249 389	263 167	307 202	351 371	321 626	424 791	534 183	370 945	568 538	563 986
Scooters	935 279	987 498	1 021 013	943 944	1 074 933	1 161 276	1 494 409	2 144 765	2 659 340	3 026 512	3 676 193
Motor cycles	4 355 168	5 193 894	6 207 690	7 112 281	6 503 532	6 798 118	8 444 857	10 527 111	11 982 669	11 925 690	12 471 488
Mopeds	332 294	348 437	379 994	379 987	430 827	436 219	571 070	704 575	785 523	791 954	732 210
Electric two-wheelers	NR	NR	NR	30 454	17 389	24 179	2 567	NR	NR	NR	NR
Total two-wheelers	5 622 741	6 529 829	7 608 697	8 466 666	8 026 681	8 419 792	10 512 903	13 376 451	15 427 532	15 744 156	16 879 891
Three-wheelers	356 223	374 445	434 423	556 126	500 660	497 020	619 194	799 553	879 289	839 748	830 120
Grand total	7 243 564	8 467 853	9 743 503	11 087 997	10 853 930	11 172 275	14 057 064	17 916 035	20 382 026	20 647 611	21 481 526

LCVs – light commercial vehicles; M&HCVs – medium and heavy commercial vehicles; MUVs – multi-utility vehicles; NR – not reported

Source MoRTH (Ministry of Road Transport and Highways). 2015. ***Road Transport Year Book 2013/14***. New Delhi: MoRTH, Transport Research Wing, Government of India

TABLE 9.22 Total and surfaced road length (kilometre) by categories in India as on 31 March 2013

	Total/Surfaced	2000	2004	2005	2006	2007	2008	2009	2010	2011	2012	2013
National highways	Total	52 010	65 569	65 569	66 590	66 590	66 754	70 548	70 934	70 934	76 818	79 116
	Surfaced	51 952	65 358	65 358	66 590	66 590	66 754	70 548	70 934	70 934	76 818	79 116
State highways	Total	132 797	133 177	144 396	148 090	152 235	154 522	158 497	160 177	163 898	164 360	169 227
	Surfaced	130 592	131 262	142 898	146 325	150 713	152 738	156 715	158 202	161 920	162 950	167 219
Other PWD roads	Total	730 680	719 257	786 230	803 669	835 003	863 241	962 880	977 414	998 895	1 022 287	1 066 747
	Surfaced	601 512	597 866	643 705	664 652	689 935	719 383	832 534	853 441	880 553	914 101	921 353
Rural roads	Total	1 948 043	2 140 569	2 266 439	2 308 125	2 393 488	2 450 559	2 629 165	2 692 535	2 749 805	2 838 220	2 259 639
	Surfaced	545 378	678 533	681 761	714 326	761 429	810 258	933 051	996 221	1 042 536	929 789	1 400 988
Project and urban roads	Total	462 235	562 935	546 522	554 177	569 085	574 516	650 419	681 379	700 379	763 709	757 193
	Surfaced	244 366	287 749	261 576	266 791	276 091	286 930	331 654	354 014	368 737	431 729	419 124
All-India	Total	3 325 765	3 621 507	3 929 439	4 003 930	4 140 544	4 109 592	4 471 509	4 582 439	4 683 911	4 865 394	4 331 922
	Surfaced	1 573 800	1 760 768	1 846 629	1 910 792	1 997 323	2 036 063	2 324 502	2 432 812	2 524 680	2 515 387	2 987 800

PWD – Public Works Department

Note In 2013, excluding rural roads (900 000 km) constructed under Jawahar Rozgar Yojana.

Source MoRTH (Ministry of Road Transport and Highways). 2015. ***Basic Road Statistics of India 2012/13***. New Delhi: MoRTH, Transport Research Wing, Government of India

TABLE 9.23 Total road length (kilometre) and percentage share of each category of road (1951–2013)

Year (as on 31 March)	National highways	State highways	PWD roads	Panchayat Raj roads	JRY and PMGSY roads (b)	Urban roads	Project roads	Total
1951	19 811	a	173 723	206 408	0	0	0	399 942
(%)	4.95	0	43.44	51.61	0	0	0	100
1961	23 798	a	257 125	197 194	0	46 361	0	524 478
(%)	4.54	0	49.02	37.6	0	8.84	0	100
1971	23 838	56 765	276 833	354 530	0	72 120	130 893	914 979
(%)	2.61	6.2	30.26	38.75	0	7.88	14.31	100.01
1981	31 671	94 359	421 895	628 865	0	123 120	185 511	1 485 421
(%)	2.13	6.35	28.4	42.34	0	8.29	12.49	100
1991	33 650	127 311	509 435	931 288	329 142	186 799	209 737	2 327 362
(%)	1.45	5.47	21.89	40.01	14.14	8.03	9.01	100
2001	57 737	132100	736 001	1 068 020	903 996	252 001	223 665	3 373 520
(%)	1.71	3.92	21.82	31.66	26.8	7.47	6.63	100.01
2002	58 112	137 711	695 335	1 134 326	926 697	250 295	224 124	3 426 600
(%)	1.70	4.02	20.29	33.1	27.04	7.3	6.54	99.99
2003	58 112	134 807	696 960	1 155 491	926 697	297 259	259 328	3 528 654
(%)	1.65	3.82	19.75	32.75	26.26	8.42	7.35	100
2004	65 569	133 177	719 257	1 189 058	951 511	301 310	261 625	3 621 507
(%)	1.81	3.68	19.86	32.83	26.27	8.32	7.22	99.99
2005	65 569	144 396	786 230	1 299 463	966 976	286 707	259 815	3 809 156
(%)	1.72	3.79	20.64	34.11	25.39	7.53	6.82	100
2006	66 590	148 090	803 669	1 318 258	989 867	291 991	262 186	3 880 651
(%)	1.72	3.82	20.71	33.97	25.51	7.52	6.76	100.01
2007	66 590	152 235	835 003	1 372 867	1 020 621	300 580	268 505	4 016 401
(%)	1.66	3.79	20.79	34.18	25.41	7.48	6.69	100
2008	66 754	154 522	863 241	1 388 750	1 061 809	304 327	270 189	4 109 592
(%)	1.62	3.76	21.01	33.79	25.84	7.41	6.57	100
2009	70 548	158 497	962 880	1 514 952	1 114 213	373 802	276 617	4 471 509
(%)	1.58	3.54	21.53	33.88	24.92	8.36	6.19	100

Contd...

TABLE 9.23 | *Contd...*

Year (as on 31 March)	National highways	State highways	PWD roads	Panchayat Raj roads	JRY and PMGSY roads (b)	Urban roads	Project roads	Total
2010	70934	160 177	977 414	1 518 205	1 174 330	402 448	278 931	4 582 439
(%)	1.55	3.5	21.33	33.13	25.63	8.78	6.09	100
2011	70934	163 898	998 895	1 530 366	1 219 438	411 679	281 628	4 676 838
(%)	1.52	3.5	21.36	32.72	26.07	8.8	6.02	100
2012	76 818	164 360	1 022 287	1 587 787	1 250 433	464 294	299 415	4 865 394
(%)	1.58	3.38	21.01	32.63	25.7	9.54	6.15	100
2013 (P)	79 116	169 227	1 066 747	1 725 318	1 434 321	446 238	310 955	5 231 922
(%)	1.51	3.23	20.39	32.98	27.41	8.53	5.94	100

JRY – Jawahar Rozgar Yojana; P – provisional; PMGSY – Pradhan Mantri Gram Sadak Yojana; PWD – Public Works Department

[a] included in other Public Works Department (PWD) roads

[b] includes roads constructed under Pradhan Mantri Gram Sadak Yojana launched in December 2000 and Jawahar Rozgar Yojana launched in 1989/90

Source MoRTH (Ministry of Road Transport and Highways). 2015. ***Basic Road Statistics of India 2012/13***. New Delhi: MoRTH, Transport Research Wing, Government of India

TABLE 9.24 | Length (kilometre) of national highways in India: 2005–13 (as on 31 March)

State/union territory	2005	2006	2007	2008	2009	2010	2011	2012	2013
State									
Andhra Pradesh	4 472	4 472	4 472	4 472	4 537	4 537	4 537	4 537	5 022
Arunachal Pradesh	392	392	392	392	1 992	1 992	1 992	2 027	2 027
Assam	2 836	2 836	2 836	2 836	2 836	2 836	2 836	2 940	2 940
Bihar	3 537	3 642	3 642	3 642	3 642	3 642	3 642	4 105	4 168
Chhattisgarh	2 184	2 184	2 184	2 184	2 184	2 184	2 184	2 289	2 289
Goa	269	269	269	269	269	269	269	269	269
Gujarat	2 871	3 245	3 245	3 245	3 245	3 245	3 245	4 032	3 828
Haryana	1 468	1 512	1 512	1 512	1 512	1 518	1 518	1 633	1 633
Himachal Pradesh	1 208	1 208	1 208	1 208	1 409	1 409	1 409	1 506	1 506
Jammu and Kashmir	823	1 245	1 245	1 245	1 245	1 245	1 245	1 245	1 695
Jharkhand	1 805	1 805	1 805	1 805	1 805	1 805	1 805	2 170	2 374
Karnataka	3 843	3 843	3 843	3 843	4 396	4 396	4 396	4 396	4 642
Kerala	1 440	1 440	1 440	1 457	1 457	1 457	1 457	1 457	1 457
Madhya Pradesh	5 200	4 670	4 670	4 670	4 670	5 027	5 027	5 064	5 116
Maharashtra	4 176	4 176	4 176	4 176	4 176	4 191	4 191	4 257	4 498
Manipur	959	959	959	959	959	959	959	1 317	1 317
Meghalaya	810	810	810	810	810	810	810	1 171	1 171
Mizoram	927	927	927	927	927	927	927	1 027	1 027
Nagaland	494	494	494	494	494	494	494	494	494
Odisha	3 704	3 704	3 704	3 704	3 704	3 704	3 704	3 704	4 416
Punjab	1 557	1 557	1 557	1 557	1 557	1 557	1 557	1 557	1 557
Rajasthan	5 585	5 585	5 585	5 585	5 585	5 585	5 585	7 130	7 180
Sikkim	62	62	62	62	62	62	62	149	149
Tamil Nadu	4 183	4 462	4 462	4 462	4 832	4 832	4 832	4 943	4 943
Tripura	400	400	400	400	400	400	400	400	400
Uttarakhand	1 991	1 991	1 991	1 991	2 042	2 042	2 042	2 042	2 042
Uttar Pradesh	5 599	5 874	5 874	5 874	6 774	6 774	6 774	7 818	7 818
West Bengal	2 325	2 377	2 377	2 524	2 578	2 578	2 578	2 681	2 681
Union territory									
Andaman and Nicobar Islands	300	300	300	300	300	300	300	300	300
Chandigarh	24	24	24	24	24	24	24	24	24
Delhi	72	72	72	72	72	80	80	80	80
Puducherry	53	53	53	53	53	53	53	53	53
All-India	65 569	66 590	66 590	66 754	70 548	70 934	70 934	76 817	79 116

Source MoRTH (Ministry of Road Transport and Highways). 2015. ***Basic Road Statistics of India 2012/13***. New Delhi: MoRTH, Transport Research Wing, Government of India

HOUSEHOLD ENERGY 10

AT A GLANCE

- As per the 68th round of National Sample Survey (NSS), 72.7% of rural households and 96.1% of urban households depend on electricity for lighting.
- According to the 68th round of NSS, 15% of rural households and 68.4% of urban households depend on liquefied petroleum gas for cooking.
- As on 31 May 2015, 96.7% of the inhabited census villages were electrified.
- More than 5 million slum inhabitants depend on kerosene as the primary energy source for lighting.
- More than 5.5 million slum households depend on fuel sources such as firewood, crop residue, cow dung, coal/lignite/charcoal, and kerosene for cooking.

Introduction

In the era of sustainable development goals,[1] access to energy is fundamental for the development and prosperity of the poor. Two widely popular indicators that demonstrate the challenges of energy access in a country are (i) the number of people living without access to electricity and (ii) those who are still dependent on cooking technology based on traditional solid biomass. With a total population of 1.2 billion, India is home to a large section of energy-poor communities that lack access to clean and modern energy services. The International Energy Agency (IEA) estimates that India accounts for nearly one-fourth of the world's population that lacks access to electricity and close to one-third of the population that is dependent on the traditional use of solid biomass for cooking (Figure 1).

At the household level, the primary use of energy can be classified into two forms: basic consumptive use and productive use. This chapter focuses on the primary consumption of various sources of energy for lighting and cooking in addition to discussing the progress being made on energy access. Also, relevant policies and programmes of the Government of India will be briefly discussed to highlight the efforts being made in facing the challenges to energy access.

Status and progress of energy access in India

There is a wide gap between rural India and urban India in terms of progress in energy access for both cooking and lighting. The latest census reveals that

[1] Sustainable Development Goals are the 17 global commitments that the world leaders have agreed to at the UN Sustainable Development Summit held from 25–27 September 2015 in New York, USA.

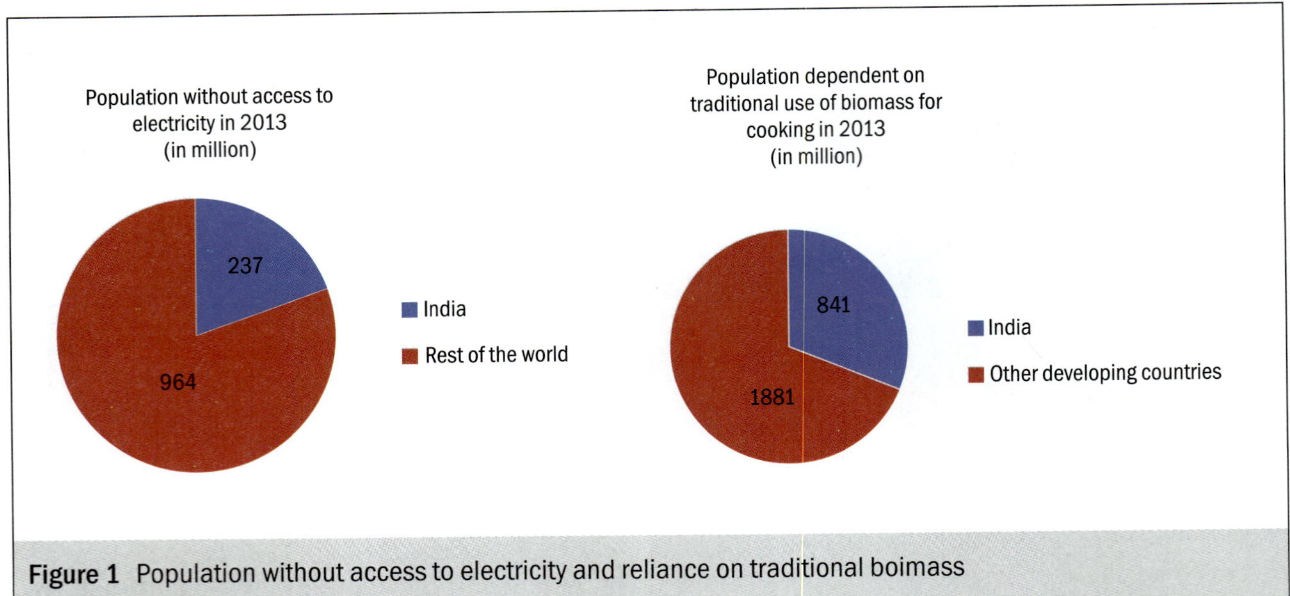

Figure 1 Population without access to electricity and reliance on traditional boimass
Source IEA (2015a)

less than 60% of the total rural households depend on electricity for lighting, as against more than 90% of the urban households that have access to cleaner forms of energy for lighting (Table 1). Similarly, the estimates from Census of India 2011 data for cooking suggest that over 85% of rural households and 26.2% of urban households depend on solid biomass as the primary fuel (Table 2).

As per the Census of India 2011 data, at the national level, 67.3% of households use electricity for lighting and 28.6% of households depend on liquefied petroleum gas (LPG) for cooking (Tables 10.1a and 10.1b).[2] Disaggregation of the state-level data outlines that households in rural areas of underdeveloped states are considerably less reliant on electricity for lighting and are relatively more

Table 1 Distribution of households on the basis of energy source for lighting

Energy source	Census of India 2001		Census of India 2011	
	Percentage of rural households dependent on energy source	Percentage of urban households dependent on energy source	Percentage of rural households dependent on energy source	Percentage of urban households dependent on energy source
Electricity	43.6	87.6	55.3	92.7
Kerosene	55.6	11.6	43.2	6.5
Solar energy	0.3	0.2	0.5	0.2
Any other source	0.3	0.2	0.5	0.3
No lighting	0.3	0.4	0.5	0.3

Note Census Info 2011a (version 2.0) reports that 43.6% of rural households depend on electricity. However, the official Census website reports this figure as 43.5% of households.

Sources Details available at <www.censusindia.gov.in/2011census/hlo/Data_sheet/India/HLO_Indicators.pdf>, last accessed on 17 November 2015; Details available at <www.devinfolive.info/censusinfodashboard/website/index.php/pages/source_lighting/total/Households/IND>, last accessed on 10 February 2015

[2] As per the 68th round of National Sample Survey, 72.7% of rural households and 96.1% of urban households depend on electricity for lighting. It is also estimated that 15% of rural households and 68.4% of urban households depend on liquefied petroleum gas for cooking.

Table 2 Source of energy for domestic cooking in India

Energy source	Census of India 2001		Census of India 2011	
	Percentage of rural households	Percentage of urban households	Percentage of rural households	Percentage of urban households
Firewood	64.1	22.7	62.5	20.1
Crop residue, cow dung, coal, lignite, charcoal	26.9	8.7	24	6.1
Kerosene	1.6	19.2	0.7	7.5
LPG/PNG	5.7	48	11.4	65
Any other	1.4	0.9	1.1	0.8

LPG – liquefied petroleum gas; PNG – piped natural gas

Sources Details available at <www.censusindia.gov.in/2011census/hlo/Data_sheet/Source%20of%20Lighting.pdf>, last accessed on 17 November 2015; Details available at <www.devinfolive.info/censusinfodashboard/website/index.php/pages/kitchen_fuelused/total/Households/IND>, last accessed on 10 February 2015

dependent on traditional biomass for cooking when compared with the urban households of the state (Tables 10.2a and 10.2b). A large number of energy poor households live in slums,[3] which indicates urban energy poverty (Tables 10.3a and 10.3b).

Owing to the prioritization of energy access in the overall development agenda, progress in cleaner sources/methods of cooking and lighting has been registered over the years (Tables 3a and 3b).

According to a recent UN report, more than 50 million people in India gained access to electricity because of the progress made from 2010 to 2012 (SE4ALL 2015). Based on the data from the World Health Organization, the report further suggests that only 12% of the total rural population has access to non-solid fuels, in contrast to 72% of the urban population. As per estimates presented in the report,[4] 791 million people still do not have access to modern cooking methods.

In terms of distribution of households based on occupation, households engaged in casual labour in both rural and urban areas appear to be poor relative to all other household types (NSSO 2015). Disaggregation of data[5] from the 68th National Sample Survey (NSS) round reveals that the incidence of use of electricity for lighting is lowest for household type "casual labour" in urban areas and "non-agricultural casual labour" in rural areas. Similarly, among the different household types, the incidence of use of firewood and chips is highest (80.4%) for casual labour in agriculture (rural households) and casual labour in urban India. This affirms the fact that the poorest of the poor are also the most energy poor in the country (Figures 2a and 2b, Tables 4a and 4b).

Energy consumption

One of the indicators that determine/reflect progress is the total energy consumption at the household level. This is because an increase in energy consumption is linked to higher income levels, improved access, and better quality of life among others. This section looks at consumption data and levels for various energy sources.

Over the years, the total residential energy consumption in India has been rising. Between 2010 and 2012, energy balance on a net calorific value basis increased by 3.7% (Figure 3). This can be attributed to various factors such as increase in the number of consumers, enhanced use of electrical/electronic appliances by households, rapid urbanization, and stacking of multiple fuel sources by households for cooking.

Based on the IEA's Balances report, India's total residential consumption of energy for 2012 was close to 8.77% of the world's total residential energy

[3] In Census of India 2011, slums have been defined as residential areas where dwellings are unfit for human habitation by reasons of dilapidation, overcrowding, faulty arrangements and design of such buildings, narrowness or faulty arrangement of street, lack of ventilation, light, or sanitation facilities, or any combination of these factors that are detrimental to safety and health. Slums have been classified into (a) notified slums; (b) recognized slums; and (c) identified slums (Chandramouli 2011).

[4] On the basis of household type

[5] As per Census 2001

Table 3a Percentage distribution of households by primary energy source for cooking (2001/02–2011/12)

NSS round	Period	Coke, coal, and charcoal (%)	Firewood and chip (%)	LPG (%)	Dung cake (%)	Kerosene (%)	No cooking/ others (%)
Rural							
68	2011/12	1.10*	67.30	15	9.60	0.90	6.20**
66	2009/10	0.8	75.9	12.1	6.1	0.8	4.3
64	2007/08	0.8	77.6	9.1	7.4	0.6	4.5
63	2006/07	0.8	75.4	8.9	9.1	0.8	5
62	2005/06	1.1	74	9.3	9	1	5.6
61	2004/05	0.8	75	8.6	9.1	1.3	5.2
59	2003	0.9	74.9	9.1	9.3	1.9	3.9
57	2001/02	1.2	73.4	8.1	10.5	2	4.8
Urban							
68	2011/12	2.10	14	68.4	1.3	5.7	8.4
66	2009/10	2.2	17.6	64.6	1.4	6.4	7.8
64	2007/08	2.1	20.1	61.8	1.4	7.6	7.1
63	2006/07	2.3	22.1	59.2	1.7	7.5	7.1
62	2005/06	3.5	20.9	57.1	1.7	9.2	7.6
61	2004/05	2.8	21.7	57.1	1.7	10.2	6.5
59	2003	3.3	20	55.4	1.8	13	6.6
57	2001/02	3	23.3	49.9	1.6	15.3	7.1

LPG – liquefied petroleum gas; NSS – National Sample Survey
*Includes coke and coal
**Includes households not reporting primary source of energy for cooking and households using gobar gas, charcoal, electricity, and others
Source Compiled from various NSS rounds

consumption. Electricity is one of the key constituents of the energy consumed at the household level. According to the latest edition of *Energy Statistics 2015* published by the Government of India, the domestic sector accounted for 22% of the total electricity consumed during 2013/14 (Figure 4). As per the 68th round of NSS, electricity constitutes 50% of the household fuel expenditure for urban households and 22% for rural households.

While all the 5161 towns (MoPNG 2015b) across all states and union territories are reported to be electrified, village electrification is still under implementation (CEA 2014). Electrification of 19 706 villages is yet to be carried out (Table 5) (GoI 2015)

During 2012/13, per capita ultimate consumption at the all-India level was 150.9 kWh, and more than 183 700 GWh of electrical energy was sold by utilities to the end consumers (Table 6). Bihar registered lowest per capita ultimate consumption of 28.64 kWh, followed by Assam (51.94 kWh) and Manipur (81.82 kWh). The highest per capita ultimate consumption was registered in Delhi at 575.98 kWh (CEA 2014). Overall, at the all-India level, the annual consumption of electricity by the domestic sector is increasing (Figure 5).

Similarly, in case of natural gas, the domestic sector is reported to have consumed 8.6% of the total 33.96 billion cubic metres of natural gas consumed during 2013/14. In the past few years, with the continued efforts of the government and enterprises, the number of domestic LPG consumers of public sector oil-marketing companies has been increasing.

Table 3b Percentage distribution of households by primary energy source for lighting (2001/02–2011/12)

NSS round	Period	Kerosene (%)	Electricity (%)	Other/no lighting arrangement (%)
Rural				
68	2011/12	27.97	71.24	0.78
66	2009/10	33.54	65.61	0.86
64	2007/08	38.6	60.2	1.2
63	2006/07	42.3	56.1	1.6
62	2005/06	42	56.3	1.5
61	2004/05	44.4	54.9	0.6
59	2003	46.6	51.6	1.7
57	2001/02	47.2	51.9	0.9
Urban				
68	2011/12	3.27	96.29	0.45
66	2009/10	4.9	93.8	1.3
64	2007/08	5.1	93.8	1.2
63	2006/07	6.4	92.7	0.9
62	2005/06	7.2	92	0.8
61	2004/05	7.1	92.3	0.6
59	2003	8.3	90.8	1
57	2001/02	7.8	91.4	0.8

NSS – National Sample Survey

Source Compiled from various NSS rounds

As per the *Indian Petroleum and Natural Gas Statistics 2014-15*[6] of the Government of India, the consumption of superior kerosene oil (SKO) is declining and the consumption of LPG for domestic use is rising (Table 7).

The government offers subsidies to end users to facilitate the shift towards modern fuels, such as LPG. The total subsidy under recoveries reported at the end of 2014/15 amounted to ₹63 902 crore for public distribution system (PDS), SKO, and LPG (MoPNG 2015b).

Key policies, programmes, and regulations for modern energy access in households

- **Electricity Act, 2003:** The Electricity Act, 2003 deals with domains, such as generation, transmission, distribution, and trade of power. This Act extends to the whole of India except the state of Jammu and Kashmir. The Act was amended twice, and the Electricity (Amendment) Act, 2007 came into force on 15 June 2007.
- **National Electricity Policy, 2005:** The Central Government, in compliance with Section 3 of the Electricity Act, 2003, notified the National Electricity Policy early in 2005. The policy recognized electricity as a critical infrastructure for the socio-economic development of the country.
- **Rural Electrification Policy, 2006:** In compliance with Sections 4 and 5 of the Electricity Act, 2003, the Central Government notified the Rural Electrification Policy in 2006.
- **Remote Village Electrification Programme:** The government launched the Remote Village Electrification Programme to advance basic lighting facilities in

[6] The programme is also applicable to institutional users such as Anganwadi kitchens of Mid-day Meal Scheme of the government.

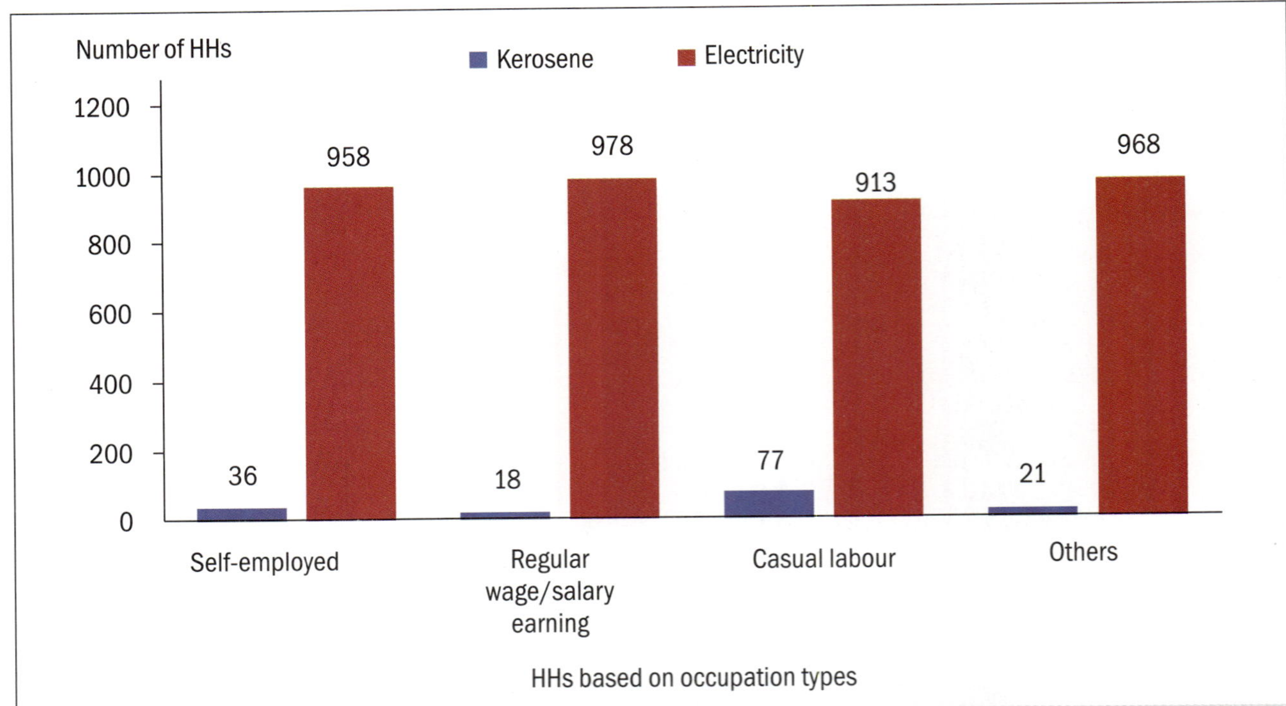

Figure 2a Distribution of urban households (per 1000) of each household type by the primary source of energy for lighting
HHs – households
Source NSSO (2015)

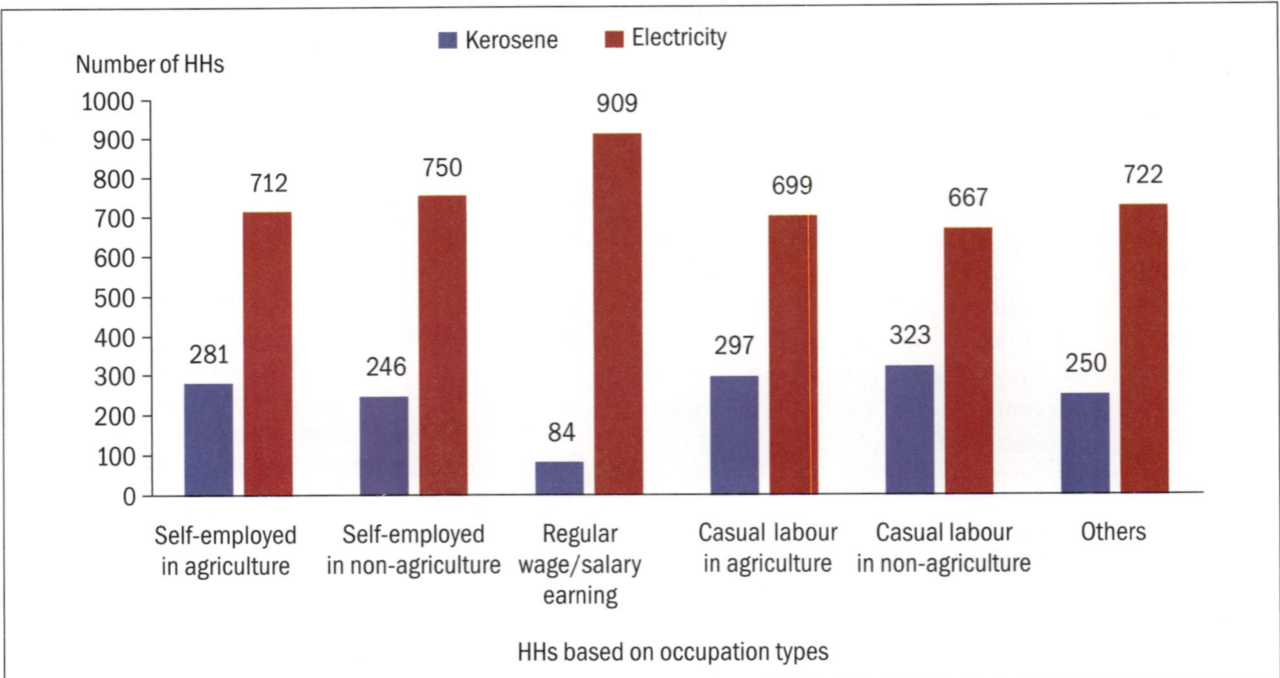

Figure 2b Distribution of households (per 1000) of each rural household type by the primary source of energy for lighting
HHs – households
Source NSSO (2015)

Table 4a Distribution of rural households (per 1000) of each household type by primary source of energy for cooking

All-India	Coke and coal	Firewood and chips	Liquefied petroleum gas	Gobar gas	Dung cake	Charcoal	Kerosene	Electricity	Others	No cooking arrangement
Self-employed in agriculture	8	706	108	5	128	0	3	0	42	0
Self-employed in non-agriculture	23	586	245	1	87	1	12	1	40	3
Regular wage/ salary earning	11	414	445	2	59	0	20	3	16	30
Casual labour in agriculture	6	804	46	0	58	0	7	0	79	1
Casual labour in non-agriculture	16	729	85	0	118	0	12	0	33	5
Others	10	489	220	1	83	0	14	1	42	141

Source NSSO (2015)

Table 4b Distribution of urban households (per 1000) of each household type by primary source of energy for cooking

All-India	Coke and coal	Firewood and chips	Liquefied petroleum gas	Gobar gas	Dung cake	Charcoal	Kerosene	Electricity	Others	No cooking arrangement
Self-employed	24	152	735	0	21	1	45	2	11	9
Regular wage/ salary earning	17	66	766	0	3	0	62	4	12	69
Casual labour	42	412	389	0	27	1	102	1	8	18
Others	6	66	569	0	7	1	29	8	8	306

Source NSSO (2015)

un-electrified census villages and hamlets that were not covered under the Rajiv Gandhi Grameen Vidyutikaran Yojana (RGGVY) and where electrification through grid extension was technically/financially not feasible.

- **National Biogas and Manure Management Programme**: The government also launched the National Biogas and Manure Management Programme, which has been active since 1981/82. It caters mainly to setting up family-type biogas plants.
- **Jawaharlal Nehru National Solar Mission:** The ambitious Jawaharlal Nehru National Solar Mission (JNNSM) or the National Solar Mission is one of the eight missions under the National Action Plan on Climate Change. The implementation of JNNSM comprises three phases: Phase 1 (2010–13), Phase 2 (2013–17), and Phase 3(2017–22).
- **Solar City:** In 2008, the government launched a programme on "Development of Solar Cities". The programme was revised in 2014 for implementation during the Twelfth Five-year Plan. The programme aims a 10% reduction in the projected demand for conventional energy by the end of the plan period.
- **Akshay Urja Shops:** To facilitate the availability of and easy after-sales services for solar products, the government started promoting the institutionalization of Aditya Solar Shops in 1995, which under present scheme has been renamed as Akshay Urja shops.

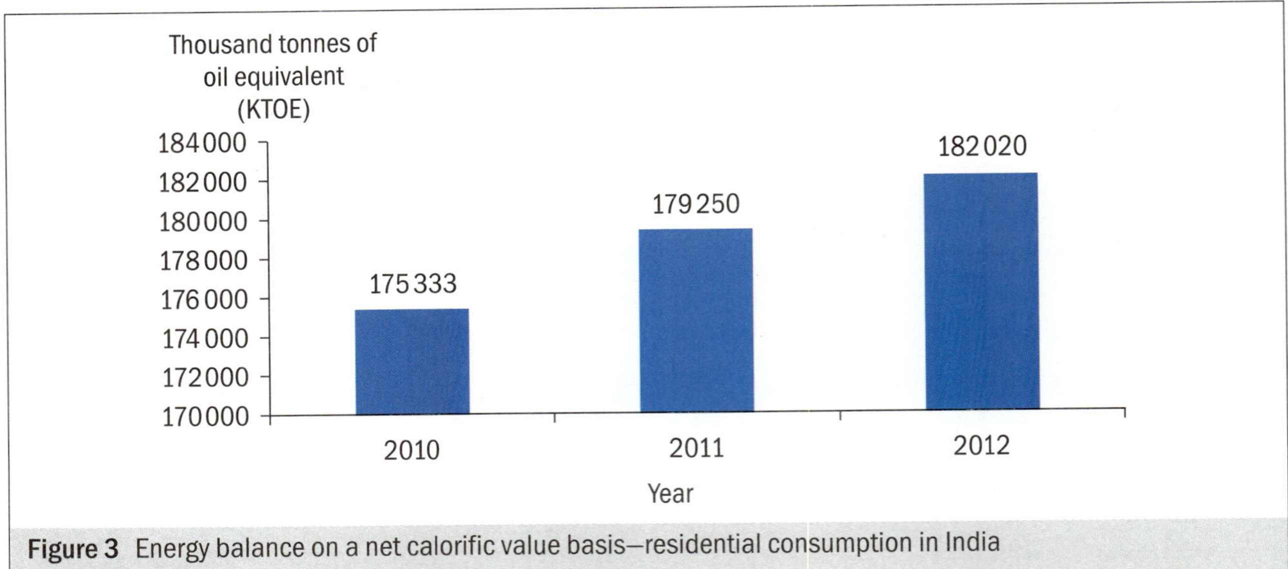

Figure 3 Energy balance on a net calorific value basis—residential consumption in India
Source IEA (2015b)

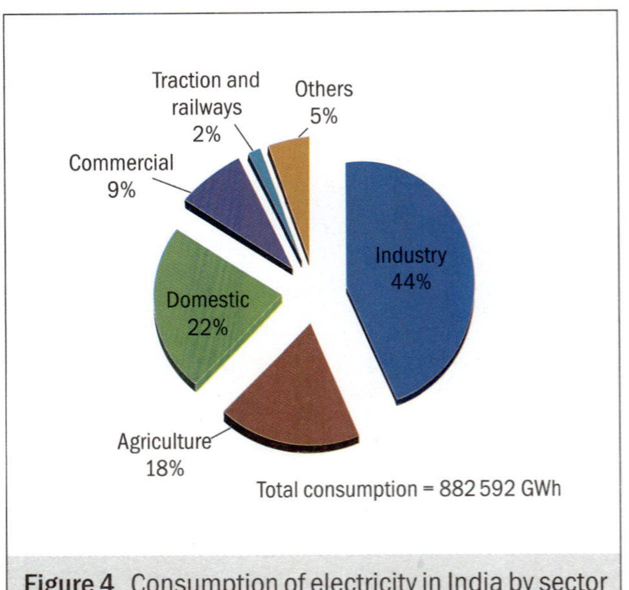

Figure 4 Consumption of electricity in India by sector (2013/14)
Source MoSPI (2015)

- **National Biomass Cookstove Programme:** Based on inputs from the National Biomass Cookstove Initiative, launched in 2009, the government launched the National Biomass Cookstove Programme for implementation in the Twelfth Five-year Plan.
- **Rajiv Gandhi Gramin LPG Vitaran Yojana:** To enhance the penetration of LPG, the government launched the Rajiv Gandhi Gramin LPG Vitaran (RGGLV) Yojana in 2009. The yojana aims at setting up small-sized LPG distribution agencies to reach remote and low potential areas [locations having potential of 600 cylinders (refill sales) per month].
- **Subsidy on PDS SKO and domestic LPG**: The Government of India provides SKO for cooking and lighting through PDS in all states and union territories on a quarterly basis.
- **Capping of LPG supply:** To reduce the subsidy burden, the government introduced a capping policy in September 2012. As per the existing scheme, the number of domestic cylinders for which subsidy can be availed has been capped at 12 refills per LPG connection per financial year.
- **Portability for selecting LPG distributor:** To facilitate customer experience to customers, the government introduced portability. A pilot scheme was launched in Chandigarh in 2013, which allowed consumers to select their distributor. Later, intercompany portability was launched in 24 cities in October 2013.
- **Direct Benefit Transfer of LPG Scheme:** The Cabinet Committee on Political Affairs approved the Direct Benefit Transfer Scheme in May 2013.
- **Unnat Chulha Abhiyan**: The Government of India launched the Unnat Chulha Abhiyan to catalyse the development and deployment of improved biomass cookstoves in India.

Table 5 Progress in village electrification in India

State/union territory	Total inhabited villages as per Census 2011	Villages electrified as on 30 March 2015 (P)	Villages electrified as on 30 March 2015 (P) (%)	Cumulative achievement as on 31 May 2015	Villages electrified as on 31 May 2015 (%)	Unelectrified villages as on 31 May 2015
Andhra Pradesh	16 158	16 158	100	16 158	100	0
Arunachal Pradesh	5 258	3 694	70.3	3 694	70.3	1 564
Assam	25 372	24 569	96.8	24 569	96.8	803
Bihar	39 073	37 316	95.5	37 316	95.5	1 757
Chhattisgarh	19 567	19 124	97.7	19 125	97.7	442
Goa	320	320	100	320	100	0
Gujarat	17 843	17 843	100	17 843	100	0
Haryana	6 642	6 642	100	6 642	100	0
Himachal Pradesh	17 882	17 828	99.7	17 828	99.7	54
Jammu and Kashmir	6 337	6 224	98.2	6 224	98.2	113
Jharkhand	29 492	27 328	92.7	27 387	92.9	2 105
Karnataka	27 397	27 363	99.88	27 363	99.9	34
Kerala	1 017	1 017	100	1 017	100	0
Madhya Pradesh	51 929	50 471	97.2	50 471	97.2	1 458
Maharashtra	40 956	40 920	99.9	40 920	99.9	36
Manipur	2 379	2 061	86.6	2 061	86.6	318
Meghalaya	6 459	5 176	80.1	5 176	80.1	1 283
Mizoram	704	659	93.6	659	93.6	45
Nagaland	1 400	1 271	90.8	1 271	90.8	129
Odisha	47 677	43 799	91.9	43 799	91.9	3 878
Punjab	12 168	12 168	100	12 168	100	0
Rajasthan	43 264	39 098	90.4	39 098	90.4	4 166
Sikkim	425	425	100	425	100	0
Tamil Nadu	15 049	15 049	100	15 049	100	0
Tripura	863	837	97	837	97	26
Telangana	10 128	10 128	100	10 128	100	0
Uttar Pradesh	97 813	96 515	98.7	96 515	98.7	1 298
Uttarakhand	15 745	15 638	99.3	15 638	99.3	107
West Bengal	37 463	37 461	99.99	37 461	99.99	2
Total (states)	596 780	577 102	96.7	577 162	96.7	19 618
Andaman and Nicobar Islands	396	308	77.8	308	77.8	88
Chandigarh	5	5	100	5	100	0
Dadra and Nagar Haveli	65	65	100	65	100	0
Daman and Diu	19	19	100	19	100	0

Contd...

Table 5 *Contd...*						
State/union territory	Total inhabited villages as per Census 2011	Villages electrified as on 30 March 2015 (P)	Villages electrified as on 30 March 2015 (P) (%)	Cumulative achievement as on 31 May 2015	Villages electrified as on 31 May 2015 (%)	Unelectrified villages as on 31 May 2015
Delhi	103	103	100	103	100	0
Lakshadweep	6	6	100	6	100	0
Puducherry	90	90	100	90	100	0
Total (union territories)	684	596	87.1	596	87.1	88
Total	597 464	577 698	96.7	577 758	96.7	19 706

P – provisional
Source GoI (2015)

Table 6 Electrical energy sales to ultimate consumers (utilities): 2012/13

State/union territory	Domestic
Northern region	
Haryana	5 554.81
Himachal Pradesh	1 616.58
Jammu and Kashmir	2 469.39
Punjab	9 284.90
Rajashthan	8 155.85
Uttar Pradesh	19 843.95
Uttarakhand	1 799.56
Chandigarh	686.54
Delhi	1 1116.35
Subtotal	60 427.93
Western region	
Gujarat	10 604.54
Madhya Pradesh	7 771.65
Chhattisgarh	3 807.42
Maharashtra	22 594.63
Goa	835.00
Daman and Diu	77.31
Dadra and Nagar Haveli	71.80
Subtotal	45 762.35
Southern region	
Andhra Pradesh	16 049.63
Karnataka	9 191.70

Contd...

Table 6 *Contd...*	
State/union territory	Domestic
Kerala	8 382.68
Tamil Nadu	17 834.00
Puducherry	589.81
Lakshadweep	27.13
Subtotal	52 074.96
Eastern region	
Bihar	2 858.03
Jharkhand[+]	3 780.00
Odisha	4 299.83
West Bengal[+]	11 070.87
Andaman and Nicobar Islands	100.20
Sikkim	140.58
Subtotal	22 249.51
North-eastern region	
Assam	1 619.00
Manipur	203.74
Meghalaya	403.63
Nagaland	220.50
Tripura	403.59
Arunachal Pradesh	137.65
Mizoram	197.59
Subtotal	3 185.70
Total (All-India)	183 700.45

[+]Includes Damodar Valley Corporation sales to ultimate consumers in Jharkhand and West Bengal
Source CEA (2014)

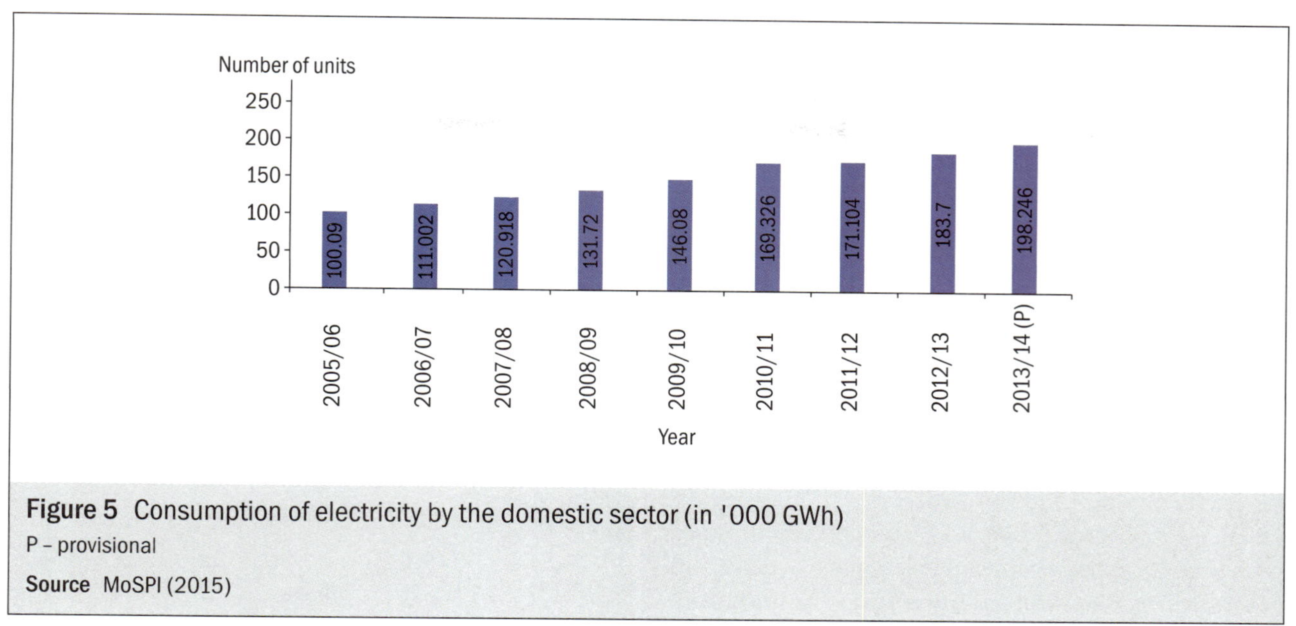

Figure 5 Consumption of electricity by the domestic sector (in '000 GWh)

P – provisional

Source MoSPI (2015)

Table 7 Consumption of LPG and SKO by the domestic sector

Domestic consumption of petroleum products		
Financial year	Consumption in '000 tonnes	
	LPG domestic distribution	SKO
2008/09	10 637.46	9 131.2
2009/10	11 364.33	9 101.0
2010/11	12 368.68	8 722.0
2011/12	13 295.91	8 045.0
2012/13	13 568.03	7 349.04
2013/14	14 411.60	7 008.86
2014/15 (P)	16 040.45	6 917.34

LPG – liquefied petroleum gas; P – provisional; SKO – superior kerosene oil

Source MoPNG (2015b)

For individual households, the programme is applicable only in rural areas.[7] Beyond deployment, the programme activities include research and development, support for test centres, development of standards and guidelines for testing of cookstoves, creation of enabling environment for mass production of processed biomass fuel, network of dealers, training of entrepreneurs, and supply chain mechanism. The programme has a target of 2.4 million family-type or household cookstoves during the Twelfth Five-year Plan.

- **Deendayal Upadhyaya Gram Jyoti Yojana:** The Government of India subsumed the RGGVY for rural electrification into its new scheme Deendayal Upadhyaya Gram Jyoti Yojana. The yojana aims to separate agriculture and non-agriculture feeders to facilitate a more judicious rostering of electricity supply in rural areas. It also aims to strengthen and augment the sub-transmission and distribution infrastructure and involve the metering of feeders, distribution transformers, and consumers. In addition, it aims to realize targets that were approved for implementation under the Twelfth and Thirteenth Five-year Plans.

[7] The programme is also applicable to institutional users such as Anganwadi kitchens for Mid-day Meal Scheme of the government.

Shyama Prasad Mukherji Rurban Mission

The Union Cabinet chaired by the Prime Minister of India announced the ambitious Shyama Prasad Mukherji Rurban Mission (SPMRM) in September 2015. This mission has an outlay of ₹5142.08 crore and aims to develop a cluster of "smart villages" across states and union territories of the country. The Mission aims to create 300 "rurban" growth clusters over the next 3 years. The 14 suggested components of development in each village include components for energy access too. The Mission will leverage funding for cluster development through various government schemes. Moreover, an additional funding support of up to 30% of the project cost per cluster as critical gap funding will be provided as centre's share to enable the development of such "rurban" clusters. The SPMRM includes an innovation budget for facilitating research, development, and capacity building.

References

CEA (Central Electricity Authority). 2014. ***All-India Electricity Statistics. General Review 2014***. New Delhi: CEA, Government of India

Chandramouli, C. 2011. Housing stock, amenities and assets in slums. Details available at https://censusindia.gov.in/2011-Documents/On_Slums-2011Final.ppt, last accessed on 22 December 2015

GoI (Government of India). 2015. Progress on village electrification. Details available at https://data.gov.in/catalog/progress-report-village-electrification, last accessed on 20 October 2015

IEA (International Energy Agency). 2015a. ***World Energy Outlook 2015***. Paris: IEA

IEA (International Energy Agency). 2015b. India: Balances for 2012. Details available at www.iea.org/statistics/statisticssearch/report/?country=INDIA&product=balances&year=2012, last accessed on 20 October 2015

MoPNG (Ministry of Petroleum and Natural Gas). 2015a. State-wise LPG domestic consumers. Details available at http://ppac.org.in/content/148_1_Marketing.aspx, last accessed on 20 October 2015

MoPNG (Ministry of Petroleum and Natural Gas). 2015b. ***India Petroleum and Natural Gas Statistics 2014–15***. New Delhi: Economics and Statistics Division, MoPNG, Government of India

MoSPI (Ministry of Statistics and Programme Implementation). 2015. ***Energy Statistics 2015***. New Delhi: MoSPI, Government of India

NSSO (National Sample Survey Organization). 2015. ***Energy Sources of Indian Households for Cooking and Lighting, 2011–12. NSS 68th Round: JULY 2011–JUNE 2012***. New Delhi: MoSPI, Government of India

SE4ALL (Sustainable Energy for All). 2015. ***Progress towards Sustainable Energy: Global Tracking Framework. Summary Report 2015***. Vienna: SE4ALL

TABLE 10.1a State-wise distribution of households based on source of energy for lighting

State code	India/state/union territory	Total number of households	Electricity	Kerosene	Solar energy	Other oil	Any other	No lighting
1	2	3	4	5	6	7	8	9
00	India	246 692 667	67.3	31.4	0.4	0.2	0.2	0.5
01	Jammu and Kashmir	2 015 088	85.1	9.7	1.0	0.2	2.0	2.0
02	Himachal Pradesh	1 476 581	96.8	2.8	0.1	0.1	0.1	0.1
03	Punjab	5 409 699	96.6	2.2	0.1	0.2	0.3	0.7
04	Chandigarh	235 061	98.4	1.2	0.1	0.1	0.1	0.2
05	Uttarakhand	1 997 068	87.0	11.1	1.2	0.2	0.2	0.3
06	Haryana	4 717 954	90.5	8.1	0.2	0.3	0.5	0.5
07	NCT of Delhi	3 340 538	99.1	0.7	0.1	0	0.1	0.1
08	Rajasthan	12 581 303	67.0	30.9	0.6	0.3	0.3	0.8
09	Uttar Pradesh	32 924 266	36.8	61.9	0.5	0.3	0.3	0.2
10	Bihar	18 940 629	16.4	82.4	0.6	0.3	0.3	0.1
11	Sikkim	128 131	92.5	6.6	0.2	0.1	0.1	0.5
12	Arunachal Pradesh	261 614	65.7	18.5	2.9	0.3	2.1	10.5
13	Nagaland	399 965	81.6	15.6	0.3	0.2	1.1	1.1
14	Manipur	507 152	68.4	25.1	1.9	0.4	3.6	0.6
15	Mizoram	221 077	84.2	13.5	1.3	0.3	0.3	0.3
16	Tripura	842 781	68.4	29.1	1.9	0.2	0	0.3
17	Meghalaya	538 299	60.9	37.0	0.8	0.3	0.2	0.8
18	Assam	6 367 295	37.1	61.8	0.8	0.1	0.1	0.2
19	West Bengal	20 067 299	54.5	43.5	1.2	0.2	0.1	0.5
20	Jharkhand	6 181 607	45.8	53.1	0.7	0.2	0.1	0.1
21	Odisha	9 661 085	43.0	55.3	0.4	0.1	0.1	1.1
22	Chhattisgarh	5 622 850	75.3	23.2	0.9	0.2	0.1	0.3
23	Madhya Pradesh	14 967 597	67.1	32.1	0.3	0.2	0.1	0.2
24	Gujarat	12 181 718	90.4	8.1	0.1	0.2	0.2	1.0
25	Daman and Diu	60 381	99.1	0.8	0	0	0.1	0.1
26	Dadra and Nagar Haveli	73 063	95.2	4.4	0	0	0	0.3
27	Maharashtra	23 830 580	83.9	14.5	0.2	0.2	0.3	0.9
28	Andhra Pradesh	21 024 534	92.2	6.9	0.3	0.2	0.1	0.4
29	Karnataka	13 179 911	90.6	8.6	0.2	0.1	0.1	0.4

Contd...

TABLE 10.1a | *Contd...*

State code	India/state/union territory	Total number of households	Electricity	Kerosene	Solar energy	Other oil	Any other	No lighting
1	2	3	4	5	6	7	8	9
30	Goa	322 813	96.9	2.4	0.2	0.1	0.1	0.4
31	Lakshadweep	10 703	99.7	0.2	0	0	0.1	0
32	Kerala	7 716 370	94.4	5.2	0.2	0.1	0.1	0
33	Tamil Nadu	18 493 003	93.4	5.9	0.1	0.2	0.1	0.4
34	Puducherry	301 276	97.7	2.1	0	0.1	0	0.2
35	Andaman and Nicobar Islands	93 376	86.1	12.9	0.2	0.3	0.1	0.5

Source Details available at <www.censusindia.gov.in/2011census/hlo/Data_sheet/India/HLO_Indicators.pdf>, last accessed on 17 November 2015

TABLE 10.1b State-wise distribution of households by type of fuel used for cooking

State code	India/state/union territory	Households (excluding institutional households)	Firewood	Crop residue	Cow dung cake	Coal, lignite, charcoal	Kerosene	LPG	Electricity	Biogas	Any other	No cooking
1	2	3	4	5	6	7	8	9	10	11	12	13
00	India	246 692 667	49.0	8.9	8.0	1.5	2.9	28.6	0.1	0.4	0.5	0.3
01	Jammu and Kashmir	2 015 088	58.9	2.5	4.2	0	1.3	31.6	0.4	0.8	0.2	0.2
02	Himachal Pradesh	1 476 581	57.5	1.1	0.2	0	2.1	38.6	0.2	0.1	0	0.3
03	Punjab	5 409 699	13.4	6.5	20.4	0.2	3.2	54.5	0	1.4	0.1	0.3
04	Chandigarh	235 061	4.6	0.3	0.2	0.1	21.9	71.6	0	0.1	0.1	1.1
05	Uttarakhand	1 997 068	48.7	1.3	3.2	0.1	1.8	44.2	0	0.5	0	0.3
06	Haryana	4 717 954	26.1	14.1	14.2	0.1	1.0	44.0	0	0.3	0.1	0.2
07	NCT of Delhi	3 340 538	3.4	0.3	0.6	0.1	5.3	89.9	0	0.1	0.1	0.3
08	Rajasthan	12 581 303	61.8	11.0	3.0	0.1	0.9	22.8	0	0.1	0.1	0.2
09	Uttar Pradesh	32 924 266	47.7	8.7	23.1	0.3	0.7	18.9	0.1	0.2	0.1	0.2
10	Bihar	18 940 629	34.7	32.5	21.7	1.0	0.3	8.1	0.1	0.3	1.2	0.2
11	Sikkim	128 131	52.5	0.6	0.2	0.1	4.4	41.3	0.3	0.1	0	0.6
12	Arunachal Pradesh	261 614	68.7	0.7	0.1	0	0.7	29.2	0.1	0.1	0.1	0.3
13	Nagaland	399 965	77.9	0.8	0.1	0	0.6	20.0	0.1	0.1	0.1	0.2
14	Manipur	507 152	65.7	1.1	0.2	2.1	0.2	29.7	0.1	0.2	0.6	0.1
15	Mizoram	221 077	44.5	0.3	0.1	0.4	1.8	52.6	0.2	0.1	0.1	0.1
16	Tripura	842 781	80.5	0.8	0.1	0.1	0.6	17.6	0	0.1	0.1	0.1
17	Meghalaya	538 299	79.0	0.9	0.3	2.3	3.7	11.9	1.6	0.2	0.1	0.2
18	Assam	6 367 295	72.1	6.4	0.9	0.1	0.6	19.0	0.1	0.1	0.4	0.4
19	West Bengal	20 067 299	33.1	25.6	10.0	7.9	2.1	18.0	0.1	0.3	2.7	0.3
20	Jharkhand	6 181 607	57.6	4.0	7.2	18.1	0.2	11.7	0.3	0.1	0.6	0.1
21	Odisha	9 661 085	65.0	10.2	9.4	1.6	1.1	9.8	0.4	0.2	2.0	0.3
22	Chhattisgarh	5 622 850	80.8	0.9	3.7	2.3	0.5	11.2	0.1	0.2	0.1	0.2
23	Madhya Pradesh	14 967 597	66.4	5.6	7.7	0.2	1.3	18.2	0.1	0.4	0.1	0.2
24	Gujarat	12 181 718	44.0	5.7	2.6	0.5	7.6	38.3	0	0.9	0.1	0.4
25	Daman and Diu	60 381	10.8	1.5	0.2	0.2	30.8	53.0	0.1	0.9	0.1	2.4
26	Dadra and Nagar Haveli	73 063	40.4	0.4	0.2	0.1	17.8	39.8	0	0.4	0	0.8
27	Maharashtra	23 830 580	42.6	4.5	1.2	0.2	6.5	43.4	0.1	0.7	0.1	0.8
28	Andhra Pradesh	21 024 534	56.8	1.4	0.6	0.3	3.9	35.8	0.1	0.7	0.1	0.3
29	Karnataka	13 179 911	57.5	2.9	0.2	0.1	5.4	32.5	0.1	0.9	0.1	0.3
30	Goa	322 813	20.7	0.9	0.2	0.1	4.1	72.7	0.1	0.4	0.1	0.7
31	Lakshadweep	10 703	54.8	10.7	0.1	0.1	13.7	16.6	1.2	0.2	0	2.5
32	Kerala	7 716 370	61.9	0.8	0.1	0.1	0.4	35.8	0	0.6	0	0.3

Contd...

TABLE **10.1b** *Contd...*

State code	India/state/union territory	Households (excluding institutional households)	Firewood	Crop residue	Cow dung cake	Coal, lignite, charcoal	Kerosene	LPG	Electricity	Biogas	Any other	No cooking
1	2	3	4	5	6	7	8	9	10	11	12	13
33	Tamil Nadu	18 493 003	43.5	0.6	0.2	0.1	6.9	47.9	0.1	0.3	0.1	0.4
34	Puducherry	301 276	18.0	0.3	0.1	0	10.3	70.5	0.1	0.1	0	0.6
35	Andaman and Nicobar Islands	93 376	33.8	0.4	0	0	19.8	44.5	0	0	0.1	1.4

LPG – liquefied petroleum gas

Source Details available at <www.censusindia.gov.in/2011census/hlo/Slum_table/hl-slum/SHH3410-crc.pdf>, last accessed on 17 November 2015

TABLE 10.2a Source of energy for domestic lighting in rural India

State code	India/state/union territory	Rural									
		Total households		Percentage of households having							
				Electricity		Kerosene		Other sources		No lighting	
		2011	2001	2011	2001	2011	2001	2011	2001	2011	2001
1	2	13	14	15	16	17	18	19	20	21	22
00	India	167 826 730	138 271 559	55.3	43.5	43.2	55.6	1.0	0.6	0.5	0.3
01	Jammu and Kashmir	1 497 920	1 161 357	80.7	74.8	12.6	19.2	4.1	5.4	2.6	0.6
02	Himachal Pradesh	1 310 538	1 097 520	96.6	94.5	3.0	4.9	0.3	0.4	0.1	0.2
03	Punjab	3 315 632	2 775 462	96.5	89.5	2.9	8.9	0.7	0.5	0.9	1.1
04	Chandigarh	6 785	21 302	97.3	97.4	2.4	2.1	0.2	0.3	0.1	0.2
05	Uttarakhand	1 404 845	1 196 157	83.1	50.3	14.5	46.7	2.1	2.7	0.3	0.3
06	Haryana	2 966 053	2 454 463	87.2	78.5	11.3	20.6	1.0	0.5	0.5	0.4
07	NCT of Delhi	79 115	169 528	97.8	85.5	1.4	13.0	0.6	0.9	0.2	0.6
08	Rajasthan	9 490 363	7 156.703	58.3	44.0	39.3	54.7	1.5	0.8	1.0	0.5
09	Uttar Pradesh	25 475 071	20 590 074	23.8	19.8	75.0	79.5	1.1	0.6	0.1	0.1
10	Bihar	16 926 958	12 560 007	10.4	5.1	88.4	94.5	1.2	0.4	0.1	0
11	Sikkim	92 370	91 721	90.2	75.0	8.7	24.3	0.5	0.3	0.6	0.4
12	Arunachal Pradesh	195 723	164 501	55.5	44.5	23.6	37.9	7.0	7.1	14.0	10.5
13	Nagaland	284 911	265 334	75.2	56.9	21.1	37.5	2.2	2.4	1.5	3.2
14	Manipur	335 752	296 354	61.2	52.5	32.2	45.1	5.9	1.2	0.7	1.1
15	Mizoram	204 874	79 362	68.8	44.1	26.9	52.8	3.9	2.3	0.5	0.7
16	Tripura	607 779	539 680	59.5	31.8	37.7	67.6	2.5	0.4	0.4	0.2
17	Meghalaya	422 197	329 678	51.6	30.3	45.9	68.2	1.6	0.8	0.9	0.8
18	Assam	5374 553	4 220 173	28.4	16.5	70.4	83.1	1.1	0.3	0.2	0.1
19	West Bengal	13 717 186	11 161 870	40.3	20.3	57.8	79.2	1.5	0.5	0.4	0.1
20	Jharkhand	4 685 965	3 802 412	32.3	10.0	66.4	89.6	1.2	0.3	0	0
21	Odisha	8 144 012	6 782 879	35.6	19.4	62.8	79.8	0.6	0.5	1.1	0.4
22	Chhattisgarh	4 384 112	3 359 078	70.0	46.1	28.2	52.9	1.5	0.6	0.3	0.4
23	Madhya Pradesh	11 122 365	8 124 795	58.3	62.3	40.9	37.2	0.6	0.3	0.2	0.2
24	Gujarat	6 765 403	5 885 961	85.0	72.1	12.8	26.2	0.8	0.7	1.4	1.0

Contd...

TABLE 10.2a | *Contd...*

State code	India/state/union territory	Rural									
		Total households		Percentage of households having							
				Electricity		Kerosene		Other sources		No lighting	
		2011	2001	2011	2001	2011	2001	2011	2001	2011	2001
1	2	13	14	15	16	17	18	19	20	21	22
25	Daman and Diu	12 750	22 091	98.3	97.5	1.5	2.0	0.1	0.2	0.2	0.4
26	Dadra and Nagar Haveli	35 408	32 783	91.7	82.6	7.6	16.0	0.1	0.3	0.6	1.0
27	Maharashtra	13 016 652	10 993 623	73.8	65.2	23.9	33.6	1.0	0.7	1.3	0.6
28	Andhra Pradesh	14 246 309	12 676 218	89.7	59.7	9.2	39.7	0.6	0.4	0.5	0.3
29	Karnataka	7 864 196	6 675 173	86.7	72.2	12.3	27.2	0.5	0.3	0.5	0.3
30	Goa	124 674	140 173	95.6	92.4	3.4	6.9	0.3	0.3	0.6	0.4
31	Lakshadweep	2 523	5 351	99.8	99.7	0.2	0.1	0	0.1	0	0
32	Kerala	4 095 674	4 942 550	92.1	65.5	7.4	33.8	0.4	0.7	0	0
33	Tamil Nadu	9 563 899	8 274 790	90.8	71.2	8.3	25.2	0.3	0.3	0.6	0.4
34	Puducherry	95 133	72 199	95.8	81.0	3.6	18.6	0.2	0.2	0.5	0.1
35	Andaman and Nicobar Islands	59 030	49 653	79.4	68.1	19.3	29.9	0.8	1.2	0.6	0.8

Source Details available at <www.censusindia.gov.in/2011census/hlo/Data_sheet/Source%20of%20Lighting.pdf>, last accessed on 17 November 2015

TABLE 10.2b Source of energy for domestic lighting in urban India

State code	India/state/ union territory	Urban									
		Total households		Percentage of households having							
				Electricity		Kerosene		Other sources		No lighting	
		2011	2001	2011	2001	2011	2001	2011	2001	2011	2001
1	2	23	24	25	26	27	28	29	30	31	32
00	India	78 865 937	53 692 376	92.7	87.6	6.5	11.6	0.5	0.5	0.3	0.4
01	Jammu and Kashmir	517 168	390 411	98.0	97.9	1.2	1.6	0.6	0.4	0.2	0.1
02	Himachal Pradesh	166 043	143 113	98.1	97.4	1.6	2.2	0.2	0.3	0.1	0.1
03	Punjab	2 094 067	1 489 694	98.3	96.5	1.0	2.6	0.3	0.4	0.3	0.5
04	Chandigarh	228 276	180 576	98.4	96.7	1.1	2.9	0.2	0.2	0.2	0.2
05	Uttarakhand	592 233	390 164	96.5	90.9	2.9	8.4	0.4	0.4	0.3	0.3
06	Haryana	1 751 901	1 075 179	96.2	92.9	2.7	6.1	0.7	0.5	0.4	0.4
07	NCT of Delhi	3 261 423	2 384 621	99.1	93.4	0.6	5.7	0.2	0.7	0.1	0.2
08	Rajasthan	3 090 940	2 185 591	93.9	89.6	5.2	9.6	0.5	0.4	0.5	0.4
09	Uttar Pradesh	7 449 195	5 170 527	81.4	79.9	17.2	19.3	0.9	0.5	0.4	0.3
10	Bihar	2 013 671	1 322 583	66.7	59.3	32.2	39.9	0.8	0.7	0.2	0.2
11	Sikkim	35 761	13 015	98.7	97.1	0.9	2.8	0.1	0.1	0.3	0
12	Arunachal Pradesh	65 891	48 114	96.0	89.4	3.2	9.4	0.4	0.5	0.4	0.7
13	Nagaland	115 054	66 716	97.4	90.3	2.1	8.3	0.3	0.5	0.2	0.9
14	Manipur	171 400	101 302	82.4	82.0	11.2	17.3	5.9	0.5	0.5	0.3
15	Mizoram	116 203	81 604	98.1	94.4	1.5	5.2	0.3	0.2	0.1	0.2
16	Tripura	235 002	122 343	91.6	86.4	7.0	13.0	1.2	0.4	0.2	0.2
17	Meghalaya	116 102	90 568	94.9	88.1	4.4	10.9	0.3	0.5	0.3	0.5
18	Assam	992 742	715 185	84.1	74.3	15.2	25.0	0.5	0.6	0.3	0.2
19	West Bengal	6 350 113	4 554 045	85.1	79.6	12.7	19.5	1.6	0.6	0.6	0.3
20	Jharkhand	1 495 642	1 060 178	88.0	75.6	11.4	23.8	0.5	0.5	0.1	0.1
21	Odisha	1 517 073	1 087 248	83.1	74.1	15.3	24.3	0.4	0.7	1.2	1.0
22	Chhattisgarh	1 238 738	789 440	93.7	82.9	5.7	16.5	0.3	0.3	0.2	0.3
23	Madhya Pradesh	3 845 232	2 794 858	92.7	92.3	6.6	7.1	0.4	0.3	0.2	0.2
24	Gujarat	5 416 315	3 758 028	97.2	93.4	2.1	5.5	0.2	0.5	0.5	0.7
25	Daman and Diu	47 631	12 251	99.3	98.3	0.6	1.3	0.1	0.1	0.1	0.3

Contd...

TABLE 10.2b *Contd...*

State code	India/state/ union territory	Urban									
		Total households		Percentage of households having							
				Electricity		Kerosene		Other sources		No lighting	
		2011	2001	2011	2001	2011	2001	2011	2001	2011	2001
1	2	23	24	25	26	27	28	29	30	31	32
26	Dadra and Nagar Haveli	37 655	11 190	98.5	95.8	1.4	3.8	0	0.3	0.1	0
27	Maharashtra	10 813 928	8 069 526	96.2	94.3	3.1	5.1	0.3	0.3	0.4	0.3
28	Andhra Pradesh	6 778 225	4 173 639	97.3	90.0	2.0	9.2	0.5	0.5	0.2	0.3
29	Karnataka	5 315 715	3 556 960	96.4	90.5	3.1	8.8	0.3	0.4	0.2	0.3
30	Goa	198 139	138 461	97.7	54.7	1.8	4.6	0.3	0.3	0.3	0.4
31	Lakshadweep	8 180	3 889	99.7	0.2	0.2	0.3	0.1	0	0	0
32	Kerala	3 620 696	1 652 656	97.0	84.3	2.8	15.1	0.2	0.5	0	0
33	Tamil Nadu	8 929 104	5 898 836	96.1	88.0	3.4	11.1	0.2	0.4	0.3	0.4
34	Puducherry	206 143	135 456	98.5	91.4	1.3	8.2	0.1	0.2	0.1	0.1
35	Andaman and Nicobar Islands	34 346	23 409	97.7	95.2	1.9	4.3	0.1	0.2	0.2	0.2

Source Details available at <www.censusindia.gov.in/2011census/hlo/Data_sheet/Source%20of%20Lighting.pdf>, last accessed on 17 November 2015

TABLE 10.3a Source of energy for lighting in slum households

Total number of households	Electricity	Kerosene	Solar energy	Other oil	Any other	No lighting
13 749 424	12 448 161	1 133 954	37 543	22 848	33 469	73 449

Source Details available at <www.censusindia.gov.in/2011census/hlo/Slum_table/hl-slum/SHH2507-crc.pdf>, last accessed on 17 November 2015

TABLE 10.3b Source of energy for cooking in slum households

Total number of households	Firewood	Crop residue	Cow dung cake	Coal/ lignite/ charcoal	Kerosene	LPG/PNG	Electricity	Biogas	Any other	No cooking	
13 749 424	3 542 928	222 330	282 285	541 205	1 922 158	7 047 692	16 744	62 262	37 489	74 331	Total
12 935 856	3 050 854	191 830	252 715	502 090	1 871 239	6 958 016	15 988	60 877	32 247	–	Cooking inside house
8 976 710	1 580 914	100 524	96 548	204 395	1 092 519	5 825 229	11 507	50 503	14 571	–	Has kitchen
3 959 146	1 469 940	91 306	156 167	297 695	778 720	1 132 787	4 481	10 374	17 676	–	Does not have kitchen
739 237	492 074	30 500	29 570	39 115	50 919	89 676	756	1 385	5 242	–	Cooking outside house
278 215	167 706	15 395	12 690	15 130	18 769	45 731	390	769	1 635	–	Has kitchen
461 022	324 368	15 105	16 880	23 985	32 150	43 945	366	616	3 607	–	Does not have kitchen
74 331	–	–	–	–	–	–	–	–	–	74 331	

LPG – liquefied petroleum gas; PNG – petroleum and natural gas

Source Details available at <www.censusindia.gov.in/2011census/hlo/Slum_table/hl-slum/SHH3410-crc.pdf>, last accessed on 17 November 2015

LOCAL AND GLOBAL ENVIRONMENT

- ENVIRONMENT
- CLIMATE CHANGE

LOCAL AND GLOBAL ENVIRONMENT

ENVIRONMENT 11

AT A GLANCE

- The Ministry of Environment, Forests, and Climate Change is in the process of revising solid waste management, plastic waste, biomedical waste, and e-waste rules.
- Particulate matter concentration exceeded the prescribed limit in 83% of the monitored locations in 2012.
- The Government of India launched the "Clean India Campaign" or "Swachh Bharat Abhiyan" on 2 October 2014. The campaign aims to develop a clean, green, and hygienic India by involving the government and the people in the mission.
- The Central Ground Water Board has estimated the average annual utilizable water in India at 1123 billion cubic metres (BCM), of which 433 BCM is groundwater.
- India's forest cover is 701 673 km^2, which is 21.34% of the total geographical area of the country.
- Compared to 2014, wildlife sanctuaries have increased from 525 to 531, conservation reserves have increased from 60 to 66, and community reserves have seen an exponential increase from 4 to 26 in the country. The number of national parks has remained the same.

Introduction

One of the biggest challenges for human beings in the twenty-first century is coping with the impacts of environmental pollution. The human race has gone through some major lifestyle changes in the last century owing to growing industrial production, increased use of automobiles, and other developments. Although these changes have added to the quality of life, they have also altered the environment almost irreversibly.

Figure 1 shows the percentage of the total population living in urban areas in 2014. It can be seen from the figure that 53% of the world population was living in urban areas in 2014. While 85% of people in Brazil lived in urban settlement, the corresponding figure for India is just 32%.

It is clear from Figure 1 that more than half of the world population live in urban areas, and the number is expected to rise in the future. This will have a direct impact on the environment. Thus, to have a sustainable future, it is most important to frame policies and develop technologies that incorporate environmental needs along with the ever-increasing human needs. This chapter discusses the current status of environmental pollution in India and the steps taken by the government and other agencies to abate it.

Air

Clean air is undeniably one of the key reasons for life on earth. In order to live a healthy life, human beings need clean air. However, the situation is

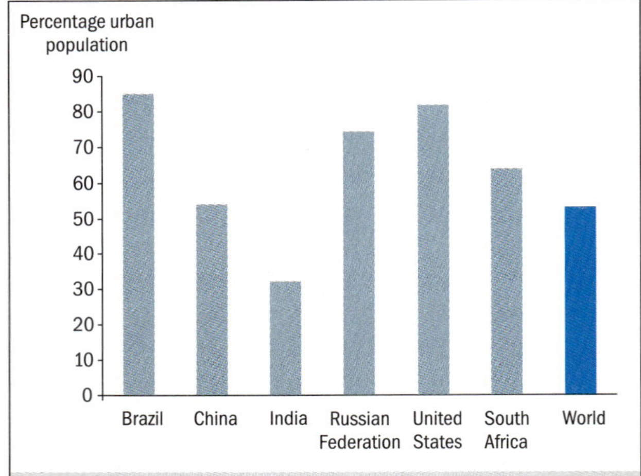

Figure 1 Urban population as a percentage of the total population of some selected countries
Source World Bank (2014)

gradually worsening. According to the Air Act, 1981, air pollution is defined as "the presence in the atmosphere of any air pollutant", where "air pollutant" is "any solid, liquid, or gaseous substance (including noise) present in the atmosphere in such concentration as may be or tend to be injurious to human beings or other living creatures or plant or property or environment". Metropolitan cities in developing countries such as India and China are severely suffering from air pollution. Particulate matter, sulphur dioxide (SO_2), and oxides of nitrogen (NO_x) are some of the key pollutants that have polluted cities globally. Sources of air pollution may vary according to the subject area. While the major sources of air pollution are vehicular and industrial emissions in urban areas, biomass used for cooking and space heating and unregulated burning of refuse in the agricultural sector are the major sources of pollution in rural areas.

Ambient air pollution has been identified as the fifth biggest cause of mortality in India (Atkinson, Cohen, Mehta, *et al.* 2011). As per the 2013 Global Burden of Disease Report, ambient air pollution is the fifth largest killer after high blood pressure, indoor air pollution, tobacco smoking, and poor nutrition, accounting about 620 000 premature deaths in India every year. Thus, air pollution has been identified as a national problem. As a remedial action, the Central Pollution Control Board (CPCB) initiated National Air Quality Monitoring Programme (NAMP) in 1984. As on 31 March 2011, there were 456 monitoring stations spread across 190 cities in 26 states and 5 union territories. The key idea behind NAMP is to understand and obtain knowledge about environmental pollution, its status and trends, natural dilution and dispersion phenomenon, and identification of non-attainment cities. This knowledge is eventually used to develop preventive and corrective measures. Under NAMP, the major air pollutants monitored regularly are SO_2, NO_x, and respirable suspended particulate matter (RSPM, PM_{10} and $PM_{2.5}$). Additional parameters such as respirable lead, ozone, carbon monoxide (CO), and benzene are also being monitored in a few cities.

Status of air quality

Details of the ambient air quality in various cities of India in terms of SO_2, NO_x, and RSPM concentrations for 1995–2012 are provided in Table 11.1. As shown in Figure 2, of the major criteria pollutants, RSPM is of prime concern as it exceeds the ambient air quality standards at 83% of the locations (CPCB 2014). The table also shows that the annual average levels of SO_2 are well within the annual average National Ambient Air Quality Standards (NAAQS) prescribed by the CPCB (50 $\mu g/m^3$) at most of the locations, while NO_x concentrations are found to be on the verge of violation (40 $\mu g/m^3$) at a majority of locations and violating the standards at other locations. Table 11.2 shows the monitoring stations, depicting the percentage violation of prescribed standards for SO_2, NO_x, and SPM during 2001–12.

In 2012, out of all the stations in residential/industrial/rural/commercial areas, 99% of stations indicated low concentrations of SO_2, and 14% and 83% showed violations of NAAQS for NO_2 and PM_{10}, respectively (Figure 2).

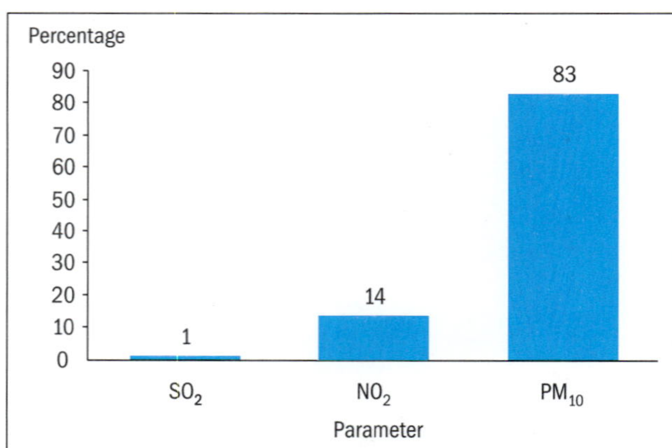

Figure 2 Percentage of locations in residential/industrial/commercial/rural areas exceeding NAAQS
Source CPCB (2014)

Figure 3 shows the NO_2 concentrations of top 10 cities. The highest level is 87 µg/m^3 in Dombivali. The annual average levels of NO_2 in all 10 cities are above the NAAQS (40 µg/m^3, represented by the red line in the figure).

The annual average NAAQS for PM_{10} is 60 µg/m^3 (red line in Figure 4). The figure clearly shows that all the top 10 cities have violated these standards. The PM_{10} level in Allahabad is almost five times the prescribed standards.

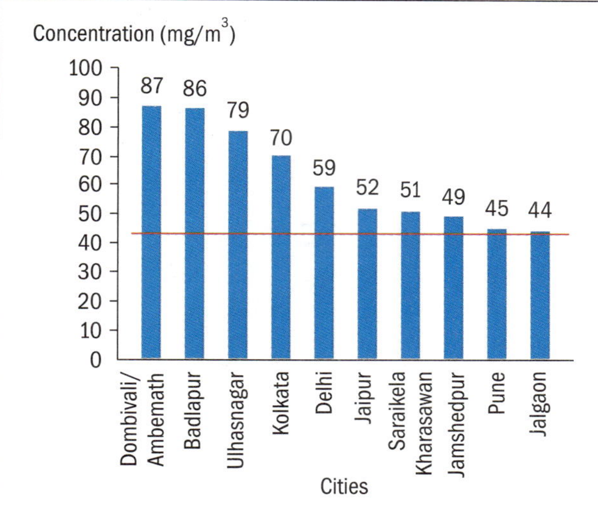

Figure 3 Top 10 cities with highest NO_2 values (annual average) in 2012
Source CPCB (2014)

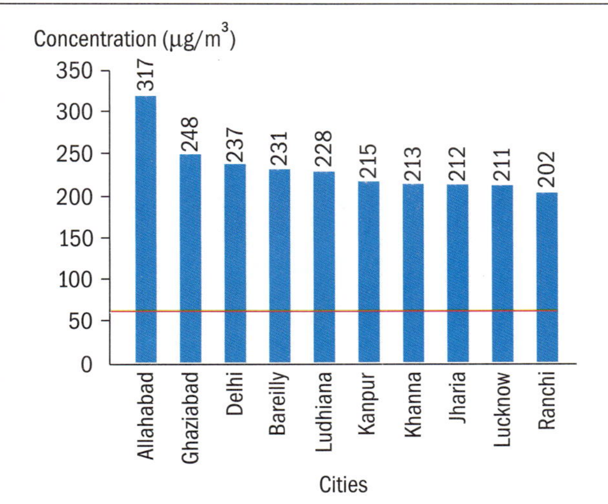

Figure 4 Top 10 cities with the highest PM_{10} values (annual average) in 2012
Source CPCB (2014)

It is evident from the observed data for various pollutants that the rising concentration of particulate matter is a major concern at the national level, and there is an urgent need to address this PM_{10} problem. The government must take early initiatives to curb the growth of this pollutant in the atmosphere because of the major health impacts of particulate matter.

Impacts of air pollution

Air quality is not just an environmental issue but also a public health issue. According to the Global Burden Disease Study 2010, air pollution is the fifth leading cause of death in India. Studies suggest that elevated concentrations of PM_{10} in the ambient air could result in the decreased lung function, high mortality (especially cardiovascular and respiratory), increased incidence of respiratory symptoms, an increased number of patients in hospitals, and increased lung cancer cases (Pope 3rd 1995). While the direct impacts of NO_2 on human health are still being investigated, certain toxicological studies reveal that NO_2 exposure can potentially lead to pulmonary inflammation and decreased respiratory defence mechanism. NO_2 can also be correlated with reduced lung function and worsening symptoms among asthma patients (Mukulai, Pekkaneni, Tiittaneni, *et al.* 2001). Around 400 000–550 000 premature deaths in India are due to only indoor air pollution (Dey, Girolamo, Donkelaar, *et al.* 2012). A recent study by Lakshmi, Virdi, Sharma, *et al.* (2013) using India's 2003/04 District Level Household Survey II data concluded that the usage of biomass and kerosene cooking fuels at home was related to stillbirths in India. Nearly 12% of these stillbirths could be prevented if cleaner fuels are used (Lakshmi, Virdi, Sharma, *et al.* 2013). Figure 5 shows the prevalence of active tuberculosis among persons aged 20 years or older, acute respiratory infection (ARI) among children under 3 years of age, and partial/complete blindness among women aged 30 years or older by type of cooking fuel used in India.

However, to improve air quality and reduce the burden of diseases, several interventions have been adopted.

Interventions

- **Environmental standards:** Revised ambient air quality standards were notified in 2009, under which 12 pollutants were measured compared to six in the previous standards (Table 11.4).

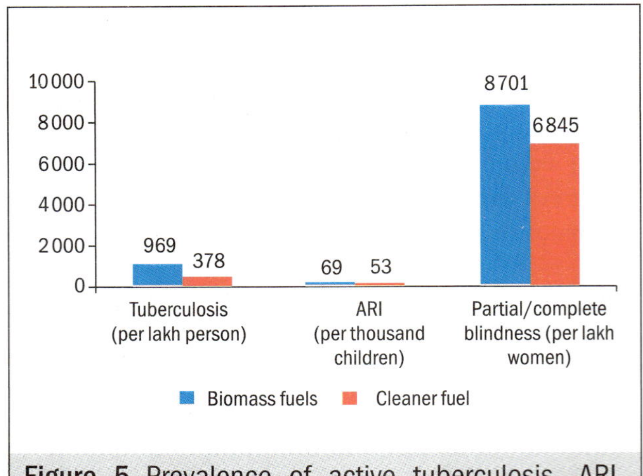

Figure 5 Prevalence of active tuberculosis, ARI, and partial/complete blindness in India by type of cooking fuel used

ARI – acute respiratory infection

Source Mishra, Retherford, and Smith (2002)

- **Vehicular pollution control:** Vehicular emission standards were first set in India in 1986 by the CPCB for both petrol- and diesel-driven vehicles, which were later revised in 1987 and 1989. To further curb the vehicular pollution, the Ministry of Petroleum and Natural Gas (MoPNG) has adopted the Auto Fuel Policy and recommended a road map for emission norms for new vehicles and fuel quality (more details are given in Chapter 9). The Bharat Stage (BS) IV norms have been implemented in 13 cities[1] since April 2010, and BS III norms have been implemented in the entire India since October 2010. Compressed natural gas (CNG) has been introduced in many cities of the country. From 41 cities at present, India is posed to have at least 150 cities with authorized city gas distribution networks (CNG and piped gas) by 2015, and the number is expected to increase to 250 cities by 2020 (Ayush 2009). Table 11.5 lists the various interventions adopted to control vehicular emissions during 1991–2007. To meet the mobility demand of people, public transport infrastructure is being strengthened in some cities. An ultra-modern metro rail system is now operational in Delhi and expanding its network to meet the mobility demands of city's residents. It has now been proposed in 26 other Indian cities, including Ahmedabad, Amritsar, Bengaluru, Chandigarh, Chennai, Goa, Hyderabad, Jalandhar, Kochi, Ludhiana, Noida, Pune, and Thane. In 2015, pollution tax was introduced on commercial vehicles entering Delhi.
- **Noise pollution standards:** Ambient noise standards were notified by CPCB in 1987 under Section 20 of the Air (Prevention and Control of Pollution) Act, 1981. Noise standards for different areas (residential, commercial, industrial, and silence zones) are notified under the Environment (Protection) Act, 1986. Noise limits have been specified for automobiles, domestic appliances, and construction equipment at the manufacturing stage. Source-specific standards for portable generator sets and for 5–1000 kVA generator sets/stationary diesel generator sets have been evolved. A notification to control noise pollution from loudspeakers and public address systems was issued in February 2000. Apart from these, standards have been evolved for noise limits in cinema halls. The noise standards for fire crackers were notified by the Environment (Protection) Rules, 1999 in October 1999.
- **Indoor pollution control:** Promoting the use of easily available fuels such as biomass and cow dung for cooking can help in reducing indoor air pollution. Thus, after the failure of the National Programme on Improved Chulha, launched in 1983, the government has initiated the National Project on Biogas Development. However, the rate of progress is very slow. In 2008/09, around 101 529 biogas plants were installed in comparison to a target of 124 000 (MNRE 2009). Also, the government has introduced programmes such as the Rajiv Gandhi Grameen Vidyutikaran Yojana (Ministry of Power, Government of India) and the Village Energy Security Programme and the Remote Village Electrification Programme to provide clean lightning energy.

Water

Living organisms need water for survival. Unfortunately, although more than three-quarters of earth's surface is covered with water, human beings have a limited amount of drinking water. Thus, water pollution further aggravates the problem.

[1] Agra, Ahmedabad, Bengaluru, Chennai, Hyderabad, Kanpur, Kolkata, Mumbai, Delhi, Pune, Lucknow, Solapur, and Surat

The major sources of inland water pollution could be both anthropogenic and natural. Contamination of groundwater with arsenic in India (Bengal) and Bangladesh is an example of water pollution through natural sources. The contamination of water by industrial effluent and the drainage of untreated sewage into rivers are examples of water pollution through anthropogenic sources.

Estimates indicate that the total amount of surface and groundwater resources in India is around 1869 billion cubic metres (BCM). Of this, 40% is not available for use owing to geological and topographical reasons. Around 4000 BCM of freshwater is available from precipitation in the form of rain and snow, most of which returns to the sea via rivers (Table 11.6).

India has 12 major river systems along with a number of smaller rivers and streams (MoEF 2009). More than 70% of India's rivers drain in the Bay of Bengal, mostly as part of the Ganga–Brahmaputra system, while the Arabian Sea receives 20% of the total drainage from the Indus system as well as a number of smaller rivers down the western coast. The remaining 10% drain into the interior basins and a few natural lakes spread across the country (MoEF 2009). The total replenishable groundwater potential of the country has been estimated by the Ministry of Water Resources as 431 000 m^3 per year (MoEF 2009).

The assessed average annual utilizable water in India is 1123 BCM, out of which 433 BCM is from groundwater (CGWB 2010). Groundwater meets the water requirements of nearly 55% of irrigational, 85% of rural, and 50% of urban and industrial activities. The net annual groundwater availability is 396 BCM.

Status of water quality

The exponential increase in population and rapid urbanization and an increase in industries and agricultural activities have adversely influenced the quality of water in India (CPCB 2013). Under the National Water Quality Monitoring Programme, 2500 water quality monitoring stations are spread across the country covering 445 rivers, 154 lakes, 12 tanks, 78 ponds, 41 creeks/sea water, 25 canals, 45 drains, 807 wells, and 10 water treatment plants (raw water). Figure 6 shows the BOD trends of the monitoring stations across the country. BOD is the biochemical oxygen demand of water. It represents the amount of oxygen required by the microorganisms in water to biologically decompose organic matter. Higher BOD levels indicate the presence of organic matter in water. The figure shows that there had been an increase in the number of stations with less than 3 mg/L of BOD till 2007, followed by a gradual decrease in the number. This shows a slight deterioration in the water quality.

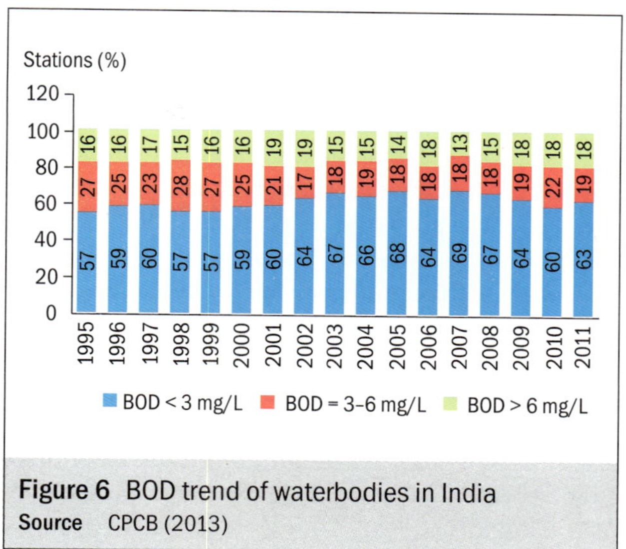

Figure 6 BOD trend of waterbodies in India
Source CPCB (2013)

Figures 7 and 8 show the total coliform (TC) and faecal coliform (FC) trends, respectively, in waterbodies across India. TC and FC are the indicators of the presence of harmful disease-causing microorganisms, that is, pathogens. Both the figures show a decrease in the number of stations monitoring high FC and TC, which implies the improvement in water quality.

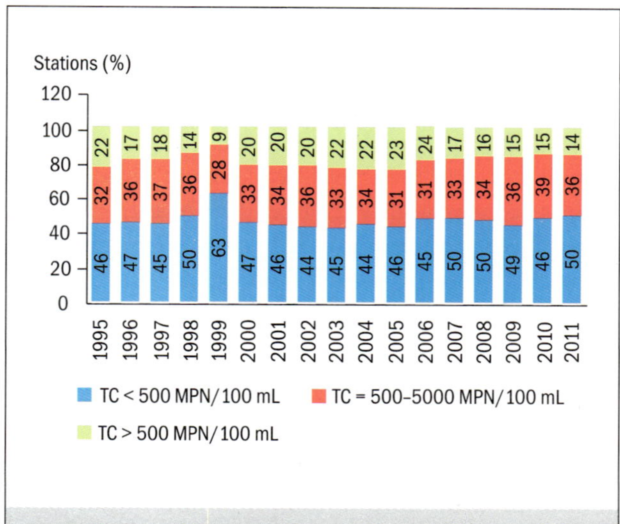

Figure 7 Total coliform trend of waterbodies in India
Source CPCB (2013)

Impacts

Water pollution refers to presence of harmful materials in water in concentrations that make the water unfit for use. Water contamination weakens or destroys natural ecosystems that support human health, food production, and biodiversity (CAG 2012). Water pollution is a bane of India today, with almost 70% of its surface water resources contaminated by biological, toxic, organic, and inorganic pollutants (MoEF 2009). These sources have been rendered unsafe for human consumption and for the purpose of irrigation and industrial uses (MoEF 2009). FC levels, BOD, and chemical oxygen demand have been reported to be very high in most waterbodies across India.

Water pollution has created a serious shortfall of freshwater. With the exception of Ganga in certain stretches, all other rivers continue to be plagued by high levels of organic pollution, a low level of oxygen availability for aquatic organisms, and the presence of bacteria, protozoa, and viruses, which have faecal origin and cause illnesses (CAG 2012). While many lakes have disappeared because of drying up of their catchment areas and have been reclaimed for urbanization. Most lakes in India are under threat from nutrient overloading, leading to eutrophication and subsequent choking due to the proliferation of weeds in the nutrient-rich water (CAG 2012).

Groundwater contamination is a serious problem in India because a large section of country's population relies on groundwater. Inland salinity in groundwater is prevalent mainly in the arid and semi-arid regions of Andhra Pradesh, Delhi, Gujarat, Haryana, Karnataka, Maharashtra, Punjab, Rajasthan, Tamil Nadu, and Uttar Pradesh (CGWB 2010). Coastal salinity problems have also been observed in most coastal areas of the country, with salinity ingress conspicuously noticed in Minjur area of Tamil Nadu and Mangrol–Chorwad–Porbandar belt along the Saurashtra coast (CGWB 2010). Fluoride levels beyond the permissible limit of 1.5 mg/L have been observed in 19 states in India, while arsenic contamination beyond the permissible limit of 0.05 mg/L has been observed beyond Assam, Bihar, West Bengal, and in new areas of Chhattisgarh and Uttar Pradesh (CGWB 2010). Groundwater contamination by iron (more than 1 mg/L) has been reported from Andhra Pradesh, Assam, Bihar, Chhattisgarh, Goa, Gujarat, Haryana, Jammu and Kashmir, Jharkhand, Karnataka, Kerala, Madhya Pradesh, Maharashtra,

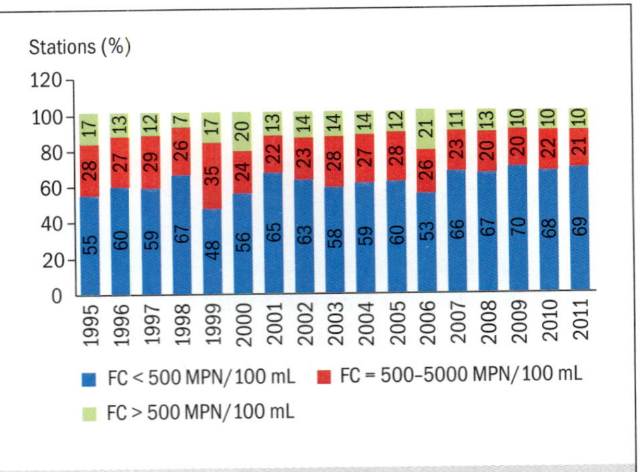

Figure 8 Faecal coliform trend of waterbodies in India
Source CPCB (2013)

Manipur, Meghalaya, Odisha, Punjab, Rajasthan, Tamil Nadu, Tripura, Uttar Pradesh, West Bengal, and the union territory of Andaman and Nicobar Islands. Also, a high nitrate concentration in groundwater in India has been found in almost all hydrogeological formations. Groundwater salinity has increased considerably in several areas. Increased nitrate levels in groundwater have been reported consistently from various parts of the country. Moreover, over-extraction of groundwater has led to a sharp decline in the water table across several parts of the country, including Andhra Pradesh, Maharashtra, and Punjab.

Interventions

- **National River Conservation Plan:** Initiated by the Ministry of Environment and Forests (MoEF, now renamed Ministry of Environment, Forests, and Climate Change [MoEFCC]) in 1995, a host of initiatives have been undertaken by various states under this plan. In 1996, it was merged with the second phase of the Ganga Action Plan. Financial and technical aid has been extended to about 20 states so far.
- **National Lake Conservation Plan:** Under this plan, funds have been released for about 59 lake restoration projects across 14 states in India. Launched in 2001, the programme exclusively aims at restoring lakes in different parts of the country. The scheme operates on a funding pattern of 70:30 cost sharing between the central and the participating state governments.

- **City Sanitation Plans:** To develop capacities for wastewater treatment, funding is being provided for cities under the Jawaharlal Nehru National Urban Renewal Mission (JNNURM). The first stage of this activity is the development of City Sanitation Plans (CSPs), following which states are allocated funds. Till now, 158 cities across 13 states in India have submitted CSPs.
- **Groundwater Aquifer Atlas of India:** Developed by CGWB, the atlas was released in 2012. This is intended to provide a base work for the development of "strategies for ensuring long-term sustainability of groundwater in identified aquifers by adopting the supply-side and demand-side management".
- **The Wetlands (Conservation and Management) Rules, 2010:** Wetland conservation has been accorded a high priority in India. Since 1987, the National Wetlands Conservation Programme has been in action and 115 wetlands have been identified for conservation and management till date. India is also a party to the Ramsar Convention under which 25 wetlands from India are included in the list of wetlands of international importance. The Wetlands (Conservation and Management) Rules, 2010 are a positive step towards conservation of wetlands in India. It is the first time that legally enforceable rules are being notified for such eco-sensitive areas in our country.
- **National Mission for Clean Ganga:** This mission is a registered society formed by the MoEFCC in 2011. Its objective is to "ensure effective abatement of pollution and rejuvenation of River Ganga by adopting a river basin approach to promote inter-sectoral co-ordination for comprehensive planning and management and to maintain minimum ecological flows in the River Ganga with the aim of ensuring water quality and environmentally sustainable development".
- **National Water Mission:** Launched in 2009 under the National Action Plan for Climate Change, the mission's objective is "conservation of water, minimizing wastage and ensuring its more equitable distribution both across and within states through integrated water resources development and management" (MoWR 2009).
- **National Water Policy, 2012:** The policy was formed in 1987 and reviewed in 2002 and 2012 with the aim of better utilization of water resources by proper governance and planning.

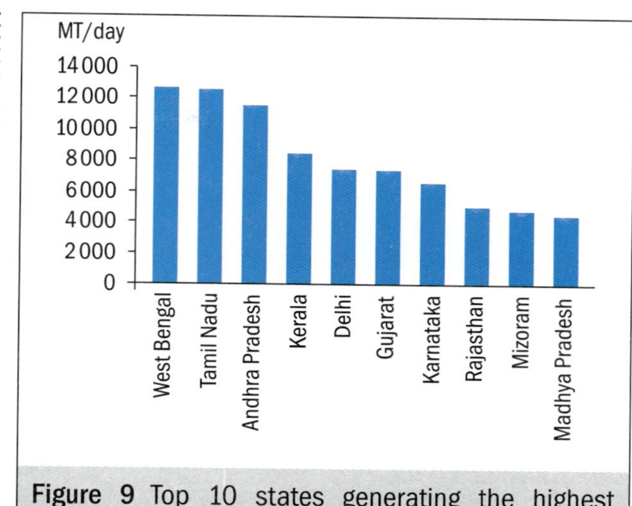

Figure 9 Top 10 states generating the highest municipal solid waste per day (2009–12)

Solid waste

Solid waste generated can create a huge problem if it is not disposed properly. Most of the cities in India are facing this problem. Waste generation is quantified in terms of three categories: municipal solid waste (MSW), plastic waste, and hazardous waste. Hazardous waste is further divided into subcategories such as biomedical, industrial, and e-waste.

Figure 9 shows the solid waste generated per day in 10 states of India. The solid waste generated in these states is the highest among all other Indian states and union territories. West Bengal tops the list, generating 12557 MT of solid waste per day, followed by Tamil Nadu and Andhra Pradesh, with respective figures being 12504 MT per day and 11500 MT per day (CPCB 2012).

The amount of solid waste generated increases with urbanization. The consumption levels and, hence, waste generation are higher in cities compared to rural areas. According to the CPCB, Delhi and Mumbai generate 6800 tonnes and 6500 tonnes, respectively, of MSW every day. Delhi and Mumbai are the most urbanized cities in the country and, thus, generate more solid waste compared to any other city. Thus, while framing policies related to solid waste management, it must be understood that urbanization plays a crucial role in the dynamics of solid waste generation.

Hazardous waste is defined as a waste that causes danger or is likely to cause harm to health or environment, whether alone or in combination with other wastes or substances, by virtue of its physical, chemical, reactive, toxic, flammable, explosive, or corrosive characteristics. Until February 2009, there were about 36 165 hazardous waste generating industries, producing about 6.2 MT of hazardous waste every year. The quantum of hazardous waste ending up for final disposal in secured landfill is about 2.7 MT (43.78%), and 0.4 MT (6.67%) is incinerated. About 3.1 MT (49.55%) of the total hazardous waste in India is recyclable.

Impacts

Urbanization has a huge impact on solid waste management. Mumbai and Delhi, the most urbanized cities in the country, produce the highest amount of solid waste in the country. Before the introduction of Municipal Solid Waste (Management and Handling) Rules in 2000, solid waste was particularly a problem in urban areas. Owing to the absence of properly defined laws, unsegregated biomedical and electronic wastes (often hazardous) used to end up in one corner of the city. This was a serious threat to the health of people living in the close vicinity of the dumping site. For instance, villages near the dumpsites in Jalandhar have reported health problems, indicating a strong to moderate health impact on the resident population owing to the dumping of solid waste in their vicinity (Puri, Kumar, and Johal 2008).

Unscientific management of waste and laxity in enforcement of laws also pose challenges. Open burning of solid waste and landfill fires in Mumbai emit nearly 22 000 tonnes of pollutants per year into the air. These pollutants include CO, hydrocarbons (HC), particulate matter, NO_x, and SO_2, as well as an estimated 10 000 g toxicity equivalent quotient (TEQ) of dioxins/furans. Open burning was found to be the largest polluter in Mumbai, among the activities that do not contribute any economic value to the city. Since open burning happens at the ground level, the resultant emissions enter the breathing zone of the atmosphere, thus causing direct exposure to humans.

Interventions

- **Swachh Bharat Abhiyan:** A nationwide campaign was introduced by the Prime Minister of India on 2 October 2014. The mission not only aims to bring behavioural change about cleanliness in citizens, but also has an objective to eliminate open defecation and manual scavenging, besides promoting modern scientific methods for waste management and strengthening the urban local bodies (Swachhbharaturban 2015).
- **Draft Solid Waste Management Rules, 2015:** The MoEFCC is in the process of revising solid waste, plastic waste, biomedical waste, and e-waste management rules. A draft version was published in 2015, and soon the revised rules are expected to be enforced.
- **Sustainable Habitat Mission:** The mission focuses on solid waste management along with energy efficiency in buildings and cleaner transportation technologies. For the mission, ₹1000 crore has been sanctioned, out of which ₹50 crore was incurred in the Eleventh Five-year Plan, and the remaining ₹950 crore shall be incurred in the Twelfth Five-year Plan (Planning Commission 2012).
- **National Urban Sanitation Policy:** The policy was prepared by the Ministry of Urban Development in 2008. The objective of the policy is "to transform urban India into community-driven, totally sanitized, healthy, and liveable cities and towns". The policy stresses upon awareness and behaviour change, open defecation free cities, and integration of sanitation in all the other aspects of cities (MoUD 2015).
- **National Clean Energy Fund:** The fund was set up in 2010 to finance projects, programmes, and policies promoting clean technologies in waste management, renewable energy, and green transport (Planning Commission 2012).
- **Waste Minimization Circles:** These circles, sponsored by the MoEFCC and implemented with the assistance of the National Productivity Council, assist small-scale industries in minimizing the waste produced by them.
- **Municipal Solid Wastes (Management and Handling) Rules, 2000:** Notified by the MoEFCC in 2000, the rules made it mandatory for municipal authorities to set up waste processing and disposal facilities, identify sanitary landfill sites, and improve existing dumpsites by December 2003.
- **Waste management initiatives at the municipal level:** JNNURM is used as one of the mechanisms to finance waste management initiatives at the municipal level. The total number of projects sanctioned so far is 46, with 10 projects having been completed. More than ₹1 000 crore has been sanctioned for these projects.

- **Initiative by local governments:** Local governments are exhorting residents to contribute to waste management. Also, the informal sector is being co-opted to help in sorting waste. For instance, the Bruhat Bangalore Mahanagara Palike started an initiative to rope in ragpickers and municipal workers to segregate solid waste at the source.
- **Hazardous waste management:** A strong legislative regime governs hazardous wastes. Strict action has been taken in several states such as Punjab and Gujarat, against hospitals and nursing homes that violated rules regarding biomedical waste, even threatening their closure.
- **Biomedical waste management:** In 1998, the MoEFCC promulgated the Biomedical Waste (Management and Handling) Rules under Sections 6, 8, and 25 of the Environment (Protection) Act, 1986, which were amended in 2003.

  Most of the state government initiatives are aimed at setting up of common biomedical waste treatment facilities (CBWTFs). The establishment of CBWTFs has been on the rise, from 157 in 2006/07 to 165 in 2007/08. At present, as many as 170 CBWTFs exist in the country for the management of biomedical waste. The total number of health care facilities using CBWTF is 49 971, and the total number of health care facilities that have applied for authorization is 47 750 (CPCB 2008).
- **Plastic waste:** The Recycled Plastics Manufacture and Usage Rules, 1999 and Amendment 2002 have been notified by the MoEFCC under Sections 3 and 25 of the Environment (Protection) Act, 1986.
- **E-waste:** The E-waste (Management and Handling) Rules, 2011, effective from May 2012, have been notified by the MoEFCC under Sections 6, 8, and 25 of the Environment (Protection) Act, 1986.

Forest and biodiversity

Forests are important for human well-being since people derive many ecosystem services from forests, including provisioning, regulating, cultural, and supporting services (IIFM 2013). India has a forest cover of 701 673 km^2, which is 21.34% of the total geographical area of the country. In India, the Forest Survey of India (FSI), a premier national organization under the MoEFCC, is responsible for regularly assessing and monitoring forest resources.

As per the *India State of Forest Report 2015* by the FSI, an increase of 3775 km^2 of forest cover has been reported as against the 2013 assessment, when India's forest cover was 697 898 km^2. This increase can be attributed to efficient conservation and management interventions.

As a mega-biodiversity nation, India has 16 major forest types and 251 subtypes based on climatic and edaphic features. For assessment, forest cover is classified into predefined classes on the basis of tree canopy density: very dense forest, moderately dense forest, and open forest. Scrub, which is shown separately, is not counted in the forest cover. The country-level forest cover in the absolute area and as a percentage of the geographical area is summarized in Table 1, while the forest cover map of the country is shown in Map 1.

Table 1 Forest cover of India in 2013

Class	Area (km^2)	Percentage share of geographical area
Forest cover		
• Very dense forest	85 904	2.61
• Moderately dense forest	315 374	9.59
• Open forest	300 395	9.14
Total forest cover*	701 673	21.34
Scrub	41 362	1.26
Non-forest	2 544 228	77.40
Total geographical area	3 287 263	100.00

*Includes 4740 km^2 under mangroves

Source FSI (2015)

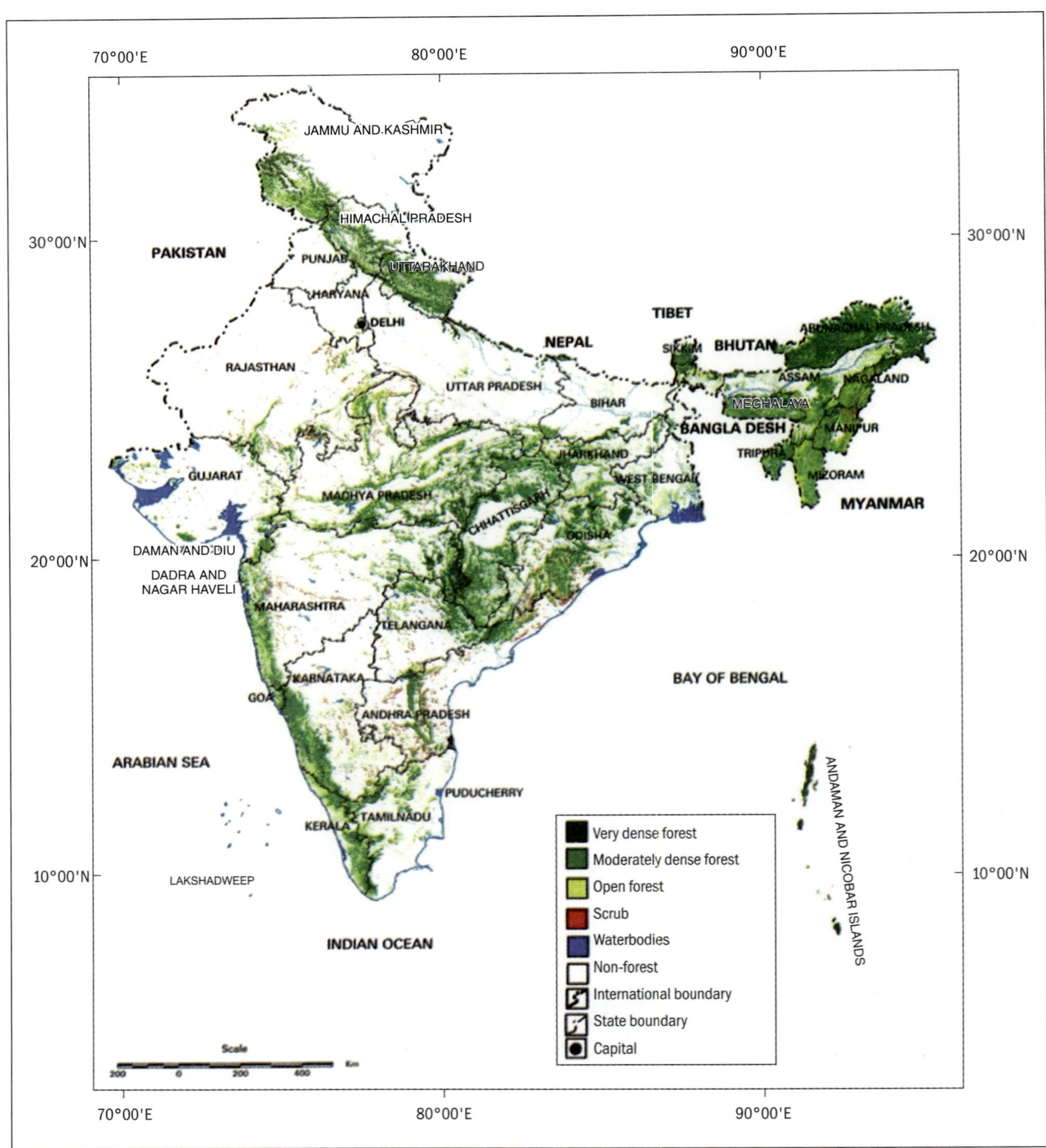

Map 1 Forest cover of India
Source India State of Forest Report (2015)

Tree cover

India's tree cover has been estimated to be 92 572 km^2, which constitutes 2.82% of the country's geographical area. The highest tree cover has been observed in Central Highlands (11 004 km^2), followed by East Deccan (10 120 km^2), and Western Himalaya (9835 km^2). The state with the maximum tree cover is Maharashtra (9558 km^2), followed by Jammu and Kashmir (8354 km2), Rajasthan (8269 km^2) and Gujarat (7914 km^2) (FSI 2015).

Protected areas for biodiversity conservation

India has been generously endowed with nature. Four of the 34 global biodiversity hotspots are in India: the Himalayas, the Western Ghats, the North-East, and the Nicobar Islands. India is among the few countries that have developed a biogeographic classification for conservation planning and mapped biodiversity, rich areas. To realize the objectives of the United Nations Convention on Biological Diversity, 1992, India also enacted the Biological Diversity Act in 2002.

With 2.4% of the world's land area, India is home to 7%–8% of recorded species in the world, which include 46 000 species of plants and 91 000 species of animals (PIB 2015a). About 70% of world's tiger population is in India (MoEFCC 2015). India has significant endemism; about 4045 species of flowering plants (angiosperms) endemic to India are distributed among 141 genera belonging to 47 families. In terms of endemism in vertebrates, India is ranked tenth in birds (with 69 species), fifth in reptiles (with 156 species), and seventh in amphibians (with 110 species) (MoEF 2014). As per the International Union for Conservation of Nature (IUCN), 1039 species are threatened in India. Table 2 gives the number of threatened species in each major group of organisms in India.

Table 2 Number of threatened species in India

Species	Number
Mammals	96
Birds	82
Reptiles*	53
Amphibians	75
Fishes*	214
Molluscs*	7
Other invertebrates*	128
Plants*	384
Total	1039

*For these groups, there are still many species that have not yet been assessed for the IUCN Red List and therefore their status is not known (that is, these groups have not yet been completely assessed). Therefore, the figures presented in the chapter for these groups should be interpreted as the number of species known to be threatened within those species that have been assessed to date, and not as the overall total number of threatened species for each group.

Source IUCN (2015)

Conservation of wild flora and fauna has been an integral part of Indian history. From a network of 54 national parks covering 21 003 km^2 and 373 sanctuaries covering 88 649 km^2, giving a combined coverage of 109 652 km^2 or 3.34% of the country's geographical area in 1988, the network has grown steadily. In October 2015, there were 726 protected areas in India covering 160 499.31 km^2, which is 4.88% of the country's geographical area (Table 3). Based on India's submission and the World Database on Protected Areas, the United Nations' list of protected areas in India in 2014 is shown in Map 2.

Compared to the previous year, the number of wildlife sanctuaries has increased from 525 to 531, while there has been no change in the number of national parks. Compared to the previous year, conservation reserves have increased from 60 to 66 and community reserves have witnessed an exponential increase from 4 to 26 in the country.

Table 3 Details of protected areas in India (as on 6 October 2015)

Type of protected areas	Number	Area (km^2)	Geographical area of India (%)
National parks	103	40 500.13	1.23
Wildlife sanctuaries	531	117 607.72	3.58
Conservation reserves	66	2 344.53	0.07
Community reserves	26	46.93	0.001
Protected areas	726	160 499.31	4.88

Source WII (2015)

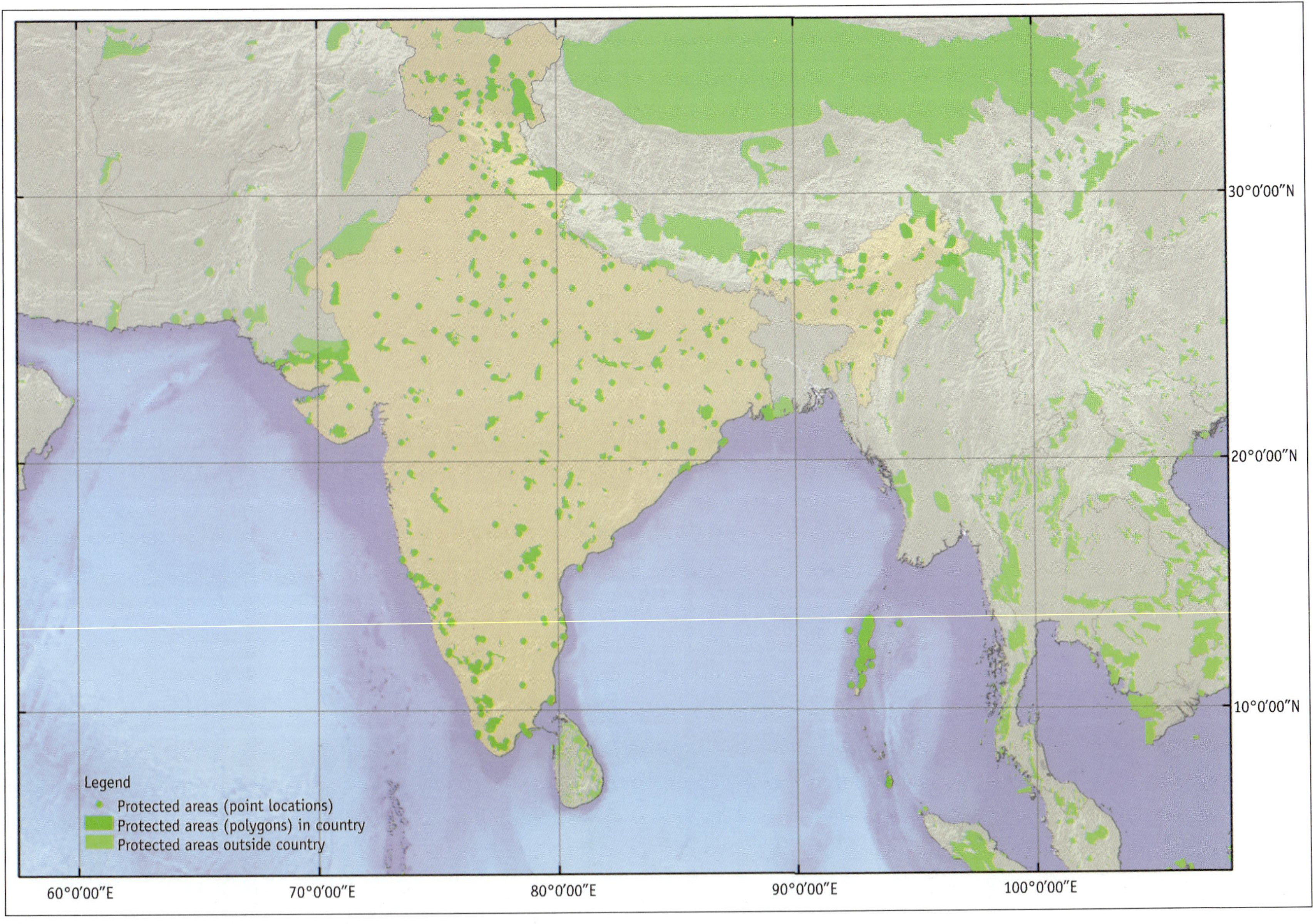

Map 2 United Nations' list of protected areas of India in 2014

Source Based on details available at <http://blog.protectedplanet.net/post/102481051829/2014-united-nations-list-of-protected-area, last accessed on 5 January 2016

Interventions at the sub-national level for biodiversity conservation

Some of the recent interventions by the Government of India and state governments for conservation and sustainable management of biodiversity are as follows:

Economic Valuation of Tiger Reserves in India: The National Tiger Conservation Authority, Ministry of Environment, Forests, and Climate Change, Government of India, assigned Indian Institute of Forest Management (IIFM), a first-of-its kind study to provide conservative estimates of the economic value of six selected tiger reserves in India: Jim Corbett, Kanha, Kaziranga, Periyar, Ranthambore, and Sundarbans. The study findings indicate that the monetary value of flow benefits emanating from selected tiger reserves ranges from ₹8.3 billion to ₹17.6 billion annually. In terms of the unit area, this translates to ₹50 000–190 000 per hectare per year.

Source IIFM (2015)

Launch of an initiative by Ministry of Environment and Forests to highlight economic consequences of the loss of biodiversity: The Government of India launched The Economics of Ecosystems and Biodiversity (TEEB)–India Initiative (TII), which aims to highlight the economic consequences of the loss of biological diversity and the associated decline in ecosystem services. This initiative focuses on three ecosystems, namely, forests, inland wetlands, and coastal and marine ecosystems.

Source PIB (2015b)

Creation of an additional carbon sink: In its Intended Nationally Determined Contribution (INDC), India has communicated its intent to create an additional carbon sink of 2.5–3 billion tonnes of CO_2 equivalent through additional forest and tree cover by 2030. Further, the Government of India has communicated that it will launch new initiatives for full implementation of the Green India Mission (GIM) and other programmes of afforestation.

Source GoI (2015)

Approval of Green India Mission plans of four states: The National Executive Council of the National Mission for a Green India has approved the Perspective Plans and Annual Plan of Operations submitted by Mizoram, Manipur, Jharkhand, and Kerala with a total financial outlay of ₹90 202.68 lakh for a plan period of 5–10 years. Under the plan, the total forest and non-forest area in these four states under GIM during the total plan period will be 108 335 ha, out of which 81 939 ha will be for improving the density of existing forests and 16 396 ha will be new areas.

Source PIB (2015c)

Establishment of Centre for World Natural Heritage Management and Training for Asia and Pacific Region at Wildlife Institute of India, Dehradun: In association with UNESCO, a Centre for World Natural Heritage Management and Training for Asia and Pacific Region (Category 2 Centre) at the Wildlife Institute of India (WII), Dehradun, is being established. The centre will function in the areas of management and training in the field of natural heritage of the world. It will function as an integral part of the WII.

Source PIB (2015d)

References

Atkinson, R. W., A. Cohen, S. Mehta, and H. R. Anderson. 2011. Systematic review and meta-analysis of epidemiological time-series studies on outdoor air pollution and health in Asia. ***Air Quality, Atmosphere and Health*** 383-391

Ayush, G. 2009. Preparedness to handle emergency in city gas distribution networks. In ***Proceedings of the 4th 2009 Pipeline Technology Conference***, Ostend, Belgium, 12-14 October 2009, organized by the Laboratorium Soete and the Technological Institute vzw of the Royal Flemish Society of Engineers (TI-KVIV)

CAG (Comptroller and Auditor General). 2012. ***Performance Audit of Water Pollution in India, Report No. 21 of 2011/12***. New Delhi: CAG, Ministry of Environment, Forests, and Climate Change

CGWB (Central Groundwater Board). 2010. ***Groundwater Scenario of India, 2009/10***. New Delhi: CGWB, Ministry of Water Resources, Government of India

CPCB (Central Pollution Control Board). 2008. Highlights 2008. ***Biomedical Waste Management***. Details available at http://cpcb.nic.in/Highlights/2008/chapter-10.pdf

CPCB (Central Pollution Control Board). 2010. ***National Ambient Air Quality Status and Trends in India-2010***. New Delhi: CPCB

CPCB (Central Pollution Control Board). 2012. ***Status Report on Solid Waste Management***. New Delhi: CPCB

CPCB (Central Pollution Control Board). 2013. ***Annual Report 2011-12***. New Delhi: CPCB

CPCB (Central Pollution Control Board). 2014 ***National Ambient Air Quality Status and Trends in India-2012***. New Delhi: CPCB

Dey, S., L. D. Girolamo, A. Donkelaar, S. Tripathi, T. Gupta, and M. Mohan. 2012. Variability of outdoor fine particulate ($PM_{2.5}$) concentration in the Indian subcontinent: a remote sensing approach. ***Remote Sensing of Environment*** 127: 153-161

FSI (Forest Survey of India). 2015. ***India State of Forest Report 2015***. Dehradun: FSI

GoI (Government of India). 2015. India's Intended Nationally Determined Contribution: working towards climate justice. Details available at http://www4.unfccc.int/submissions/INDC/Published%20Documents/India/1/INDIA%20INDC%20TO%20UNFCCC.pdf, last accessed on 6 November 2015

IIFM (Indian Institute of Forest Management). 2013. ***Revision of Rates of NPV Applicable for Different Class/Category of Forests***. Bhopal: IIFM

IIFM (Indian Institute of Forest Management). 2015. ***Economic Valuation of Tiger Reserves in India: A Value Approach***. Bhopal: IIFM

IUCN (International Union for Conservation). 2015. ***IUCN Red List Version 2015.2: Table 5***. Gland: International Union for Conservation

Lakshmi, P. V., N. K. Virdi, A. Sharma, J. P. Tripathy, K. R. Smith, M. N. Bates, and R. Kumar. 2013. Household air pollution and stillbirths in India: analysis of DLHS-II National Survey. ***Environmental Research*** 121: 17-22

Mishra, V., R. D. Retherford, and K. R. Smith. 2002. Indoor air pollution: the quiet killer. ***Asia-Pacific Issues*** 63

MNRE (Ministry of New and Renewable Energy). 2009. ***Annual Report 2008/09***. New Delhi: MNRE

MoEF (Ministry of Environment and Forests). 2009. ***State of Environment Report, India***. New Delhi: MoEF, Government of India

MoEF (Ministry of Environment and Forests). 2014. ***India's Fifth National Report to the Convention on Biological Diversity***. New Delhi: MoEF, Government of India

MoEFCC (Ministry of Environment, Forests, and Climate Change). 2015. ***Development without Destruction: A Saga of Turnaround***. New Delhi: MoEFCC, Government of India

MoUD (Ministry of Urban Development). 2015. ***National Urban Sanitation Policy***. New Delhi: MoUD

MoWR (Ministry of Water Resources). 2009. ***National Water Mission***. New Delhi: MoWR

Mukulai, K., J. Pekkaneni, P. Tiittaneni, S. Alm, S. Saloneni, and J. Tuomisto. 2001. Personally measured weekly exposure to NO_2 and respiratory health among preschool children. ***European Respiratory Journal*** 13: 1411-1417

PIB (Press Information Bureau). 2015a. Environment minister's message on International Day for Biological Diversity. Details available at http://pib.nic.in/newsite/PrintRelease.aspx?relid=121919, last accessed on 6 November 2015

PIB (Press Information Bureau). 2015b. Environment m nistry launches TEEB-India Initiative to highlight economic consequences of the loss of biological diversity. Details available at http://pib.nic.in/newsite/PrintRelease.aspx?relid=123641, last accessed on 5 November 2015

PIB (Press Information Bureau). 2015c. Establishment of Centre for World Natural Heritage Management and Training for Asia and Pacific Region at Wildlife Institute of India, Dehradun. Details available at http://pib.nic.in/newsite/PrintRelease.aspx?relid=124513, last accessed on 6 November 2015

PIB (Press Information Bureau). 2015d. Green India Mission Plans of four states approved. Details available at http://pib.nic.in/newsite/PrintRelease.aspx?relid=128649, last accessed on 6 November 2015

Planning Commission. 2012. ***Twelfth Five Year Plan (2012-2017): faster, more inclusive and sustainable growth***. New Delhi: Planning Commission

Pope 3rd, C. A. 1995. Particulate pollution and health: a review of the Utah valley experience. ***Journal of Exposure Analysis and Environmental Epidemiology*** 6: 23-34

Puri, A., M. Kumar, and E. Johal. 2008. Solid-waste management in Jalandhar city and its impact on community health. ***Indian Journal of Occupational and Environmental Medicine*** 12: 76-81

Swachhbharaturban. 2015. Details available at https://swachhbharaturban.gov.in/writereaddata/Mission_objective.pdf

WII (Wildlife Institute of India). 2015. Protected areas of India. Protected Areas. Details available at http://wiienvis.nic.in/Database/Protected_Area_854.aspx, last accessed on 6 November 2015

World Bank. 2014. Details available at http://data.worldbank.org/indicator/SP.URB.TOTL.IN.ZS/countries/1W-IN-US-CN-RU-BR-ZA?display=graph, last accessed on 8 October 2014

TABLE 11.1a Ambient SO_2 levels in highly polluted cities (µg/m³): 1995–2012

City	SO_2																	
	1995	1996	1997	1998	1999	2000	2001	2002	2003	2004	2005	2006	2007	2008	2009	2010	2011	2012
Agra	-	20.5	16.6	17.4	7.4	8.2	11	5.5	7	5.8	9.5	5.3	5.3	6.2	0	5	6	5
Akola																	9	9
Ahmedabad	32	25	15.1	-	12.2	8.4	10	12	16	16	13	10	13	12.3	16	15	13	12
Aizawal														BDL	2		2	2
Alappuzha																2	2	2
Allahabad													20	8	3	4	6	5
Alwar	27.2	24	13.5	14.1	15.1	16.5	16	16	11	6.3	7.7	6.6	8	8	8	8	12	11
Amona																6	11	10
Amravati													10	9.7		12	11	12
Amritsar													14	15		14	14	15
Angul	21.2	14.5	16.8	5.2	5.7	6	9	16	3	2.5		5.7	5.5	7	7	7	7	7
Ankleshwar	86.7	86	31.5		54.3			19	20	24	22	24	22	19	19	16	16	18
Anpara	48.7	47	55.2	58.5	59.2	64.2	52	30	19	18	18	21	20	18.5	18	17	16	18
Asansol										6	6	5.2	7	9		8	7	10
Assanora																5	9	9
Aurangabad													5	8.3	7	6	9	10
Baddi														BDL	4	3	3	2
Badlapur																32	39	6
Balasore														-		2	2	3
Bengaluru	-	21	28	20	38	21	20	13	12	8.5	8.7	19	17	15	16	12.6958	17	14.1429
Barajamda																21		
Batala																		
Barrackpore																12	21	13
Bareilly																	14	13
Bathinda													14	8		9	10	
Behrampur														BDL	2	2	2	2
Belgam														BDL	2	2	2	2
Bhilai	19.8	19.3	31.8	31.4	30.9	18.7	19	18	16	15	16	17	17	17.7	15	9	9	9
Bhopal	12	15	16.3	17.3	16.3	20.5	22	15	6	6	5.3	4.3	6.3	7	7	9	3	2

Contd...

TABLE 11.1 a | *Contd...*

City	SO_2																	
	1995	1996	1997	1998	1999	2000	2001	2002	2003	2004	2005	2006	2007	2008	2009	2010	2011	2012
Bhubaneshwar	0.5						4				0	2		BDL	2	2	2	2
Bicholim																9	10	10
Bilaspur																8	8	6
Bongaigaon													4	5	6	6	6	7
Byrnihat																83	74	35
Chandrapur	-	14.7	28.8	50.6	28.3	23.6	21	14	19	19	23	27	38	34	36	26	22	10
Chandigarh	11	6.5	6.6	5.3	6.5			7.8	2	6	BDL	2		BDL	2	2	2	2
Champhai																	2	2
Chennai	21.7	8.1	15.9	12.6	11.9	12.5	17	20	15	12	12	11	11	9.5	9	9	13	13
Chittoor				31.7	34.9									-		4	4	4
Codli																10	9	16
Coimbatore	1.7	1.5	8.2	13.4	4.9		7	8.5	10	8.7	7.7	11	6.7	6	6	5	5	3
Curchorem																32	7	15
Cuttack														BDL	2	2	2	2
Dadra and Nagar Haveli																7		
Daman															7	7	7	9
Damtal		0.5	6.3	2.9		0.4	3	3.7	2		5.5	2.1		BDL	3	2	2	2
Daranga																5	6	3
Dawki																2	3	2
Dehradun	28.4			16.5	17.1	18.6	20	18	18	19	22	25	24	23.5		28	22	26
Delhi	23.5	17.3	16.3	15.4	17.5	15.2	13	11	10	9.9	9.1	9.6	6.4	6.6	6		5	6
Derabassi													12	8		10	9	9
Devanagere																	10	5
Dewas													17	17.5	13	12	21	18
Dhanbad	31.1	38	37.3	65.6	84.4	18.8	17	15	16	12	16	21	20	19		14	16	17
Dibrugarh													5	5	5	6	6	3
Dharwad										8				-				
Dimapur											0	2.2		BDL		2	2	2

Contd...

TABLE 11.1a *Contd...*

City	SO_2																	
	1995	1996	1997	1998	1999	2000	2001	2002	2003	2004	2005	2006	2007	2008	2009	2010	2011	2012
Dombivali		20.5			33.1						38		31	-				52
Durgapur										7.7	7.3	7	7.3	8.3		8	7	13
Faridabad	39.1	34.5	37.8	35.7	33.7		23	13	10	9	9.5	10	12	13		18	21	12
Firozabad										22	21	19	20	22.3	21	14	12	12
Gajraula	22	24.5	28.6	17.6	21.8	22.9	31	36	39	36	28	28	28			16	15	20
Ghaziabad					33.1		24			18	13	12	20	20.5	30	30	31	30
Gobindgarh		17.5	19.9		13.2	13.1	12	10	11	13	17	11	12	11	10	18	9	9
Gorakhpur																	23	18
Golaghat													5	4		6	6	2
Gulbarga														BDL		9	4	3
Guntur																2	4	5
Guwahati	4.3	2.5				6.8		3.2	0	2	5.3	7	8.8	8	8	7	6	3
Gwalior													9	8.5		12	12	13
Hailakandi													5	6				
Haridwar																	23	28
Haldia	28		11.1	43.9		29.5	15	7.6	6	7	7.5	8.8	9.5	9.5		14	15	13
Haldwani																	17	
Hassan													4	4	4	5	5	5
Hissar													7	7.5		8	6	6
Honda																	13	16
Howrah	73		82.1			12.7	11	13	7	14	16	8.7	7.5	10.3	14	12	13	13
Hubli-Dharwad										15	4.5	2.6		BDL	3	5	7	5
Hyderabad	17.2	16.8	16.4	11.8	14	12.4	10	7.3	6	5.6	5.5	4.9	5	5.5	5	4	5	5
Imphal														BDL				
Indore	6	6	7.2	14.2	24.4	24.8	20.6	12	13	13	8.7	6.2	8.7	10	6	14	12	12
Jabalpur	1.7	2.5	6		33.5	14.6		2			0	2		BDL	2	2	2	2
Jaipur	13.4	15	8.5	10.1	12.9	19.8	11	11	8	9.3	6.2	6.2	5.3	6.2	6	6	7	9
Jalna																	7	10
Jalandhar		15	23		50.7	21.3	20	13	10	11	11	11	13	12.5	12	11	12	13

Contd...

TABLE 11.1a | *Contd...*

City	SO_2																	
	1995	1996	1997	1998	1999	2000	2001	2002	2003	2004	2005	2006	2007	2008	2009	2010	2011	2012
Jalgaon																17	17	19
Jammu														-			5	6
Jamnagar										10	9	11	12	10	14	12	12	12
Jamshedpur	62.8	63	40.7	41.2	50.7	49.9	44	43	44	42	40	37	38	37	36	35	16	37
Jhansi													8.5	8.5	21	9	7	8
Jharia	40.9	42	45	76.4	80.1	22.3	16	17	16	14	18	20	20	19	17	16	16	17
Jodhpur	43.7	31.7	8.1	4.3	10		9	7.9	9	6.3	8.2	6.8	6.3	6.5	6	6	5	6
Kadappa																3		
Kakinada																	4	
Kala Amb														BDL	2	3	3	3
Kanpur	14.1	14	14.1	16.9	16.9	18.9	17	10	7	9.4	8.6	6.6	7	7	8	7	8	8
Kashipur																	25	
Khanna													9.5	9.5	9	9	11	10
Khurja													45	39.5		29	28	26
Kochi	10.3	7.3	7.7	9.4	14.8	20	14	13	8	14	14	12	4.5	4.3	4	4	3	3
Kohima																2	2	2
Kolasib																	2	2
Kolhapur													7	11.7	12	15	16	20
Kalinganagar																		3
Kolkata	35.7	21.3		34.3	44.5	17.4	18	11	17	9.3	9.3	8.1	8.6	7.7	16	11	13	12
Kollam																	4	5
Korba	10.4	12	15.1	15.1	13.9	13.6	13	13	13	13	13	13	13	13.3	13	13	13	12
Kota	7.2	9	16.7	10	4.4	6.7	7	5.5	6	6.3	8	8.4	8	9	7	10	7	8
Kothagudem																2	4	5
Kottayam	2.5	0.5	6.2	1.8			2	2			4.5	5.3	5.5	6	6	6	5	6
Kozhikode	12.2	2	6.3	2.4	2.5	2.4	2	2			0			BDL	2	2	2	2
Kurnool													4	4	4	4	4	4
Kundaim																		9
Lakhimpur																6	6	3

Contd...

TABLE 11.1a | Contd...

City	SO_2																	
	1995	1996	1997	1998	1999	2000	2001	2002	2003	2004	2005	2006	2007	2008	2009	2010	2011	2012
Latur																8	6	8
Lote													19	25		60	44	25
Lucknow	29.8	29.2	31.4	30	27.7		28	23	18	16	13	10	8.6	8.2	8	8	8	8
Ludhiana		14.6	19.6		20.5	11.6	12	11	12	20	21	16	10	9.8	9	9	11	11
Lunglei																	2	3
Madurai				6.1	5.3	4.8	6	8.9	11	13	12	11	9.3	10	10	11	11	14
Mahad																19		
Mallapuram																2	2	2
Manali																	2	4
Mandya																	11	10
Mangalore													7	7	7	6	7	6
Mapusa																		7
Margao																	15	9
Margherita																	9	5
Mathura																	20	
Meerut													11	10		8	5	4
Mormugao											4	2.7		BDL	4	32	4	6
Mizoram																2		
Mumbai	31.1	18	25.1	11.5	14.9	12.1	16	9.1	8	6.7	7.3	24	10	8.7	6	4	5	5
Muradabad																11	13	13
Mysore	33.5	36	32.4	36	31	30.7	23	21	11	11	20	17	13	15.5	15	10	12	11
Nagaon																6	6	4
Nagda	46.3	57.7	66	46.4	24.4	42.9	38	27	26	25	25	23	21	23.3	22	19	21	27
Nagpur	8.4	7.5	8.6	7.0	6.6	7.2	9	8.4	8	6.8	5	6.4	7.7	BDL	6	7	8	10
Nahan															2			
Nalagarh															4	3	2	2
Nalbari																7	6	3
Nalgonda																5	4	4
Nanded																	28	30

Contd...

TABLE 11.1 a | *Contd...*

City	SO_2																	
	1995	1996	1997	1998	1999	2000	2001	2002	2003	2004	2005	2006	2007	2008	2009	2010	2011	2012
Nasik	-	33.7	38.6	37.8	33.9	32.1	26	26	27	31	36	34	43	29.7	23	21	25	24
Navi Mumbai													16	18	15	20	17	17
Naya Nangal													9.5	8	8	7	7	7
Nellore																2	4	5
Noida										26	29	23	13	11		11	10	9
Palakkad											0	9.7		3	3	3	3	3
Panjim											0	4.3		BDL		4	4	9
Paonta Sahib	4	0.5		3.7		0.5	2	2			0	2.1		BDL	2	3		
Parwanoo														BDL	3	3	2	2
Patancheru	-	0.5	6.5	1.4		1.7	4	3.5			4	2.6	17	11	17	11	6	6
Pathanamthitta																2	2	2
Pathankot																7	7	7
Patiala																7	8	5
Patna	26.2	21	21	17.1	12.7	14	13	13	14	12	13	10	10	7			5	7
Ponda	8.4	9	11.2	9.8	7.8	3.6	4							-				9
Puducherry	45.8	51.7	31.5		17.6	17.9	14	17	20	22	16	8	4.3	4.7		6	7	8
Pune	22.6	42.7	54	49.5	39.9	43.5	43	41	34	30	23	25	20	22	23	29	32	22
Raebareli																	8	11
Raichur																		
Raipur	16.5	16	8.8	9.4	9.4	10.4	10	8.7	8	11	11	12	14	19.5	19	15	15	14
Rajkot	15.5	19						14		12	11	12	13	10.5	12	13	13	13
Ramagundam													4	4	4	5	6	6
Ranchi										16	19		22	18	18	19	18	18
Rayagada	18.2	12.5	12.9	12.5			9	9.5	2	0	0	2.9	0	BDL	3	2	3	3
Raniganj																8	7	14
Rishikesh																	21	22
Roha																16	21	
Rourkela	39.3	25	27	33.8	15.6	14.1	10	7.4	5	4.5	5	5.1	5.5	5.5	6	5	5	5

Contd...

TABLE 11.1a | *Contd...*

City	SO_2																	
	1995	1996	1997	1998	1999	2000	2001	2002	2003	2004	2005	2006	2007	2008	2009	2010	2011	2012
Rudrapur																	15	
Sagar													5	3	4	4	4	3
Salem				19.2	4		4	5.3	8	7	7	7.9	8	8	9	8	8	9
Saharanpur																	29	11
Sambalpur														3	4	3	4	3
Sangli																14	10	11
Sanguem																		14
Sankrail																10	9	
Saraikela Kharsawan																35	35	39
Satna	15.3	13.5	13.3	11.3	11.5	12.4	12	8.1		2.5	4	3.3	2	4		3	3	
Shillong	-		6.3			1.2	4	4		2.5	0	3	2	BDL	2	2	3	3
Shimla	4	3	6.4	2.7	3.1	3.5	4	6.1	3	3	0	5.9	2	BDL	3	3	3	3
Sibsagar																	6	4
Silchar															6	6	6	4
Silvassa															9.0			9
Sindri	34	41	41	64.5	69.3		17	16	15	12	16	19	18	18	16	16	16	17
Singrauli																15		24
Sivasagar													5	5	5	6		
South Suburban																	6	8
Solapur	-		19.5	17.4	18.1	19.2	20	20	20	18	18	16	17	17.5	18	17	17	17
Sunder Nagar																	2	2
Surat	85.8	51	54.9		33.4			20	17	22	20	24	19	17.7	19	16	17	17
Talcher	17	20.5	11	10.1	5.8	5.6	7	9.7		5	5.3	8.8	9	12	12	12	9	8
Tarapur													25	31				
Tezpur													4	5	5	6	5	25
Thane										8	6.3	11	12	12	13	14	16	20
Tilamol																	18	15
Trichy																		11
Tirupati										7	5	4	4	4		4	4	4

Contd...

TABLE 11.1a | *Contd...*

City	SO_2																	
	1995	1996	1997	1998	1999	2000	2001	2002	2003	2004	2005	2006	2007	2008	2009	2010	2011	2012
Thiruvananthapuram	14.3	17.3	12.2	13.5	17.5	14.5	11	12	15	11	10	9.2	9	-	11	9	10	6
Thrissur																2	2	2
Tinsukia																7	6	4
Tura																2	4	2
Tuticorin	8.4	6	14.5	14.6	5.7			11	17	21	19		28	-	23	12	10	14
Udaipur	18.1	10	18.4	26.6	31.1	10.7	7	7.7	7	6.7	7.7	7.1	9.3	6.7	5	6	6	6
Ujjain													12	11.3	11	12	14	13
Ulhasnagar																31	37	46
Unnao																	10	10
Usgao																6	15	10
Vadodara	69.7	67.3	38.9		41.5			22	19	16	18	17	14	13.7	16	17	17	17
Vapi	39.7	53	32.4		68.5			25	24	26	25	23	23	16.5	16	16	16	20
Varanasi	23.6			-	-			11	14	17	17	15	16	16		18	17	18
Vasco	8.9	9	9.7	8.6	7.7	3.5	5	2.9			4	2		BDL	4	6	4	7
Vijaywada													7.5	5.5	5	4	7	6
Visakhapatnam	18.8	23.3	31.2	12.8	16.8	16.4	14	8	6	10	12	11	8.7	10.3	13	9	13	13
Warangal																4	6	3
Wayanad																2	2	2
West Singhbum																	24	19
Yamuna Nagar	34.3	31	27.8	32.2	9.8	18.9	22	29	28	29	19	16	12	17		12	11	

BDL - below detection limit

Sources

1. CPCB (Central Pollution Control Board). 1997. Ambient air quality: status and statistics 1995. In ***Ambient Air Quality Monitoring Series: NAAQMS/8/1996-97.*** New Delhi: Central Pollution Control Board
2. CPCB (Central Pollution Control Board). 1998. Ambient air quality: status and statistics 1996. In ***Ambient Air Quality Monitoring Series: NAAQMS/10/1998-99.*** New Delhi: Central Pollution Control Board
3. CPCB (Central Pollution Control Board). 1999. Ambient air quality: status and statistics 1997. In ***National Ambient Air Quality Monitoring Series: NAAQMS/11/1999-00.*** New Delhi: Central Pollution Control Board
4. CPCB (Central Pollution Control Board). 2000. National Ambient Air Quality: status and statistics 1998. In ***National Ambient Air Quality Monitoring Series: NAAQMS/15/2000-2001.*** New Delhi: Central Pollution Control Board
5. CPCB (Central Pollution Control Board). 2001. National Ambient Air Quality: status and statistics 1999. In ***National Ambient Air Quality Monitoring Series: NAAQMS/21/2001-2002***, pp. 148. New Delhi: Central Pollution Control Board
6. CPCB (Central Pollution Control Board). 2002. National ambient air quality: status and statistics 2000. In ***National Ambient Air Quality Monitoring Series: NAAQMS/22/2001-2002.*** New Delhi: Central Pollution Control Board
7. CPCB (Central Pollution Control Board). 2004a. National ambient air quality: status 2001. In ***National Ambient Air Quality Monitoring Series: NAAQMS/23/2003-2004.*** New Delhi: Central Pollution Control Board
8. CPCB (Central Pollution Control Board). 2004b. National ambient air quality: status 2002. In ***National Ambient Air Quality Monitoring Series: NAAQMS/24/2003-2004.*** New Delhi: Central Pollution Control Board
9. CPCB (Central Pollution Control Board). 2005. National air quality: status 2003. In ***National Ambient Air Quality Monitoring Series: NAAQMS/26/2004-2005.*** New Delhi: Central Pollution Control Board
10. CPCB (Central Pollution Control Board). 2006a. National air quality: status 2004. In ***National Ambient Air Quality Monitoring Series: NAAQMS/27/2006-2007.*** New Delhi: Central Pollution Control Board
11. CPCB (Central Pollution Control Board). 2006b. National air quality: status 2005. In ***National Ambient Air Quality Monitoring Series: NAAQMS/30/2006-07.*** New Delhi: Central Pollution Control Board
12. CPCB (Central Pollution Control Board). 2007. National air quality: status 2006. In ***National Ambient Air Quality Monitoring Series: NAAQMS//2007-2008.*** New Delhi: Central Pollution Control Board
13. CPCB (Central Pollution Control Board). 2008. National air quality: status 2007. In ***National Ambient Air Quality Monitoring Series: NAAQMS//2008-09.*** New Delhi: Central Pollution Control Board
14. CPCB (Central Pollution Control Board). 2009. National air quality: status 2008. In ***National Ambient Air Quality Status: NAAQMS//2009-10.*** New Delhi:Central Pollution Control Board
15. CPCB (Central Pollution Control Board). 2014a. National air quality: status 2011. In ***National Ambient Air Quality Status NAAQMS// 2013-14.*** New Delhi: Central Pollution Control Board
16. CPCB (Central Pollution Control Board). 2014b. National air quality: status 2012. In ***National Ambient Air Quality Status NAAQMS// 2013-14.*** New Delhi: Central Pollution Control Board

TABLE 11.1b Ambient NO_X levels in highly polluted cities (μg/m³): 1995–2012

City	NO_x																	
	1995	1996	1997	1998	1999	2000	2001	2002	2003	2004	2005	2006	2007	2008	2009	2010	2011	2012
Agra	0	18	15.1	14.9	6.8	7.5	10	11	10	25	27	20	28	23	0	20	22	23
Akola																	10	10
Ahmedabad	18.8	14.8	20	-	33	28.6	39	32	25	24	26	22	22	20	21	21	25	24
Aizawl													0	BDL	6		5	10
Alappuzha																5	5	5
Allahabad													40	37	24	24	23	32
Alwar	73	82.7	78.1	77.3	75.9	77.5	72	56	33	12	6.3	16	23	31	23	24	22	23
Amona																18	16	12
Amravati													13	13		14	13	13
Amritsar													33	35		36	36	39
Angul	12.3	13	15.6	6.6	6	25.6	25	16	14	8.5	16	18	18	20	20	21	20	20
Ankleshwar	25.1	34	22.9		42.7			26	31	31	29	31	29	26	27	24	26	27
Anpara	46.1	47.5	55.1	60.9	61.9	67.4	58	42	31	29	31	31	30	29	29	28	26	31
Asansol										46	50	54	57	74		66	55	38
Assanora																16	14	12
Aurangabad													19	20	25	20	31	32
Baddi													13	11	13	16	15	23
Badlapur																73	63	86
Balasore													11	-		12	12	14
Bengaluru	-	28	20	25	27	40	23	25	35	52	44	37	39	41	40	31	30	29
Barajamda																32		
Batala																		
Barrackpore																74	46	48
Bareilly																	56	24
Bathinda													30	20		21	22	
Behrampur													15	13	20	13	15	18
Belgam													16	14	15	15	14	19
Bhilai	32.9	33.7	35.3	32.7	33.6	31.5	31	28	26	25	26	25	24	25	25	22	22	23
Bhopal	19.2	24	26.2	26.2	24.1	24.1	25	20	13	13	12	14	16	21	18	18	18	21

Contd...

TABLE **11.1b** | *Contd...*

City	NO_x																	
	1995	1996	1997	1998	1999	2000	2001	2002	2003	2004	2005	2006	2007	2008	2009	2010	2011	2012
Bhubaneshwar	7.4						19			11	15	15	14	19	18	19	25	18
Bicholim																50	16	18
Bilaspur																19	21	20
Bongaigaon													9.5	11	15	15	14	14
Byrnihat																15	17	20
Chandrapur	-	32.7	46.1	44	45.4	42.9	43	36	37	29	34	36	48	55	33	26	28	19
Chandigarh	32.3	19	4.8	9.4	14.8		0	9.5	28	30	15	13	15	15	15	16	16	19
Champhai																	6	5
Chennai	17.5	9	13	16.7	10.7	14.4	18	18	26	17	19	19	17	15	17	15	19	26
Chittoor				70.1	68.3									-		9	9	9
Codli																27	15	21
Coimbatore	5.8	2.5	13	16.8	14.2		40	52	51	46	43	37	29	30	29	27	33	27
Curchorem																17	16	20
Cuttack													30	20	24	21	18	17
Dadra and Nagar Haveli																18		
Daman															19	18	20	20
Damtal	4.6	9.8	10.4	10.8	0	2.5	15	17	17	17	21	15	19	18	12	11	10	13
Daranga																14	14	13
Dawki																6	7	8
Dehradun	19.9			14.4	16.8	18.4	21	21	23	24	26	28	26	25		30	24	29
Delhi	47.2	39.7	34	33.9	35.7	39.9	37	37	42	46	47	47	50	57	49		57	57
Derabassi													30	22		23	22	28
Devanagere																	15	10
Dewas													23	25	20	18	26	23
Dhanbad	39.1	51	47.5	57.5	53.6	17.7	32	48	66	58	47	56	52	44		38	35	40
Dibrugarh													11	11	13	14	13	13
Dharwad														-				
Dimapur											15	14	14	14		7	7	7

Contd...

TABLE 11.1b | *Contd...*

City	NO$_x$																	
	1995	1996	1997	1998	1999	2000	2001	2002	2003	2004	2005	2006	2007	2008	2009	2010	2011	2012
Dombivali		25			28.2						41		35	-				88
Durgapur										42	47	50	56	72		66	57	48
Faridabad	14.7	12.5	14.6	13.3	12		15	27	29	23	22	22	25	25		29	43	39
Firozabad										31	29	31	31	32	32	34	40	31
Gajraula	17.8	25		-	11.9	10.4	14	14	16	16	14	16	18			22	22	29
Ghaziabad					28.9		24			25	13	18	15	16	31	37	39	34
Gobindgarh		41	45.6		34.6	31	34	31	34	39	31	30	29	26	29	35	25	35
Gorakhpur																	42	53
Golaghat													12	11		15	15	14
Gulbarga													14	14		13	12	11
Guntur																11	10	11
Guwahati								19	23	15	18	17	16	17	16	15	14	14
Gwalior													17	16		20	20	27
Hailakandi													12	13				
Haridwar																	24	
Haldia	44.9		16.8	57.7		68.6	50	26	25	27	38	44	43	50		52	55	41
Haldwani																	24	
Hassan													19	20	22	22	24	17
Hissar													0	BDL		8	7	8
Honda																	19	21
Howrah			84.3			51	60	81	65	92	96	71	72	77	81	75	62	40
Hubli-Dharwad										12	0	5.6	0	12	13	13	13	15
Hyderabad	37.8	25	30.7	30.8	24.3	25.2	31	25	26	30	30	27	24	26	22	19	29	28
Imphal													19	19				
Indore	10.2	11	10	14.2	19.4	18.4	13.9	25	30	23	18	14	17	19	11	18	14	20
Jabalpur	61.6	13.5	11.9	12.8	14	18.2	17	15	15	13	18	22	24	25	24	25	25	24
Jaipur	27.3	41.3	14.9	21.5	37.8	43.4	31	34	24	24	26	29	28	35	36	37	42	52
Jalna																	25	30
Jalandhar		26.3	38.7		57.3	29.2	26	24	27	29	31	30	31	30	29	29	25	27

Contd...

TABLE 11.1b | *Contd...*

City	NO_x																	
	1995	1996	1997	1998	1999	2000	2001	2002	2003	2004	2005	2006	2007	2008	2009	2010	2011	2012
Jalgaon																45	42	44
Jammu													0	27			12	12
Jamnagar										22	22	23	27	28	28	27	25	25
Jamshedpur	10.4	10	51.4	51.4	57.3	49.4	52	56	56	54	52	50	52	51	49	48	48	49
Jhansi													28	28	29	25	24	21
Jharia	54.2	52	57	53.7	51.3	18.5	36	70	63	67	49	56	52	45	41	37	39	40
Jodhpur	32.8	30.3	18.7	15.2	23.8		18	16	16	19	21	21	23	24	23	22	23	24
Kadappa																15		
Kakinada																	9	
Kala Amb													13	13	17	18	16	16
Kanpur	15.9	16.2	19.1	23.1	22.6	25.6	26	23	18	20	19	21	23	24	31	34	33	34
Kashipur																	29	
Khanna													37	39	36	31	31	26
Khurja													31	33		27	28	25
Kochi	14.8	18.7	12.8	8.1	16.6	20.6	15	10	6	8.7	8	8.6	6.7	15	12	11	13	10
Kohima																5	7	5
Kolasib																	5	5
Kolhapur													14	20	14	19	21	29
Kalinganagar																		9
Kolkata	29.9	29.3		32	30.5	34.8	74	82	71	60	39	58	60	64	56	62	66	69
Kollam																	20	20
Korba	7.9	16	20.4	20.6	19.1	18.9	21	20	20	20	21	21	21	21	21	20	20	19
Kota	64.9	44	48.4	25.8	18.3	28.1	27	20	19	24	21	24	21	25	24	29	30	32
Kothagudem																11	9	11
Kottayam	35.6	19.5	24	26.8	27.7		15	8.4	8	11	19	19	18	19	19	19	17	16
Kozhikode	11.4	16	8.6	7	3.5	9.8	9	8.2			0		5	9	9	9	8	8
Kurnool													17	11	9	9	9	9
Kundaim																		16

Contd...

TABLE 11.1b | Contd...

City	NO_x																	
	1995	1996	1997	1998	1999	2000	2001	2002	2003	2004	2005	2006	2007	2008	2009	2010	2011	2012
Lakhimpur																15	13	2
Latur																16	16	20
Lote													21	27		30	32	14
Lucknow	28.5	29.6	34.7	32.1	29.8		31	29	28	33	27	29	32	36	36	34	33	32
Ludhiana		41	45.1		39.5	30.4	32	33	36	43	39	36	37	40	37	32	29	26
Lunglei																	5	5
Madurai				27.6	29.2	28.6	31	31	26	23	23	28	22	23	25	25	24	30
Mahad																38	38	
Mallapuram																5	5	5
Manali																	9	12
Mandya																	22	24
Mangalore													0	BDL	6	8	8	7
Mapusa																		9
Margao																	23	17
Margherita																	20	15
Mathura																	24	29
Meerut													44	42		47	45	43
Mormugao										10	0	5.1		9	9	18	19	19
Mizoram																6		
Mumbai	34.2	35.3	34.3	19.5	29.6	25.5	23	17	21	18	21	45	37	39	42	19	33	20
Muradabad																21	23	23
Mysore	31.1	32	29.9	32	27.1	27.3	31	28	18	20	32	26	18	21	30	28	22	23
Nagaon																14	13	13
Nagda	31.6	32.7	37.7	30.8	22.1	34.1	36	32	30	34	31	26	24	25	24	24	34	27
Nagpur	14.2	15.8	18.3	15.8	19.1	20.2	18	15	19	21	23	25	26	33	30	33	35	32
Nahan															17			
Nalagarh															13	18	13	23
Nalbari																16	15	15
Nalgonda																23	18	15

Contd...

TABLE 11.1b | Contd...

City	NO_x																	
	1995	1996	1997	1998	1999	2000	2001	2002	2003	2004	2005	2006	2007	2008	2009	2010	2011	2012
Nanded																	29	30
Nasik	-	29.7	32.7	33.8	25.7	21.6	21	20	23	30	29	27	35	25	29	26	27	27
Navi Mumbai													29	39	42	39	44	43
Naya Nangal													28	21	20	16	17	20
Nellore																12	10	11
Noida										48	51	55	50	43		46	45	35
Palakkad											0	11	0	BDL	7	6	9	7
Panjim										11	0	11	11	13		17	17	14
Paonta Sahib	7.9	7.5	7.8	7.1		7.2	8	8.5	5	4.5	17	15	13	13	16	17	9	16
Parwanoo													5.5	10	11	12	9	8
Patancheru	12.2	17	14	11.6	16	10.1	15	15	14	11	22	10	29	24	26	23	22	22
Pathanamthitta																13	13	15
Pathankot																14	14	12
Patiala																20	19	13
Patna	29.0	27.5	24.7	20.5	15.3	17	15	16	29	26	38	41	50	40			48	36
Ponda	8.8	12	20.5	18.3	12.9	13.1	18							-				16
Puducherry	64.1	36.3	32.3		25	22.9	17	13	16	18	15	10	0	11		13	16	14
Pune	23.5	47.3	62.2	58.6	54	58.4	79	72	62	47	34	42	45	37	40	39	58	45
Raebareli																	11	16
Raichur																		
Raipur	20.8	24	25.9	32.3	36.1	37.4	39	37	38	38	36	33	35	43	46	43	42	40
Rajkot	8.3	13						32		25	19	18	18	14	19	17	18	18
Ramagundam													12	18	15	9	11	11
Ranchi										36	30		44	33	32	35	35	35
Rayagada	29.3	27	17.3	19.6			15	16	5	8	13	11	12	20	21	20	22	22
Raniganj																63	52	45
Rishikesh																	22	24
Roha																34	35	
Rourkela	38.7	16.3	22.7	31.9	17.7	19.7	18	12	6	4.5	11	10	10	11	11	11	11	11

Contd...

TABLE **11.1b** | *Contd...*

City	NO_x																	
	1995	1996	1997	1998	1999	2000	2001	2002	2003	2004	2005	2006	2007	2008	2009	2010	2011	2012
Rudrapur																	20	
Sagar													13	17	18	17	15	13
Salem				16.1	24.9		9	21	34	35	33	31	25	25	25	26	21	22
Saharanpur																	20	30
Sambalpur													11	14	16	15	14	15
Sangli																27	35	40
Sanguem																		19
Sankrail																65	50	
Saraikela Kharsawan																46	45	51
Satna	0	18	15.1	14.9	6.8	7.5	10	11	10	25	27	20	28	23	0	20	22	23
Shillong																	10	10
Shimla	18.8	14.8	20	-	33	28.6	39	32	25	24	26	22	22	20	21	21	25	24
Sibsagar													0	BDL	6		5	10
Silchar																5	5	5
Silvassa													40	37	24	24	23	32
Sindri	73	82.7	78.1	77.3	75.9	77.5	72	56	33	12	6.3	16	23	31	23	24	22	23
Singrauli																18	16	12
Sivasagar													13	13		14	13	13
South Suburban													33	35		36	36	39
Solapur	12.3	13	15.6	6.6	6	25.6	25	16	14	8.5	16	18	18	20	20	21	20	20
Sunder Nagar	25.1	34	22.9		42.7			26	31	31	29	31	29	26	27	24	26	27
Surat	46.1	47.5	55.1	60.9	61.9	67.4	58	42	31	29	31	31	30	29	29	28	26	31
Talcher										46	50	54	57	74		66	55	38
Tarapur																16	14	12
Tezpur													19	20	25	20	31	32
Thane													13	11	13	16	15	23
Tilamol																73	63	86
Trichy													11	-		12	12	14

Contd...

TABLE 11.1b | Contd...

City	NO_x																	
	1995	1996	1997	1998	1999	2000	2001	2002	2003	2004	2005	2006	2007	2008	2009	2010	2011	2012
Tirupati	-	28	20	25	27	40	23	25	35	52	44	37	39	41	40	31	30	29
Thiruvananthapuram																32		
Thrissur																		
Tinsukia																74	46	48
Tura																	56	24
Tuticorin													30	20		21	22	
Udaipur													15	13	20	13	15	18
Ujjain													16	14	15	15	14	19
Ulhasnagar	32.9	33.7	35.3	32.7	33.6	31.5	31	28	26	25	26	25	24	25	25	22	22	23
Unnao	19.2	24	26.2	26.2	24.1	24.1	25	20	13	13	12	14	16	21	18	18	18	21
Usgao	7.4						19			11	15	15	14	19	18	19	25	18
Vadodara																50	16	18
Vapi																19	21	20
Varanasi													9.5	11	15	15	14	14
Vasco																15	17	20
Vijaywada	-	32.7	46.1	44	45.4	42.9	43	36	37	29	34	36	48	55	33	26	28	19
Visakhapatnam	32.3	19	4.8	9.4	14.8		0	9.5	28	30	15	13	15	15	15	16	16	19
Warangal																	6	5
Wayanad	17.5	9	13	16.7	10.7	14.4	18	18	26	17	19	19	17	15	17	15	19	26
West Singhbum				70.1	68.3									-		9	9	9
Yamuna Nagar																27	15	21

BDL - below detection limit

Sources

1. CPCB (Central Pollution Control Board). 1997. Ambient air quality: status and statistics 1995. In ***Ambient Air Quality Monitoring Series: NAAQMS/8/1996-97.*** New Delhi: Central Pollution Control Board
2. CPCB (Central Pollution Control Board). 1998. Ambient air quality: status and statistics 1996. In ***Ambient Air Quality Monitoring Series: NAAQMS/10/1998-99.*** New Delhi: Central Pollution Control Board
3. CPCB (Central Pollution Control Board). 1999. Ambient air quality: status and statistics 1997. In ***Ambient Air Quality Monitoring Series: NAAQMS/11/1999-00.*** New Delhi: Central Pollution Control Board
4. CPCB (Central Pollution Control Board). 2000. National ambient air quality: status and statistics 1998. In ***National Ambient Air Quality Monitoring Series: NAAQMS/15/2000-2001.*** New Delhi: Central Pollution Control Board
5. CPCB (Central Pollution Control Board). 2001. National ambient air quality: status and statistics 1999. In ***National Ambient Air Quality Monitoring Series: NAAQMS/21/2001-2002.*** New Delhi: Central Pollution Control Board
6. CPCB (Central Pollution Control Board). 2002. National ambient air quality: status and statistics 2000. In ***National Ambient Air Quality Monitoring Series: NAAQMS/22/2001-2002.*** New Delhi: Central Pollution Control Board
7. CPCB (Central Pollution Control Board). 2004a. National ambient air quality: status and statistics 2001. In ***National Ambient Air Quality Monitoring Series: NAAQMS/23/2003-2004.*** New Delhi: Central Pollution Control Board
8. CPCB (Central Pollution Control Board). 2004b. National ambient air quality: status and statistics 2002. In ***National Ambient Air Quality Monitoring Series: NAAQMS/24/2003-2004.*** New Delhi: Central Pollution Control Board
9. CPCB (Central Pollution Control Board). 2005. National air quality: status 2003. In ***National Ambient Air Quality Monitoring Series: NAAQMS/26/2004-2005.*** New Delhi: Central Pollution Control Board
10. CPCB (Central Pollution Control Board). 2006a. National air quality: status 2004. In ***National Ambient Air Quality Monitoring Series: NAAQMS/27/2006-2007.*** New Delhi: Central Pollution Control Board
11. CPCB (Central Pollution Control Board). 2006b. National air quality status 2005. In ***National Ambient Air Quality Monitoring Series: NAAQMS/30/2006-2007.*** New Delhi: Central Pollution Control Board
12. CPCB (Central Pollution Control Board). 2007. National air quality: status 2006. In ***National Ambient Air Quality Monitoring Series: NAAQMS//2007-2008.*** New Delhi: Central Pollution Control Board
13. CPCB (Central Pollution Control Board). 2008. National air quality: status 2007. In ***National Ambient Air Quality Monitoring Series: NAAQMS//2008-2009.*** New Delhi: Central Pollution Control Board
14. CPCB (Central Pollution Control Board). 2009. National air quality: status 2008. In ***National Ambient Air Quality Monitoring Series: NAAQMS//2009-10.*** New Delhi: Central Pollution Control Boardi
15. CPCB (Central Pollution Control Board). 2014a. National air quality status: 2011. In ***National Ambient Air Quality Status NAAQMS// 2013-14.*** New Delhi: Central Pollution Control Board
16. CPCB (Central Pollution Control Board). 2014b. National air quality status: 2012. In ***National Ambient Air Quality Status NAAQMS// 2013-14.*** New Delhi: Central Pollution Control Board

TABLE 11.1c Ambient RSPM levels in highly polluted cities (μg/m³): 1999–2012

City	RSPM													
	1999	2000	2001	2002	2003	2004	2005	2006	2007	2008	2009	2010	2011	2012
Agra						197	197	203	212	198	0	185	165	196
Akola													137	139
Ahmedabad	161	237	198	169	154	152	111	100	80	76	95	95	79	83
Aizawl									39	37	38		46	54
Alappuzha												44	41	50
Allahabad									159	181	60	218	260	317
Alwar			104	129	171	139	118	122	118	144	153	225	214	152
Amona												55	79	90
Amravati									62	66		118	100	100
Amritsar												219	200	
Angul		79	83	86	70	78	84	91	100	108	115	110		106
Ankleshwar				141	126	126	132	123	100	93	89	77	91	100
Anpara				153	159	134	133	146	128	124	136	128	140	135
Asansol						120		132	112	135		141	145	111
Assanora												71	78	84
Aurangabad									70	75	86	75	83	80
Baddi									204	99	63	105	98	99
Badlapur												112	105	124
Balasore									62			73	76	82
Bengaluru		90	68	64	76	69	66	75	80	100	122	89	94	121
Barajamda												302		
Batala														
Barrackpore												121	98	130
Bareilly													218	232
Bathinda												216		
Behrampur									62	64	87	58	75	80
Belgam									39	33	30	33	41	72
Bhilai				119	117	128	76	126	113	109	112	106	106	103
Bhopal			201	107	58	57	60	74	79	102	115	133	175	173

Contd...

TABLE 11.1c *Contd...*

City	RSPM													
	1999	2000	2001	2002	2003	2004	2005	2006	2007	2008	2009	2010	2011	2012
Bhubaneshwar						59	62	85	71	86	75	62	121	81
Bicholim												89	111	119
Bilaspur														
Bongaigaon									48	66	81	58	53	46
Byrnihat												175	178	138
Chandarpur				62	94	79	110	102	164	170	107	152	153	148
Chandigarh					106	108	95	111	101	95	81	92	102	110
Champhai													65	62
Chennai	72	65	78	75	86	60	61	65	63	63	70	59	65	57
Chittoor												39	39	40
Codli												90	76	121
Coimbatore			54	52	48	68	48	61	63	75	42	78	86	68
Curchorem												38	87	112
Cuttack									82	81	83	74	70	68
Dadra and Nagar Haveli												39		
Daman											57	35	24	
Damtal									68	65	66	68	67	87
Daranga												58	56	56
Dawki												71	63	44
Dehradun	138	137	170		177	146	148	120	92	110		145	156	188
Delhi	172	155	146	158	151	149	140	171	170	214	243		222	237
Derabassi												162	105	129
Devanagere													67	75
Dewas									27	84	113	85	84	92
Dhanbad						129	121	139	107	131		127	202	178
Dibrugarh									59	56	44	38	42	56
Dharwad						41								
Dimapur						69	95	73	68	72		76	88	90

Contd...

TABLE 11.1c | *Contd...*

City	RSPM													
	1999	2000	2001	2002	2003	2004	2005	2006	2007	2008	2009	2010	2011	2012
Dombivali		80					96		105					115
Durgapur						104	116	136	121	133		141	166	108
Faridabad					216	164	159	164	159	150		164	177	184
Firozabad									185	222	187	207	271	212
Gajraula			246	95	76	65	69	64	64			87	90	158
Ghaziabad			82			158	339	252	230		236	290	231	248
Gobindgarh						256	241	231	243	216	206	224	166	201
Gorakhpur													67	123
Golaghat									67	71		73	63	55
Gulbarga									80	102		65	63	65
Guntur												81	74	75
Guwahati			121	141	136	135	112	117	99	71	119	105	93	92
Gwalior									166	123		308	309	329
Hailakandi									47	163				
Haridwar														149
Haldia				56	56	67	81	81	60	66		57	163	238
Haldwani													125	137
Hassan									62	62	46	45	46	36
Hissar									114	50		96	102	112
Honda													184	135
Howrah		62	115	119	102	135	113	114	112	112	115	118	131	186
Hubli-Dharwad						157	62	61	118	104	107	92	73	77
Hyderabad	127	98	69	71	64	71	76	83	75	85	82	39	86	79
Imphal									56	84				
Indore			321	188	138	138	138	117	120	196	114	120	132	143
Jabalpur			81	85	84	69	82	82	107	136	136	135	73	75
Jaipur	119	156	109	111	133	125	104	139	111	127	151	164	147	187
Jalna													116	109
Jalandhar					287	209	186	173	170	150	140	144	142	140

Contd...

TABLE 11.1c | Contd...

City	RSPM													
	1999	2000	2001	2002	2003	2004	2005	2006	2007	2008	2009	2010	2011	2012
Jalgaon												122	118	130
Jammu									68	72			107	119
Jamnagar				150		130	128	94	103	95	106	104	104	100
Jamshedpur					206	168	167	154	166	172	172	153	152	150
Jhansi									165	130	192	118	118	110
Jharia						202	159	176	180	200	261	221	223	212
Jodhpur			89	124	118	111	108	127	129	148	164	181	168	189
Kadappa												47		
Kakinada													59	58
Kala Amb									172		141	166	170	165
Kanpur	212	211	208	197	179	186	190	194	190	212	211	203	189	215
Kashipur													291	180
Khanna									215	247	249	231	237	213
Khurja									205	231		173	160	168
Kochi	118	88	127	69	58	60	59	67	46	43	40	36	39	70
Kohima												66	77	82
Kolasib													94	57
Kolhapur									66	84	82	83	86	110
Kalinganagar														52
Kolkata	140	145	117	128	121	134	122	129	106	103	126	99	115	135
Kollam													53	41
Korba				148	183	172	76	90	100	107	117	98	94	81
Kota		112	81	101	96	96	108	125	118	125	91	132	139	156
Kothagudem												62	81	63
Kottayam		54	52	87	53	48	40	45	48	46	45	47	57	56
Kozhikode		31	25	22		57	68		26	34	32	42	47	56
Kurnool									80	71	79	85	82	74
Kundaim														63
Lakhimpur												76	64	45
Latur												106	116	116

Contd...

TABLE 11.1c Contd...

City	RSPM													
	1999	2000	2001	2002	2003	2004	2005	2006	2007	2008	2009	2010	2011	2012
Lote									77	76		119	86	40
Lucknow		187	171	170	169	173	190	190	189	190	197	204	189	211
Ludhiana			215	202	238	253	236	236	212	271	254	214	221	228
Lunglei													41	42
Madurai			50	59	89	94	51	38	42	42	42	47	45	48
Mahad												93	207	
Mallapuram												30	30	36
Manali													110	85
Mandya													43	49
Mangalore									62	60	41	48	55	31
Mapusa														82
Margao													90	67
Margherita													53	54
Mathura													206	208
Meerut									117	115		166	158	129
Mormugao						60	42	24		54	67	59	72	112
Mizoram												42		
Mumbai	115	107	67	69	70	78	82	85	93	127	109	97	119	117
Muradabad												195	146	166
Mysore		63	81	60	55	54	47	44	42	49	48	44	48	55
Nagaon												103	86	79
Nagda			106	122	116	123	112	94	86	101	105	95	101	103
Nagpur	178	113	86	80	67	72	76	82	111	114	99	113	106	103
Nahan											153			
Nalagarh											70	86	89	89
Nalbari												68	95	82
Nalgonda												85	75	79
Nanded													39	53
Nasik			75	62	82	89	102	68	45	79	89	76	101	95
Navi Mumbai									92	110	111	113	129	120

Contd...

TABLE 11.1c | *Contd...*

City	RSPM													
	1999	2000	2001	2002	2003	2004	2005	2006	2007	2008	2009	2010	2011	2012
Naya Nangal									94	155	162	94	76	89
Nellore												65	63	62
Noida				179		109	99	180	165	148		132	138	136
Palakkad		58		42		107	145	121	68	30	49	32	23	37
Panjim							52	63	45	57		85	87	67
Paonta Sahib								93	129	132	120	135	72	154
Parwanoo									73	80	82	84	87	80
Patancheru		74	65	79	84	74	49	76	107	97	87	76	76	108
Pathanamthitta												27	22	23
Pathankot												76	69	68
Patiala												143	97	63
Patna		108		131	86	82	109	113	123	105			174	167
Ponda														60
Puducherry				46	53	52	51	57	50	50		38	42	42
Pune		87	121	132	142	116	99	131	109	103	82	82	113	92
Raebareli													133	163
Raichur														
Raipur				211	243	284	204	153	142	192	260	289	293	268
Rajkot				166		220	145	117	100	105	105	96	98	99
Ramagundam									65	87	106	52	65	82
Ranchi						138	95		136	173	182	172	170	202
Rayagada			110	108	61	65	81	57	56	64	60	58	58	54
Raniganj												159	153	126
Rishikesh													94	109
Roha												89	133	
Rourkela		100	112	95	86	92	93	102	101	102	106	105	104	98
Rudrapur													150	158
Sagar									66		124	66	77	120
Salem			31	37	35	36	41	41	59	78	82	85	65	60

Contd...

TABLE 11.1c | Contd...

City	RSPM													
	1999	2000	2001	2002	2003	2004	2005	2006	2007	2008	2009	2010	2011	2012
Saharanpur													42	186
Sambalpur									49	50	54	50	52	53
Sangli												62	76	80
Sanguem														100
Sankrail												100	114	
Saraikela Kharsawan												171	216	160
Satna			144	159	136	182	189	215	205	128		135	99	
Shillong		37	51	51	54	57	60	67	67	73	55	79	73	65
Shimla					20	12	22	50	51	54	68	58	54	58
Sibsagar													99	109
Silchar											57	81	78	91
Silvassa											71		24	
Sindri						142	92	128	128	136	166	174	214	170
Singrauli												59		63
Sivasagar									90	81	89	59		
South Suburban													96	119
Solapur	178	192	187	181	167	142	124	107	90	79	71	66	78	83
Sunder Nagar													85	95
Surat				160	132	124	127	124	92	87	91	76	94	97
Talcher		75	89	89	70	79	68	90	82	95	104	116	109	116
Tarapur									85	77				
Tezpur									66	76	83	68	60	11
Thane		55		60		47	48	52	51	62	58	50	56	73
Tilamol													150	114
Trichy														75
Tirupati						52	41	34	26	96		37	37	37
Thiruvananthapuram		143	136	128	128	104	85	75	72	34	65	56	58	55
Thrissur												31	33	73
Tinsukia												58	56	57

Contd...

TABLE 11.1c *Contd...*

City	RSPM													
	1999	2000	2001	2002	2003	2004	2005	2006	2007	2008	2009	2010	2011	2012
Tura												63	59	52
Tuticorin				37	32	39	64		90	67	108	119	130	134
Udaipur			69	86	91	97	83	78	95	88	85	116	171	155
Ujjain									109	102	86	80	98	80
Ulhasnagar												106	114	111
Unnao													132	98
Usgao												245	125	121
Vadodara				173	167	146	112	139	101	77	86	93	88	102
Vapi				186	131	136	130	112	72	84	78	80	89	100
Varanasi			95	157	97	133	102	107	114	106		127	127	138
Vasco					65	52	74	59	51	50	73	61		84
Vijaywada									89	96	80	84	90	97
Visakhapatnam	71	67	84	89	92	88	92	101	91	81	97	72	80	65
Warangal												53	61	49
Wayanad												46	28	52
West Singhbum													231	153
Yamuna Nagar					98	62	128	130	99	375		261	172	

RSPM – respirable suspended particulate matter

Sources

1. CPCB (Central Pollution Control Board). 1997. Ambient air quality: status and statistics 1995. In *Ambient Air Quality Monitoring Series: NAAQMS/8/1996-97.* New Delhi: Central Pollution Control Board
2. CPCB (Central Pollution Control Board). 1998. Ambient air quality: status and statistics 1996. In *Ambient Air Quality Monitoring Series: NAAQMS/10/1998-99.* New Delhi: Central Pollution Control Board
3. CPCB (Central Pollution Control Board). 1999. Ambient air quality: status and statistics 1997. In *Ambient Air Quality Monitoring Series: NAAQMS/11/1999-00.* New Delhi: Central Pollution Control Board
4. CPCB (Central Pollution Control Board). 2000. National ambient air quality: status and statistics 1998. In *National Ambient Air Quality Monitoring Series: NAAQMS/15/2000-2001.* New Delhi: Central Pollution Control Board
5. CPCB (Central Pollution Control Board). 2001. National ambient air quality: status and statistics 1999. In *National Ambient Air Quality Monitoring Series: NAAQMS/21/2001-2002.* New Delhi: Central Pollution Control Board
6. CPCB (Central Pollution Control Board). 2002. National ambient air quality: status and statistics 2000. In *National Ambient Air Quality Monitoring Series: NAAQMS/22/2001-2002.* New Delhi: Central Pollution Control Board
7. CPCB (Central Pollution Control Board). 2004a. National ambient air quality: status and statistics 2001. In *National Ambient Air Quality Monitoring Series: NAAQMS/23/2003-2004.* New Delhi: Central Pollution Control Board
8. CPCB (Central Pollution Control Board). 2004b. National ambient air quality: status and statistics 2002. In *National Ambient Air Quality Monitoring Series: NAAQMS/24/2003-2004.* New Delhi: Central Pollution Control Board
9. CPCB (Central Pollution Control Board). 2005. National air quality status 2003. In *National Ambient Air Quality Monitoring Series: NAAQMS/26/2004-2005.* New Delhi: Central Pollution Control Board
10. CPCB (Central Pollution Control Board). 2006a. National air quality status 2004. In *National Ambient Air Quality Monitoring Series: NAAQMS/27/2006-2007.* New Delhi: Central Pollution Control Board
11. CPCB (Central Pollution Control Board). 2006b. National air quality status 2005. In *National Ambient Air Quality Monitoring Series: NAAQMS/30/2006-2007.* New Delhi: Central Pollution Control Board
12. CPCB (Central Pollution Control Board). 2007. National air quality status 2006. In *National Ambient Air Quality Monitoring Series: NAAQMS//2007-2008.* New Delhi: Central Pollution Control Board
13. CPCB (Central Pollution Control Board). 2008. National air quality status 2007. In *National Ambient Air Quality Monitoring Series: NAAQMS//2008-2009.* New Delhi: Central Pollution Control Board
14. CPCB (Central Pollution Control Board). 2009. National air quality status 2008. In *National Ambient Air Quality Monitoring Series: NAAQMS//2009-10.* New Delhi: Central Pollution Control Board
15. CCPCB (Central Pollution Control Board) 2014a. National air quality: status 2011. In *National Ambient Air Quality Status NAAQMS// 2013-14.* New Delhi: Central Pollution Control Board
16. CCPCB (Central Pollution Control Board) 2014b. National air quality: status 2012. In *National Ambient Air Quality Status NAAQMS// 2013-14.* New Delhi: Central Pollution Control Board

TABLE 11.2 Number of air monitoring stations in India violating the 24 hourly standards for a particular per cent of times in a year (2001–10)

	Residential						Industrial				
	0%-2%	2%-5%	5%-25%	>=25%	TSN		0%-2%	2%-5%	5%-25%	>=25%	TSN
2001						2001					
SO_2	91	0	0	0	91	SO_2	77	0	1	0	78
NO_x	83	2	4	2	91	NO_x	73	0	3	0	76
SPM	15	3	9	50	77	SPM	39	10	16	4	69
2002						2002					
SO_2	97	0	0	0	97	SO_2	68	1	0	0	69
NO_x	87	3	5	1	96	NO_x	68	0	2	1	71
SPM	11	7	10	58	86	SPM	34	9	15	5	63
2003						2003					
SO_2	105	0	0	0	105	SO_2	76	0	0	0	76
NO_x	94	4	4	2	104	NO_x	72	0	2	0	74
SPM	14	5	17	55	91	SPM	36	6	23	3	68
RSPM	11	6	19	60	96	RSPM	18	5	15	31	69
2004						2004					
SO_2	113	0	0	0	113	SO_2	83	0	0	0	83
NO_x	103	2	7	2	114	NO_x	79	1	2	1	83
SPM	19	6	22	59	106	SPM	44	7	20	5	76
RSPM	15	5	24	66	110	RSPM	19	5	17	39	80
2005						2005					
SO_2	136	2	0	0	138	SO_2	91	0	0	0	91
NO_x	131	4	1	3	139	NO_x	88	1	1	1	91
SPM	15	13	23	87	138	SPM	49	10	19	6	84
RSPM	18	12	19	84	133	RSPM	28	4	21	33	86
2006						2006					
SO_2	177	1	0	0	178	SO_2	106	4	0	0	110
NO_x	162	6	6	3	177	NO_x	105	3	3	0	111
SPM	20	10	35	99	164	SPM	64	11	17	10	102
RSPM	23	5	34	100	162	RSPM	29	9	23	42	103
2007						2007					
SO_2	195	0	0	0	195	SO_2	118	0	0	0	118
NO_x	178	7	8	4	197	NO_x	112	3	3	0	118
SPM	30	9	34	112	185	SPM	69	13	15	11	108
RSPM	27	14	42	110	193	RSPM	33	7	22	49	111

Contd...

TABLE 11.2 | *Contd...*

	Residential						Industrial				
	0%-2%	2%-5%	5%-25%	>=25%	TSN		0%-2%	2%-5%	5%-25%	>=25%	TSN
2008						2008					
SO_2	198	0	0	0	198	SO_2	116	0	0	0	116
NO_x	173	11	8	6	198	NO_x	117	2	3	0	122
SPM	38	12	26	114	190	SPM	65	11	20	11	107
RSPM	32	10	42	94	178	RSPM	28	12	20	55	115
	0%-2%	2%-5%	6%-25%	26%-50%	21-75%	> 75%*					
2009											
SO_2	343	6	2	0	0	0					
NO_x	304	11	18	17	1	0					
SPM	51	20	48	74	54	104					
2010											
SO_2	352	4	3	1	0	0					
NO_x	313	14	20	15	0	0					
SPM	50	19	64	83	56	87					

* including residential/industrial/others

RSPM – respirable suspended particulate matter; TSN – total stations violating the standards at least one time

Sources

1. CPCB (Central Pollution Control Board). 2002. ***National Ambient Air Quality Monitoring Series: NAAQMS/22/2001-02***. New Delhi: CPCB, Ministry of Environment and Forests
2. CPCB (Central Pollution Control Board). 2003. ***National Ambient Air Quality Monitoring Series: NAAQMS/23/2003-04***. New Delhi: CPCB, Ministry of Environment and Forests
3. CPCB (Central Pollution Control Board). 2004. National ambient air quality: status 2002. In ***National Ambient Air Quality Monitoring Series: NAAQMS/24/2003-2004***. New Delhi: CPCB, Ministry of Environment and Forests
4. CPCB (Central Pollution Control Board). 2005. National ambient air quality: status 2003. In ***National Ambient Air Quality Monitoring Series: NAAQMS/26/2004-2005***. New Delhi: CPCB, Ministry of Environment and Forests
5. CPCB (Central Pollution Control Board). 2006a. National ambient air quality: status 2004. In ***National Ambient Air Quality Monitoring Series: NAAQMS/27/2006-2007***. New Delhi: CPCB, Ministry of Environment and Forests
6. CPCB (Central Pollution Control Board). 2006b. National ambient air quality : status 2005. In ***National Ambient Air Quality Monitoring Series: NAAQMS/30/2006-07***. New Delhi: CPCB, Ministry of Environment and Forests
7. CPCB (Central Pollution Control Board). 2008. National ambient air quality : status 2007. In ***National Ambient Air Quality Status: NAAQMS//2008-09***. New Delhi: CPCB, Ministry of Environment and Forests
8. CPCB (Central Pollution Control Board). 2009. National ambient air quality : status 2008. In ***National Ambient Air Quality Status: NAAQMS//2009-10***. New Delhi: CPCB, Ministry of Environment and Forests
9. CPCB (Central Pollution Control Board). 2011. National ambient air quality : status 2009. In ***National Ambient Air Quality Status: NAAQMS//2010-11***. New Delhi: CPCB, Ministry of Environment and Forests
10. CPCB (Central Pollution Control Board). 2012. National ambient air quality : status 2010. In ***National Ambient Air Quality Status: NAAQMS//2011-12***. New Delhi: CPCB, Ministry of Environment and Forests

TABLE 11.3 Concentration of pollutants as a result of burning biofuels

Location and type of stove	Total suspended particulate matter (mg/m³)		Carbon monoxide (parts per million)	
	Sample size (number)	Average	Sample size (number)	Average
Traditional				
Nepal	22	3.1	27	300
India				
Gujarat	45	6.4	–	–
Haryana	51	3.3	51	6
Karnataka	38	3.2	38	17
Central and South India	129	4.6	129	51
Pauri Garhwal	95	5.6	100	21
Tehri Garhwal	50	1.2	–	–
Pune, urban poor[a]	–	1.1	–	8
Ahmedabad, urban poor	20	2.6	20	165
Delhi slums[a]	80	1.2	80	12
Improved				
Nepal	27	1.1	26	67
India				
Gujarat	14	4.6	–	–
Haryana	36	2.9	36	5
Karnataka	38	2.5	38	8

[a]respirable particulate matter

Sources

1. Reid, H, K. R. Smith, and B. Sherchand. 1986. Indoor smoke exposures from traditional and improved cookstoves: comparisons among rural Nepali women. ***Mountain Research and Development*** 6(4): 293–304
2. Smith, K. R., A. L. Aggarwal, and R. M. Dave. 1983. Air pollution and rural biomass fuels in developing countries: a pilot village study in India and implications for research and policy. ***Atmospheric Environment*** 17(11): 2343–2362
3. Ramakrishna, J. 1988. Patterns of domestic air pollution in rural India. ***PhD Dissertation,*** University of Hawaii
4. Menon, P. 1988. Indoor spatial monitoring of combustion generated pollutants (total suspended particulates, carbon monoxide, and BAP) by Indian cookstoves. ***PhD Dissertation***, University of Hawaii
5. Saksena, S., R. Prasad, R. C. Pal, and V. Joshi. Patterns of daily exposure to total suspended particulates and carbon monoxide in the Garhwal Himalaya. ***Atmospheric Environment*** 26(11): 2125–2134
6. TERI (Tata Energy Research Institute). 1995. ***Biomass Fuels, Indoor Air Pollution and Health: A Multi-disciplinary, Multi-center Study*** (phase 1B). Final report submitted to Biomass Users Network, USA. New Delhi: TERI
7. Raiyani, C. V., S. H. Shah, N. M. Desai, K. Venkaiah, J. S. Patel, D. J. Parikh, and S. K. Kashyap. 1993. Characterization and problems of indoor pollution due to cooking stove smoke. ***Atmospheric Environment*** 27A(11): 1643–1655
8. TERI (Tata Energy Research Institute). 1995. ***Health risks and cost-effective interventions for indoor air pollution***. Final report submitted to The Catholic University of Louvain, Belgium. New Delhi: TERI
9. Smith, K. R., M. G. Apte, M. A. Yuqing, W. Wongsekiarttirat, and A. Kulkarni. 1994. Air pollution and the energy ladder in Asian cities. ***Energy*** 19(5): 587–600

TABLE **11.4** New National Ambient Air Quality Standards (NAAQS) for India

Pollutant	Residential, industrial, rural, and other areas		Ecologically sensitive areas (notified by Central Government)	
	24 hourly standard (mg/m³)	Annual standard (mg/m³)	24 hourly standard (mg/m³)	Annual standard (mg/m³)
SO_2	80	50	80	20
NO_x	80	40	80	30
PM_{10}	100	60	100	60
$PM_{2.5}$	60	40	60	40
Ozone	180*	100**	180*	100**
Lead	1	0.5	1	0.5
CO	4000*	2000**	4000*	2000**
NH_3	400	100	400	100
As (ng/m³)		6		6
Benzene		5		5
BaP (ng/m³)		1		1
Ni (ng/m³)		20		20

*Annual arithmetic mean of 104 measurements in a year at a particular site taken twice a week 24 hourly at uniform intervals.

**24 hourly or 8 hourly or 1 hourly monitored values, as applicable, shall be complied with 98% of the time in a year. 2% of the time, they may exceed the limits but not on two consecutive days of monitoring.

Source Details available at http://www.cpcb.nic.in, last accessed on 1 April 2010

TABLE 11.5 Interventions adopted to control vehicular emissions in the period 1991–2010

Vehicular pollution		1991	1996	1997	1998	1999	2000	2001	2002	2003	2004	2005	2006	2007	2008	2009	2010
Emission norms		• First emission standards were set up • First idle emission norms (PUC) checks for on-road vehicles came into force	• Emission norms made stringent		• Emission norms for catalytic converters made stringent in NCT • Notification for specifications of 2-stroke engine	• Applicability of Euro I norms in NCR	• Applicability of Euro I norms all over the India • Applicability of Bharat stage II norms in NCR • Emission norms for CNG notified	• Bharat stage II emission norms introduced in Mumbai, Kolkata, and Chennai for new vehicles (except two- and three-wheelers) • Emission norms for LPG notified • Notification of new emission norms for CNG				• Bharat stage II norms introduced in whole country • Bharat stage III norms in selected 11 cities					
		(AFP)	(AFP, SOE)		(MoEF 97–98)	(SOE)	(AFP)	(AFP)									
Fuel quality	• Eradication of gasoline lead (Annual report, MoEF)		• Lead phased out from gasoline in metros	• Low leaded gasoline (0.15 g/l) in entire India	• Unleaded or very low leaded gasoline in NCT and other state capitals and major cities	• Unleaded gasoline in NCR	• Lead phased out from gasoline in the entire country										
	• Sulphur content reduction (Annual reports, MoEF)		• Low sulphur (0.5%) in metros	• Low sulphur (0.25%) in Delhi	• Low sulphur (0.25%) in metros	• Low sulphur (0.25%) in entire India	• 0.05% sulphur for all vehicles in Mumbai and private vehicles in NCR	• 0.05% sulphur for all vehicles in NCR	• 0.05% sulphur for all vehicles in metros			• Bharat stage III fuel introduced in 11 selected cities • Sulphur content in diesel and gasoline reduced to 0.035% and 0.015%, respectively					• Bharat stage IV norms introduced in 13 major cities and Bharat stage III implemented in rest of India

Contd...

TABLE 11.5 *Contd...*

Vehicular pollution	1991	1996	1997	1998	1999	2000	2001	2002	2003	2004	2005	2006	2007	2008	2009	2010
Others			• Fuel quality specifications laid down by BIS for gasoline and diesel for 2000–05 and beyond 2005 for the country • Specifications laid down by BIS for gasoline and diesel for 2000–05 and beyond 2005 for the country	• Gasoline benzene reduced to 3% in Delhi • Ban on 2T oil in petroleum in NCT		• Petrol with benzene 1% made available in NCT • Investment of ₹ 10 000 crore was made by oil companies for upgradation of fuel quality	• Petrol with benzene 1% made available in NCR	• Ethanol blended petrol introduced in some states as phase 1	• Auto Fuel Policy introduced by the Ministry of Petroleum and Natural Gas		• Bharat stage II fuel introduced in 11 selected cities					
			(SOE)	(Annual report, MoEF)		(Annual report, MoEF, AFP)	(Annual report, MoEF)	(Annual report, MoEF)								
Other measures		• Fitting of catalytic converters/ CNG Kits for in-use government vehicles made mandatory in Delhi • CNG for automobiles through nine retail outlets • Air monitoring at traffic intersections also (CPCB air quality report)	• Authorization of 305 pollution checking centres for petrol and 32 for diesel in NCT • 17.7 lakh vehicles were issued PUC in NCT	• Phasing out 15 years old vehicles from Delhi • New ISBT in Delhi • Fitting of catalytic converters/ CNG Kits for in-use government vehicles made mandatory in 45 cities	• Public surveys conducted for their opinion about traffic management, encroachment, cleaner fuels,	• New taxis and autos to be registered with CNG only in NCT • Establishment of 1 Fuel Testing Lab • 60 CNG stations established in Delhi	• Conversion of buses to CNG started in Delhi and Mumbai • 87 CNG stations established in Delhi	• 4 cities identified for CNG conversion as Delhi • 9 cities identified for immediate improvement in air quality			• Phase I of metro rail project completed in Delhi	• Work for Bus Rapid Transit (BRT) project started in Delhi				
			(Annual report, MoEF)	(Annual report, MoEF)	(Annual report, MoEF)	(Annual report, MoEF)	(Annual report, MoEF)	(Annual report, MoEF)								

TABLE 11.6 Available water resources: summary for India and world

India	Value
Annual rainfall (2005)	1 208 mm
Major river basins	12
Catchment area >20 000 km^2	253 Mha
Medium river basins	46
Catchment area <20 000 km^2	24.6 Mha
Average annual precipitation	4 000 BCM
Average precipitation during monsoon (June–September)	3 000 BCM
Natural run-off	1 986.5 BCM
Estimated utilizable surface water resources	690 BCM
Total utilizable groundwater resources	433 BCM
Total annual utilizable water resources	1 123 BCM
Per capita water availability	1 720.29 m^3
World	
Sources of water (approximately)	Volume (million BCM)
Salt water in oceans	1 348
Freshwater	37.5
Sources of freshwater (approximately)	Volume ('000 BCM)
Polar ice and glaciers	28 200
Groundwater < 800 m deep	3 740
800–4000 m deep	4 710
Lakes and rivers	127
Others (soil moisture and atmospheric vapours)	704

Source Details available at <http://cwc.nic.in/main/webpages/statistics.html#1>, last accessed on 15 September 2010

TABLE 11.7 Municipal solid waste generation in India (2009–12)

State/union territory	Municipal solid waste (MT/day)
Andaman and Nicobar Islands	50
Andhra Pradesh	11 500
Arunachal Pradesh	93.8
Assam	1 146.28
Bihar	1 670
Chandigarh	380
Chhattisgarh	1 167
Daman and Diu and Dadra and Nagar Haveli	41
Delhi	7 384
Goa	193
Gujarat	7 378.78
Haryana	536.85
Himachal Pradesh	304.3
Jammu and Kashmir	1 792
Jharkhand	1 710
Karnataka	6 500
Kerala	8 338
Lakshadweep	21
Maharashtra	19.2
Manipur	112.9
Meghalaya	284.6
Mizoram	4 742
Madhya Pradesh	4 500
Nagaland	187.6
Odisha	2 239.2
Puducherry	380
Punjab	2 793.5
Rajasthan	5 037.3
Sikkim	40
Tamil Nadu	12 504
Tripura	360
Uttar Pradesh	11.58
Uttarakhand	752
West Bengal	12 557
Total	127 485.11

Source CPCB (Central Pollution Control Board). 2012. ***Status Report on Municipal Solid Waste Management***. New Delhi: Central Pollution Control Board

CLIMATE CHANGE 12

AT A GLANCE

- The annual mean temperature of India during 2014 was +0.53°C above the 1961–90 average. The year 2014 was the fifth warmest year on record since 1901.
- Rainfall activity over the country was below normal (88% of long period average) during 2014.
- Barring some parts of southern peninsular India, the regional climate models show a rise in the future monsoon rainfall in the range 9%–16% towards the end of twenty-first century.
- As per the *Fifth Assessment Report* of the Intergovernmental Panel on Climate Change, the current trajectories for emissions point towards a likely increase of 3.2–5.4°C in temperature above the pre-industrial levels by 2100.
- CO_2 emissions in India was approximately in the range of 2088–2187 $MtCO_2$ in 2014.
- The per capita emissions in India are reported to be 1.6 tonnes/capita, which is well below the global average of 5.1 tonnes/capita.
- India doubled the clean energy cess on coal from ₹100/tonne in 2013/14 to ₹200/tonne during 2014/15.
- Renewable energy targets for India have been increased to 175 GW, of which the target for solar energy is 100 GW by 2022. Moreover, a target has been set to achieve 40% cumulative electric power installed capacity by 2030 from non-fossil energy resources.

Introduction

The *Fifth Assessment Report* (AR5) of the Intergovernmental Panel on Climate Change (IPCC) observes that climate change is unequivocal and essentially driven by anthropogenic factors. Emissions in the past 40 decades have risen at a much higher rate than the increase in emissions before that period. The decade post-2000 recorded the fastest growth in emissions worldwide, including India. The current emissions trajectories in the IPCC AR5 point towards a "likely" increase of 3.2–5.4°C in the mean temperature above the pre-industrial levels by 2100. However, latest studies after the submissions of the Intended Nationally Determined Contributions (INDCs) have pointed towards a rise of 2.7–3.6°C in physical temperature by 2100.

Against this backdrop, the chapter explains the observed climate trends over India in the past 100 years and presents a synthesis of temperature and precipitation trends over the country. It also discusses the trends in greenhouse gas (GHG) emissions globally and in India during 2014/15. The measures adopted by the Government of India in 2014/15 to deal with climate change domestically and internationally are discussed in the following sections.

Physical climate science: observed climate trends over India

This section explains the temperature and precipitation trends observed over the Indian region in the last 100 years. The analysis presents a synthesis of various studies covering temporal and spatial dimensions.

Temperature

The all-India annual mean temperatures have shown a significant warming trend of 0.51°C per 100 years (Kothawale, Munot, and Krishna Kumar 2010). The all-India annual mean temperature anomalies for 1901–2014 based on the 1961–1990 average (Figure 1) clearly show a warming trend of +0.62°C per 100 years (IMD 2014). This warming has been predominant in recent years because both maximum and minimum temperatures have been steadily rising over large parts of the Indian region since 1990, with a higher increase in minimum temperatures than in maximum temperatures (IMD 2012). The year 2014 has been the fifth warmest year on record for India since 1901 (IMD 2014).

The analysis of temperature data for 1971–2007 shows the accelerated warming owing to the intense warming during 1998–2007. The rates of warming in winter and post-monsoon seasons have increased by 0.80°C and 0.82°C, respectively, in the past 100 years. The all-India maximum and minimum temperatures have, respectively, increased by 0.71°C and 0.27°C per 100 years for the analysis period of 1901–2007 (MoEF 2010) (Figure 2).

Figure 3 demonstrates that spatially also the annual mean temperature shows positive trends (warming) over most parts of India, barring parts of Rajasthan, Gujarat, and Bihar, where negative trends (cooling) have been observed for 1901–2014 (IMD 2014).

A seasonal analysis of observed data suggests a maximum rise of 0.77°C in the annual mean temperature during the post-monsoon season (October–December), which is followed by 0.70°C during the winter season (January–February). The pre-monsoon season (March–May) has shown an increase of 0.64°C, which is followed by the monsoon season (June–September) with 0.33°C rise. The rise in minimum temperatures during winter over northern plains has been attributed to the rise in pollution levels, which leads to frequent occurrences of fog (Attri and Tyagi 2010).

The regional climate model projections over India for the future time period indicate an overall warming climate over the Indian subcontinent, with a projected increase of 3.5–4.3°C in the mean surface air temperature by the end of the century. The daily extremes in temperature are also projected to increase in the future, with the highest maximum temperature showing an increase of 1–4°C towards 2050s and an increase of 4.5°C for night-time temperatures (MoEF 2012).

Precipitation

Although the Indian monsoon does not show any significant trend, it depicts a well-defined epochal variability, with each epoch of approximately three decades. All-India monsoon rainfall series based on the period 1871–2009 indicates that the mean rainfall is 848 mm with a standard deviation of 83 mm.

Rainfall during the principal rainy season [southwest (summer) monsoon season (June–September)] for the country was below normal (88% of long period average) (IMD 2014).

Studies have shown that all-India monsoon rainfall and all-India annual rainfall do not show any trend for 1950–80 (Parthasarthy 1984) or for even longer periods from 1901–09 (Attri and Tyagi 2010), though many regional level trends have been observed

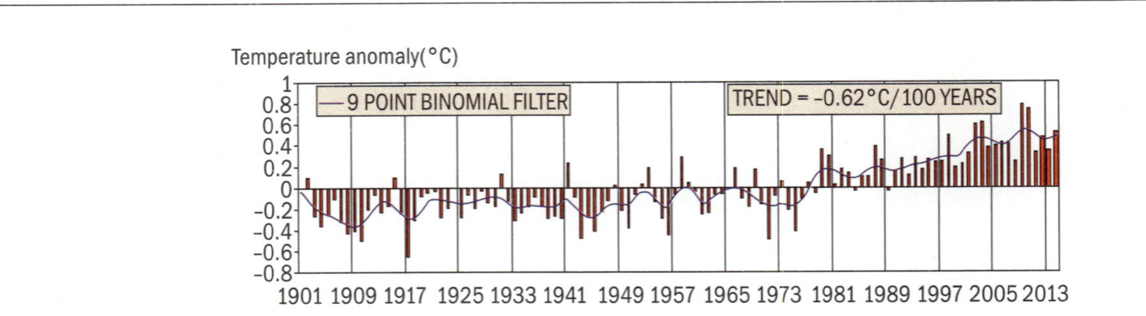

Figure 1 All-India annual mean temperature anomalies for 1901–2013 (based on the 1961–90 average)

Source IMD (2014)

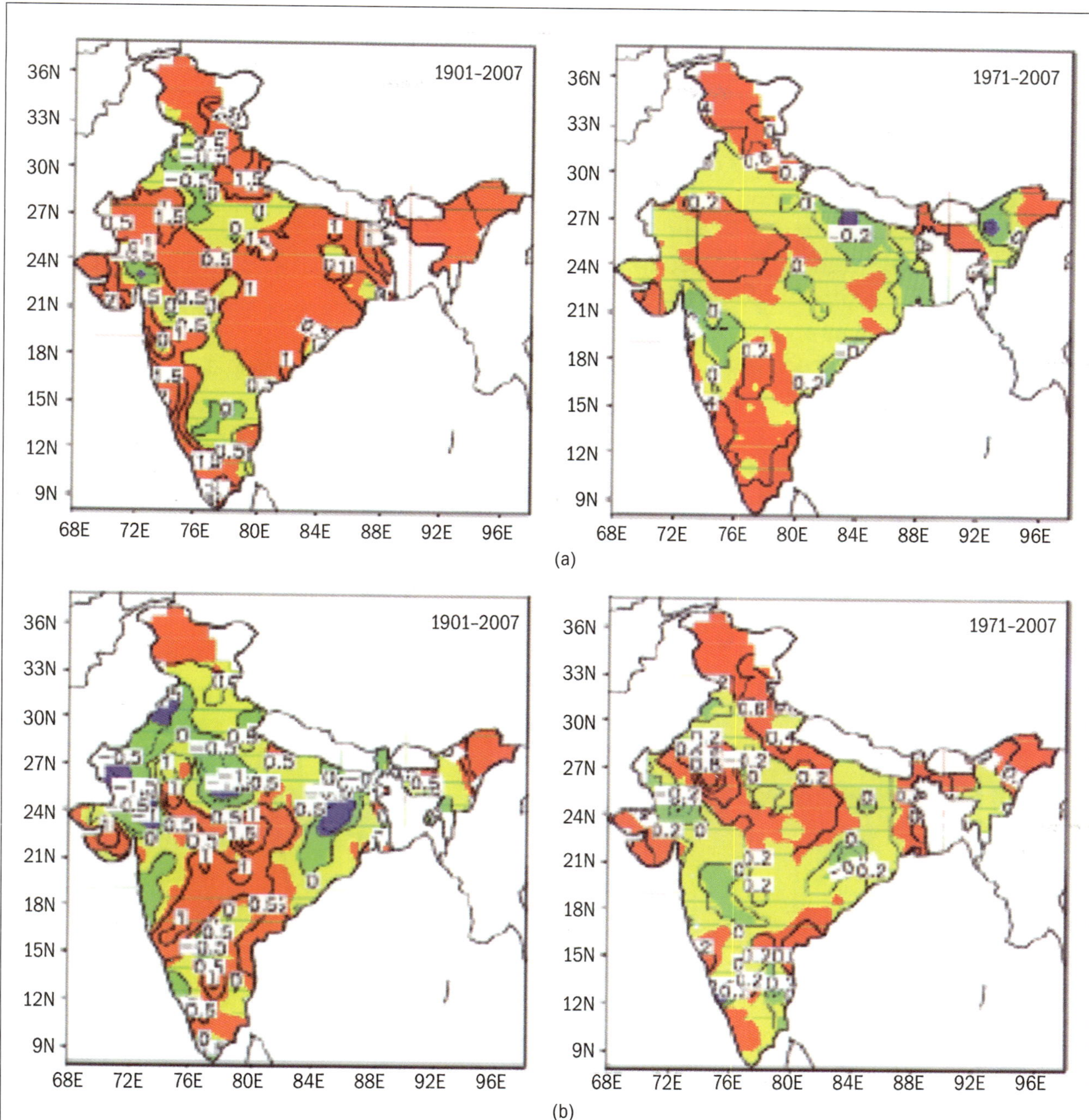

Figure 2 Spatial patterns of linear trends of (a) maximum and (b) minimum temperatures

Note Red—warming trend at 5%; blue—cooling trend at 5%; green—cooling trend; yellow—warming trends

Source INCCA (2010)

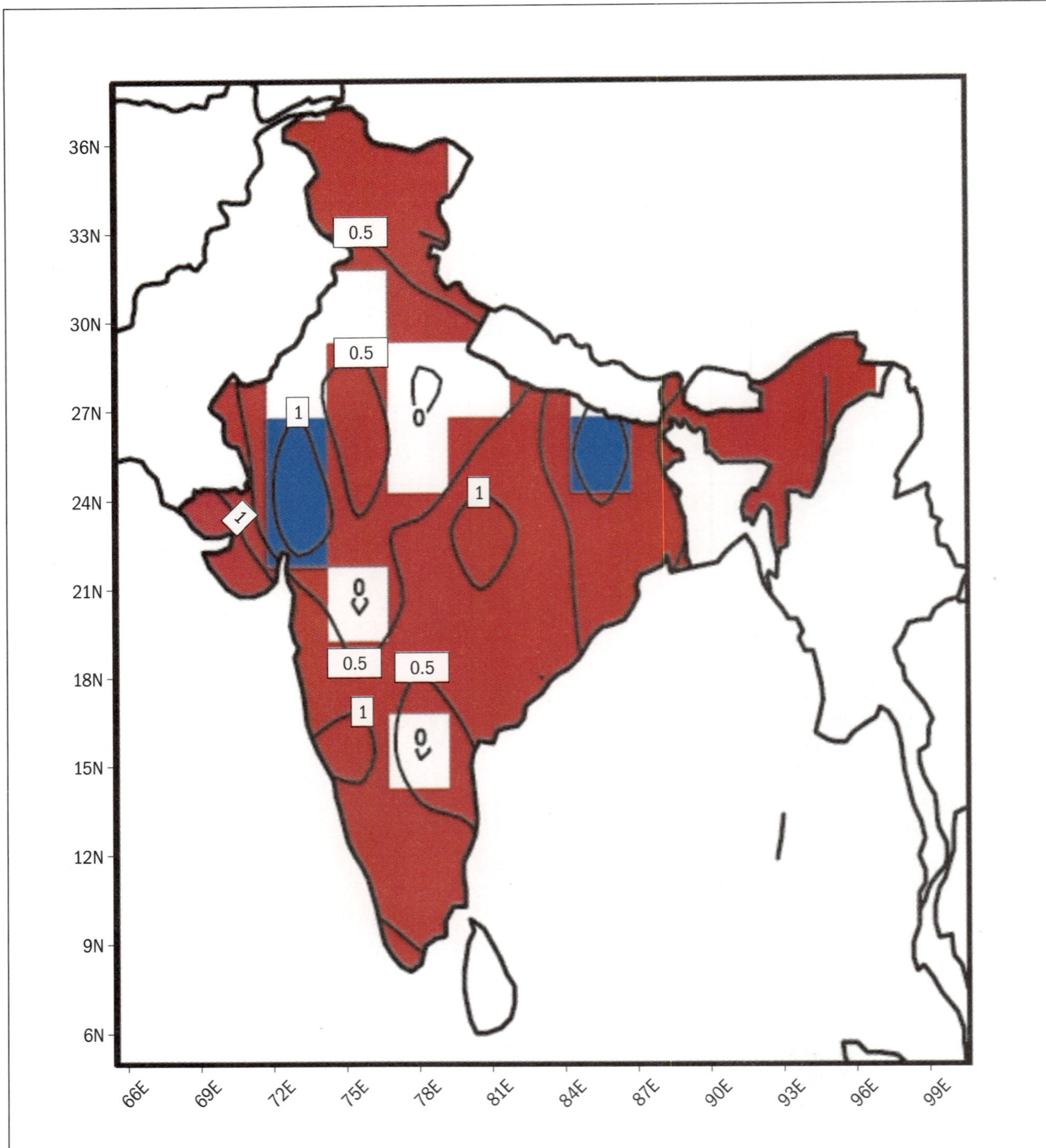

Figure 3 Spatial patterns of trends (°C per 100 years) in mean annual temperature anomalies (1901–2014)

Note Areas having significant trends are shaded: red denotes warming and blue denotes cooling.

Source IMD (2014)

(Kumar, Kumar, Pant 1994). Sub-divisional trends in seasonal rainfall and monsoon rainfall (Figure 4) also show that no clear trend for India as a whole can be established. Rajeevan, Jyoti, Kale, *et al.* (2006) have also shown no significant trends in the number of break and active days of the south-west monsoon season (June–September) during the analysis period of 1951–2003 (Figure 5). However, an overall increase in extreme rainfall events and their intensities during 1910–2000 over the entire country was reported by Sen Roy and Balling (2004).

Barring some parts of southern peninsular India, the regional climate models show a rise in the future monsoon rainfall in the range of 9%–16% towards the end of the twenty-first century (MoEF 2012). The number of rainy days is projected to decrease, whereas the intensity of rainfall is projected to increase in the future over most parts of North and Central India, indicating a rise in extreme precipitation events in the future.

Cyclonic storms

In 2014, three cyclonic storms formed over Indian waters (one each over Arabian Sea in June and October and one over Bay of Bengal in October). Hudhud, the strongest of them, was very severe, which crossed the Andhra Pradesh coast and caused heavy rainfall and widespread damage to crops and property in Andhra Pradesh and Odisha. The cyclonic storm Nanauk in June and very severe cyclonic storm Nilofar in October, which formed over Arabian Sea, dissipated in situ over the sea itself without making landfall (IMD 2014).

Models have projected a decrease in the frequency of cyclonic disturbances over the coasts of India by the end of the twenty-first century. These models project that although the frequency might decrease, the intensity of cyclones occurring in the future would be higher than the present level. No significant changes in the tracks of cyclones have been reported for the future climate simulations (MoEF 2012).

Emission trends

GHG emission pathways from IPCC's *Fifth Assessment Report*

For the temperature changes caused by anthropogenic GHG emissions to be kept below 2°C relative to the pre-industrial levels, atmospheric concentrations of GHGs must be less than 450 ppm CO_2-eq in the year 2100 (IPCC AR5, likely scenario, high confidence). The latest report of IPCC AR5 used a set of emission scenarios called the Representative Concentration Pathway (RCP) to project climate models.

The report suggests that the global surface temperature change by the end of the twenty-first century is likely to exceed 1.5°C relative to the period 1850–1900 for all RCP scenarios except RCP2.6. Warming will continue beyond the year 2100 under all RCP scenarios except RCP2.6 and continue to exhibit interannual-to-decadal variability and it will not be regionally uniform. In all RCPs, atmospheric CO_2 concentrations are higher in the year 2100 relative to the present day as a result of further increase in cumulative emissions of CO_2 during this century (Figure 6). Only for the RCP2.6 scenario, the global mean temperature is likely to remain below 2°C during the twenty-first century. Table 1 gives the temperature change over the twenty-first century under the relative RCP scenarios.

The IPCC notes that the increase in global temperature is proportional to the build-up of long-lasting GHGs in the atmosphere, particularly CO_2. Thus, to limit the warming caused by anthropogenic CO_2 emissions alone (>66% confidence) to less than 2°C since the pre-industrial period, cumulative CO_2 emissions from all anthropogenic sources need to stay below 2900 $GtCO_2$ (790 GtC). Since emissions began growing rapidly after the 1970s, the world has already emitted around 1900 $GtCO_2$ by 2012 and used up a large part of this budget. The absolute decadal increase in CO_2 emissions was larger from 2000 to 2010 compared to that from 1970 to 2000 (IPCC 2014). The current trajectories for emissions thus point towards a 3.2–5.4°C "likely" increase in temperature above the pre-industrial levels by 2100 (IPCC 2014, Figure 6).

Globally observed emissions

In 2013, approximately 36 $GtCO_2$ was emitted to the atmosphere globally. This represents a 2.1%–2.3% increase in emissions over the previous year and approximately 66% increase from the Kyoto period (1990 level) (GCP 2014). These levels of CO_2 emissions were expected to increase further by 2.5% in 2014; however, the actual estimates are not yet available (IEA 2014). The increase in the level of CO_2 emissions till 2013, as estimated by different agencies, is given in Table 2. Annex 1 countries produced 36%

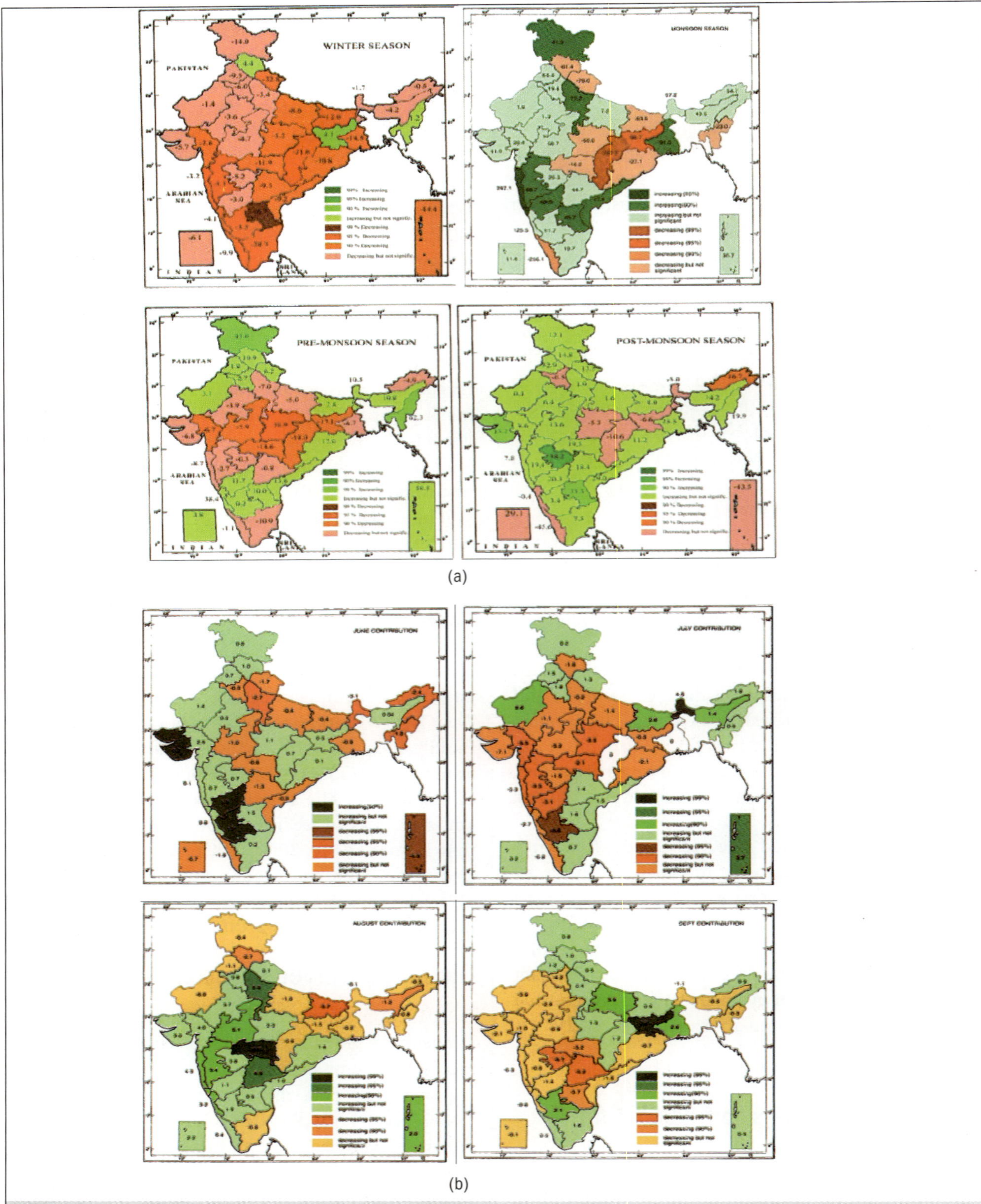

Figure 4 Sub-divisional trends of (a) seasonal and (b) monsoon rainfall for 1901–2003

Note Different colours show different levels of significance.

Source IMD (2010)

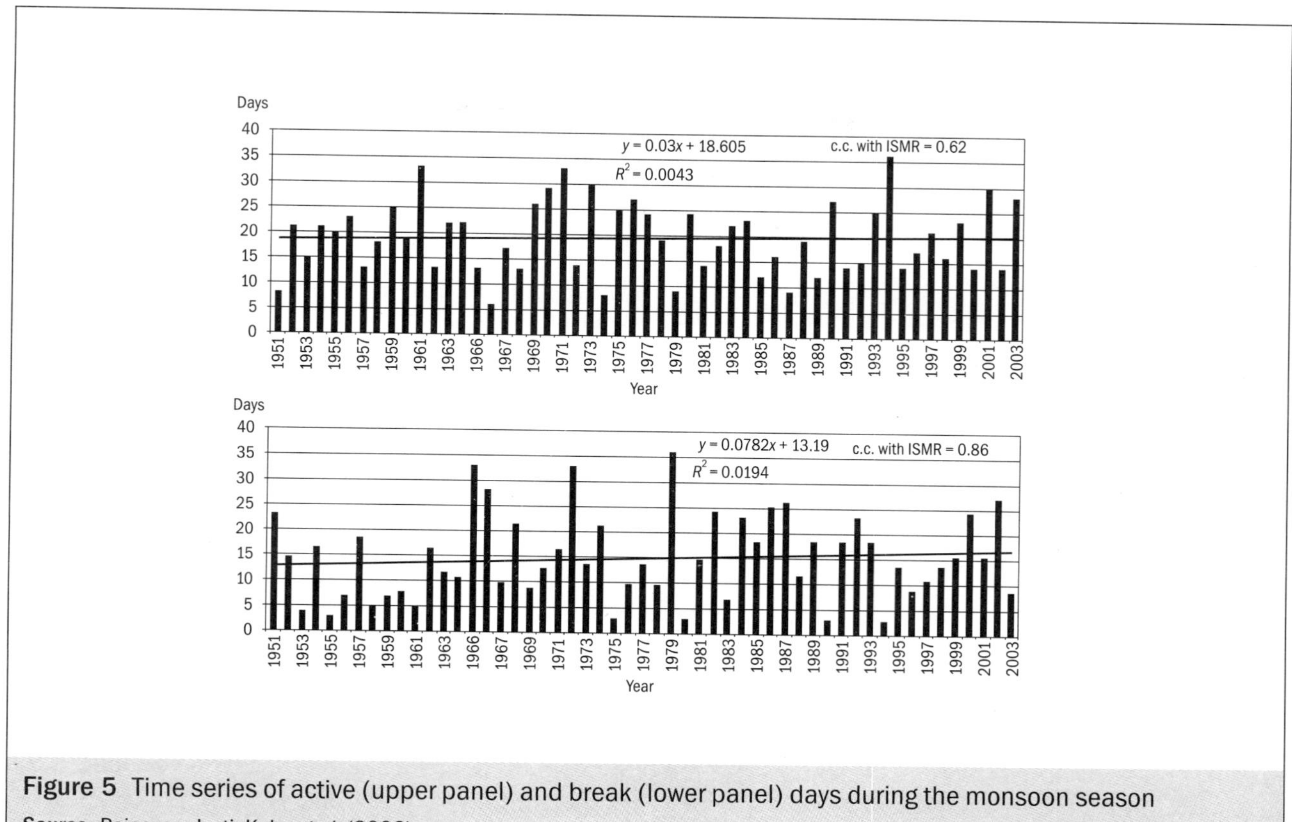

Figure 5 Time series of active (upper panel) and break (lower panel) days during the monsoon season
Source Rajeevan, Jyoti, Kale, *et al.* (2006)

Table 1 Probability statements regarding RCP scenarios

Relative position of RCPs	CO_2-eq concentration in 2100 (concentration range)	Cumulative CO_2 emissions ($GtCO_2e$) 2011-2100	2100 temperature change (°C)
RCP2.6	430-480	630-1180	1.5-1.7
	480-530 (no overshoot of 530 ppm CO_2-eq)	960-1430	1.7-1.9
	530-580 (no overshoot of 580 ppm CO_2-eq)	1240-2240	2.0-2.2
RCP4.5	580-650	1870-2440	2.3-2.6
	650-720	2570-3340	2.6-2.9
RCP6.0	720-1000	3620-4990	3.1-3.7
RCP8.5	>1000	5350-7010	4.1-4.8

ppm – parts per million; RCP – Representative Concentration Pathway
Source IPCC (2014)

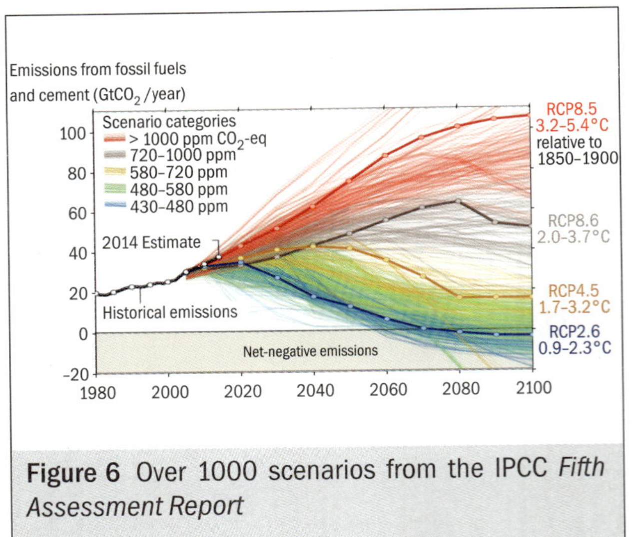

Figure 6 Over 1000 scenarios from the IPCC *Fifth Assessment Report*

Source IPCC (2014)

of these emissions, while non-Annex 1 countries emitted 58%. Table 12.1 gives the disaggregated data for CO_2 emissions in countries worldwide.

With the rise in absolute CO_2 emissions, the global average of per capita emissions also rose from 4.88 tonnes per capita in 2012 to 5.10 tonnes per capita in 2013.

Table 3 shows that the per capita emission in the top three major emitting countries is far more than the global average in 2013, while it is well below the global average for India, at 1.6 tonnes CO_2 per capita. The disaggregated emission trends in India are discussed in the next section.

Bridging the global emission gap

To limit warming up to 2°C, the global economy needed to reduce carbon intensity by 6% per year but was able to achieve only a reduction of 1.3% in 2013. However, only Australia recorded a decarbonization of 7.2% in 2013. UK, Italy, and China achieved a decarbonization rate between 4% and 5%, while the five countries, that is, France, USA, India, Germany, and Brazil, increased their carbon intensity in 2013 though the average change in carbon intensity over 2008–13 decreased for these countries except for Brazil (Table 4) (PwC 2014).

The United Nations Framework Convention on Climate Change (UNFCCC) invited all parties to submit their INDCs ahead of Conference of Parties (COP) 21, and by 1 October 2015, 119 INDCs were submitted. Many developing country INDCs also have a conditional component which is contingent upon the provision of financial, technological and capacity building support by developed country parties. Table 5 gives the global emissions and emissions gap under the implementation of unconditional INDCs and inclusion of conditional INDCs.

Global GHG emissions for unconditional INDCs are projected to reach 54 $GtCO_2e$ in 2025, and for conditional INDCs, the projection is 53 $GtCO_2e$ in 2025. The emissions gap between the full implementation of unconditional INDCs and the least cost emission level pathway to stay below 2°C is estimated to be 7 $GtCO_2e$ (UNEP 2015).

Table 2 Level of CO_2 emissions and per capita CO_2 emissions in the world

Year	CO_2 emissions (billion tonnes of CO_2)				Per capita emissions (tonnes CO_2/capita)
	International Energy Agency	US Energy Information Agency	BP Statistical Review	WRI/CAIT	
2014	32.20	–	35.50	–	–
2013	32.190	–	35.09	–	5.10
2012	31.491	–	34.46	33.84	4.88
2011	31.293	32.57	33.74	33.37	4.87
2010	29.838	31.50	32.28	32.33	4.79
2009	28.321	30.23	31.10	30.71	–
2008	–	30.25	31.54	31.14	–
2007	–	29.73	31.19	30.91	–

CAIT – Climate Data Explorer; WRI – World Resources Institute

Source Compiled from EDGAR (2014), IEA (2015), BP (2015), GCP (2014), and CAIT (2015)

Table 3 Emission trend across four major CO_2-emitting countries

Country	Share of global CO_2 emissions in 2013 (%)	Per cent change in emissions from 2012/13	Rate of growth of GDP from 2012/13 (%)	Per capita emissions in 2013	Historical cumulative emissions share (1870–2013) (%)
USA	14	20% increase	2.9	16.4	26
China	28	58% increase	4.2	7.2	11
EU28	10	11% decrease	−1.8	6.8	23
India	7	17% increase	5.1	1.6–1.9	3

Source GCP (2014)

Table 4 Emissions related to energy

Country	Change in energy-related emissions	Real GDP growth (PPP)	Carbon intensity (tonnes CO_2/ 2013$m)	Change in carbon intensity (2012/13)	Annual average change in carbon intensity (2008–13)
Australia	−4.7%	2.7%	338	−7.2%	−4.6%
UK	−3.2%	1.7%	206	−4.8%	−2.9%
Italy	−5.9%	−1.9%	172	−4.1%	−2.9%
USA	2.5%	1.9%	326	0.6%	−2.4%
China	3.4%	7.7%	561	−4.0%	−1.6%
EU	−2.5%	0.1%	209	−2.5%	−2.3%
France	0.5%	0.2%	145	0.3%	−2.0%
Brazil	8.1%	2.5%	157	5.5%	1.8%
Germany	3.3%	0.4%	224	2.9%	−0.9%
India	6.0%	5.0%	271	0.9%	−0.7%
World	1.8%	3.1%	323	−1.2%	−0.6%

Source PwC (2014)

Table 5 Global emissions and emission gap under the implementation of INDCs

INDC	Global GHG emissions ($GtCO_2e$)		Emissions gap ($GtCO_2e$)	
	2025	2030	2025	2030
Implementation of unconditional INDCs	54 (range 53–58)	56 (range 54–59)	7 (range 5–10)	14 (range 12–17)
Inclusion of conditional INDCs	53 (range 52–56)	54 (range 52–57)	5 (range 4–8)	12 (range 10–15)

GHG – greenhouse gas; INDC – Intended Nationally Determined Contribution
Source UNEP (2015)

From Table 6, it is clear that more efforts are required, but these INDCs can be considered a step towards a successful climate agreement.

India's emission trends

Cumulative emission trends

For emission estimates in India at the national level, a comprehensive GHG emission assessment with a bottom-up approach was undertaken for submission of India's National Communication (NATCOM) to the UNFCCC. The Second National Communication was submitted in 2010, which gave emission estimates for the year 2000. Further, the Indian Network for Climate Change Assessment (INCCA), a network of 127 research institutions supported by the Indian Government, released an updated GHG inventory report for 2007 in 2010.[1] Although the next round of an updated inventory of emissions in India is yet to be completed, CO_2 emissions data till 2013 are available from the estimates of other energy institutions.

The estimates on emissions for India differ slightly across the different sources of data, primarily owing to the methodology of calculations. The approximate increase in emissions over 2013 has been estimated to be more than 100 $MtCO_2$-eq. The total CO_2 emission estimates for India till 2013 are given in Table 6.

Emissions trends by sector and fuel type

Carbon dioxide emissions in India are mainly attributed to the emissions from the energy sector. Of this, fossil fuel combustion comprises more than 90% of the total emissions, mainly because of the burning of coal for energy. In the transportation sector, road transport accounts for the largest share of CO_2 emissions followed by the aviation sector, mostly because these sectors depend on energy from oil. In the residential, industrial, and commercial sectors, emissions are mainly attributed to the use of energy sources for cooking, lighting, heating, and operating household appliances. The distributed share of emissions across different subsectors for India is given in Table 7.

Currently, India's clean energy capacity is 33 GW, while its dependence on thermal sources is 69.73% (CEA 2015). Coal will still account for 40% of India's energy mix in 2022, and all fossil fuels together will constitute 75% of the total energy demand (Figure 7). In the INDC scenario, India's energy-related CO_2 emissions will be 30% higher in 2022 as compared to 2013, reaching 2.4 GT, and, it will exceed 3.0 GT by 2030. Though the per capita emissions will increase through 2030, it will still be only half of the global average (WEO 2015).

Table 6 CO_2 emissions in India (in $MtCO_2$)

CO_2 emissions for India ($MtCO_2$)	International Energy Agency	US Information Agency (EIA)/ (International Energy Statistics)	BP Energy Review	NATCOM and INCCA assessments	Enerdata (Global Energy Intelligence)	EDGAR 2014
2014			2088	–	2187	
2013	–	–	1931.1	–	2000	2071.51
2012	1954.02	1830.9	1823.2	–	1909	1983.43
2011	1745.04	1752.76	1798.0	–	1787	1821.71
2010	1625.79	1695.6	1682.7	–	1710	1776.37
2009	1585.8	1622.7	1584.9	–	1629	1686.86
2007	1357.2	1368.4	1327.1	1497.06	1350	1476.59
2000	972.5	–	1001.2	1027.02	927	1059
1994	718.4	–	700.4	743.82	684	800

EDGAR – Emission Database for Global Atmospheric Research; INCCA – Indian Network on Climate Change Assessment; NATCOM – National Communication

Source Compiled from IEA (2014), US EIA (2014), BP (2015), NATCOM, INCAA Assessments, Enerdata (2015), and EDGAR (2014)

[1] NATCOM methodology follows the IPCC Revised Guidelines 1996, supported by the IPCC Good Practice Guidance (GPG) 2000 and 2003.

Table 7 CO_2 emissions within energy subsectors		
CO_2 emissions by sector (in million tonnes)	2013	2012
Total CO_2 emissions from fuel combustion	1868.624818	1954
Electricity and heat production	944.58028	1044.2
Manufacturing industries and construction	492.931974	473.2
Transport	222.322521	216.2
Other energy industry own use	43.0867	67.2

Sources IEA (2015) and IEA (2014a)

Emission trends by gases

The Government of India conducted the bottom-up GHG inventorization for 2000 and 2007. The trend indicates that CO_2 constitutes more than 90% of the total GHG emissions in India, primarily owing to fuel combustion. Methane (CH_4), nitrous oxide (N_2O), and other fluorides have contributed negligibly to India's emission share, both in 2000 and 2007, as per the national estimates. Emission estimates by the International Energy Agency (IEA) for 2000 and 2010 are given in Table 8.

Future emission scenario

Under the current policy scenario, with the assumption of 6.4% GDP (real) growth per year, the Climate Action Tracker projects GHG emissions [excluding land use, land-use change and forestry (LULUCF)] to reach 3.5 $GtCO_2e$ in 2020 and 5.0–5.1 $GtCO_2e$ in 2030. This is a 40% increase in emissions from the 2010 levels by 2020 and twice the 2010 levels by 2030. This growth in emissions is in line with India's 2020 intensity pledge and the recently submitted INDC target.

With population projected to increase by 24%, reaching 1.53 billion people (United Nations 2015) during 2010–30, the projected per capita emission is estimated to increase by around 63%–66% to 3.4–3.5 tCO_2e per capita (excluding LULUCF) by 2030. Although this represents a considerable increase, India's 2030 per capita emissions will still be far below the world average of 6.7 tCO_2e per capita (CAT 2015).

Response to climate change

International responses

India is committed to engaging actively in multilateral negotiations under the UNFCCC in an effective, cooperative, and forward-looking manner.

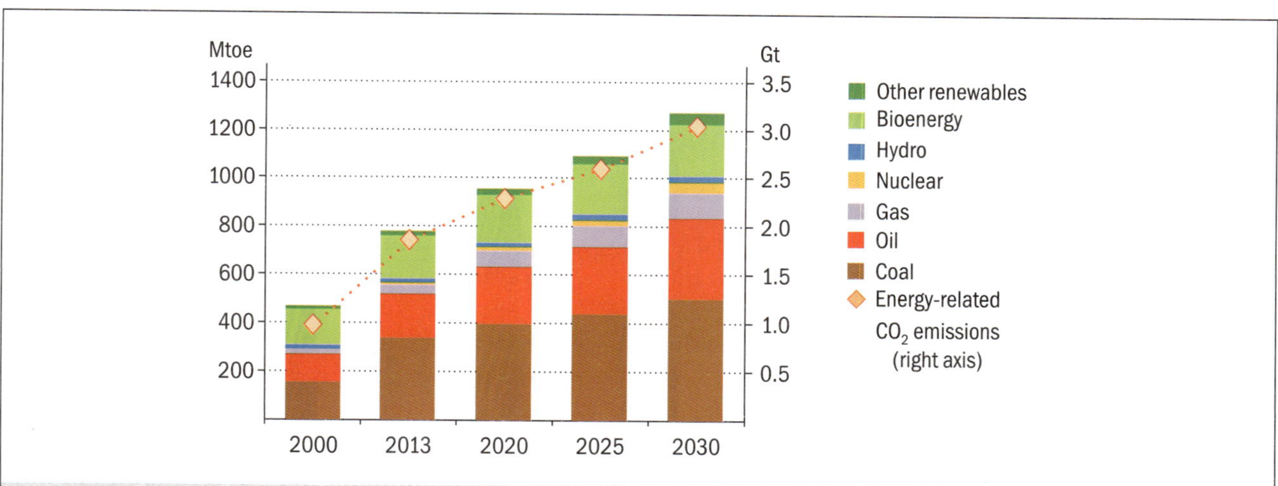

Figure 7 India's primary energy demand by fuel type and related CO_2 emissions in the INDC scenario

Source WEO (2015)

Table 8 GHG emission trends in India by gases (million tonnes of CO_2 equivalent)

		National estimates		IEA estimates	
Gas	Process	NATCOM 2000	INCCA 2007	IEA 2000	IEA 2010
CO_2	Fuel combustion	952	992	978.1	1749.3
	Fugitive	2	–	7.7	32.7
	Industrial processes	72	405	42.0	120.1
	Other	236	98	57.3	36.1
	Total	1027	1497.0	1085.1	1938.3
CH_4	Energy	8	4.3	82.1	116.1
	Agriculture	14	13.77	376.0	377.6
	Waste	–	2.5	101.1	125.3
	Other	2.8	0.90	2.4	2.5
	Total	19.9	20.5	561.6	621.5
N_2O	Energy	0.038	0.056	23.4	28.8
	Industrial process	0.012	0.020	1.6	0.3
	Agriculture	0.192	0.146	149.9	170.6
	Other	0.013	0.015	24.6	34.5
	Total	0.264	0.239	199.5	234.1
HFC	Industrial processes	Negligible	–	8.1	13.4
PFC	Industrial processes	Negligible	–	2.0	1.7
SF_6	Industrial processes	Negligible	–	3.4	5.8
Total		1301.2	1727.77	1859.7	2814.9

IEA – International Energy Agency; INCCA – Indian Network on Climate Change Assessment; NATCOM – National Communication

Source Compiled from MoEF (2010), MOEF (2012), and IEA (2014)

India's Intended Nationally Determined Contribution

In response to COP decisions 1/CP.19 and 1/CP.20, India communicated its INDC to the UNFCCC Secretariat on 2 October 2015. India's INDC builds on India's Copenhagen pledge to reduce GHG intensity by 20%–25% till 2020 from the 2005 levels (emissions from the agriculture sector excluded from the assessments of emissions intensity) submitted to the UNFCCC Secretariat in January 2010. It has put forward the following actions in response:

- To put forward and further propagate a healthy and sustainable way of living based on traditions and values of conservation and moderation.
- To adopt a climate-friendly and a cleaner path than the one followed hitherto by others at the corresponding level of economic development.
- To reduce the emissions intensity of its GDP by 33%–35% till 2030 from the 2005 level.
- To achieve about 40% cumulative electric power installed capacity from non-fossil fuel-based energy resources by 2030 with the help of transfer of technology and low-cost international finance, including Green Climate Fund.
- To create an additional carbon sink of 2.5–3.0 billion tonnes of CO_2 equivalent through additional forest and tree cover by 2030.
- To better adapt to climate change by enhancing investments in development programmes in sectors vulnerable to climate change, particularly agriculture, water resources, Himalayan region, coastal regions, and health and disaster management.
- To mobilize domestic and additional funds from developed countries to implement the aforementioned mitigation and adaptation actions in view of the resources required and the resource gap.

- To build capacities and create domestic framework and international architecture for quick diffusion of cutting-edge climate technology in India and for joint collaborative R&D for such future technologies.

It is clarified that India's INDCs do not bind the country to any sector-specific mitigation obligation or action, including the agriculture sector. India aims to reduce the overall emission intensity and improve energy efficiency of its economy over time and at the same time protect the vulnerable segments of the economy and society. The successful implementation of INDCs is contingent upon an ambitious global agreement, including additional means of implementation to be provided by developed country parties, technology and capacity building following Articles 3.1 and 4.7 of the UNFCCC Convention.

To achieve the aforementioned contributions, India is determined to continue with its ongoing interventions, enhance the existing policies as detailed in the previous sections, and launch new initiatives in the following priority areas:

- Introducing new, more efficient, and cleaner technologies in thermal power generation
- Promoting renewable energy generation and increasing the share of alternative fuels in the overall fuel mix
- Reducing emissions from the transportation sector
- Promoting energy efficiency in the economy, notably in industry, transportation, buildings, and appliances
- Reducing emissions from waste
- Developing climate-resilient infrastructure
- Completely implementing Green India Mission and other programmes of afforestation
- Planning and implementating actions to enhance climate resilience and reduce vulnerability to climate change

Clean Development Mechanism

Since 2005, India has been an active participant in the Clean Development Mechanism (CDM) under the Kyoto Protocol and represents a significant component of the global CDM market, registering the second highest number of projects for any country. The National CDM Authority[2] (NCDMA) of the Ministry of Environment, Forests, and Climate Change has so far accorded Host Country Approval to more than 2938 projects in India. Among the submissions, energy industries scored the highest with 2309 projects, followed by manufacturing industries and the energy demand sector (NCDMA 2015).

In the second commitment period that commenced in 2013, the number of CDM projects worldwide has come down drastically. In 2013, India registered 115 projects, which came down to 70 in 2014 and to 18 in 2015 (January–October 2015). Table 9 gives the sector-wise distribution of approved CDM projects in India in 2015.

The low commitments of emissions reduction by developed countries and the withdrawal of some countries from the second commitment period of the Kyoto Protocol have dampened the prospects of CDM in recent years. Currently, proposals to revive the price of the carbon credits and the third commitment period cycle of negotiations are being discussed.

Domestic responses

National and state action plans on climate change

India's National Action Plan on Climate Change (NAPCC) was released in June 2008. It outlined India's strategy to meet developmental goals concurrent to meeting the challenges of climate change. The NAPCC established eight national missions running through 2017 and 2022. Several fiscal incentives and policy provisions were made

Table 9 Total approved CDM projects in 2015

Sector	Number of projects	Number of CERs (annual)	Number of CERs (2012)
Energy industries (renewable/non-renewable sources)	16	561 097	8 936
Waste handling and disposal	1	400 000	0
Afforestation and reforestation	1	4 267	0
Total	18	965 364	8 936

CERs – certified emission reductions
Source NCDMA (2015)

[2] Details available at http://www.cdmindia.gov.in/reports_new.php

in the NAPCC mission documents to promote climate change mitigation and adaptation strategies in India. The interim progress and achievements in the NAPCC missions during 2013/14 along with their financial outlay for the plan period are given in Table 10. Report suggests that the indicative fund requirement for the fulfilment of the mission activities during the Twelfth Five-year Plan went from ₹230 000 crore (Planning Commission 2013) to ₹256 800 crore in 2014.

Consistent with the objectives of the NAPCC, in August 2009, the Prime Minister of India urged all state governments to develop their respective state action plans on climate change. The objective of this appeal was to set the stage for stimulating action on climate change at the state level and to effectively ensure that the national objectives are aligned with regional development priorities and the local environment. Until May 2015, 30 states and union territories have completed the draft plan; the ministry has endorsed 19 plans and the Expert Committee on Climate Change has considered 3 (Table 11). The remaining states are likely to get their plans approved by the end of this year.

On 5 November 2014, the Union Government reconstituted the Prime Minister's Council on Climate Change (PMCCC 2014) to formulate and coordinate the National Action Plan for Assessment, Adaptation, and Mitigation of Climate Change. The council has 18 members, which includes 13 government representatives and 5 non-government members. Among the 13 government representatives, 8 are ministers. The council will review the progress made by the eight national missions and provide suggestions regarding the launch of new missions in the field of wind energy, health, waste-to-energy conservation, and coastal management under NAPCC (GoI 2015). The final review of all eight missions under the NAPCC will be undertaken by the newly constituted PMCCC in 2017.

Enhanced renewable energy targets

In the first half of 2015, targets for renewable energy sources were raised to a cumulative installed capacity of 175 GW by the year 2022. This target includes 100 GW solar, 60 GW wind, 10 GW biomass, and 5 GW small-scale hydro (MNRE 2015). Wind power is supported via Generation Based Incentive scheme, while state-level feed-in tariffs apply for all renewables. These are regulated by the Electricity Act, 2003 and the National Tariff Policy, 2006. Renewable Energy Certificates (RECs) are in place that promote renewable energy and facilitate Renewable Purchase Obligations (RPOs), which legally mandate a percentage of electricity to be produced from renewable energy sources. In April 2015, the RPOs were revised upwards, from 3% by 2022 to 8% by 2019.

The Ministry of Power announced in April 2015 that every new coal-fired power plant would have to be accompanied by a renewable power plant of at least 10% of the generating capacity. Given the supporting mechanisms and policies in place, it is expected that India will meet its ambitious renewable energy targets (CAT 2015).

National Clean Energy Fund

In 2010, India introduced a cess on coal at $0.8 per tonne of coal. Recently, the cess has been quadrupled to $3.2 per tonne of coal. The coal cess translates into a carbon tax equivalent, using the emission factor for coal, of around $2 per tonne. This forms a corpus for the National Clean Energy Fund to a total of $2.7 billion till 2014/15. This is used for financing 46 clean energy projects worth $2.6 billion in clean energy, technologies, and project-related activities.

National Adaptation Fund

India has set up a National Adaptation Fund with an initial allocation of $55.6 million to meet adaptation needs in sectors such as agriculture, water, and forestry, in addition to sectoral spending by respective ministries.

Implicit carbon tax on petrol and diesel

India resorted to an implicit carbon tax policy of $140 for petrol and $64 for diesel at the time when global crude price fell drastically during 2014/15. Estimates suggest that these measures will help India achieve a net reduction of 11 million tonnes of CO_2 emissions in less than a year (GoI 2015).

Tax-free Infrastructure Bond

The Department of Economic Affairs, Ministry of Finance, has in principle approved the issuance of tax-free infrastructure bonds of ₹5000 crore in the financial year 2015/16 by the Central Public Sector Enterprises (CPSEs) for funding renewable energy

Table 10 National Action Plan on Climate Change

Mission	Aim	Date of implementation/ approval	Implementing agency	Financial outlay	Achievements till 2014
National Solar Mission	To increase the share of solar energy in the total energy mix	Launched by prime minister on 11 January 2010	Ministry of New and Renewable Energy	Total estimated cost of ₹ 8759 crore in Twelfth Five-year Plan	• Government of India raised the ambition to increase solar energy capacity addition targets from 20 GW to 100 GW by 2022 • 522 MW grid-connected solar power projects, 22.7 MW off-grid solar power projects, 5 solar heating systems • 70 lakh m^2 of collector area
National Mission for Enhanced Energy Efficiency	To promote energy efficiency through demand-side management programme and supply-side energy efficiency shifts	This mission is under implementation from 1 April 2010	Ministry of Power/ Bureau of Energy Efficiency	Total estimated cost of ₹ 190 crore in Twelfth Five-year Plan	• Commencement of first commitment period of PAT • Expansion of energy efficiency financing platform through MoU with public sector banks • Implementation of compact fluorescent lamp programme • Operationalization of the Partial Risk Guarantee Fund/Venture Capital Fund for energy efficiency
National Mission for Sustainable Habitat	Promote energy efficiency in buildings, management of solid waste, and GHG mitigation in the transportation sector	Approved by PM's Council on 18 June 2010	Ministry of Urban Development/ Ministry of Housing and Urban Poverty/ Alleviation	Total estimated cost of ₹ 950 crore in Twelfth Five-year Plan	• NSH standards developed for six subsectors: (a) solid waste management, (b) water and sanitation, (c) storm water drainage, (d) urban planning, (e) energy efficiency, and (f) urban transport • Energy Conservation Building Code 2007 made mandatory for both new and old buildings and incorporated in the CPWD general specifications for electrical works in 2013 • Green Building norms made mandatory for the CPWD since 2009 and incorporated in the CPWD Works Manual 2012
National Water Mission	Conservation of water, minimizing wastage, and ensuring equitable distribution between states	Approved by PM's Council on 28 May 2010. The Union Cabinet approved its implementation framework on 6 April 2011	Ministry of Water Resources	Total estimated cost of ₹ 89 101 crore till 2017. Proposals for ₹ 196 crore only have been approved	• Preparation of state-specific action plans on climate change underway. • A pilot study of basin-wise water conducted for two basins: Godavari and Brahmani–Baitarani. The studies are being extended to all basins. • An MoU signed between the Ministry of Water Resources and Asian Development Bank for technical assistance with the objective of undertaking operationally relevant research for identifying and testing integrated flood mitigation and floodplain management strategies.

Contd...

Table 10 *Contd...*

Mission	Aim	Date of implementation/ approval	Implementing agency	Financial outlay	Achievements till 2014
National Mission for Sustaining the Himalayan Ecosystems	Management measures for sustaining and safeguarding the Himalayan glacier and mountain ecosystem	–	Ministry of Science and Technology	Total estimated cost of ₹1500 crore in Twelfth Five-year Plan. Proposals for ₹500 crore have been approved	• Four networking and strengthening knowledge institutions set up • Four new centres created, which are relevant to climate change in existing institutions in Himalayan states • Developed one observational network for monitoring the health of Himalayan ecosystem • Developed a regional cooperation on Glaciology in a neighbouring country
National Mission for a Green India	Enhancing ecosystem services and carbon sinks through afforestation on degraded forestland and expansion of forest and tree cover	Approved by PM's Council on Climate Change on 23 February 2011	Ministry of Environment, Forests, and Climate Change	Total fund requirement of ₹45 800 crore in Twelfth Five-year Plan. Of this, ₹13 000 crore has been approved for activities.	• About 10 states have already submitted their perspective plans to be taken up in 33 landscapes over a working area of 0.85 lakh hectares
National Mission for Sustainable Agriculture	Ensuring food security and protecting land, water, biodiversity, and genetic resources for sustainable production of food	Approved by PM's Council on 23 September 2010	Ministry of Agriculture and Cooperation/DARE	Total fund requirement of ₹108 000 crore in Twelfth Five-year Plan and proposals for ₹13 034 crore have been approved so far	Main achievements have been in the area of horticulture area expansion activities, seed processing, agriculture market creation of storages to enhance agriculture supply chain management, increase in fish production
National Mission on Strategic Knowledge on Climate Change	Research and technology development into various aspects of climate change	–	Ministry of Science and Technology	Total fund requirement of ₹2500 crore in Twelfth Five-year Plan	The mission has led to set up of 2 thematic knowledge networks in the country, 3 regional climate models, 25 specially trained high-quality climate change professionals, and 2 technology watch groups

CPWD – Central Public Works Department; DARE – Department of Agriculture, Research, and Education; GHG – greenhouse gas; NSH – National Sustainable Habitat; PAT – Perform, Achieve, and Trade

Source Compiled from MoF (2014) and Planning Commission (2011)

Table 11 Current status of state action plans on climate change

	Name of state/union territory
Endorsed by National Steering Committee on Climate Change	Andaman and Nicobar Islands, Andhra Pradesh, Arunachal Pradesh, Chhattisgarh, Himachal Pradesh, Jammu and Kashmir, Lakshadweep, Madhya Pradesh, Manipur, Meghalaya, Mizoram, Nagaland, Odisha, Puducherry, Punjab, Rajasthan, Sikkim, Tripura, West Bengal
Considered by Expert Committee on Climate Change	Haryana, Jharkhand, Karnataka

Source MoEF (2014a)

projects aimed towards achieving 175 GW capacity by 2022 as announced in Budget 2015/16. Accordingly, the Ministry of New and Renewable Energy has submitted a proposal to the Ministry of Finance for the allocation of tax-free bonds to five CPSEs:

(i) Indian Renewable Energy Development Agency—₹2000 crore
(ii) Indian Infrastructure Finance Company ₹1000 crore
(iii) Power Finance Corporation—₹1000 crore
(iv) Rural Electrification Corporation—₹1000 crore
(v) Solar Energy Corporation of India—₹100–200 crore

Tax-free infrastructure bonds of $794 million are being introduced for funding renewable energy projects during 2015/16 (MoF 2015).

India's Declaration on International Solar Alliance at COP 21 in Paris

Indian Prime Minister Shri Narendra Modi launched an "International Solar Alliance" in a joint declaration with the French President H. E. Francois Hollande on 30 November 2016 at UNFCCC Conference at Paris. The declaration is backed by 120 governments and it aims to promote solar energy and unlock investments to drive the transformation of solar technology, particularly in developing countries. The alliance will provide a common platform for cooperation among solar-rich countries and for undertaking efforts to mobilize more than $1 trillion worth of investments by 2030. It recognizes that developing countries will need technology and capacity building to quickly take solar energy to scale and that initial public financing is critical for this purpose.

Collective vision of the alliance is to bring down the cost of capital and the cost of technology for immediate deployment of competitive solar generation assets in all 120 countries and pave the way for technological innovations which are adapted to country-specific circumstances, with primary focus on the solar-rich countries between the Tropic of Cancer and Tropic of Capricorn. According to the declaration, an International Steering Committee will also be set up (potentially in India) to provide necessary guidance, direction and advice to establish the international solar alliance.

Sources UNFCCC (2015); Details available at http://newsroom.unfccc.int/clean-energy/international-solar-energy-alliance-launched-at-cop21/and MNRE (2015)

References

Attri, S. D. and A. Tyagi. 2010. ***Climate Profile of India***. Met Monograph No. Environment Meteorology-01/2010

BP. 2015. BP Statistical Review of World Energy. Details available at www.bp.com/en/global/corporate/energy-economics/statistical-review-of-world-energy/downloads.html

CAIT Climate Data Explorer. 2015. Details available at http://cait.wri.org

CAT (Climate Action Tracker). 2015. Details available at http://climateactiontracker.org/countries/india

CEA (Central Electricity Authority). 2015. ***Executive Summary Power Sector***. New Delhi: CEA

EDGAR (Emission Database for Global Atmospheric Research). 2014. European Commission, Joint Research Centre (JRC)/PBL Netherlands Environmental Assessment Agency. Details available at <http://edgar.jrc.ec.europa.eu/index.php

Enerdata. 2015. ***Global Energy Statistical Yearbook 2015***. Details available at www.enerdata.net

GCP (Global Carbon Project). 2014. ***Global Carbon Budget Report***. Details available at www.globalcarbonproject.org

GoI (Government of India). 2015. India's Intended Nationally Determined Contribution (INDC). Details available at http://www4.unfccc.int/submissions/INDC/Published%20Documents/India/1/INDIA%20INDC%20TO%20UNFCCC.pdf

IEA (International Energy Agency). 2014a. ***CO_2 Emissions from Fossil Fuel Combustion***. France: IEA

IEA (International Energy Agency). 2014a. ***World Energy Statistics 2014***. France: IEA

IEA (International Energy Agency). 2015. Energy and climate change. ***World Energy Outlook Special Report***. Details available at https://www.iea.org

IEA (International Energy Agency). 2015. CO_2 ***Emissions from Fossil Fuel Combustion***. France: IEA

IMD (India Meteorological Department). 2012. ***Annual Climate Summary 2012***. Pune: National Climate Centre. Details available at http://www.imdpune.gov.in/research/ncc/climatebulletin/Annual_2012.pdf

IMD (India Meteorological Department). 2012. ***Annual Climate Summary 2014***. Pune: National Climate Centre. Details available at www.imd.gov.in/doc/report2014.pdf

INCCA 2010: Indian Network for Climate Change Assessment report 2. Climate Change and India: a 4 × 4 Assessment, (2010). Ministry of Environment and Forests, Government of India

IPCC (Intergovernmental Panel on Climate Change). 2013. Summary for policymakers. In ***Climate Change 2013: The Physical Science Basis. Contribution of Working Group I to the Fifth Assessment Report of the Intergovernmental Panel on Climate Change,*** edited by T. F. Stocker, D. Qin, G. K. Plattner, M. Tignor, S. K. Allen, J. Boschung, A. Nauels, Y. Xia, V. Bex, and P. M. Midgley. Cambridge, UK and New York, USA: Cambridge University Press

IPCC (Intergovernmental Panel on Climate Change). 2014. Summary for policymakers. In: ***Climate Change 2014: Mitigation of Climate Change. Contribution of Working Group III to the Fifth Assessment Report of the Intergovernmental Panel on Climate Change,*** edited by O. Edenhofer, R. Pichs-Madruga, Y. Sokona, E. Farahani, S. Kadner, K. Seyboth, A. Adler, I. Baum, S. Brunner, P. Eickemeier, B. Kriemann, J. Savolainen, S. Schlomer, C. von Stechow, T. Zwickel, and J. C. Minx. Cambridge, UK and New York, USA: Cambridge University Press

Kothawale, D. R., A. A. Munot, and K. Krishna Kumar. 2010. Surface air temperature variability over India during 1901–2007 and its association with ENSO, ***Climate Research*** 42: 89-104

MNRE (Ministry of New and Renewable Energy). 2015. ***Schemes for Installing Large Solar Power Plants***. New Delhi: MNRE

MoEF (Ministry of Environment and Forests). 2010. ***India: Green House Gas Emissions 2007***. New Delhi: Indian Network for Climate Change Assessment (INCCA), MoEF, Government of India. Details available at http://moef.nic.in/downloads/public-information/Report_INCCA.pdf

MoEF (Ministry of Environment and Forests). 2012. ***India's Second National Communication to the United Nations Framework Convention on Climate Change***. New Delhi: MoEF, Government of India. Details available at http://unfccc.int/resource/docs/natc/indnc2.pdf

MoEF (Ministry of Environment and Forests). 2014a. ***State Action Plan on Climate Change***. Details available at http://envfor.nic.in/ccd-sapcc

MoEF (Ministry of Environment and Forests). 2014b. Statement by Hon'ble Environment Minister at the High Level Segment of UNFCCC COP-20 on 9 December 2014.

MoF (Ministry of Finance). 2014. ***Economic Survey 2013–2014***. New Delhi: MoF, Government of India. Details available at http://indiabudget.nic.in/survey.asp

NCDMA (National CDM Authority). 2015. Details available at http://www.cdmindia.gov.in/reports_new.php?n=1

Parthasarthy, B. 1984. Inter-annual and long-term variability of Indian summer monsoon rainfall. ***Proceedings of the Indian Academy of Sciences (Earth and Planetary Sciences)*** 93: 371–385

Planning Commission. 2011. ***Climate Change and Twelfth Five-Year Plan: Report of Sub-group on Climate Change***. New Delhi: Planning Commission

Planning Commission. 2013. ***Twelfth Five Year Plan (2012–2017)***. New Delhi: SAGE Publications

PMCCC (Prime Minister's Council on Climate Change). 2008. ***National Action Plan on Climate Change***. New Delhi: PMCCC

PwC. 2014. Two Degrees of Separation: Ambition and Reality. Details available at www.pwc.co.uk/assets/pdf/low-carbon-economy-index-2014.pdf

Rajeevan, M., B. Jyoti, J. D. Kale, and B. Lal. 2006. High resolution daily gridded rainfall data for the Indian region: analysis of break and active monsoon spells. ***Current Science*** 91(3): 296–306

Rupa Kumar, K., K. Krishna Kumar, and G. B. Pant. 1994. Diurnal asymmetry of surface temperature trends over India. ***Geophysical Research Letter*** 21: 677–680

Sen Roy, S. and R. C. Balling. 2004. Trends in extreme daily precipitation indices in India. ***International Journal of Climatology*** 24: 457–466

UN. 2015. World Population Prospects: The 2015 Revision. New York: United Nations

UNEP (United Nations Environment Programme). 2014. ***The Emissions Gap Report 2014***. Details available at http://www.unep.org/emissionsgapreport2014/

UNEP (United Nations Environment Programme). 2015. ***Emissions Gap Report: Executive Summary***. Details available at http://uneplive.unep.org/media/docs/theme/13/EGR_2015_ES_English_Embargoed.pdf

United Nations. 2015. ***World Population Prospects: The 2015 Revision***. New York: United Nations

US EIA (US Energy Information Agency). 2014. CO_2 emissions from energy consumption database. Details available at www.eia.gov/cfapps/ipdbproject/IEDIndex3.cfm?tid=90&pid=44&aid=8

Ziesing, H-J. 2014. Updated development of global greenhouse gas emissions 2013. "Low Carbon Markets and the Legacy of Nuclear Power" 19th REFORM Group Meeting, Schloss Leopoldskron Salzburg, 01-05 September 2014

TABLE 12.1 Carbon dioxide emissions across world economies

CO_2 (MT)	2000	2001	2002	2003	2004	2005	2006	2007	2008	2009	2010	2011	2012	2013	2014	Change in 2014 over 2013 (%)	2014 share of total (%)
USA	6377.0	6248.4	6293.4	6343.5	6473.3	6495.0	6412.8	6521.5	6332.1	5908.2	6142.7	6001.3	5786.0	5941.4	5994.6	0.9	16.9
Canada	593.6	592.7	605.3	631.6	629.3	636.0	629.1	640.9	633.4	588.1	611.9	623.4	613.1	619.8	620.5	0.1	1.7
Mexico	382.7	383.7	400.1	422.5	433.0	471.0	482.4	479.7	472.0	473.9	493.1	514.2	504.6	507.3	499.9	-1.5	1.4
Total North America	7353.3	7224.8	7298.7	7397.6	7535.5	7602.0	7524.3	7642.1	7437.5	6970.2	7247.7	7138.9	6903.6	7068.5	7114.9	0.7	20.0
Argentina	134.8	127.2	120.8	131.8	142.4	152.2	157.9	171.0	174.4	170.1	181.1	187.1	192.6	201.6	199.4	-1.1	0.6
Brazil	356.2	361.7	359.8	356.6	375.4	383.5	388.8	412.3	442.1	423.6	480.4	501.2	518.7	559.1	581.7	4.1	1.6
Chile	59.6	57.7	57.7	59.1	64.7	64.8	70.7	78.2	77.5	75.9	76.3	85.8	88.0	91.8	88.0	-4.1	0.2
Colombia	56.6	54.8	52.6	54.2	53.4	58.0	57.8	58.0	62.8	65.0	71.7	74.6	78.0	80.8	84.3	4.3	0.2
Ecuador	20.2	20.1	21.2	22.2	22.9	24.9	26.5	27.0	27.7	28.5	32.7	33.4	34.9	36.8	38.5	4.8	0.1
Peru	25.3	23.9	24.8	24.0	27.2	28.0	27.2	30.7	34.7	35.6	40.9	45.3	46.9	49.3	50.9	3.1	0.1
Trinidad and Tobago	24.7	27.3	29.2	31.7	34.7	38.2	49.0	51.4	50.6	52.1	54.8	54.5	51.7	52.4	51.5	-1.8	0.1
Venezuela	132.9	145.7	146.6	127.4	139.9	144.9	164.2	168.0	176.7	175.4	161.9	164.9	176.4	185.0	182.1	-1.6	0.5
Other South and Central America	175.8	185.2	187.4	199.4	201.7	200.1	205.3	207.6	199.0	197.4	200.8	205.2	207.7	207.6	210.3	1.3	0.6
Total South and Central America	986.2	1003.6	1000.0	1006.4	1062.4	1094.5	1147.4	1204.3	1245.6	1223.5	1300.7	1352.0	1394.9	1464.4	1486.6	1.5	4.2
Austria	67.8	72.5	73.5	79.6	78.3	79.9	79.5	75.4	75.9	70.2	75.7	72.6	70.2	70.2	65.9	-6.2	0.2
Azerbaijan	30.5	28.2	27.2	29.5	31.4	34.7	34.0	30.8	30.4	26.7	25.6	29.5	30.9	32.1	33.5	4.3	0.1
Belarus	61.0	59.2	61.0	61.0	65.4	64.4	69.0	66.7	67.3	65.0	67.3	67.8	74.7	75.5	76.7	1.5	0.2
Belgium	157.6	154.9	153.7	160.7	160.0	156.7	156.5	155.6	163.0	147.5	160.1	145.4	138.6	140.8	138.1	-1.9	0.4
Bulgaria	45.0	48.1	44.9	48.5	47.6	48.7	49.6	52.8	51.4	43.4	44.6	49.9	45.2	40.2	42.8	6.6	0.1
Czech Republic	125.0	128.5	125.0	127.3	131.7	130.7	133.1	132.9	127.0	116.9	120.6	117.3	113.4	109.0	107.5	-1.4	0.3
Denmark	57.8	57.9	56.8	62.0	56.1	53.3	61.5	57.0	54.4	50.6	51.6	47.1	42.1	44.1	40.6	-7.9	0.1
Finland	58.3	62.8	65.8	74.6	68.4	58.5	68.3	66.3	59.6	57.5	64.9	56.6	50.3	51.0	47.7	-6.5	0.1
France	430.3	430.1	421.6	430.8	435.1	433.4	426.8	421.2	419.2	399.5	403.9	379.1	379.5	380.6	347.5	-8.7	1.0
Germany	904.1	917.9	905.3	910.0	898.5	880.0	898.4	863.9	853.9	798.5	835.4	811.4	826.5	846.0	798.6	-5.6	2.2
Greece	101.1	103.0	102.3	100.7	107.1	105.7	107.8	109.4	106.6	102.0	93.7	92.8	87.7	79.7	74.8	-6.1	0.2
Hungary	58.9	60.0	59.0	61.6	62.1	65.8	66.7	64.7	63.5	56.5	57.8	54.0	50.1	46.7	45.0	-3.6	0.1
Republic of Ireland	43.6	46.8	46.2	44.8	45.1	47.3	47.2	48.3	47.4	42.7	42.5	38.6	38.9	37.1	36.8	-0.8	0.1
Italy	473.8	475.1	476.0	492.0	497.0	498.4	495.8	488.7	475.7	432.2	441.4	430.3	406.8	377.8	347.1	-8.1	1.0

Contd...

TABLE 12.1 *Contd...*

CO_2 (MT)	2000	2001	2002	2003	2004	2005	2006	2007	2008	2009	2010	2011	2012	2013	2014	Change in 2014 over 2013 (%)	2014 share of total (%)
Kazakhstan	124.1	125.7	131.7	142.4	146.2	149.4	169.4	177.3	188.2	168.9	168.5	190.0	199.3	191.5	188.6	-1.5	0.5
Lithuania	13.4	14.7	14.3	14.7	15.2	16.2	16.5	17.2	17.1	14.5	15.7	16.2	16.2	14.8	14.1	-4.7	-
Netherlands	244.5	255.4	254.6	257.3	264.9	269.0	267.1	267.5	258.7	252.7	263.0	251.5	243.7	237.8	225.2	-5.3	0.6
Norway	41.2	42.8	41.6	43.9	43.7	43.5	43.8	44.5	43.8	43.4	44.2	44.4	44.2	45.0	44.2	-1.7	0.1
Poland	312.8	314.0	309.3	317.2	320.8	318.0	330.6	332.7	330.8	313.6	338.1	337.0	329.0	329.1	316.8	-3.7	0.9
Portugal	68.8	67.3	71.3	67.6	69.6	72.5	66.5	65.8	63.2	61.6	57.1	56.9	54.8	54.8	53.0	-3.3	0.1
Romania	96.5	97.9	103.9	105.0	107.2	104.3	107.7	105.9	103.5	86.1	83.3	89.7	87.0	75.2	75.4	0.3	0.2
Russian Federation	1 557.9	1 571.5	1 569.3	1 605.6	1 605.9	1 594.5	1 661.5	1 663.0	1 689.3	1 582.1	1 646.1	1 710.2	1 716.8	1 683.0	1 657.2	-1.5	4.7
Slovakia	41.1	42.0	41.5	41.7	40.6	42.4	40.8	39.0	40.1	37.2	39.2	37.5	35.1	36.1	32.0	-11.3	0.1
Spain	333.2	338.5	355.7	363.6	383.4	394.8	387.7	400.5	374.7	336.4	313.9	329.0	325.4	287.6	285.7	-0.7	0.8
Sweden	58.7	59.7	62.2	66.2	66.2	64.3	64.9	63.7	62.3	57.8	62.3	57.6	55.5	54.7	54.0	-1.3	0.2
Switzerland	43.7	46.7	44.4	43.7	43.8	44.7	45.5	41.6	44.2	44.5	42.6	40.7	41.9	43.9	39.3	-10.5	0.1
Turkey	216.6	200.4	208.9	222.8	230.3	241.1	270.5	295.9	294.9	297.6	304.5	324.1	337.2	324.7	348.5	7.3	1.0
Turkmenistan	38.1	38.1	40.5	45.3	47.4	49.8	54.1	60.9	61.9	57.4	65.4	68.1	74.6	67.5	78.1	15.7	0.2
Ukraine	342.2	344.4	338.6	348.1	344.4	336.5	343.0	336.2	336.2	282.4	300.5	318.0	312.0	296.9	243.3	-18.0	0.7
United Kingdom	591.2	598.4	581.9	594.9	601.6	603.6	605.0	587.6	578.5	530.5	547.2	508.0	528.3	514.8	470.8	-8.5	1.3
Uzbekistan	122.4	129.9	132.0	121.8	119.6	111.0	111.3	116.8	122.4	101.9	101.6	115.4	113.7	118.5	120.5	1.7	0.3
Other Europe and Eurasia	182.3	189.8	200.1	211.1	221.0	218.5	209.5	219.1	222.0	211.5	219.1	226.9	215.7	216.7	208.4	-3.8	0.6
Total Europe and Eurasia	7 043.8	7 122.2	7 120.3	7 295.7	7 355.7	7 331.4	7 489.4	7 468.7	7 426.8	6 889.6	7 097.4	7 113.3	7 085.2	6 923.2	6 657.5	-3.8	18.8
Iran	352.6	371.2	397.3	403.8	439.4	470.7	511.2	546.1	575.9	601.4	594.5	620.0	620.9	633.2	650.4	2.7	1.8
Israel	66.3	67.1	68.4	70.3	70.7	72.2	72.3	74.6	75.5	73.1	75.2	77.0	81.8	75.5	74.2	-1.7	0.2
Kuwait	56.5	59.5	60.4	71.2	79.7	85.7	80.7	80.4	85.3	88.9	97.0	98.6	104.9	107.5	110.6	2.8	0.3
Qatar	26.6	30.5	32.7	35.3	42.1	51.5	55.4	57.3	59.8	71.2	83.6	104.8	111.5	115.3	125.8	9.1	0.4
Saudi Arabia	329.5	342.7	355.0	377.8	409.7	421.5	439.0	458.5	497.9	521.5	565.3	577.5	613.3	618.1	665.0	7.6	1.9
United Arab Emirates	126.7	139.6	138.6	148.4	158.9	166.1	175.3	193.1	219.5	215.7	225.8	241.0	251.4	258.5	273.2	5.7	0.8
Other Middle East	215.0	222.8	232.7	235.2	243.5	256.7	247.3	260.6	289.7	298.5	320.7	314.5	323.5	331.5	328.6	-0.9	0.9
Total Middle East	1 173.2	1 233.5	1 285.2	1 342.0	1 444.1	1 524.5	1 581.1	1 670.6	1 803.6	1 870.4	1 962.1	2 033.5	2 107.3	2 139.7	2 227.8	4.1	6.3
Algeria	69.9	72.5	75.6	78.7	81.9	85.3	89.0	93.9	99.6	105.1	102.4	108.5	118.5	125.5	135.1	7.6	0.4

Contd...

TABLE 12.1 Contd...

CO_2 (MT)	2000	2001	2002	2003	2004	2005	2006	2007	2008	2009	2010	2011	2012	2013	2014	Change in 2014 over 2013 (%)	2014 share of total (%)
Egypt	129.6	135.2	137.1	146.1	153.2	162.1	168.7	178.1	189.4	198.0	208.6	210.1	221.1	219.0	223.2	1.9	0.6
South Africa	365.6	368.2	355.1	389.2	424.3	400.1	407.7	420.2	456.1	452.8	457.5	451.5	444.0	444.6	452.2	1.7	1.3
Other Africa	249.5	264.2	271.5	274.9	298.7	312.1	306.2	318.1	338.7	330.4	358.7	346.9	367.6	376.0	384.3	2.2	1.1
Total Africa	814.6	840.1	839.3	888.9	958.1	959.6	971.6	1 010.3	1 083.8	1 086.3	1 127.1	1 117.1	1 151.3	1 165.2	1 194.8	2.5	3.4
Australia	349.5	358.6	368.6	361.6	370.8	385.4	405.1	403.9	406.3	396.6	389.4	395.6	391.6	383.7	374.9	-2.3	1.1
Bangladesh	31.4	35.8	36.9	39.0	40.3	42.9	45.8	47.5	50.1	53.0	57.3	60.9	64.8	68.0	71.5	5.1	0.2
China	3 513.7	3 673.5	4 024.6	4 722.9	5 521.1	6 326.1	6 925.9	7 518.0	7 662.9	8 036.9	8 471.9	9 206.1	9 415.4	9 674.2	9 761.1	0.9	27.5
China Hong Kong SAR	53.9	70.2	67.3	71.0	78.4	75.9	80.1	85.7	79.4	86.7	88.8	92.2	88.7	91.5	89.9	-1.8	0.3
India	952.8	959.2	1 007.2	1 040.9	1 116.3	1 180.0	1 246.5	1 341.7	1 444.3	1 570.5	1 640.1	1 704.2	1 855.0	1 931.1	2 088.0	8.1	5.9
Indonesia	287.2	305.8	321.5	351.9	348.9	358.6	371.2	402.4	392.9	408.0	453.4	495.6	524.0	529.8	548.7	3.6	1.5
Japan	1 333.1	1 328.7	1 325.9	1 381.1	1 389.5	1 405.0	1 379.3	1 395.8	1 397.7	1 230.6	1 311.9	1 313.9	1 398.4	1 386.1	1 343.1	-3.1	3.8
Malaysia	134.9	137.5	149.2	165.3	176.6	187.0	192.3	204.6	212.1	206.5	221.2	227.5	238.2	258.3	257.7	-0.2	0.7
New Zealand	35.2	37.1	37.1	37.9	37.2	38.1	38.5	37.3	38.7	35.9	36.2	35.3	37.2	37.2	38.2	2.7	0.1
Pakistan	111.3	113.0	116.5	124.7	137.0	145.7	154.1	165.7	171.6	175.1	174.1	172.9	172.7	171.1	177.4	3.7	0.5
Philippines	67.8	68.6	70.0	71.5	73.7	74.7	69.2	74.5	73.6	73.0	77.4	80.2	86.9	92.5	97.9	5.9	0.3
Singapore	114.9	128.3	123.2	115.7	129.6	140.8	154.7	166.4	177.2	190.6	205.7	215.2	214.7	220.9	226.1	2.4	0.6
South Korea	528.9	542.7	566.6	580.0	591.2	602.2	606.0	639.9	653.4	661.8	713.6	753.8	760.6	767.8	768.3	0.1	2.2
Taiwan	262.6	275.3	285.7	302.4	316.9	321.7	328.6	342.9	321.9	309.7	327.0	327.6	327.6	330.1	332.9	0.8	0.9
Thailand	182.0	188.1	208.4	227.0	246.6	256.4	258.9	269.2	274.1	284.2	303.6	312.1	334.6	334.8	346.9	3.6	1.0
Vietnam	47.3	52.0	58.3	63.9	81.9	86.8	72.5	79.0	104.3	104.0	123.1	135.3	131.9	137.7	154.6	12.3	0.4
Other Asia Pacific	123.6	127.0	124.9	130.9	131.6	140.1	144.9	136.3	139.3	140.7	141.1	129.8	133.9	136.2	139.8	2.6	0.4
Total Asia Pacific	8 130.1	8 401.3	8 892.0	9 787.7	10 787.7	11 767.3	12 473.5	13 310.8	13 599.9	13 963.8	14 735.7	15 658.4	16 176.3	16 550.9	16 817.0	1.6	47.4
Total World	25 501.2	25 825.4	26 435.6	27 718.2	29 143.5	30 279.3	31187.4	32 306.9	32 597.2	32 003.7	33 470.8	34 413.1	34 818.7	35 311.8	35 498.7	0.5	100.0

Source BP. 2015. BP Statistical Review of World Energy. Details available at http://www.bp.com/en/global/corporate/energy-economics/statistical-review-of-world-energy/downloads.htmlhttp://www.bp.com/en/global/corporate/energy-economics/statistical-review-of-world-energy/downloads.html

ENERGY AND ENVIRONMENT GOALS

- SUSTAINABLE DEVELOPMENT GOALS AND IMPLICATIONS FOR INDIA
- AIR POLLUTION AND HEALTH

SUSTAINABLE DEVELOPMENT GOALS AND IMPLICATIONS FOR INDIA 13

2030 Agenda

The Sustainable Development Goals (SDGs) resulted from the convergence of two international discourses: Millennium Development Goals (MDGs) and Sustainable Development. The global discourse on sustainable development was steered by the outcomes of many UN conferences and summits in economic, social, and environmental fields.

The key outcomes guiding the post-2015 development process include the Rio Declaration, Agenda 21, Programme for the Further Implementation of Agenda 21, United Nations Millennium Declaration, Monterrey Consensus of the International Conference on Financing for Development, Johannesburg Plan of Implementation, Johannesburg Declaration on Sustainable Development, the 2005 World Summit Outcome, Doha Declaration on Financing for Development, The Future We Want, Addis Ababa Action Agenda, and Transforming Our World: the 2030 Agenda for Sustainable Development. Figure 1 shows key milestones contributing to the sustainable development discourse.

On 2 August 2015, at the eighth session of the Intergovernmental Negotiations on the Post-2015 Development Agenda, member states adopted, by acclamation, the agenda "Transforming Our World: The 2030 Agenda for Sustainable Development". At the United Nations Sustainable Development Summit 2015 held at the UN Headquarters in New York during 25–27 September 2015, heads of state as well as government and high-ranking representatives adopted the new global SDGs.

The new sustainable development agenda with 17 goals and 169 targets will come into effect from 1 January 2016 and guide global and national development policy agendas over the next 15 years. Table 1 lists the 17 goals. Member states will implement the agenda at the regional and global levels, taking into account different national realities, capacities, and levels of development and respecting national policies and priorities.

Compared to MDGs, the SDGs have a sharper focus on social and environmental goals. SDGs have specific goals related to energy, which were missing in the MDGs. Moreover, the focus on environment was not well defined in MDGs.

Energy and environment: MDG experience and India

The MDGs did not have any specific goal on "energy". This has been addressed in the SDGs with a dedicated goal on energy. Goal 7 of SDGs focuses on ensuring access to affordable, reliable, sustainable, and modern energy for all. Apart from this, energy is also implicitly addressed in other SDGs around cities, infrastructure, and sustainable consumption and production. With respect to environmental sustainability, Goal 7 of MDGs (MDG 7),"Ensure environmental sustainability", addressed the concern for sustainable development to reverse environmental degradation and loss of biodiversity. MDG 7 had four targets and 10 indicators. The targets focused on sustainable development, biodiversity, drinking water and sanitation, and slum dwellers.

Although MDG 7 had an important role in sustainable development, its targets and indicators, by themselves, were considered inadequate for measuring the broad domain of environmental sustainability, which should also consider air quality, waste management, and climate change adaptation. The targets and indicators, thus, did not capture the essence of the goal. There were also many practical and technical data-related challenges in the use of indicators, such as involvement of many stakeholders (government and semi-government organizations) in

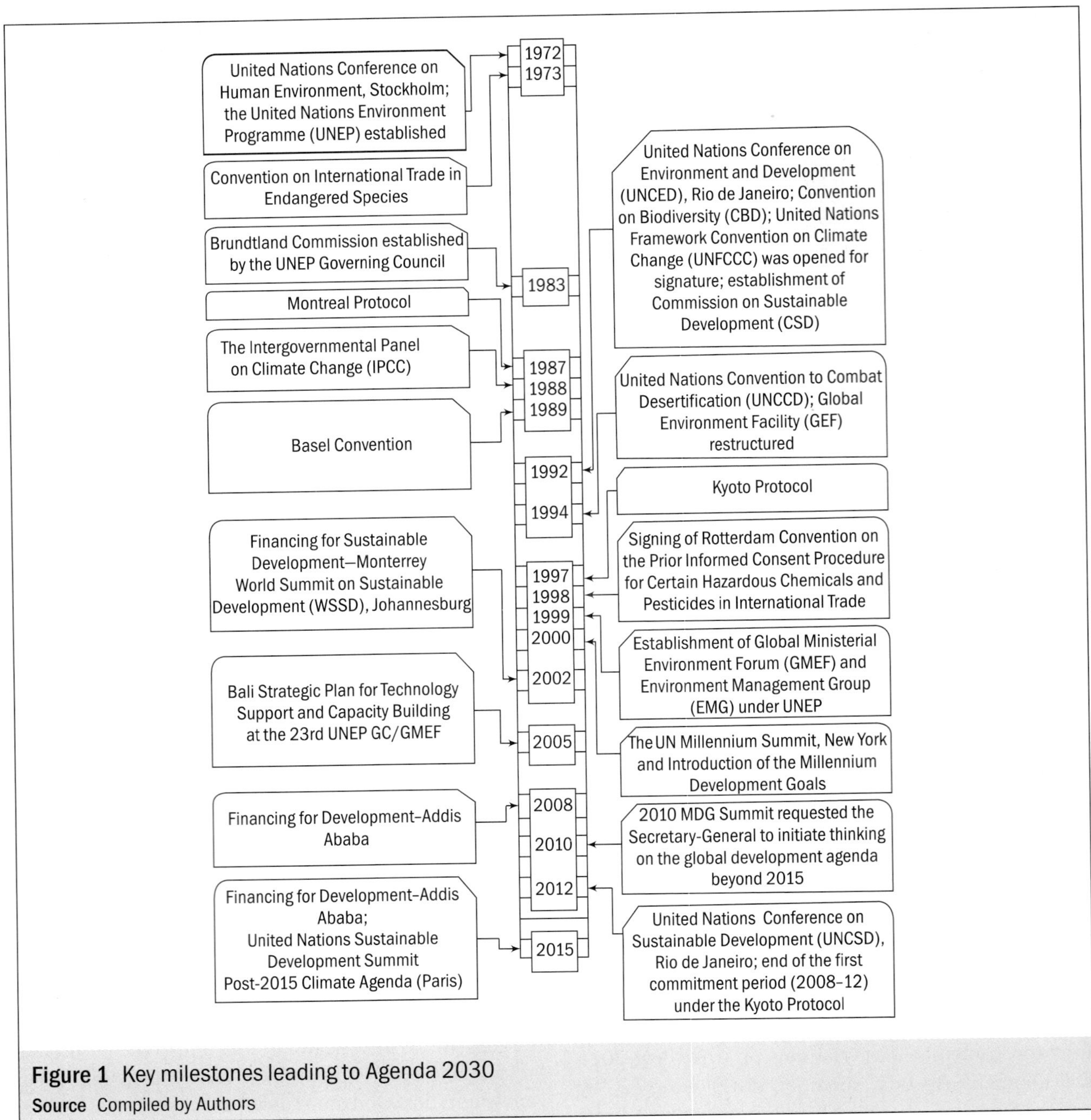

Figure 1 Key milestones leading to Agenda 2030

Source Compiled by Authors

generating data, lack of baseline for collecting data, insufficient capacity of the nodal agency to generate such data or apply the methodologies, and lack of data availability for some indicators.

In 2015, the Social Statistics Division of the Ministry of Statistics and Programme Implementation published the *Millennium Development Goals India Country Report 2015* (MoSPI 2015). On MDG 7, the report covered the eight indicators listed in Box 1.

It can be seen from Box 1 that the environmental sustainability indicators, as reported in India, did not cover aspects of water resources, threatened species, and marine biodiversity. According to MoSPI (2015), India is moderately on track with regard to MDG7. Thus, while India made significant progress in meeting goals, such as forestry and safe drinking water, it made insufficient progress on many environmental themes such as biodiversity loss and degradation of ecosystems.

Table 1 Sustainable Development Goals for Agenda 2030

Number	Goal
1	End poverty in all its forms everywhere
2	End hunger, achieve food security and improved nutrition, and promote sustainable agriculture
3	Ensure healthy lives and promote well-being for all at all ages
4	Ensure inclusive and equitable quality education and promote lifelong learning opportunities for all
5	Achieve gender equality and empower all women and girls
6	Ensure availability and sustainable management of water and sanitation for all
7	Ensure access to affordable, reliable, sustainable, and modern energy for all
8	Promote sustained, inclusive, and sustainable economic growth, full and productive employment, and decent work for all
9	Build resilient infrastructure, promote inclusive and sustainable industrialization, and foster innovation
10	Reduce inequality within and among countries
11	Make cities and human settlements inclusive, safe, resilient, and sustainable
12	Ensure sustainable consumption and production patterns
13	Take urgent action to combat climate change and its impacts*
14	Conserve and sustainably use the oceans, seas, and marine resources for sustainable development
15	Protect, restore, and promote sustainable use of terrestrial ecosystems, sustainably manage forests, combat desertification, and halt and reverse land degradation and halt biodiversity loss
16	Promote peaceful and inclusive societies for sustainable development, provide access to justice for all, and build effective, accountable, and inclusive institutions at all levels
17	Strengthen the means of implementation and revitalize the global partnership for sustainable development

*Acknowledging that the United Nations Framework Convention on Climate Change is the primary international, intergovernmental forum for negotiating the global response to climate change.

Source UN (2015)

MDG 7 had limited coverage and it omitted many important topics concerned with environment. On the contrary, SDGs cover the topics that were missing in MDGs, such as arid and semi-arid, oceanic, mountain, grassland, and arctic ecosystems and desertification. Further, SDGs cover issues such as air pollution and water pollution, accumulation of waste (chemical and solid), and unsustainable consumption of natural resources, which are the major environmental issues in most of the developing countries, including India. Finally, SDGs focus only on the most crucial aspect of environmental problem that all countries (both developed and developing) are suffering from climate change.

The biggest drawback of MDGs is the lack of synergy between MDG 7 and other MDGs. Thus, despite having a strong link with environmental sustainability, poverty and health, job creation, and gender issues, the environmental themes in MDG 7 were not linked with poverty and hunger reduction and health-related goals and targets (Goal 1 and Goal 6, respectively). Therefore, MDG 7 focused only on environmental issues without linking environment with socio-economic development. However, SDGs cover a wider range of goals and targets on environment-related issues, which are closely linked with socio-economic developmental goals and targets. The SDGs try to capture the linkages between poverty reduction, resource management, and development by collectively managing the global environmental risks.

Further, some indicators under MDG 7 included both short-term and long-term targets that were not in sync with each other. The indicator proportion of land under forest cover was a long-term goal, while reducing emission of carbon dioxide (CO_2) was both a medium-term and a long-term goal or target. SDGs try to capture the entire essence of environment by focusing on short-term, medium-term, and long-term targets that sync environmental goals with other SDGs (socio-economic development).

Box 1 Indicators on environmental sustainability

Indicators covered in MoSPI Report 2015	Official list of MDGs indicators for Goal 7
1. Proportion of land area covered by forest 2. Ratio of area protected to maintain biological diversity to surface area 3. Energy use per unit of GDP 4. Carbon dioxide emission per capita and consumption of ozone-depleting chlorofluorocarbons 5. Proportion of households using solid fuels 6. Proportion of population with sustainable access to an improved water source, urban and rural 7. Proportion of population with access to improved sanitation, urban and rural 8. Slum population as percentage of urban population	1. Proportion of land area covered by forest 2. CO_2 emissions, total, per capita, and per $1 GDP (PPP) 3. Consumption of ozone-depleting substances 4. Proportion of fish stocks within safe biological limits 5. Proportion of total water resources used 6. Proportion of terrestrial and marine areas protected 7. Proportion of species threatened with extinction 8. Proportion of population using an improved drinking water source 9. Proportion of population using an improved sanitation facility 10. Proportion of urban population living in slums

GDP – gross domestic product; PPP – purchasing power parity; MDGs – Millennium Development Goals; MoSPI – Ministry of Statistics and Programme Implementation

Sources MoSPI (2015) for Column 1; UN (2008) for Column 2

The targets of MDG 7 were difficult to measure, and some of the indicators were not properly defined or were inadequate for the Indian context. For example, for the target 7A, there is no universal format or baseline to measure to what degree India has adopted/incorporated the principles of sustainable development in its national policies and programmes. Another example is the case of reduced biodiversity loss (MDG target 7B) in which the proportion of terrestrial and marine areas under protection is increasing, yet biodiversity continues to decrease.

Thus, India needs to develop appropriate and specific national indicators for SDG targets which are measurable and accepted in the entire India. Also state-specific indicators should be developed. For example, for "desertification", there is a need to understand and examine the causes of desertification, effects of desertification on the socio-economic condition, and ways and means of resolving the issue of desertification in Indian states. SDGs focus not only on changes in the state of environment but also on the driving forces behind the changes. For example, the target for tackling the ozone layer depletion is "reducing the consumption of ozone-depleting substances".

In a nutshell, SDGs are consolidated and formulated to protect biodiversity, mitigate climate change, restore freshwater ecosystems, and promote environmental sustainability. SDGs will take on the unfinished business of current goals and targets by incorporating them in some form. A limited number of them that are important can be accelerated by being included in SDGs. Finally, SDGs are expected to stimulate the progress of existing and new targets, namely, mitigating climate change, reducing indoor air pollution, slowing down loss of biodiversity and ecosystem services, and lessening the risk of species extinction.

How will India assess energy- and environment-related goals?

Learning from the MDG experience, indicators for SDGs and targets to measure and report their progress are being designed and developed. The global indicator framework will be developed by the Inter-agency and Expert Group on SDG Indicators in coordination with the UN Statistical Commission by March 2016 and will be adopted by the Economic and Social Council and the General Assembly.

Following this, an enormous challenge lies ahead of national governments to monitor the SDGs. In terms of data for energy and environment, some efforts made in India are given in Table 2.

A concerted effort is needed on a periodic basis for collecting energy data/statistics. Multiple agencies collect energy data. Supply-side energy data can be obtained from the Ministry of Coal, Ministry of Petroleum and Natural Gas, Ministry of Power, and Ministry of New and Renewable Energy. Demand-side energy data can be collated from various agencies, especially for monitoring progress on energy

Table 2 Efforts on reporting energy and environment data/statistics

Publication name	Organization	Description
Energy Statistics	Ministry of Statistics and Programme Implementation	Compiles energy data on supply, trade, and consumption from various sources
Compendium of Environment Statistics	Ministry of Statistics and Programme Implementation	Compiles environment data on forests and biodiversity, atmosphere, land and soil, water, and human settlements from various sources
TERI Energy and Environment Data Diary and Yearbook	The Energy and Resources Institute	Compiles data from various sources on energy supply, energy demand, energy access, local environment, and climate change

Source Compiled by Authors

efficiency indicators. In the transport sector, agencies reporting data include the Ministry of Road Transport and Highways, Ministry of Railways, Ministry of Civil Aviation, Ministry of Urban Development, and Ministry of Shipping. Data for buildings and industry also need to be collected and collated. It also remains to be seen what role Niti Aayog assumes with regard to SDGs. International cooperation in development has contributed to measuring energy efficiency, especially in industry. Similar efforts are needed to improve data measurement in other energy demand sectors.

For collecting environment-related data, efforts are being made to measure forest cover and pollution in India. The network of stations for monitoring air quality and water quality needs to be expanded. Moreover, efforts are also needed to measure the resource use in different consumption activities. Another area that needs attention in the environment sector is waste management. With the recent focus on "Smart Cities" and "Clean India Mission", there is also an urgent need to focus on collecting data on waste generation, collection, and treatment.

With regard to energy and environment goals, the following key points arise:

- There is a need to recognize energy more holistically in terms of energy supply, demand, conservation, and access.
- There is an urgent need to scale up existing efforts related to the collection of environment data.
- Lessons from MDGs need to be incorporated with regard to reporting data.
- Partnerships for development need to be enhanced for strengthening statistical reporting capacities and defining baseline and methodology for collection of data and report format.
- Setting up state-specific nodal agencies that would be responsible for data collection, and reporting needs to be institutionalized.
- Cross-sectoral linkages need to be examined for better policy coherence.
- Effective monitoring of development finance for climate, biodiversity, and other environmental issues still needs attention.
- There is a need for participatory approaches for strengthening partnerships for sharing knowledge and best practices by involving communities.

References

MoSPI (Ministry of Statistics and Programme Implementation). 2015. ***Millennium Development Goals India Country Report 2015***. New Delhi: MoSPI

UN (United Nations). 2008. Official List of MDG Indicators. Details available at <http://mdgs.un.org/unsd/mdg/host.aspx?Content=indicators/officiallist.htm>, last accessed on 2 November 2015

UN (United Nations). 2015. ***Transforming Our World: The 2030 Agenda for Sustainable Development***: United Nations

AIR POLLUTION AND HEALTH 14

Introduction

Clean air is the foremost requirement of human beings and the supporting ecosystems that affect human well-being. However, pure air has become scarce since rampant industrial growth has accelerated the emission of various gases and particulate matter. Anthropogenic emissions of various kinds (called primary pollutants) generate new pollutants (called secondary pollutants) owing to chemical reactions in the atmosphere. The build-up of primary and secondary pollutants has made ambient air pollution (AAP) a prominent global threat to human health. According to the *Fifth Assessment Report* of Intergovernmental Panel on Climate Change (IPCC), "nearly all the non-CO_2 climate-altering pollutants are health damaging, either directly or by contributing to secondary pollutants in the atmosphere".

On 19 July 2014, the Open Working Group on Sustainable Development Goals submitted to the UN General Assembly a proposal with a set of goals that consider economic, social, and environmental dimensions to improve people's lives and protect the planet for future generations, at the conclusion of the group's 13th and final session. The proposal contains 17 goals with 169 targets covering a broad range of sustainable development issues, including ending poverty and hunger, improving health and education, making cities more sustainable, combating climate change, and protecting oceans and forests. The third goal aims at ensuring "healthy lives" and promoting "well-being for all at all ages". It has nine subgoals, and subgoal 3.9 targets a substantial reduction in the number of deaths and illnesses from hazardous chemicals as well as air, water, and soil pollution and contamination by 2030.

The hazardous impacts of environmental pollution must be regularly reported and monitored. Of the various kinds of pollutions, air pollution has attracted high priority and a host of environmental regulations have been passed since it directly affects human well-being by way of adverse health effects. Air quality has deteriorated in most large cities in India, a situation driven by population growth, industrialization, and increased vehicle use. Carrying out integrated air quality management, which is an evaluation and monitoring tool, is a challenge in most developing countries because of the lack of information on sources of air pollution and insufficient ambient air monitoring data available in the public domain (TERI 2009).

Urban air pollution largely results from the combustion of fossil fuels in transportation, power generation, industries, and other economic activities. Household air pollution (HAP), also known as indoor air pollution, is a serious concern in rural India since the majority of the rural population continue to depend on traditional biomass for cooking and space heating and on kerosene or other liquid fuels for lighting, all of which lead to high levels of HAP. More than 70% of the population in India depend on traditional fuels (firewood, crop residue, cow dung, coal, and lignite) for cooking and almost 32% depend on kerosene for lighting. About 3 billion people in the world (more than 40% of the global population) rely on traditional biomass for cooking, and an estimated 500 million households rely on kerosene and similar fuels for lighting (WHO 2015). In rural India, for instance, only 11.4% of households use liquefied petroleum gas (LPG) for cooking (Census of India 2011).

This chapter discusses the health impacts of AAP and HAP, and recommends policy measures to reduce these impacts. It focuses on the global burden of air pollution and recommendations to reduce air pollution in the national context.

Parameters of air quality

Air pollution consists of a complex mix of various substances in different physical and chemical states arising from various sources. Many of these substances are gravely harmful to human health. Eminent international institutions such as the World Health Organization (WHO) consider a certain set of air pollution indicators to get closer to the quantification and monitoring of air pollution since the quantification of all air pollutants has not yet been feasible on a global scale. The WHO focuses on four health-related air pollutants: particulate matter, measured as particles with an aerodynamic diameter lesser than 10 μm (PM_{10}) and lesser than 2.5 μm ($PM_{2.5}$); nitrogen dioxide; sulphur dioxide; and ozone. The focus on just these four pollutants is for the sake of monitoring the general state of air quality, and it does not mean that the other air pollutants do not impact human health and the environment (WHO 2006).

Figures 1 and 2 give the annual mean concentrations of PM_{10} and $PM_{2.5}$, respectively, in different regions of the world. For both the figures, the data represented belong to 2008–12 since these data are available for different years for different stations. The highest and lowest recorded values of PM_{10} and $PM_{2.5}$ for each region are represented in these figures. The WHO guideline values for particulate matter are 20 μg/m^3 for PM_{10}, 10 μg/m^3 for $PM_{2.5}$, 40 μg/m^3 for NO_2 (nitrogen dioxide), and 20 μg/m^3 for SO_2 (sulphur dioxide). The orange line denotes the WHO limits. The red dots plotted over each country's bar graph in both the figures represent the human development index (HDI) for that country, as reported in 2012.

In Figure 1, the highest recorded level for PM_{10} is in the Eastern Mediterranean Region (EMR) low and middle income (LMI), where the value recorded comes from Pakistan (Peshawar). The lowest recorded level is in the Western Pacific Region (WPR) high income (HI), where the value recorded comes from Australia (Illawarra). In Figure 2, the highest recorded level for $PM_{2.5}$ is in the South-East Asia Region (SEAR), where the value recorded comes from India (Delhi). The lowest recorded level is in America (AMR HI or America high income), where the value recorded

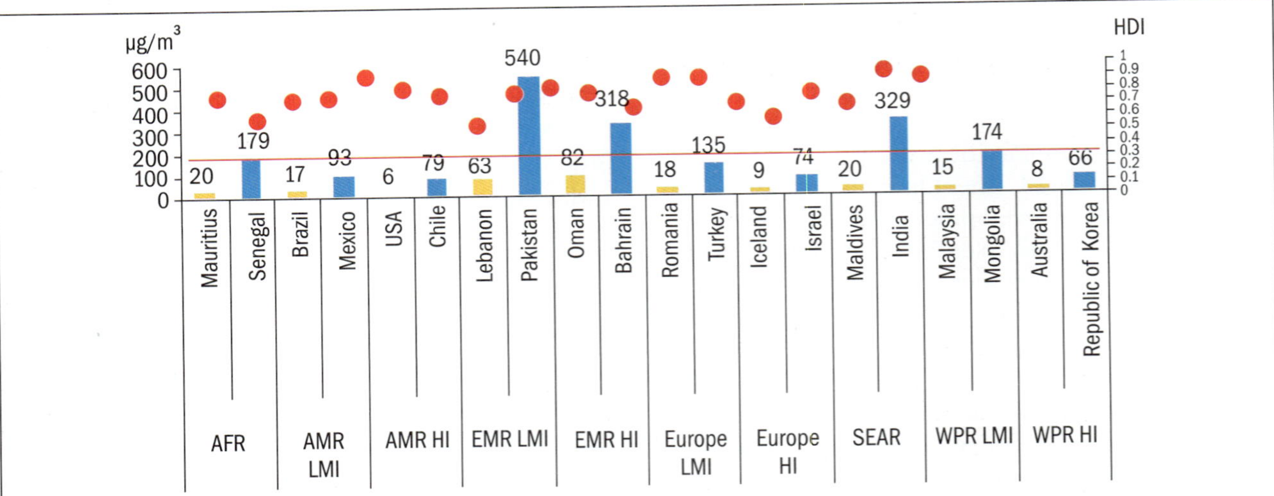

Figure 1 Annual mean concentration of PM_{10} (μg/m^3) in regions of the world

AFR – Africa; AMR – America; EMR – Eastern Mediterranean Region; HDI – human development index; HI – high income; LMI – low and middle income; SEAR – South-East Asia Region; WPR – Western Pacific Region

Notes

1. Yellow bar denotes the lowest value for the region, and blue bar denotes the highest value for the region. These are based on data for a particular region and are not average values for the region. The red line denotes the permissible limit of 20 μg/m^3. The red dots depict the HDI value.
2. City/station of specific countries whose data are presented: Mauritius (Midlands), Senegal (Dakar), Brazil (Salvador), Mexico (Mexico City), USA (Lake Havasu City-Kingman, Arizona), Chile (Rancagua), Lebanon (Beirut), Pakistan (Peshawar), Oman (Muscat), Bahrain (Hamad Town), Romania (Galati), Turkey (Igdir), Iceland (Hafnarfjordur), Israel (Ashkelon), Maldives (Male-Friday Mosque Station), India (Gwalior), Malaysia (Tanah Rata), Mongolia (Darkhan), Australia (Illawarra), and Republic of Korea (Cheongju)

Source WHO (2014)

comes from Canada (Powell River). Generally, the high-income group countries have lower PM_{10} and $PM_{2.5}$ values than the low-income group countries, except in the case of $PM_{2.5}$ concentration in Chile. The highest value of $PM_{2.5}$ in AMR HI (in Chile) is greater than the highest value in America low- and middle- income or AMR LMI (in Brazil). Overall, EMR and SEAR fare worst in terms of PM_{10} and $PM_{2.5}$ among other regions of the world. It is interesting to note that some countries that have higher PM_{10} and $PM_{2.5}$ values have lower HDI values, and some countries that have lower PM_{10} and $PM_{2.5}$ values have higher HDI values. This does not mean that there is necessarily a relation between the two, but it hints towards understanding the specific context of these countries and their respective stage of development.

Ambient air pollution has been identified as a national problem since it is the fifth largest cause of mortality in India (Atkinson, Cohen, Mehta, *et al.* 2011). The Central Pollution Control Board (CPCB) of India implements the National Air Quality Monitoring Programme (NAMP) through a network of 544 ambient air quality monitoring stations covering 224 cities/towns in 26 states[1] and 5 union territories in the country.

Under the NAMP, three major pollutants— PM_{10}, SO_2, and NO_2— have been identified for regular monitoring at all locations. The permissible levels of PM_{10}, NO_2, and SO_2 are 60, 50, and 40 μg/m³, respectively. Other parameters such as $PM_{2.5}$, carbon monoxide (CO), ammonia (NH_3), lead (Pb), ozone (O_3), benzene (C_6H_6), benzo[a] pyrene (B[a]P), arsenic (As), and nickel (Ni) are being monitored at selected locations and are slowly being added to the monitoring network under the NAMP. The CPCB gives ambient air quality data in different cities of India. More than 80% of the

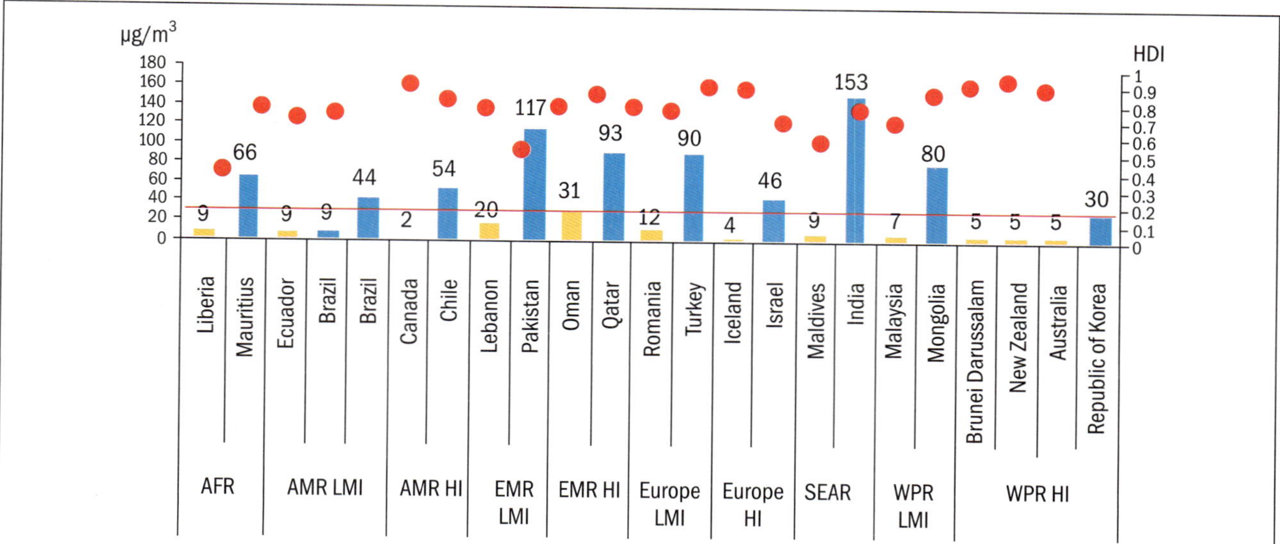

Figure 2 Annual mean concentration of $PM_{2.5}$ (μg/m³) in regions of the world

AFR – Africa; AMR – America; EMR – Eastern Mediterranean Region; SEAR – South-East Asia Region; WPR – Western Pacific Region; HI – high income; LMI – low and middle income

Notes

1. Yellow bar denotes the lowest value for the region, and blue bar denotes the highest value for the region. These are based on data for a particular region and are not average values for the region. The red line denotes the permissible limit of 30 μg/m³. The red dots depict the HDI value.
2. City/station of specific countries whose data are presented: Liberia (Buchanan rural), Mauritius (Beau Bassin/Rose Hill, Coromandel), Ecuador (Ibarra, Ambato), Brazil (Salvador, 9 μg/m³; Santa Gertrudes, 44 μg/m³), Canada (Powell River), Chile (Rancagua), Lebanon (Beirut), Pakistan (Karachi), Oman (Muscat), Qatar (Doha), Romania (Galati), Turkey (Igdir), Iceland (Hafnarfjordur), Israel (Modiín), Maldives (Male-Friday Mosque Station), India (Delhi), Malaysia (Tanah Rata), Mongolia (Darkhan), Brunei Darussalam (Brunei-Muara District), New Zealand (Lower Hutt), Australia (Melbourne), and Republic of Korea (Cheongju)

Source WHO (2014)

[1] Manipur and Sikkim are not covered by the CPCB

Indian cities violate the PM_{10} standards. Only 31 cities meet the standard of 60 μg/m^3 for PM_{10} out of 204 cities for which data exist. The levels of SO_2 in almost all cities in India, except Dombivali (52) and Ulhasnagar (46), lie below the permissible limit of 40 μg/m^3, and nine cities exceed the permissible limit of 50 μg/m^3 for NO_2. More details about the average annual concentration for different cities in India are given in Annexure 1.

Impact of air pollution on respiratory health

In 2012 alone, 7 million deaths in the world were attributed to the combined effects of ambient (3.7 million) and household (4.3 million) air pollution (WHO 2015). Figure 3 shows the leading causes of global deaths and premature deaths in the world (as of 2010). Ischaemic heart disease (IHD) is at the forefront, and chronic obstructive pulmonary disease (COPD), lower respiratory infections, and lung cancer are also among the top five causes of deaths worldwide.

Figure 4 shows the total number of deaths in 2012 that are attributable to HAP and AAP across all regions of the world. As part of an update on WHO (2010), it was found that essentially five kinds of respiratory health conditions are behind the deaths. About 36% of the deaths are attributable to IHD, 33% to stroke, 17% to COPD, 8% to acute lower respiratory disease (ALRI), and 6% to lung cancer. The update also quantifies the effect on people from different gender and age groups. According to it, 49% of the 7 million victims in 2012 were men (aged 25 years or above), 42% were women (aged 25 years or above), and 9% were children (under the age of 5 years) (WHO 2014). The number of deaths in the Western Pacific LMI region is the highest in the world, followed by South-East Asia and Africa.

Table 1 presents data for every 5 years from 1990 to 2010 for the percentage of total global deaths attributed to air pollution categorized under three main diseases. From the data, it can be stated that the total global deaths owing to all the major respiratory diseases seem to have reduced systematically for the time period in consideration. This is also due to the improvements made over these years (1990–2010) in the quality, accessibility, and availability of health services and medicines in the geographical areas where the illnesses are prevalent. There are insufficient data to state that the percentage of total global deaths attributable to air pollution has reduced over these years due to improvements in global air quality.

Global life expectancy for both sexes increased from 65.3 years in 1990 to 71.5 years in 2013, while the number of deaths increased from 47.5 million to 54.9 million over the same interval. This means that improving health care is pushing the average

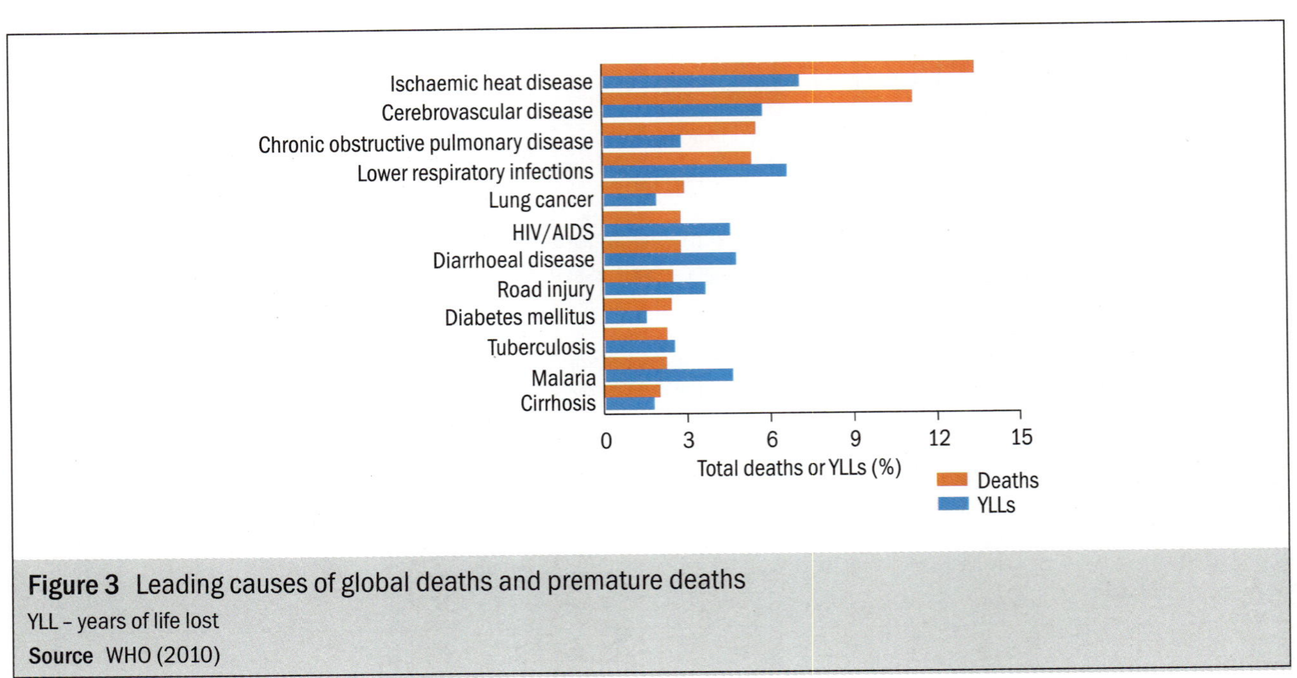

Figure 3 Leading causes of global deaths and premature deaths

YLL – years of life lost

Source WHO (2010)

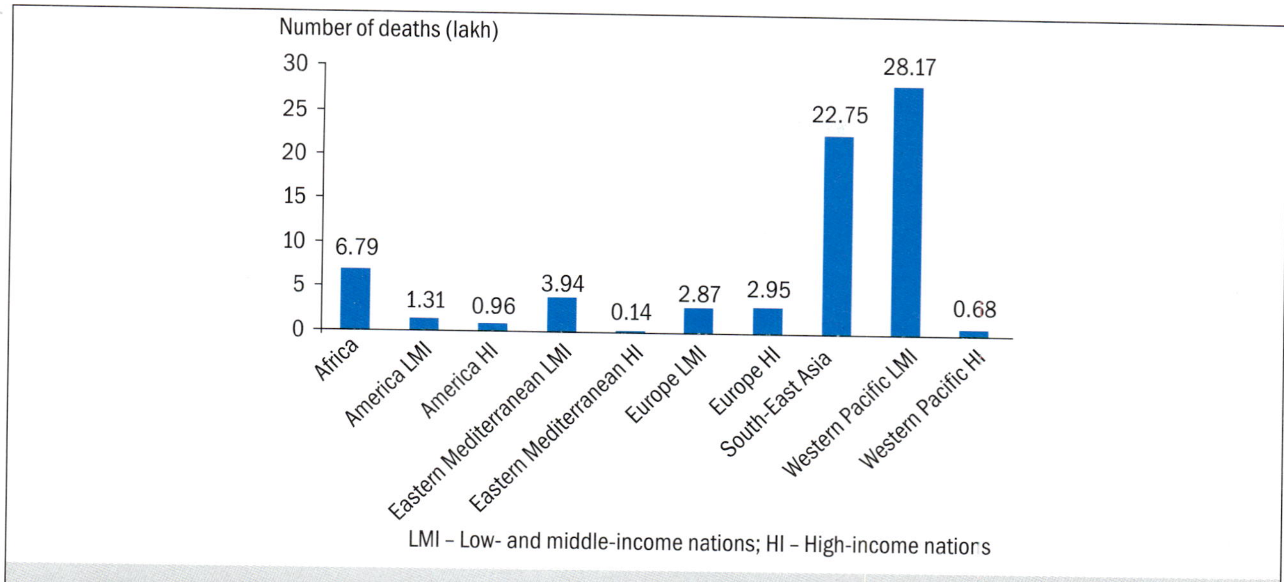

Figure 4 Total global deaths attributable to household and ambient air pollution in 2012 (by region)
Source WHO (2014)

Table 1 Percentage of total global deaths due to causes attributed to air pollution

Year	Chronic obstructive pulmonary disease	Lower respiratory infections	Pneumoconiosis	Other chronic respiratory diseases	Total
1990	6.7	7.3	0.4	0.6	15
1995	6.3	6.7	0.3	0.6	13.9
2000	5.7	6.1	0.3	0.5	12.6
2005	5.4	5.5	0.2	0.5	11.6
2010	5.5	5.3	0.2	0.6	11.6

Source WHO (2015)[2]

age bar of the global population, also leading to the inference that we have less young population now than we had in earlier years. Respiratory infections have a major impact on child health, especially of those children less than 5 years of age, since AAP and HAP begin to affect child health even before birth and during the prime years of growth. Lower respiratory infection remains one of the top five causes of death in children under 5 years (IHME 2013). Figure 5 illustrates the significant decline in the number of global deaths of children aged between 1 month and 4 years in the last decade, while the number of deaths in children aged less than 27 days has barely changed during the same time period.

Table 2 gives a detailed account of the total global number of years of life lost (YLL) from mortality, years lost due to disability (YLD), and disability-adjusted life years (DALYs) for male (M) and female (F) of all age groups for major communicable and non-communicable diseases that are believed to be attributed to air pollution.

Indian cities today are among the most polluted in the world, and it is estimated that outdoor air pollution leads to approximately 670 000 deaths annually (Lim, Vos, Flaxman, *et al.* 2010). Current standards for particulate matter set by the CPCB are much higher than those recommended by the WHO (Krzyzanowski and Cohen 2008). In addition, unlike other countries (Bell, Samet and Dominici 2003; Dominici, Peng, Zeger, *et al.* 2007), the CPCB does not take into account the findings from the health literature when deciding on air quality standards (Balakrishnan, Ganguli, Ghosh, *et al.* 2011). A periodic review of epidemiological evidence enables policymakers to assess current

[2] Institute for Health Metrics and Evaluation. Details available at <www.who.int/gho/en/>, last accessed on 1 June 2015

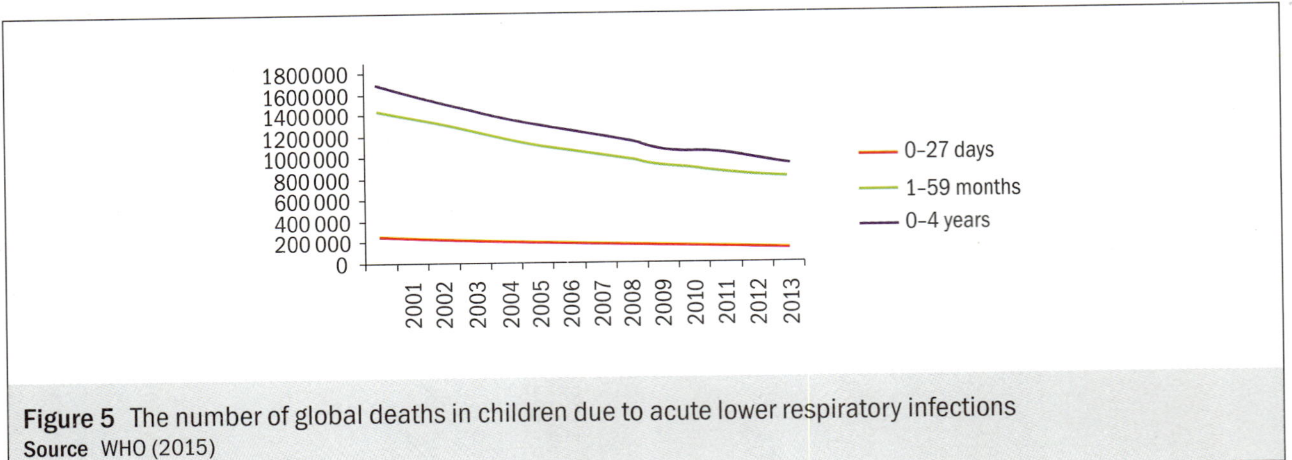

Figure 5 The number of global deaths in children due to acute lower respiratory infections
Source WHO (2015)

Table 2 Total number of YLL, YLD, and DALY (in million) for 2000 and 2012 for male and female of all age groups for the diseases attributable to air pollution

	2000						2012					
	YLL		YLD		DALY		YLL		YLD		DALY	
	M	F	M	F	M	F	M	F	M	F	M	F
Communicable diseases												
Respiratory infections	110.85	95.97	4.33	4.02	83.83	70.34	79.03	65.91	4.80	4.43	83.83	70.34
Lower respiratory infections	110.50	95.66	1.10	1.08	79.98	66.89	78.76	65.69	1.22	1.20	79.97	66.88
Upper respiratory infections	0.22	0.20	0.80	0.79	1.02	1.01	0.17	0.14	0.89	0.88	1.06	1.01
Non-communicable diseases												
Ischaemic heart disease	80.79	54.24	3.82	3.38	98.58	67.14	93.62	62.87	4.96	4.27	98.56	67.12
Respiratory diseases	51.38	39.23	21.13	20.61	76.73	60.38	50.54	35.36	26.19	25.02	76.72	60.37
COPD	36.47	28.46	12.70	11.79	52.19	40.19	36.14	25.49	16.05	14.70	52.18	40.18
Asthma	6.91	4.90	5.92	6.48	13.40	11.80	6.63	4.44	6.77	7.36	13.40	11.80
Other respiratory diseases	7.99	5.88	2.51	2.34	11.14	8.39	7.78	5.43	3.36	2.97	11.14	8.39

COPD – chronic obstructive pulmonary disease; F – female; M – male
Note YLL is the years of life lost from premature mortality, YLD is the years of life lived with any short-term or long-term health loss, and disability-adjusted life year (DALY) is the sum of years lost due to premature death (YLL) and years lived with disability (YLD). DALYs are also defined as years of healthy life lost.
Source WHO (2015)

health risks associated with air pollution and sets the agenda towards finding a balance between reducing health impacts and the costs of implementing further air pollution controls (Dholakia, Bhadra, and Garg 2014). An interesting finding of the analysis is that cities such as Ahmedabad and Mumbai that have higher levels of pollution experience a relatively lower increase in mortality for every 10 mg/m^3 increase in PM_{10}. In contrast, the percentage increase in mortality is highest for Shimla, which is among the cleanest cities. The percentage increase in mortality associated with a 10 mg/m^3 increase in PM_{10} is highest for Shimla (1.36%) and the least for Ahmedabad (0.16%). Bangalore and Mumbai showed similar results with a 0.22% and 0.20% mortality increase, respectively, at a 95% confidence interval.

Findings suggest that a small reduction in pollution in cleaner cities will yield large health benefits, whereas in less clean cities, even a large reduction in pollution may yield only modest health benefits in a relative sense. The WHO findings (WHO 2006, 2013) state that all-cause daily mortality is estimated to increase from 0.2%

to 0.6% per 10 μg/m^3 increase in PM_{10}, and long-term exposure to $PM_{2.5}$ is associated with an increase in the long-term risk of cardiopulmonary mortality from 6% to 13% per 10 μg/m^3 increase in $PM_{2.5}$. The Harvard Six Cities Study found that a 2.5 μg/m^3 decrease in the annual average level of $PM_{2.5}$ was associated with a 3.5% reduction in all-cause mortality in the United States (Lepeule, Laden, Dockery, *et al.* 2012). The study was conducted on a group of people living in six cities in the United States and whose lifestyle was followed from 1974 to 2009 to estimate the effects of air pollution on health. The study also concluded that the high exposure to $PM_{2.5}$ led to increased mortality due to cardiovascular illnesses and lung cancer. Consequently, the study also found that the reduced levels of $PM_{2.5}$ could quite immediately bring out positive differences in the health of people and reduce morbidity and mortality due to illnesses arising from air pollution.

Figure 6 presents the predicted average gain in life expectancy (in months) for people aged 30 years for a reduction in average annual levels of $PM_{2.5}$ down to the WHO Air Quality Guideline annual mean level of 10 μg/m^3 in 25 European cities participating in the Aphekom project conducted between 2008 and 2011 (WHO 2013). The blue line denotes the air quality guideline limit for $PM_{2.5}$. The gain in life expectancy is predicted after calculating the effect of reduced $PM_{2.5}$ levels (down to 10 μg/m^3) in each of the 25 cities. The study concluded that the higher the $PM_{2.5}$ levels, the greater is the gain in life expectancy could be achieved if the existing levels were brought down to 10 μg/m^3. For instance, the gain in life expectancy was found to be zero simply because the $PM_{2.5}$ levels were under the WHO limits. Similarly, the gain in life expectancy was highest in Bucharest after the $PM_{2.5}$ levels were modelled and brought down below the WHO limits.

Ambient air pollution has been identified as the fifth largest cause of mortality in India (Lim, Vos, Flaxman, *et al.* 2010). India registered an increase of about 12% in the number of deaths and about 3% in YLL between 2005 and 2010. Figure 7 shows the alarming increase in the death toll in India from 2005 to 2010 (72 000 more deaths). Figure 8 gives the percentage distribution of deaths by diseases attributable to ambient particulate matter pollution in India.

Many studies across the world and also in India prove that outdoor and indoor air pollutions are serious environmental risk factors that cause or aggravate acute and chronic diseases. Comparing of city-level air pollution measures with child-level data from the National Family Health Survey (2005/06) for six cities in India shows that an increase in AAP significantly increases child morbidity (Ghosh and Mukherjee 2010). The six cities considered were Chennai, Delhi, Hyderabad, Indore, Kolkata, and Nagpur. The study found that a rise in AAP significantly increases the likelihood of a child suffering from cough and fever in the past week. However, the type of cooking fuel used at home is not significantly related to child morbidity after accounting for AAP and other child- and household-level control variables. Thus, while bad air is harmful to child health,

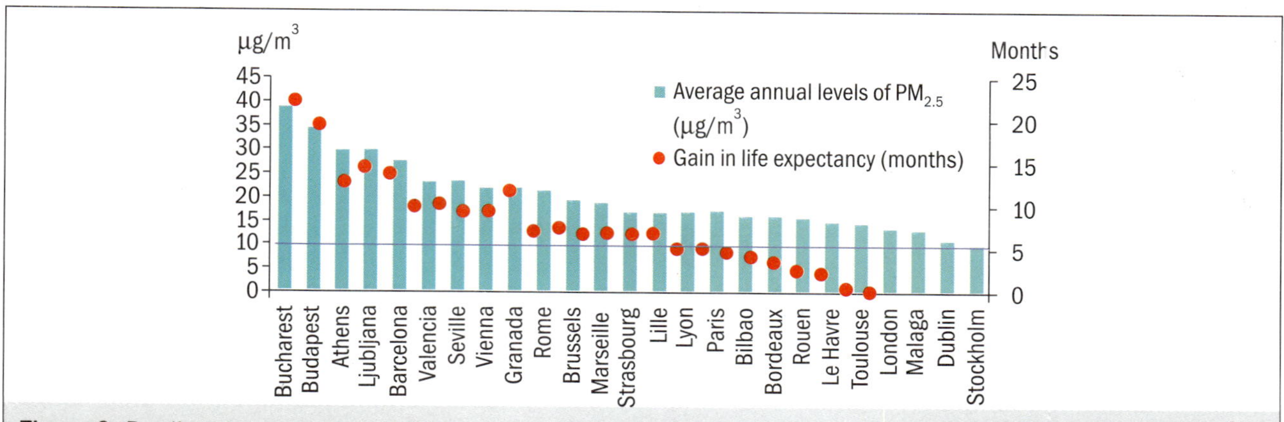

Figure 6 Predicted average gain in life expectancy with a reduction in average annual levels of $PM_{2.5}$ 10 μg/m^3 in 25 European cities

Source WHO (2013)

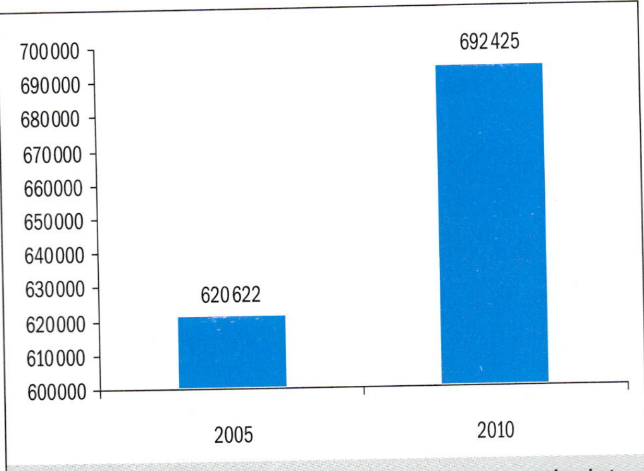

Figure 7 Total deaths from ambient particulate matter and ozone pollution in India
Source OECD (2014)

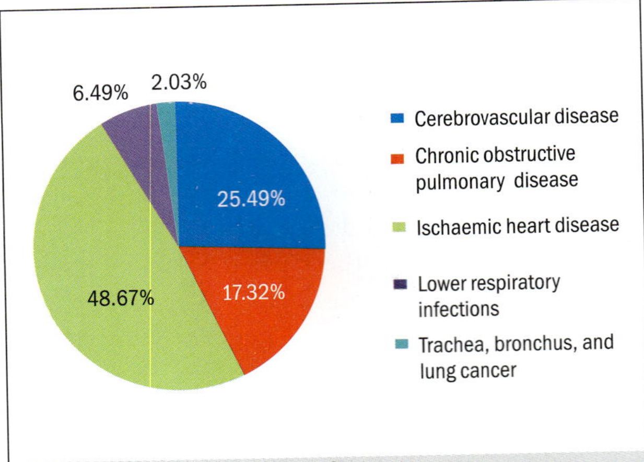

Figure 8 Percentage distribution of deaths from ambient particulate matter pollution in India
Source Lim, Vos, Flaxman, *et al.* (2010)

AAP is a more significant determinant of the child health outcomes. A significant correlation has been observed between the two child morbidity outcomes: fever and cough. Controlling citywide air pollution could significantly lower child morbidity and should receive greater emphasis in urban planning and infrastructure development.

Indoor air quality has emerged as one of the most important issues of environment and health worldwide. HAP due to biomass cooking fuel is an important risk factor for a range of diseases, especially among adult women who are primary cooks in India (Lim, Vos, Flaxman, *et al.* 2010). India has recently identified HAP as one of the key indicators in its National Monitoring Framework for Prevention and Control of Non-communicable Diseases (MoHFW 2012). Biomass fuels remain a widely used energy source in rural India where nearly 80% of households use them as the primary cooking fuel. In contrast, the majority of urban households use LPG as the primary cooking fuel; however, about 19% of urban households use biomass fuel for cooking. Poverty, inaccessibility to improved cooking fuel, and lack of awareness about the harms of biomass emissions are among the major factors that drive their widespread use (MoSPI 2012). The use of biomass fuel leads to harmful health effects owing to the emission, during its incomplete combustion, of a large number of air pollutants such as CO, SO_2, respirable particulate matter ($PM_{2.5}$ and PM_{10}), polycyclic aromatic hydrocarbons, benzene, and metals such as lead and copper.

It has been estimated that about 2.4 million of 5.6 million cases of chronic bronchitis, 0.3 million of 0.76 million cases of tuberculosis, 5 million of 51.4 million cases of cataract among adult Indian women, and 0.02 million of 0.15 million stillbirths across India are attributable to HAP owing to biomass cooking fuel (Sehgal, Rizwan, and Krishnan 2014).

Several studies have linked biomass cooking fuel to adverse pregnancy outcomes such as preterm births, low birthweight, and post-neonatal infant mortality, but very few have studied the associations with the cooking fuel independent of other factors associated with stillbirths. Using India's 2003/04 District Level Household Survey II data, it is found that the usage of biomass and kerosene cooking fuels by women is related to stillbirths. Nearly 12% of these stillbirths can be prevented by providing access to cleaner cooking fuels (Lakshmi, Virdi, Sharma, *et al.* 2013).

Annually, 400 000–550 000 premature deaths have been attributed to indoor air pollution in India (Smith, Samet, Romieu, *et al.* 2010). There is consistent evidence that indoor air pollution increases morbidity and mortality from respiratory tract symptoms in childhood (Hertz-Picciotto, Baker, Yap, *et al.* 2007). Indoor air pollution measured in terms of $PM_{2.5}$ is an important cause of respiratory illness in children. Increased $PM_{2.5}$ levels and biomass fuel use number are found to be associated with an increased occurrence of respiratory illness in children (Kumar, Goel, Gupta, *et al.* 2013).

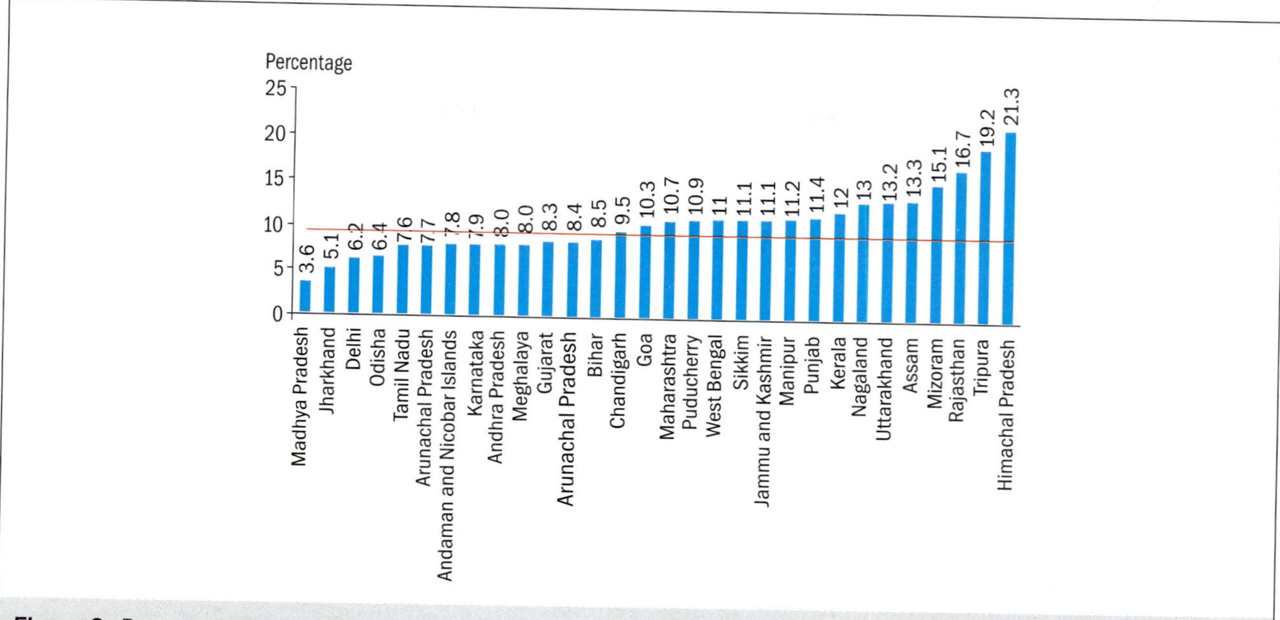

Figure 9 Percentage by state of medically certified deaths due to diseases of the respiratory system in 2011
Note The red line in the graph shows the all-India value of 9.71%.
Source CBHI (2013)

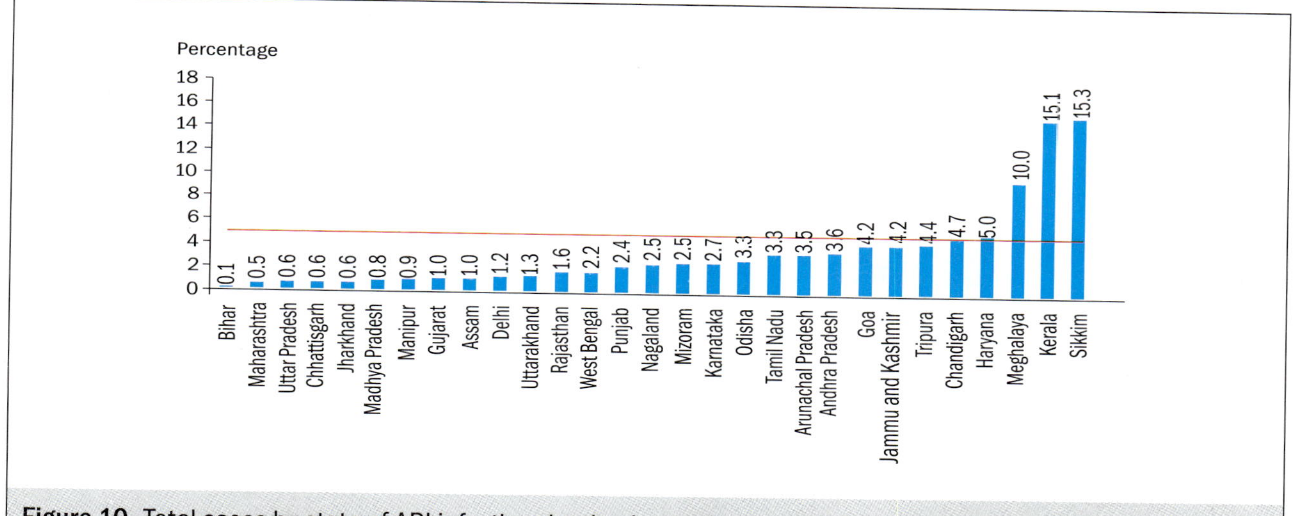

Figure 10 Total cases by state of ARI infection deaths due to diseases of the respiratory system in 2011
Note The red line in the graph shows the all-India value of 4.77%.
Source CBHI (2013)

Figure 9 shows the percentage by state of medically certified deaths due to diseases of the respiratory system. Figure 10 shows the cases of acute respiratory infection (ARI) as a percentage of total population in the state. The red lines in both the graphs depict the all-India value.

Indian metropolitan cities remain exposed to high levels of air pollutants mainly owing to the high vehicular movement and poor roads. Although the concentrations of these pollutants vary according to the traffic density, types of vehicles, and time of day, some people by virtue of their occupation are more exposed to high levels of traffic-related air pollutants. These people include filling station workers, traffic policemen, professional drivers, and tollbooth workers, who are more exposed to high emissions from vehicle idling, deceleration, and acceleration (Sehgal, Suresh, Sharma, *et al.* 2014). A recently conducted study of air quality monitoring ($PM_{2.5}$, CO, NO_x, SO_2, and EC/OC) at highway toll plazas, municipality toll plazas, and control sites, which were 23 in number, found that there was a high level of air pollution at almost all locations with $PM_{2.5}$ values exceeding the national permissible limit (60 $\mu g/m^3$) except at a few control sites. The study found that pollutant concentrations were highest at municipality toll plazas with minimum protective work areas. The observed reduction in lung function indices was significant over the years of occupational exposure, even after making adjustments for age, among non-smoking outdoor workers.

Economic costs of air pollution

Air pollution leads to a significant amount of mortality and morbidity. The economic cost of this mortality and morbidity is necessary to estimate in order to determine the health costs arising from air pollution. Such cost estimates can serve as effective indicators of the economic burden of air quality-related health impact on both people and the national exchequer. The cost of mortality is estimated using the value of a statistical life (VSL) as a measure derived from individuals' valuation of their willingness to pay to reduce the risk of dying. India, with its rapid growth rate during this five-year period, has registered a 60% increase in VSL.

Table 3 gives the estimated economic cost of health impacts of AAP in the Organisation for Economic Co-operation and Development (OECD) countries, India, and China for 2005 and 2010. Mortality and morbidity costs are reflected in million US dollars. Thirty-four OECD countries together accounted for 478,104 deaths in 2010 owing to AAP, while India alone accounted for 692 425 deaths. This trend was similar in 2005 as well. It is interesting to note, however, that the difference in morbidity and mortality costs in 2010 between all the OECD countries together ($1.7 trillion) and India alone ($0.4 trillion) is a revealing fact to understand the loss India incurs not only by the burden of disease but also economically. India's gross domestic product (GDP) in 2010 was $1.7 trillion, which means that India was spending about 23% of its GDP on dealing with the economic costs of health impacts of AAP. The costs in OECD countries are much higher than those calculated for India (less than four times) even when the

Table 3 Indicative estimate of the economic costs of health impacts of ambient air pollution in OECD countries, India, and China in 2005 and 2010

	2005					2010				
	Number of deaths	VSL 2005 (in million USD)	Mortality costs (in million USD)	Mortality + morbidity (health) costs (in million USD) if add-on ~10%	Health costs as a percentage of GDP	Number of deaths	VSL 2010 (in million USD)	Mortality costs (in million USD)	Mortality + morbidity costs (health) (in million USD) if add-on ~10%	Health costs as a percentage of GDP
OECD countries	497 958	–	1 470 487	~1 617 536	4.44%	478 104	–	1 571 170	~1 728 287	3.98%
India	620 622	0.375	232 736	~256 010	30.68%	692 425	0.602	416 704	~458 734	26.85%
China	1 215 180	0.61	741 019	~815 121	36.11%	1 278 890	0.975	1 246 713	~1 371 384	23.12%

GDP – gross domestic product; OECD – Organisation for Economic Co-operation and Development; USD – US dollar; VSL – value of a statistical life

Note GDP values are in current USD.

Source OECD (2014, 2015); World Bank Data for GDP[3]

[3] Details available at <http://data.worldbank.org>, last accessed on 1 June 2015

absolute deaths in OECD countries together were fewer by more than 120 000 than those in India. A similar comparison holds for OECD countries and China. Among the three, China has the maximum number of mortalities due to AAP. Mortalities have increased in both India and China, while they have reduced in OECD countries. The mortality and morbidity costs have changed linearly with the mortality numbers in these countries.

According to a World Bank study in 2009, about ₹1100 billion (1.7% of GDP) and more than ₹800 billion (1.3% of GDP) were the estimated annual costs of environmental damage caused by AAP and HAP, respectively. This means about 52% of the relative share of damage cost in the environment category was due to AAP and HAP put together (World Bank 2013).

Data from the country's apex environmental regulator, the CPCB, reveal that 77% of Indian urban agglomerations exceeded the National Ambient Air Quality Standard (NAAQS) for respirable suspended particulate matter PM_{10} in 2010.[4] Estimates from the WHO suggest that 13 of the 20 cities in the world with the worst fine particulate $PM_{2.5}$ air pollution are in India, including Delhi, the worst ranked city.[5]

Air pollution also leads to a reduction in life expectancy. Using a combination of ground-level in situ measurements and satellite-based remote sensing data, it has been estimated that 660 million people, over half of India's population or nearly every Indian (1204 million people or 99.5% of the population), live in areas that exceed the NAAQS for fine particulate pollution. Reducing pollution in these areas to achieve the standard would increase life expectancy by 3.2 years on an average for a total of 2.1 billion life years (Greenstone, Nilekani, Pande, *et al.* 2015).

Recommendations

Various strategies must be urgently adopted to improve urban and indoor air quality. To improve air quality and reduce the burden of diseases, several interventions have been adopted in India. In order to reduce vehicular pollution, CPCB set vehicular emission standards in India in 1986 for both petrol- and diesel-driven vehicles. The standards were later revised in 1987 and 1989. To curb further vehicular pollution, the Ministry of Petroleum and Natural Gas adopted the Auto Fuel Policy in 2002, which laid down a roadmap for the introduction of cleaner fuels and vehicles in the country up to 2010. The Bharat Stage (BS) IV norms were implemented in 13 cities from April 2010 based on the roadmap, and BS III norms were implemented in the entire India from October 2010. Compressed natural gas has been introduced in many cities of the country. Heavy-duty trucks still remain on BS III standards all across the country, although they are registered in cities where BS IV norms have been introduced.

The Auto Fuel Vision Committee was set up in 2013 to recommend the future roadmap for the advancement of fuel quality and vehicular emission standards up to 2025. The committee has recommended the introduction of BS IV and BS V norms across the country by 2017 and 2020, respectively. BS-VI emission norms are recommended to be introduced by 2024. The roadmap recommended by the Auto Fuel Vision 2025 Committee puts India 10 years behind the United States and European countries (TERI 2014). It is, therefore, necessary that the Government of India should introduce and adopt BS V fuels other than those recommended in the Auto Fuel Vision (AFV 2025) Report, which will necessitate and enable the Indian refineries to leapfrog from BS III to BS V fuels by 2018 (ICAMP 2014).

In order to control the indoor air pollution through the provision of clean energy access, the government has introduced various schemes that would facilitate clean energy access, such as the Rajiv Gandhi Grameen Vidyutikaran Yojana, the Village Energy Security Programme, and the Remote Village Electrification Programme. The Ministry of New and Renewable Energy (MNRE) is promoting the setting up of biogas plants in all states and union territories of the country. Biogas plants are being promoted under the scheme National Biogas and Manure Management Programme (NBMMP) for setting up of family type biogas plants. According to the MNRE, about 827 003 family type biogas plants were set up till March 2014 against the projected target of 106 000 biogas plants (TERI 2015).

[4] Details available at <http://cpcb.nic. in/National_ Ambient_Air_Quality_Standards.php>, last accessed on 1 June 2015

[5] Details available at <www.who.int/entity/quantifying_ehimpacts/national/countryprofile/AAP_PM_database_May2014,xls?ua=1>, last accessed on 1 June 2015

Issues of acceptability, affordability, and poor awareness about alternative cooking fuels or modern technologies are the major hurdles. On the supply side, pricing and distribution of different cooking fuels vary geographically and across demographic segments. Government planning without community participation affects the implementation and success of such schemes. Awareness generation among public and policymakers would help improve the path each community could take towards cleaner fuels.

More epidemiological studies should be taken up to determine how AAP and HAP affect people's health. Quantification of impacts is particularly insightful to provide policy with the necessary push and can act as a benchmark to understand the need for particular tools for air quality monitoring. Data gaps in the quantification of the impact need to be addressed strongly and promptly, so the research undertaken research can perform the necessary analysis and provide the most suitable policy recommendations. These data gaps also hinder further into the matter and limit the understanding of trends in pollutant concentration levels and spread of diseases.

Given the huge economic costs of the health effects of air pollution, the benefits of reducing the burden could easily outweigh the monetary cost of investments in more ambitious programmes to reduce pollution (OECD 2014). In 2010, the best available estimate of the economic costs of the health impacts of outdoor air pollution in China and India combined was larger than the OECD total—about $1.4 trillion in China and about $0.5 trillion in India. There should be advances in instrumentation to adopt continuous information systems for monitoring ambient concentrations and pollution from high-emitting industries or sectors (transport). Increased monitoring can play an important role as a health advisory system and as a means of increasing pressure on polluters to comply with existing regulations. There is an urgent need to develop a sustainability index to benchmark and develop strong environmental governance guidelines.

Annexure 1 Ambient air quality data in different cities in India (2012)

State/union territory	City	SO_2	NO_2	PM_{10}
Andhra Pradesh	Chittoor	4	9	40
	Guntur	5	11	75
	Hyderabad	4	28	79
	Kakinada	5	11	63
	Kothagudem	4	9	74
	Kurnool	4	15	79
	Nalgonda	5	11	62
	Nellore	6	22	108
	Patancheru	6	11	82
	Ramagundam	4	9	37
	Tirupati	6	12	97
	Vijayawada	3	11	49
	Warangal	12	13	65
	Visakhapatnam	5	11	63
Assam	Darrang	5	13	56
	Dibrugarh	6	13	56
	Guwahati	6	14	92
	Margherita	6	15	54
	Lakhimpur	2	2	45
	Nagaon	6	13	79
	Nalbari	6	15	82
	Sibsagar	7	15	109
	Silchar	6	14	91
	Tezpur	3	9	11
	Tinsukia	5	12	57
Bihar	Patna	6	36	166
Chandigarh	Chandigarh	2	19	110
Chhattisgarh	Bhillai	8	22	103
	Bilaspur	6	20	NA
	Korba	12	19	81
	Raipur	14	40	NA
Dadra and Nagar Haveli	Silvassa	8	20	NA
Daman and Diu	Daman	8	20	NA
Delhi	Delhi	5	59	237
Goa	Panaji	9	14	67

Contd...

Annexure 1 *Contd...*

State/union territory	City	SO_2	NO_2	PM_{10}
	Mormugao	6	19	112
	Vasco	7	20	84
	Curchorem	15	20	112
	Codli	16	21	121
	Bicholim	10	18	119
	Amona	10	12	90
	Assanora	9	12	84
	Usgao	10	18	121
	Margao	9	17	67
	Tilamol	15	20	114
	Mapusa	7	9	82
	Sanguem	14	19	100
	Ponda	9	16	60
	Kundaim	9	16	63
Gujarat	Ahmedabad	12	24	83
	Ankleshwar	18	27	99
	Jamnagar	12	25	101
	Rajkot	13	17	99
	Surat	16	26	97
	Vadodara	16	33	102
	Vapi	19	30	100
Haryana	Faridabad	12	38	184
	Hissar	6	8	111
	Yamunanagar	NA	NA	NA
Himachal Pradesh	Baddi	2	23	99
	Damtal	2	12	97
	Kala Amb	2	16	165
	Nalagarh	2	23	89
	Parwanoo	2	8	79
	Paonta Sahib	2	15	153
	Una	NA	NA	69
	Sunder Nagar	2	11	94
Jammu and Kashmir	Jammu	6	12	119
Jharkhand	Dhanbad	17	40	178
	Jamshedpur	37	49	149
	Jharia	17	40	212
	Ranchi	18	35	202

Contd...

Annexure 1 *Contd...*

State/union territory	City	SO_2	NO_2	PM_{10}
	Saraikela Kharsawan	39	51	160
	Sindri	17	40	170
	West Singhbhum	19	27	153
Karnataka	Bengaluru	14	28	121
	Gulbarga	3	11	65
	Hassan	5	17	36
	Hubli-Dharwad	5	14	77
	Mandya	10	24	49
	Mangalore	6	7	31
	Mysore	11	23	56
	Devanagere	5	10	75
Kerala	Alappuzha	2	5	50
	Kochi	3	10	70
	Kollam	4	19	41
	Kottayam	6	16	56
	Kozhikode	2	8	56
	Malappuram	2	5	36
	Palakkad	3	7	37
	Pathanamthitta	2	15	23
	Thrissur	2	15	73
	Wayanad	2	8	33
	Thiruvananthapuram	9	22	55
Madhya Pradesh	Bhopal	3	21	173
	Dewas	18	23	92
	Gwalior	13	27	329
	Indore	12	20	143
	Jabalpur	2	24	75
	Nagda	26	27	103
	Sagar	3	13	120
	Singrauli	24	25	64
	Ujjain	12	13	80
Maharashtra	Dombivali/Ambernath	52	87	114
	Amravati	11	13	100
	Aurangabad	9	32	80
	Badlapur	6	86	124
	Chandrapur	11	19	148
	Jalgaon	19	44	130

Contd...

Annexure 1 *Contd...*

State/union territory	City	SO_2	NO_2	PM_{10}
	Latur	8	20	117
	Lote	25	14	40
	Mumbai	5	20	117
	Nagpur	10	32	103
	Nashik	24	27	95
	Navi Mumbai	17	43	120
	Pune	22	45	92
	Sangli	11	40	80
	Solapur	17	35	83
	Thane	20	12	72
	Ulhasnagar	46	79	111
	Jalna	9	30	109
	Akola	9	10	139
	Nanded	30	30	53
Meghalaya	Byrnihat	35	20	138
	Dawki	2	8	44
	Tura	2	5	52
Nagaland	Dimapur	2	7	90
Odisha	Angul	6	19	106
	Balasore	3	14	82
	Berhampur	2	18	80
	Bhubaneshwar	2	18	81
	Cuttack	2	17	68
	Rayagada	3	22	54
	Rourkela	5	11	98
	Sambalpur	3	15	53
	Talcher	8	19	116
	Kalinganagar	3	9	52
Punjab	Amritsar	15	39	202
	Dera Bassi	9	28	129
	Pathankot/Dera Baba	7	12	68
	Gobindgarh	9	35	201
	Jalandhar	13	26	136
	Khanna	10	26	213
	Ludhiana	11	27	228
	Naya Nangal	7	19	89
	Patiala	5	13	62

Contd...

Annexure 1 *Contd...*

State/union territory	City	SO_2	NO_2	PM_{10}
	Batala	8	14	42
Puducherry	Puducherry	9	52	187
Rajasthan	Jaipur	6	24	189
	Jodhpur	6	31	155
	Kota	8	14	42
	Udaipur	9	52	187
Tamil Nadu	Chennai	12	21	57
	Coimbatore	3	27	68
	Madurai	14	30	48
	Salem	9	22	60
	Tuticorin	14	14	134
	Trichy	11	17	76
Uttar Pradesh	Allahabad	4	32	317
	Anpara	17	31	134
	Bareilly	12	24	231
	Gajraula	19	29	158
	Ghaziabad	30	34	248
	Jhansi	8	21	110
	Kanpur	8	34	215
	Khurja	25	25	168
	Lucknow	8	32	211

Contd...

Annexure 1 *Contd...*

State/union territory	City	SO_2	NO_2	PM_{10}
	Meerut	4	43	129
	Muradabad	13	23	165
	Noida	9	35	136
	Varanasi	18	21	138
	Unnao	10	32	98
	Saharanpur	11	29	186
	Raebareli	11	15	163
	Mathura	23	29	208
	Gorakpur	18	35	123
Uttarakhand	Haldwani	NA	NA	137
	Haridwar	28	32	149
	Kashipur	NA	NA	180
	Rudrapur	NA	NA	158
West Bengal	Asansol	10	37	111
	Barrackpore	13	48	130
	Durgapur	13	48	108
	Haldia	13	41	238
	Howrah	13	40	186
	Kolkata	12	70	135
	Raniganj	14	45	126
	South Suburban	8	59	119

Source CPCB (2014)

References

Atkinson, R. W., A. Cohen, S. Mehta, and H. R. Anderson. 2011. Systematic review and meta-analysis of epidemiological time-series studies on outdoor air pollution and health in Asia. ***Air Quality, Atmosphere, and Health*** 5(4): 1-9

Balakrishnan, K., B. Ganguli, S. Ghosh, S. Sankar, V. Thanasekaran, V. N. Rayadu, and H. Caussy. 2011. Short-term effects of air pollution on mortality: results of a time-series analysis in Chennai, India. *Research Report* 157, Health Effects Institute

Bell, M., J. Samet, and F. Dominici. 2003. Time series studies of particulate matter. ***Annual Review of Public Health*** 25:247–80

CBHI (Central Bureau of Health Intelligence). 2013. ***National Health Profile of India***. New Delhi: Central Bureau of Health Intelligence, Directorate General of Health Services. Ministry of Health and Family Welfare. Details available at <http://cbhidghs.nic.in/index1.asp?linkid=267>

Census of India. 2011. ***Census Info India 2011: Houses, Household Amenities and Assets (Version 2)***. New Delhi: Ministry of Home Affairs, Government of India. Details available at www.devinfolive.info/censusinfodashboard/website/index.php/pages/kitchen_fuelused/total/Households/IND

CPCB (Central Pollution Control Board). 2014. ***National Ambient Air Quality Status and Trends***. New Delhi: CPCB, Ministry of Environment and Forests, Government of India

Dholakia, H., B. Bhadra, and A. Garg. 2014. Short term association between ambient air pollution and mortality and modification by temperature in five Indian cities. Ahmedabad, India: Indian Institute of Management

Dominici, F., R. D. Peng, S. L. Zeger, R. H. White, and J. M. Samet. 2007. Particulate air pollution and mortality in the United States: did the risks change from 1987 to 2000? ***American Journal of Epidemiological*** 166(8): 880-888

Ghosh, A. and A. Mukherjee. 2010. ***Air Pollution and Child Health in Urban India***. New Delhi: Indian Statistical Institute

Greenstone, M., J. Nilekani, R. Pande, N. Ryan, A. Sudarshan, and A. Sugathan. 2015. Lower pollution, longer lives: Life expectancy gains if India reduced particulate matter pollution. ***Economic and Political Weekly*** l(8)

Hertz-Picciotto, I., R. J. Baker, P. S. Yap, M. Dostál, J. P. Joad, M. Lipsett, T. Greenfield, C. E. Herr, I. Benes, R. H. Shumway, K. E. Pinkerton, and R. Srám. 2007. Early childhood lower respiratory illness and air pollution. ***Environmental Health Perspectives*** 115: 1510–1518

ICAMP (India-California Air Pollution Mitigation Program). 2014. India-California Air Pollution Mitigation Program: Options to Reduce Road Transport Pollution in India. A joint initiative by The Energy and Resources Institute (TERI) India, University of California at San Diego (UCSD), and the California Air Resources Board (CARB)

IHME (Institute for Health Metrics and Evaluation). 2013. ***The Global Burden of Disease: Generating Evidence, Guiding Policy***. Seattle: IHME

Krzyzanowski, M. and A. Cohen. 2008. Update of WHO air quality guidelines. ***Air Quality, Atmosphere, and Health*** 1:7–13

Kumar, R., N. Goel, N. Gupta, N. Singh, S. Nagar, and J. Mittal. 2013. ***Indoor Air Pollution and Respiratory Illness in Children from Rural India: A Pilot Study***. New Delhi: Vallabhbhai Patel Chest Institute, University of Delhi

Lakshmi, P. V., N. K. Virdi, A. Sharma, and J. P. Tripathy, K. R. Smith, M. N. Bates, and R. Kumar. 2013. Household air pollution and stillbirths in India: analysis of DLHS-II National Survey. ***Environmental Research*** 121: 17–22

Lepeule, J., F. Laden, D. Dockery, and J. Schwartz. 2012. Chronic exposure to fine particles and mortality: an extended follow-up of the Harvard Six Cities Study from 1974 to 2009. ***Environmental Health Perspectives*** 120: 965–970

Lim, S.S., T. Vos, A.D. Flaxman, G. Danaei, K. Shibuya, and H. Adair-Rohani. 2010. A comparative risk assessment of burden of disease and injury attributable to 67 risk factors and risk factor clusters in 21 regions, 1990-2010: a systematic analysis for the Global Burden of Disease Study 2010. ***Lancet*** 380: 2224–2260

MoHFW (Ministry of Health and Family Welfare). 2012. ***National Action Plan and Monitoring Framework for Prevention and Control of NCDs***. New Delhi: Government of India

MoSPI (Ministry of Statistics and Programme Implementation). 2012. ***Energy Sources of Indian Households for Cooking and Lighting, NSS 66th Round, July 2009–June 2010***. New Delhi: National Sample Survey Office

OECD (Organisation for Economic Co-operation and Development). 2014. ***The Cost of Air Pollution: Health Impacts of Road Transport***. Paris: OECD. Details available at <http://dx.doi.org/10.1787/9789264210448-en>

OECD (Organisation for Economic Co-operation and Development). 2015. Gross domestic product (GDP) (indicator). doi: 10.1787/dc2f7aec-en, last accessed on on 26 June 2015

Sehgal, M., R. Suresh, V. P. Sharma, and G. Gautam. 2014. Assessment of outdoor workers' exposure to air pollution in Delhi (India). ***International Journal of Environmental Studies*** 72(1): 96–116

Sehgal, M., S. Rizwan, and A. Krishnan. 2014. Disease burden due to biomass cooking-fuel-related household air pollution among women in India. ***Global Health Action*** 7. Details available at <www.globalhealthaction.net/index.php/gha/article/view/25326>

Smith, K. R., J. M. Samet, I. Romieu, N. Bruce. 2000. Indoor air pollution in developing countries and acute lower respiratory infections in children. ***Thorax*** 55(6) 518–532

TERI (The Energy and Resources Institute). 2009. ***GREEN India 2047 Renewed: Looking Back to Change Track***. New Delhi: TERI

TERI (The Energy and Resources Institute). 2014. ***Advancement of Fuel Quality and Vehicle Emission Norms to Improve Urban Air Quality in India: Policy Brief***. New Delhi: TERI

TERI (The Energy and Resources Institute). 2015. ***TERI Energy and Environment Data Diary and Yearbook 2014/15***. New Delhi: TERI

WHO (World Health Organization). 2006. ***Air Quality Guidelines: Global Update.*** Copenhagen, Denmark: WHO Regional Office for Europe

WHO (World Health Organization). 2010. Global burden of diseases. Details available at http://www.who.int/healthinfo/global burden_disease/gbd/en; last accessed on 20 June 2015

WHO (World Health Organization). 2013. ***Health Effects of Particulate Matter: Policy Implications for Countries in Eastern Europe, Caucasus and Central Asia***. Copenhagen: WHO Regional Office for Europe

WHO (World Health Organization). 2014. ***Burden of Disease from the Joint Effects of Household and Ambient Air Pollution for 2012.*** Geneva, Switzerland: WHO

WHO (World Health Organization). 2015. ***Global Health Observatory Data Repository***. Geneva, Switzerland: WHO

World Bank. 2013. World Development Indicators. Details available at <http://data.worldbank.org>

Index

100 City Domestic Efficient Lighting Programme, 111
2013 Global Burden of Disease Report, 298
21st Conference of Parties, 1

A

Aadhaar (unique identification number), 82
ABC Papers, 222
Abhishek Industries, 222
ACC Ltd, 219
Accelerated Power Development Reforms Programme (APDRP), 105
Addis Ababa Action Agenda, 369
Aditya Solar Shops, 277
Agenda 21, 369
Agriculture Demand-side Management (AgDSM) Programme, 207
Air (Prevention and Control of Pollution) Act, 1981, 298, 300
Air Corporations (Transfer of Undertakings and Repeal) Act, 1994, 240
Air Traffic Management Services, 239
Aircraft (Security) Rules, 2011, 240
Aircraft Act, 1934, 240
Aircraft Rules, 1937, 240
Airport Economic Regulatory Authority (AERA), 239, 240
Airports Authority of India (AAI), 238, 239
Airports Authority of India (Major Airports) Development Fees Rules, 2011, 240
Airports Authority of India Act, 1994, 239
Akshay Urja Shops, 277
Alternate Hydro Energy Centre, 175
Ambuja Cements Ltd, 219
Amendment of Multimodal Transportation of Goods Act, 1993, 240
American Solar, 169
AMRUT, 234
Andhra Paper Mills, 222
Andhra Pradesh Gas Distribution Corporation (APGDC), 87, 91
Andhra Pradesh Gas Infrastructure Corporation (APGIC), 87
Ankur Scientific Energy Technologies, 165
Appellate Tribunal (Forms for Application and Appeal and Fee) Rules, 2010, 240
Aryan Coal Beneficiation Pvt. Ltd, 33
Aryan Energy Pvt. Ltd, 33
Assam Gas Company Ltd (AGCL), 90
Assam Power Distribution Company Ltd, 107
Atal Mission for Rejuvenation and Urban Transformation (AMRUT), 234
ATR Solar, 169
Auto Fuel Policy, 300, 385
Auto Fuel Vision 2025 Committee, 385
Avni Energy Systems, 175

B

Bachat Lamp Yojana, 108
Bal Urja Rakshak Mission, 110
Ballarpur Industries Ltd (BILT), 222
Banaras Hindu University, 177
Bangalore Electricity Supply Company Ltd, 106
Bangladesh Power Development Board, 110
BG Group, United Kingdom, 85
Bhadla Solar Park Project, 172
Bharat Aluminium Company Ltd (BALCO), 212
Bharat Mala project, 232
Bharat Oman Refineries Ltd (BORL), 77, 79, 80
Bharat Petroleum Corporation Ltd (BPCL), 77, 79, 80, 83, 87, 94, 178
Bharat Stage (BS) IV norms, 300, 385
Bhupdevpur–Korichapar–Dharamjaigarh railway line, 36
BIMT, 222
Bio-diesel Purchase Policy, 83
Biofuel Park campus, 178
Biofuel Task Force, 178
Biological Diversity Act 2002, 307
Biomass Co-generation, 171
Biomass Energy and Cogeneration (Non-bagasse), 165
Biomedical Waste (Management and Handling) Rules, 305
Brahmaputra Valley Fertilizers Corporation Ltd, 216
Bruhat Bangalore Mahanagara Palike, 305
BS III norms, 300, 385
BS IV norms, 300, 385

BS V fuels, 385
BS-VI emission norms, 385
Bureau of Energy Efficiency (BEE), 108, 110, 207
Bureau of Indian Standards (BIS), 83
Bus rapid transit systems (BRTS), 233, 234

C
Cabinet Committee of Economic Affairs (CCEA), 2, 30, 88
Cabinet Committee on Political Affairs, 278
Calcutta Electric Supply Corporation Ltd, 108
Cansolair, 169
Carriage by Road Act, 2007, 239
Carriage by Road Rules, 2011 (1), 239
Census of India 2011, 7, 8, 10, 272, 273, 383
Central Ground Water Board, 297
Central Motor Vehicle Rules, 1989, 239
Central Pollution Control Board, 298, 300, 303, 377, 379, 385
Central Public Sector Enterprises (CPSEs), 356
Central Road Fund Act, 2000, 239
Centre for World Natural Heritage Management and Training for Asia and Pacific Region, 309
Century Textiles and Industries Ltd, 219
Chambal Fertilizers and Chemicals Ltd (CFCL), 216
Chamundeshwari Electricity Supply Corporation Ltd, Mysore, 107
Chemical Sources of Energy Programme, 177
Chennai Petroleum Corporation Ltd (CPCL), 79, 80, 87
Chettinad Cement, 219
Chhainsa–Jhajjar–Hissar Pipeline (CJPL), 90
Chhattisgarh State Power Distribution Company, 107
City Challenge Competition, 234
City Sanitation Plans, 303
Clean Development Mechanism (CDM), 108, 241, 363
Clean India Campaign, 297, 373
Clique Solar, 168
Coal Bearing Areas (Acquisition and Development) Act (CBAADA), 1957, 37
Coal Controller's Organisation, 2
Coal India Ltd (CIL), 25, 26, 28-31, 33, 34, 36-38, 40
Coal Mine Safety Legislation, 40
Coal Mines (Nationalization) Act, 1973, 37, 38
Coal Mines (Special Provision) Act, 2015, 2, 25, 28
Common biomedical waste treatment facilities (CBWTFs), 305
Conserval Engineering, 169
Control of National Highways (Land and Traffic) Act, 2002 (1), 239
Copenhagen pledge, 354
CSIR Laboratories, 177

D
Dabhol LNG, 88
Dabhol–Bangalore pipeline, 90
Dadri–Bawana–Nangal Pipeline (DBPL), 90, 91
Dadri-Panipat Natural Gas Pipeline (DPPL), 90
Dahej–Uran–Panvel Pipeline (DUPL), 90
Dahej-Vijaipur Pipeline (DVPL), 90
Dalmia Cement, 219
Deendayal Upadhyaya Gram Jyoti Yojana (DDUGJY), 82, 105, 110, 282
Defence Research and Development Organization, 177
Delhi Metro Rail Corporation (DMRC), 241
Delhi Metro Railway (Operation and Maintenance) Act, 2002, 239
Department of Atomic Energy, 177
Department of Biotechnology (DBT), 178
Department of Science and Technology (DST), 178
Development of Solar Cities, 277
Dhande Committee, 82
Direct Benefit Transfer of LPG (DBTL), 73, 82
Direct Benefit Transfer Scheme, 278
Disaster Resilient Power Strengthening System, 107
DISCOMS, 108, 111
District Level Household Survey II, 299, 382
Doha Declaration on Financing for Development, 369
Domestic Efficient Lighting Programme (DELP), 108, 111
Dominion Resources, USA, 85
Draft National Civil Aviation Policy, 2015, 239
Draft Solid Waste Management Rules, 2015, 304
DSM regulations, 108
Duliajan–Numaligarh Pipeline Ltd (DNPL), 90
Duncans Industries Limited (DIL), 216

E
Eastern Coalfields Ltd, 30, 31
Economic Survey 2013/14, 212
Economic Survey 2014/15, 41
Electricity Act, 2003, 275, 356
Electricity Department of the Government of Puducherry, 107
Eleventh Five-year Plan, 29, 105, 175, 207, 231, 235, 237, 239

Emami Paper Mills Ltd, 222
Empowered Pool Management Committee (EPMC), 92
Energy and Resources Institute, 106
Energy Clinic Programme, 110
Energy Conservation Act, 2001, 108
Energy Conservation Building Code, 108
Energy Conservation Day, 110
Energy Efficiency Services Ltd (EESL), 108, 111
Energy Information Administration, 93
Energy Statistics 2015, 274
Environment (Protection) Act, 1986, 300, 305
Environment (Protection) Rules, 1999, 305
Environmental Performance Index (EPI), 1
Environmental Protection Agency, 38
ESSAR Oil Ltd, 80
ESSAR, 79, 80, 93
ESTA International, 108
Ethanol Blended Petrol (EBP) Programme, 82, 83
E-waste (Management and Handling) Rules, 2011, 305
Exclusive Economic Zone (EEZ), 178, 181
Expert Committee on Climate Change, 356, 358

F

Feeder Renovation Programme, 208
Fertilizer Industry Coordination Committee (FICC), 92
Fifth Assessment Report of Intergovernmental Panel on Climate Change (IPCC), 1, 9, 343, 347, 375
First Demonstration Offshore Wind Power Project, 172
Floating storage re-gasification unit (FSRU), 87
Forbes Solar, 168
Forest Conservation Act, 38
Forest Survey of India (FSI), 305

G

GAIL (India) Ltd, 83, 85, 87, 88, 90, 91, 92, 94
Ganga–Brahmaputra river system, 301
Gas Authority of India Ltd, 83, 84, 87, 90, 91
Gas de France, France, 85
Gas Natural Fenosa, Spain, 85
Gas Utilization Policy, 88
Gazprom Marketing and Trading, Singapore, 85
GeoSyndicate Power Pvt. Ltd, 177
GIGL–GSPL India Gasnet Limited, 91
GITL –GSPL India Transco Ltd, 91
Give it up-campaign, 73, 82
Global Burden Disease Study 2010, 299
Godawari Green Energy Ltd, 172
Godawari Power and Ispat Ltd, 172
Golden Quadrilateral, 231
GP Green Energy Systems Pvt. Ltd, 165
Grain Processing Industries (I) Pvt. Ltd, 165
Great Eastern Energy Corporation (GEECL), 93
Green Climate Fund, 354
Green Energy Corridors, 179
Green India Mission (GIM), 309, 355
Grid-connected rooftop PV systems, 175
Groundwater Aquifer Atlas of India, 303
Gujarat Narmada Valley Fertilizers and Chemicals Limited (GNFC), 216
Gujarat Power Corporation Ltd, 172
Gujarat State Fertilizers and Chemicals Ltd (GSFC), 216
Gujarat State Petroleum Corporation Ltd (GSPCL), 85, 87, 90
Guru Gobind Singh Refinery (GGSR), 79

H

Harvard Six Cities Study, 379
Hazira LNG Pvt. Ltd, 85, 86
Hazira–Vijaipur–Jagdishpur (HVJ) pipeline, 90
Heera fields, 74
H-Energy East Coast Pvt. Ltd (HEECPL), 87
Himachal Pradesh State Electricity Board Ltd, 107
Hindalco Industries Ltd (HIL), 212, 213
Hindustan Mittal Energy Ltd (HMEL), 77, 79
Hindustan Newsprint Ltd, 222
Hindustan Petroleum Corporation Ltd (HPCL), 77, 79, 80, 83, 94
Hindustan Steel Limited, 213
Husk Power Systems, 165

I

Ib Valley Coalfield, 39
Implicit carbon tax policy, 356
India Cements Ltd, 219
India Smart Grid Forum (ISGF), 106, 108
India Smart Grid Task Force (ISGTF), 106
India State of Forest Report 2015, 305
Indian cement industry, 7, 219
Indian Energy Exchange (IEX), 105, 106
Indian Farmers Fertiliser Cooperative Ltd (IFFCO), 216
Indian Infrastructure Finance Company, 359
Indian Institute of Forest Management (IIFM), 309
Indian Institute of Science (IISc), 165
Indian Institute of Technology Roorkee, 175

Indian Iron and Steel Company (IISCO), 33, 34, 213
Indian Maritime University Act, 2008, 240
Indian Network for Climate Change Assessment (INCCA), 352, 354
Indian Oil Corporation Ltd (IOCL), 77, 79, 80, 83, 87, 90, 94
Indian Petroleum and Natural Gas Statistics 2014-15, 275
Indian Ports Act, 1908, 240
Indian Railways, 8, 36, 234-236
Indian Renewable Energy Development Agency, 41, 172, 359
Indian School of Mines, 40
Indian solar photovoltaic (SPV), 163
Indian Space Research Organization, 177
Indian Strategic Petroleum Reserves Ltd (ISPRL), 76
India–Nepal Agreement, 110
Indira Gandhi Canal Project, 172
Indira Gandhi International Airport, 241
Indo Gulf Fertilizers Ltd (IGFL), 216
Infinite Energy Pvt. Ltd, 165
Infrastructure Investment Trusts, 223
Inland Waterways Authority of India Act, 1985, 240
Integrated Scheme of Oilseeds, Pulses, Oil Palm and Maize (ISOPOM), 207
Intended Nationally Determined Contribution (INDC), 1, 309, 343, 351, 354
Inter Solar, 169
Inter-agency and Expert Group on SDG Indicators, 372
Intergovernmental Negotiations on the Post-2015 Development Agenda, 369
Intergovernmental Panel on Climate Change (IPCC), 1, 9, 10, 343, 347, 375
International Atomic Energy Agency (IAEA), 6
International Energy Agency (IEA), 8, 271-273, 278, 352-355, 358, 360-362
International Solar Alliance, 360
International Union for Conservation of Nature (IUCN), 9, 307
ITC Paperboards and Specialty Papers Division (ITC-PSPD), 222

J

J K Group, 219
J K Lakshmi Cement, 219
J K Paper Ltd, 222
Jaipur Vidyut Vitaran Nigam Ltd, 107, 110
Jal Marg Vikas project, 237
Japan Custom-cleared Crude (JCC), 74
Jawaharlal Nehru National Solar Mission (JNNSM), 163, 166, 172, 173, 208, 277
Jawaharlal Nehru National Urban Renewal Mission (JNNURM), 233, 234, 303, 304
Jaypee Cement Ltd, 219
Jharsuguda–Barpali–Saradegna railway line, 36
Johannesburg Declaration on Sustainable Development, 369
Johannesburg Plan of Implementation, 369
Jyotigram Yojana, 110, 208

K

Kanpur Fertilizers & Cement Ltd (KFCL), 216
Kerala State Electricity Board, 104, 107
Kesoram Industries, 219
KG-D6, 83, 92
KGDS Pvt. Ltd, 168
KG-DWN-98-3 block, 92
Kobe Steel, 224
Kochi–Koottanad–Bangalore–Mangalore pipeline, 88, 90, 91
Kribhco Shyam Fertilizers Ltd (KSFL), 216
Krishak Bharti Cooperative Ltd (KRIBHCO), 216
Kyoto Protocol, 355

L

Lafarge India Pvt. Ltd, 219
Lawrence Berkeley National Laboratory, 171
LNJ Bhilwara, 175

M

Madras Cements Ltd, 219
Madras Fertilizers Ltd (MFL), 216
Mahakali Treaty, 110
Maharashtra State Electricity Distribution Company Ltd, 106, 107
Maitri Thermal Power Project, 110
Major Ports Regulatory Authority Bill, 2009, 240
Major Ports Trust Act, 1963, 240
Make in India initiative, 163, 223
Mangalore Chemicals and Fertilizers Ltd (MCF), 216
Mangalore Electricity Supply Company, 106
Mangalore Refinery and Petrochemicals Ltd (MRPL), 79, 80, 87
Manohar Lal Sharma vs The Principal Secretary & Ors 2014, 29
Megawatt Solutions, 168
Merchant Shipping Act, 1958, 240
Millennium Development Goals (MDGs), 10, 369, 371, 373

Millennium Development Goals India Country Report 2015, 370
Ministry of Agriculture, 205, 208
Ministry of Civil Aviation, 373
Ministry of Coal (MoC), 4, 25, 26, 30, 36-39
Ministry of Consumer Affairs, Food and Public Distribution, 83
Ministry of Drinking Water and Sanitation, 208
Ministry of Environment and Forests, 302, 309
Ministry of Environment, Forests, and Climate Change [MoEFCC], 9, 32, 39, 297, 302, 310, 311, 312, 313, 315
Ministry of Finance (MoF), 41, 356
Ministry of New and Renewable Energy (MNRE), 110, 163, 165-172, 175, 177, 178, 200, 207, 208, 308, 357, 359, 364, 369, 370, 372, 380, 385
Ministry of Non-Conventional Energy, 178
Ministry of Petroleum and Natural Gas (MoPNG), 2, 3, 4, 8, 13-15, 74, 75, 76, 77, 79-86, 88, 91-94, 177, 190, 203, 230, 233, 237, 238, 239, 274, 275, 281, 300, 308, 372, 385
Ministry of Power, 7, 107, 108, 207, 300, 356, 357, 372
Ministry of Railways, 36, 373
Ministry of Road Transport and Highways, 373
Ministry of Science and Technology, 177, 358
Ministry of Shipping, 373
Ministry of Statistics and Programme Implementation, 168, 370, 373
Ministry of Urban Development (MoUD), 108, 233, 304, 357, 373
Mitsubishi, 177
Mobil Australia Resources Company (MARC), 86
Monterrey Consensus of the International Conference on Financing for Development, 369
Motor Vehicles (Amendment) Bill, 2012, 239
Motor Vehicles Act, 1988 (under amendment), 239
Multimodal Transportation of Goods Act, 1993, 240
Mumbai High, 74, 77, 85
Municipal Solid Waste (Management and Handling) Rules in 2000, 304

N

N R Agarwal Industries Ltd, 222
Nagarjuna Fertilizers and Chemicals Ltd (NFCL), 216
Nagarjuna Oil Corporation Ltd, 77
National Action Plan for Assessment, Adaptation, and Mitigation of Climate Change, 356
National Action Plan on Climate Change (NAPCC), 208, 277, 303, 355-357
National Adaptation Fund on Climate Change (NAFCC), 1
National Adaptation Fund, 1, 356
National Air Quality Monitoring Programme (NAMP), 298, 377
National Aluminium Company Ltd (NALCO), 212, 213
National Ambient Air Quality Standards (NAAQS), 298, 299, 385
National Bank for Agriculture and Rural Development (NABARD), 208
National Biogas and Manure Management Programme (NBMMP), 165, 277, 385
National Biomass Cookstove Programme, 278
National Clean Energy Fund (NCEF), 41, 173, 304, 356
National Commission for Integrated Water Resources Development, 203
National Committee on Plasticulture Applications, 205
National Communication (NATCOM), 352, 354
National DELP Dashboard, 111
National Electricity Policy 2005, 207, 275
National Family Health Survey, 381
National Fertilizers Ltd (NFL), 216
National Food Security Mission, 207
National Highway Authority of India Act, 1998, 239
National Highway Authority of India, 231
National Highway Development Project (NHDP), 231, 232
National Highways Act, 1956, 239
National Highways Act, 1995, 239
National Highways Fee (Determination of Rates and Collection) Rules, 2008, 239
National Highways Rules, 1957, 239
National Highways Tribunal Rules, 2003 (1), 239
National Hydrogen Energy Programme, 177
National Hydrogen Energy Road Map, 177
National Institute of Ocean Technology, Chennai, 178
National Institute of Wind Energy (NIWE), 181
National Investment and Infrastructure Fund, 223
National Lake Conservation Plan, 302
National Mission for a Green India, 309, 358
National Mission for Clean Ganga, 303
National Mission for Enhanced Energy Efficiency, 357

National Mission for Sustainable Agriculture, 358
National Mission for Sustainable Habitat, 357
National Mission on Enhanced Energy Efficiency (NMEEE), 108
National Mission on Micro Irrigation, 207
National Mission on Strategic Knowledge on Climate Change, 358
National Monitoring Framework for Prevention and Control of Non-communicable Diseases, 382
National Offshore Wind Energy Authority (NOWA), 171
National Offshore Wind Energy Policy, 171, 181
National Policy on Biofuels, 177
National Programme on Improved Chulha, 300
National Project on Biogas Development, 300
National River Conservation Plan, 302
National Sample Survey (NSS), 8, 271, 273-275
National Smart Grid Mission, 1
National Solar Mission, 6, 41, 163, 164, 166, 175, 207, 208, 277, 357
National Steering Committee on Climate Change, 358
National Tariff Policy, 2006, 173, 356
National Thermal Power Corporation (NTPC), 37, 108, 110, 172, 177
National Tiger Conservation Authority, 309
National Urban Sanitation Policy, 304
National Urban Transport Policy (NUTP), 233
National Water Mission, 303, 357
National Water Policy, 2012, 303
National Water Quality Monitoring Programme, 301
National Wetlands Conservation Programme, 303
National Wind Energy Mission, 6
Nepal Electricity Authority, 110
New Exploration Licensing Policy (NELP), 92
Nielsen (India) Pvt. Ltd, 199
Niti Aayog, 373
Niti Ayog Report 2014, 169
Nuclear Suppliers Group, 6
Numaligarh Refinery Ltd (NRL), 79, 80

O

Ocean thermal energy conversion (OTEC), 178
Off-grid and Decentralized Solar Applications scheme, 166, 200, 207, 208
Oil and Natural Gas Corporation (ONGC), 4, 74, 77, 79, 84, 85, 87, 90, 92, 93, 177
Oil India Ltd (OIL), 74, 77, 85, 92, 93
Open Working Group on Sustainable Development Goals, 10, 375
Organisation for Economic Co-operation and Development (OECD), 384-386

P

PAHAL, 82
Panx Geothermal, 175
Parliamentary Standing Committee on Coal and Steel, 40
Perform, Achieve, and Trade (PAT), 110, 213, 214, 216, 218, 219, 220, 221, 357, 358
Petroleum Planning and Analysis Cell (PPAC) 92, 199
Petronet LNG, 73, 85-88
Planning Commission, 36, 93
Planters Energy Network, 169
Posco, 224
Power Exchange Committee, 110
Power Exchange India Ltd (PXIL), 105, 106
Power Finance Corporation, 108, 172, 359
Power Grid Corporation of India Ltd (PGCIL), 106, 107, 172, 179
Power Reactor Information System (PRIS), 6
Power Trading Corporation, 172
POWERGRID, 108
Pradhan Mantri Gram Sadak Yojana (PMGSY), 232
Pradhan Mantri Gramodaya Yojana, 105
Pradhan Mantri Krishi Sinchayee Yojana, 207
Prime Minister's Council on Climate Change (PMCCC 2014), 356
Programme for the Further Implementation of Agenda 21, 369
Programme on Energy from Urban, Industrial and Agricultural Wastes/Residues, 171
PTC India Ltd, 110
Pulp and paper industry, 8, 221
Punjab State Power Corporation Ltd, 37, 107

R

Radhe Renewable Energy Development Associate, 165
Raigarh Coalfield, 36
Railway Claims Tribunal Act, 1987, 239
Railways (Amendment) Act, 2008, 239
Rainbow Papers Ltd, 222
Rajasthan Renewable Energy Corporation, 172
Rajasthan Solar Park report, 172
Rajiv Gandhi Grameen Vidyutikaran Yojana (RGGVY), 105, 277, 300, 385
Rajiv Gandhi Gramin LPG Vitaran Yojana, 278
Rajmahal mines, 30

Rama Paper Mills Ltd, 222
Ramsar Convention, 303
R-APDRP, 104, 105
Rashtriya Fertilizers and Chemicals Ltd (RCF), 216
Rashtriya Ispat Nigam Ltd (RINL), 214
Rashtriya Krishi Vikas Yojana, 110
Ratnagiri Gas and Power Pvt, 88
Recycled Plastics Manufacture and Usage Rules, 1999, 305
Reliance Gas Transportation Infrastructure Ltd (RGTIL), 90
Reliance Industries Ltd (RIL), 77, 79, 80, 83, 90, 92, 93
Remote Village Electrification Programme, 165, 275, 300, 385
Renewable energy system in India, 6, 164
Renewable Purchase Obligations (RPOs), 356
Representative Concentration Pathway (RCP), 347, 349
Reliance Gas Transportation Infrastructure Ltd (RGTIL), 90
RestructuredAcceleratedPower Development and Reforms Programme (R-APDRP), 104, 105
RIL-BP-Niko's KG-DWN, 83
Rio Declaration, 369
Rishipooja Energy and Engineering Company, 165
Risoe Laboratory, Denmark, 171
Road Transport Corporation Act, 1950, 239
Rooftop PV and Small Solar Power Generation Programme, 173
Royal Dutch Shell, 86
Rural Electrification Corporation, 108, 359
Rural Electrification Policy, 275
Rural Electrification Programme, 105

S

Sabine Pass Liquefaction LLC, USA, 85
Sagar Mala project, 232, 237
Scheffler technology, 168
Serve as Volunteer for Energy (SAVE) programme, 110
shale oil and gas, 93
Shipping Trade Practices (STP) Bill, 2010, 240
Shree Cement Ltd, 219
Shriram Fertilizers and Chemicals (SFC), 216
Shyama Prasad Mukherji Rurban Mission (SPMRM), 283
Siemens VAI, 224
Singareni Collieries Company Ltd (SCCL), 26, 28, 36
Small hydropower (SHP) plants, 1, 6, 7, 175, 179
Smart Cities, 108, 234, 373
Smart City Maturity Model, 108
Smart Mini-Grid (SMG), 106
Solar Energy Corporation of India, 359
Solar industrial process heat (SIPH), 168
Solar photovoltaic (SPV) Market, 163, 164, 166, 167, 172, 173, 175, 180, 208
Solar Pumping Programme for Irrigation and Drinking Water, 197, 200, 208
Solar water-heating systems, 167
Southern Petrochemical Industries Corporation Ltd (SPIC), 216
Special economic zone (SEZ), 34, 79
Standard and Labelling (S&L) programme, 108
Star Paper Mills, 222
State Cooperative Agricultural Rural Development Banks, 208
State Designated Agencies (SDAs), 108, 110
State electricity regulatory commissions, 104
Steel Authority of India Limited (SAIL), 33, 34, 37, 213
Strategic petroleum reserves (SPRs), 76
Sustainable development agenda, 9, 369
Sustainable Development Goals (SDGs), 9, 10, 271, 369-374
Sustainable Habitat Mission, 304
Swachh Bharat Abhiyan, 297, 304
Swan Energy Ltd (SEL), 87

T

Tamil Nadu Newsprint and Papers Ltd (TNPL), 222
Tanakpur Hydroelectric Project, 110
Tata Chemicals Ltd (TCL), 216
Tata Iron and Steel Company (TISCO), 33, 34, 213, 214
Tata Power Delhi Distribution Ltd, 106
Tata Power, 106, 110, 175
Tata Steel, 214
Tax-free Infrastructure Bond, 356, 359
Technology Mission on Cotton, 207
Technology Upgradation Fund Scheme, 223
Telangana Southern Power Distribution Company Ltd, 107
Tetra Tech, 108
The Coal Bearing Areas (Acquisition and Development) Act (CBAADA), 1957, 37
The Mines and Minerals (Development and Regulation) Amendment Act, 2015, 37
The 2030 Agenda for Sustainable Development, 369
The Airports Economic Regulatory Authority of India

Act, 2008, 240
The Economics of Ecosystems and Biodiversity (TEEB), 309
The Energy and Resources Institute (TERI), 106, 165, 373
The Future We Want programme, 369
The India Initiative (TII), 309
The Indian Railway Board Act, 1905, 239
The Inland Vessels Act, 1917, 240
The Metro Railways (Construction of Works) Act, 1978, 239
The Railways Act, 1989, 239
The Tariff Policy, 2006, 207
Thermax, 168, 175
Tori–Shivpur–Kathautia railway line, 36
Total SA, 86
Transforming Our World: the 2030 Agenda for Sustainable Development, 369
Tripura State Electricity Corporation Ltd, 104, 107
Twelfth Five-year Plan, 7, 35, 36, 88, 99, 102, 103, 104, 173, 175, 202, 207, 211, 212, 213, 231, 235, 237, 239, 278, 282, 304, 356

U

UltraTech Cement, 219
UN Statistical Commission, 372
UNESCO, 309
Unified Metropolitan Transport Authority, 234
Union Budget 2015/16, 41, 223
United Nations Convention on Biological Diversity, 1992, 307
United Nations Framework Convention on Climate Change (UNFCCC), 1, 9, 350, 352-355, 360, 371
United Nations Millennium Declaration, 369
United Nations Sustainable Development Summit 2015, 9, 369
Unnat Chulha Abhiyan, 278
Urja Samvardhan Upakram, 110
US Trade and Development Agency, 108
Uttar Gujarat Vij Company Ltd, 107
Uttar Haryana Bijli Vitran Nigam Limited, 107

V

VDPL – Vijaipur–Dadri Pipeline, 90
Vedanta Aluminium Ltd (VAL), 212
Vijaipur–Dadri Pipeline (VDPL), 90
Village Energy Security Programme, 165, 300, 385
Vinchhiya Rural Subdivision (Gujarat) study, 209
Visvesvaraya Iron and Steel Plant (VISL), 213
Vizag Steel Plant, 214

W

Waste Minimization Circles, 304
West Bengal State Electricity Distribution Company Ltd, 106, 107
Westcoast Paper, 222
Wetlands (Conservation and Management) Rules, 2010, 303
WHO Air Quality Guideline, 381
Wildlife Institute of India, Dehradun, 309
Wind Energy Technology, 171
World Bank, 209, 385
World Database on Protected Areas, 307
World Health Organization (WHO), 273, 376, 378-381, 385
World Summit Outcome 2005, 369

X

XLRI Jamshedpur, 40

Y

Yash Paper Mills, 222

Z

Zuari Industries Ltd (ZIL), 216